Digital Transmission Hierarchies and Networks

PDH, SDH and OTH

P V Sreekanth

General Manager (Operations & Maintenance)
RailTel, Secunderabad

Universities Press

UNIVERSITIES PRESS (INDIA) PRIVATE LIMITED

Registered Office
3-6-747/1/A & 3-6-754/1, Himayatnagar,
Hyderabad 500 029 (A.P.), India
e-mail: info@universitiespress.com; Web: www.universitiespress.com

Distributed by
Orient Blackswan Private Limited

Registered Office
3-6-752 Himayatnagar, Hyderabad 500 029 (A.P.) India

Other Offices
Bangalore / Bhopal / Bhubaneshwar / Chandigarh / Chennai
Ernakulam / Guwahati / Hyderabad / Jaipur / Kolkata
Lucknow / Mumbai / New Delhi / Noida / Patna

First published 2010

ISBN 978 81 7371 699 7

Typeset in Souvenir BT 10.5/13.5 *by*
Techastra Pvt Ltd., Hyderabad 500 029

Printed at
Graphica Printers
Hyderabad 500 013

Published by
Universities Press (India) Private Limited
3-6-747/1/A & 3-6-754/1, Himayatnagar,
Hyderabad 500 029 (A.P.), India

This book is dedicated to Lord Venkateswara,
the deity of Tirumala

Contents

List of Abbreviations

3GPP	Third generation partnership program
3PCC	Third party call controller
3G	Third generation mobile communication system or network
AC (in context of MPLS)	Attachment circuit
ADM	Add–drop MUX
ADSL	Asymmetric digital subscriber line
AIS	Absence of input signal
APS	Automatic protection switching
ASON	Automatically switched optical network
ATM	Asynchronous transfer mode
AU-3	Administrative unit for VC-3
AU-4	Administrative unit for VC-4
AU-AIS	Administrative unit AIS (absence of input signal)
AU-LOP	Administrative unit loss of pointer
BER	Bit error rate
BGP	Border gateway protocol
BIP Byte	Interleaved parity
B-ISDN	Broadband ISDN
BLSR	Bidirectional line switched ring
BRAS	Broadband remote access server
BRI (of ISDN)	Basic rate interface (192 kbps)
CCE	Call control element
CHEC	Core header error control field
CO	Central office (of a telecom company)
CRC	Cyclic redundancy check
CWDM	Coarse wavelength division multiplexing
DBR	Distributed Bragg reflector
DCN	Data channel for NMS
DI-MUX	Drop insert MUX
DSL	Digital subscriber line
DSLAM	Digital subscriber line access multiplexer
DSL-CPE	Digital subscriber line customer premises equipment
DXC	Digital cross-connect
DWDM	Dense wavelength division multiplexing
E1	2.048 Mbps PDH stream
E2	8.448 Mbps PDH stream
E3	34.368 Mbps PDH stream
E4	139.264 Mbps PDH stream
E5	565 Mbps PDH stream
EC	Event collector
EDFA	Erbium doped fibre amplifier

EMS	Element management system
ET	Exchange termination
FAS	Frame alignment sequence
FAW	Frame alignment word
FCS	Frame check sequence
FOA	Fibre optical amplifier
FPA	Fabry-Perot amplifier
FTTB	Fibre to the building
FTTC	Fibre to the curb
FTT Cab	Fibre to the cabinet
FTTH	Fibre to the home
FTTR	Fixed transmitter and tunable receiver
GFP	Generic framing procedure
GPON	Gigabit PON
HP-BIP	HO path BIP error
HP-PLM	HO path payload mismatch
HP-RDI	HO path remote defect indication
HP-TIM	HO path remote error indication
HP-UNEQ	HO path unequipped
IaDI	Intra-domain interface
IP	Internet protocol
IPE	IP call element
IPOA	IP over ATM
IrDI	Inter-domain interface
ISDN	Integrated services digital network
ISI	Inter symbol interference
IS-IS	Intermediate system to intermediate system
ITU	International Telecommunications Union
JB	Justification bits
JCB	Justification control bits
JOH	Justification overhead
K1, K2	Automatic protection switching (message) bytes in SDH frame
LANE	(ATM) LAN emulation
LCAS	Link capacity adjustment scheme
LCT	Local craft terminal
LDP	Label distribution protocol
LDAP	Lightweight directory access protocol
LOF	Loss of frame
LO-PLM	LO path payload label mismatch
LOS	Loss of signal
LP-BIP	LO path BIP error
LP-RDI	LO path remote defect indication
LP-REI	LO path remote error indication
LP-TIM	LO path trace identifier mismatch
LP-UNEQ	LO path unequipped

LSP	Label switched path
LSR	Label switched router
LT	Line termination
MG or MGW	Media gateway
MGC	Media gateway controller
MGCP	Media gateway control protocol
MFAS	Multi-frame alignment sequence
MFI	Multi-frame indicator
MILP	Mixed integer linear programming
MOS	Mean opinion score
MPOA	Multi-protocol over ATM
MPLS	Multi protocol label switching
MPλS	Multi protocol wavelength (lambda) switching
MS –AIS	MUX section AIS (absence of input signal)
MS-BIP	MUX section BIP(B) error
MSOH	Multiplexer section overhead
MS-RDI L	MUX section RDI (remote defect indication)
MS-REI	MUX section REI(remote error indication)
MS-SPRing	Multiplex section shared protection ring
MST	Member status field
MTIE	Maximum time interval error
MZI	Mach-Zender interferometer
NIU	Network interface unit
NAT	Network address translation
NDF	New data flag
NE	Network element
NHRP	Next hop resolution protocol
NT1	Network termination-1(w.r.t. ISDN)
NT2	Network termination-2 (w.r.t. ISDN)
NMS	Network management system
NVC	Number of virtual components
OA	Optical amplifier
OADM	Optical add–drop multiplexer
OAM&P	Operations, administration, maintenance & provisioning
OCC	Optical channel carrier
OCh	Optical channel (OCh)
OCh mesh	Optical channel mesh protection
OCh SPRing	Optical channel shared protection ring
ODN	Optical distribution network
ODTU	Optical data tributary units
ODUk	Optical channel data unit (ODUk)
O/E/O	Optical–Electrical–Optical
OFCP	Open fibre control protocol
OLT	Optical line terminal
OMS	Optical multiplex section (OMS)

OMS-DPRing	Optical multiplex section dual fibre protection ring
OMS-SPRing	Optical multiplex section shared protection ring
ONE	Optical network element
ONNI	Optical transport network node interface
ONU	Optical network unit
OOF	Out of frame
OOS	OTM overhead signal
OPU	Optical payload unit
OSC	Optical supervisory channel
OSPF	Open shortest path first
OTH	Optical transport hierarchy
OTM-0.m	OTM with reduced functionality
OTM-n.m	OTM with full functionality
OTN	Optical transport network
OTS	Optical transmission section
OTUk	Optical transport unit
OULSR	Optical unidirectional line switched ring
OXC	Optical cross-connect
PCM	Pulse code modulation
PDH	Plesiochronous digital hierarchy
PE	Provider equipment
PFI	Payload FCS (frame check sequence) indicator
PLI	Payload length indicator
PLMN	Public land mobile network
PMD	Physical media dependent
POH	Path overhead
PON	Passive optical network
PRBS	Pseudo random binary sequence
PSQM	Perceptual speech quality measurement
PRC	Primary reference clock
PRI (of ISDN)	Primary rate interface (2Mbps)
PSI	Payload structure identifier
PSTN	Public switched telephone network
PT	Payload type
PW	Pseudo wire
PXC	Photonic cross-connect
QSIG	QSIG is an ISDN based signaling protocol for signaling between private branch exchanges (PBXs) in a private integrated services network (PISN). It makes use of the connection-level Q.931 protocol and the application-level ROSE (remote operations service element protocol) protocol and ISDN proper functions as the physical.

Preface

Digital communication systems can be broadly classified as transmission, switching and networking systems. Transmission systems provide the connectivity over networks that often span several countries or regions. The evolution of digital transmission systems has been nothing short of spectacular, propelled by the advances in materials and component technology. The ITU-Study Groups have played the role of standards body, guiding the growth and deployment of technologies, as a result of which we have currently three generations of digital transmission systems in existence today, the PDH (plesiochronous digital hierarchy), SDH (synchronous digital hierarchy) and OTH (optical transport hierarchy). Each generation has a standardised hierarchy for multiplexing channels, equipment for multiplexing/ de-multiplexing, method for providing access to channels and aggregates to support various services and facilitating management and supervisory functions. Though these hierarchies are covered in a limited way in various books, there is no single book that covers all the hierarchies under one roof and explains the intricacies of hierarchical multiplexing, network management and the range of applications supported on them from the practical aspect. This book has been penned with the following objectives in mind.

1. To explain the intricacies of hierarchical multiplexing in PDH, SDH and OTH.
2. To present the evolutionary aspects in the management of digital transmission networks, namely alarm and supervisory function support of PDH, OAM&P (operations, administration, maintenance and provisioning) features of SDH, and managed light-path service features of OTH.
3. To demystify the relatively unexplored issues like justification and frame length in PDH, pointer and container operations in SDH, management overhead structures for digital wrapper and optical channel layers in OTH.
4. To cover in depth the wide range of applications including the most recent, namely,
   - conferencing, E1 branching, DSL through managed add-drop MUXes of PDH,
   - LAN-to-LAN transport, ATM, MPLS, VoIP, broadband, NGN through SDH nework,
   - MPλS, GMPLS, PON, connectivity for 3G, IPv6, NGN through OTN (optical transport networks) based on OTH.
5. To give an in-depth treatment of reliability aspects that include managing jitter/wander accumulation in PDH, SDH and OTH networks and survivability issues including 1+1 and 1:N protection, ring protection in SDH and OTH networks.
6. To describe the equipment pertaining to PDH, SDH and OTH and explain the practical aspects of bandwidth access and support to various applications.

In writing this book, I have drawn heavily from the experience I have gained from teaching this subject and related topics at the Indian Railways Institute of Signal Engineering and Telecommunication (IRISET). I have also devised and conducted some of the experiments in the Transmission Systems Laboratory, which together with the theory classes amounts to nearly 1200 hours of teaching over a period of two-and-half years. During this period, I was asked several interesting questions by the participants of these courses, which has given me an insight into the probable areas of confusion with regard to concepts. I have included few selected questions and their answers in every chapter so that readers get the benefit of the perspective presented in them. I believe that these sections should make reading of the book interesting and the learning proactive.

The book is organized into three parts: Part I features PDH systems, Part II, SDH systems and Part III, OTH systems. The book begins with a tribute to Claude E. Shannon, the father of IT. The fruits of Shannon's work are being enjoyed today by the entire humanity, either directly or indirectly.

Part I: PDH Systems

A conceptual treatment of the three theorems of IT is presented in chapter 1. The application of the information capacity theorem to determine the transmission capacity of various transmission media and the constraints that limit the realisation of the same are discussed along with the means of overcoming them. The potential that optical fibre communication system has in terms of bandwidth is highlighted. The role of multiplexing in encapsulating multiple channels onto TDM frames over transmission media is presented.

In chapter 2 the concept of multiplexing has been comprehensively developed. How transmission hierarchies evolved to meet traffic requirements of switching and networking systems; how advances in technology paved the way for this evolution and how the work done by ITU-T study groups provided direction and standardisation for the same constitute the contents of this chapter.

The entry point for digital transmission systems, i.e., PCM, forming E1 (2 Mbps stream from 30 channels) is described in chapter 3 with emphasis on alarms, supervisory, signalling and framing functions. Applications like drop–insert, programmability, derivation of $N\times64$ k bandwidth for DSL, provision of audio conferencing, etc., that ensure a niche for 30-channel primary MUX in the big world of SDH and OTH systems are described.

In chapter 4 the intricacies in PDH multiplexing and de-multiplexing, namely bit interleaving and segregation, justification and de-justification, the associated circuit logic, reasons for the choice regarding sizes and structure of frames for various hierarchical streams are explained.

Chapter 5 deals with tackling of noise and ISI in PDH networks. The ITU-T recommendations regarding control of jitter and wander at hierarchical interfaces are discussed. The setup for the measurement of jitter and wander tolerance is presented and jitter reduction through scrambler/ de-scrambler and de-jitter modules are discussed.

Part II : SDH Systems

How the inadequacies of PDH on one hand and the advances in OFC technology on the other propelled the evolution of SDH is the subject of discussion in chapter 6. SONET, the precursor of SDH has also been described.

In chapter 7 SDH is presented as a one-point support for all existing transmission structures. The framing procedures, mapping of PDH and other network traffic types onto STM (synchronous transport module) and support for OAM&P functions through frame overheads are discussed. Operation of pointers, transportation of containers and concatenation are de-mystified. Synchronisation in SDH network and processing of alarms are also explained.

The principles of SDH network element (NE) design, types of optical interfaces on SDH NE and specifications for optical interfaces are explained in chapter 8. SDH NE types (RG, DM, TM, DXC) and SDH network topology are explained. The basics of protection mechanism—multiplex layer and optical layer for point-to-point or ring topologies— are explored and UPSR, BLSR of SONET and SNCP, MS-SPRing, of SDH are presented. The chapter also looks at element management system (EMS), network management system (NMS) and implementation of DCN plan for SDH network.

In chapter 9 the concepts and practical aspects of Ethernet over SDH, VoIP over the SDH backbone, broadband connectivity over the SDH backbone and MPLS over SDH are presented. It looks at how signalling protocols for VoIP evolved and finally led to next generation switching networking (NGN). SDH support for NGN is presented. Fundamentals of MPLS and VoIP, the evolution of NGN and call processing in NGN switches are covered to ensure that readers understand these issues without prior knowledge of these topics.

Chapter 10 discusses ITU-T recommendations regarding control of jitter and wander in SDH networks. Mathematical modelling of jitter resulting in BER is presented. Based on this model, BER contribution due to jitter caused by pointer movements despite restoration through protection mechanisms from interruptions and impairments is explained along with the measures that can be taken to keep such jitter and BER under check. The design of survivable and resilient multi-ring SDH networks is also presented in this chapter.

Part III : OTH Systems

In chapter 11 the essentials of WDM are explained. It includes the constituents of a WDM system (transmitters, receivers, optical switches, optical amplifiers and wavelength converters), the ITU-T grid, WDM varieties (CWDM, DWDM) and ADM approaches in WDM.

OTH evolution, frame structures, concept of optical data unit, optical transport unit, optical channel and multiplexing in optical layers constitute the discussions of Chapter 12. Support for OAM&P as well as managed light-path services, details of framing and overheads in OTH, and maintenance signals in OTH are included.

In Chapter 13, mapping of client signals in OTH is explored. It includes the mapping of STM signals into OPUk, ATM cell streams into OPUk, GFP frames into OPUk, test signal into OPUk, and non-specific client bit streams into OPUk. The concatenation of OPUk, mapping of client signals into OPUk-Xv, mapping of ODUk signals into ODTUGjk are also explained.

The architectural features of optical transport network (OTN) and OTN NEs, viz. OLT, OADM, OA, OXC are presented in chapter 14. It covers the configurations of OXC, (electrical switch-core and optical switch-core) network survivability issues like 1+1 protection, 1:N protection and mesh protection, concepts of MP□S, GMPLS over OTN, access over passive optical networks (PON) and its variants (FTTH (fibre to the home), FTTC (fibre to the curb)). It also explores how OTN helps the cause of fixed mobile convergence(FMC) by connectivity support for 3G networks in mobile communications domain. The migration to IPv6 from IPv4, GPON (giga-bit PON) for seamless multiple broadband access and NGN are explained.

In Chapter 15, ITU-T recommendations for control of jitter and wander in OTH networks, the effect of OTN on the distribution of synchronisation via STM-N clients, network survivability through optical layer protection, dynamic control and management of optical network through ASON (automatically switched optical network) are discussed . Open fibre control protocol (OFCP) for optical safety and automatic restart is also presented.

There are three appendixes to the book which carry information that readers would find useful while seeking to expand their understanding on certain topics. In Appendix A, the architectural and functional features of STM-1 to STM-64 equipment of different makes and the interfaces supported by these equipment are discussed in brief. The material is not exhaustive, and my intention in including it is only to provide the reader with a feel of how aggregates and interfaces are tackled in these equipment. In Appendixes B and C, topics frequently referred to while covering IP and NGN are explained. For example, signalling and control protocols for IP telephony viz., SIP, MGCP, H.323 (ver.6), H.248/Megaco are presented in Appendix B, while Appendix C covers the aspect of migration from IPv4 to IPv6.

The range and depth of coverage of the topics makes the book ideal for a study by graduate and undergraduate students of communications engineering for gaining a thorough understanding of digital transmission hierarchies. Practising engineers and professionals dealing with OFC projects would also find the book useful for the insight it provides on the working of SDH, OTH and next generation networks.

Teachers can use this book to offer the following variety of courses.

1. A complete course on digital transmission hierarchies and networks (can be split over semesters/phases as per academic calendar of the institute)
2. A fast track course on SDH and its applications for participants having prior knowledge of digital multiplexing
3. A fast track course on WDM and OTH and their applications for participants having prior knowledge of SDH
4. A fast track course on SDH, OTH and their applications for field engineers having prior knowledge of digital multiplexing
5. A fast track course on SDH, OTH and their applications for senior executives
6. A fast track course on MPLS and NGN over SDH and OTH for senior executives

I have tabulated below the topics for each of the above course and the corresponding relevant sections in the textbook along with their estimated duration for coverage for the benefit of the teaching fraternity. It may please be noted that all courses are to be supplemented by practical sessions and/or field visits. For a full-time course, all the sections (almost) are to be covered in each chapter with special emphasis on the sections indicated below. For fast track courses, sections marked with special emphasis in each chapter are to be necessarily covered in the class; the other sections may be discussed briefly by the teacher from the point of view of continuity only, leaving the detailed reading to the students to do on their own.

1. A complete course on digital transmission hierarchies and networks (can be split over semesters/phases as per the calendar of the institute)

Topic	Section	Hours
The three theorems of IT and their outcome	1.1, 1.2, 1.3, 1.4	4
Multiplexing process and its evolution	2.1, 2.2, 2.3, 2.4, 2.5	2
PCM – The entry point	3.1, 3.2, 3.3, 3.4, 3.5	6
PDH MUX-DeMUX process (The following topics to be explained with regard to one hierarchy so that others are understood in the same way: bit interleaving and segregation, justification and de-justification, intricacies of frame structure and size)	4.1, 4.2, 4.3, 4.4 (4.5 to 4.8 to be left as an exercise)	6
PDH network reliability issues	5.1 to 5.9	6
Assessment	All sections	2

Evolution of SDH	6.1, 6.2, 6.5 (6.3 and 6.4 for self-study)	2
SDH process (Due emphasis to be given to sections 7.5 on 'pointers'; section 7.5.3 on exploring pointer operation; section 7.9 on concatenation and section 7.10 on synchronisation	7.1 to 7.11	12
SDH network elements and networks (Due emphasis to be given to section 8.4 on protection mechanisms; section 8.7 on EMS, NMS, DCN plan, Appendix A on equipment types	8.1 to 8.8, Appendix A	12
SDH applications (Due emphasis to be given to MPLS fundamentals in section 9.3, VoIP in section 9.6, evolution of NGN switch in section 9.7, signalling and control protocols for NGN switch in section 9.8 and SIP in Appendix B)	9.1 to 9.9, Appendix B	12
Reliability issues in SDH network	10.1 to 10.7 (10.8 as study exercise)	6
Assessment		2
WDM essentials	11.1 to 11.10	6
OTH essentials	12.1 to 12.6	6
Mapping of client signals in OTH	13.1 to 13.9	10
OTN and applications (Due emphasis to be given to OXC in section 14.6, MPλS in section 14.7, GMPLS in section 14.8, Access network (PON) in section 14.9, and migration from IPv4 to IPv6 (Appendix C)	14.1 to 14.10, Appendix-C	12
Assessment		2
	Total	**108**

2. A fast track course on SDH and applications for participants having prior knowledge of digital multiplexing

Topic	**Section**	**Hours**
Evolution of SDH	6.1, 6.2, 6.5 (6.3,6.4 for self-study)	1
SDH process (Due emphasis to be given to sections 7.5 on 'pointers', section 7.5.3 on exploring pointer operation, section 7.9 on concatenation, section 7.10 on synchronisation	7.1 to 7.11	6
SDH network elements and networks (Due emphasis to be given to section 8.4 on protection mechanisms, section 8.7 on EMS, NMS, DCN plan, Appendix A on equipment types	8.1 to 8.8, Appendix A	6
SDH applications (Due emphasis to be given to MPLS fundamentals in section 9.3, VoIP in section 9.6, evolution of NGN switch in section 9.7, signalling and control protocols for NGN switch in section 9.8 and SIP in Appendix B)	9.1 to 9.9, Appendix B	6
Reliability issues in SDH network	10.1 to 10.7 (10.8 as study exercise)	3
Assessment		2
	Total	**24**

3. A fast track course on WDM and OTH and applications for participants having prior knowledge of SDH

WDM essentials	11.1 to 11.10	4
OTH essentials	12.1 to 12.6	4
Mapping of client signals in OTH	13.1 to 13.9	6
OTN and applications		
(Due emphasis on OXC in section 14.6, MPλS in section 14.7, GMPLS in section 14.8, access network(PON) in section 14.9, and migration from IPv4 to IPv6 (Appendix C)	14.1 to 14.10, Appendix-C	6
Assessment		2
	Total	**24**

4. A fast track course on SDH, OTH and applications for field engineers having prior knowledge of digital multiplexing

Evolution of SDH	6.1, 6.2, 6.5 (6.3,6.4 for self-study)	1
SDH process (Due emphasis to be given to section 7.5 on 'pointers', section 7.5.3 on exploring pointer operation, section 7.9 on concatenation, section 7.10 on synchronisation	7.1 to 7.11	3
SDH network elements and networks (Due emphasis to be given to section 8.4 on protection mechanisms, section 8.7 on EMS, NMS, DCN plan, Appendix A on equipment types	8.1 to 8.8, Appendix A	3
SDH applications (Due emphasis to be given to MPLS fundamentals in section 9.3, VoIP in section 9.6, evolution of NGN switch in section 9.7, signalling and control protocols for NGN switch in section 9.8 and SIP in Appendix B)	9.1 to 9.9, Appendix B	4
Reliability issues in SDH network	10.1 to 10.7 (10.8 as study exercise)	2
WDM essentials	11.1 to 11.10	2
OTH essentials	12.1 to 12.6	2
Mapping of client signals in OTH	13.1 to 13.9	2
OTN and applications		
(Due emphasis on OXC in section 14.6, MPλS in section 14.7, GMPLS in section 14.8, Access network(PON) in section 14.9, and migration from IPv4 to IPv6 (Appendix C)	14.1 to 14.10, Appendix C	4
Assessment		2
	Total	**24**

5. Fast track course on SDH, OTH and applications for senior executives

SDH process (Due emphasis to be given to section 7.5 on 'pointers', section 7.5.3 on exploring pointer operation, section 7.9 on concatenation, 7.10 on synchronisation	7.1 to 7.11	2
SDH network elements and networks		
(Due emphasis to be given to section 8.4 on protection mechanisms, section 8.7 on EMS, NMS, DCN plan, Appendix A on equipment types	8.1 to 8.8, Appendix A	2
SDH applications		
(Due emphasis to be given to MPLS fundamentals in section 9.3, VoIP in section 9.6, evolution of NGN switch in section 9.7, signalling and control protocols for NGN switch in section 9.8 and SIP in Appendix B)	9.1 to 9.9, Appendix B	4
Reliability issues in SDH network	10.1 to 10.7	
	2	
WDM essentials	11.1 to 11.10	1
OTH essentials	12.1 to 12.6	1
Mapping of client signals in OTH	13.1 to 13.9	2
OTN and applications		
(Due emphasis on OXC in section 14.6, MPλS in section 14.7, GMPLS in section 14.8, access network(PON) in section 14.9, and migration from IPv4 to IPv6 (Appendix C)	14.1 to 14.10, Appendix C	3
Assessment		1
	Total	**18**

6. Fast track course and MPLS and NGN over SDH/OTH for senior executives

Topic	**Sections**	**Hr**
MPLS & SDH support	9.3	2
VoIP, NGN switch	9.7	2
SIP MGCP, H.323 ver.6,H.248/Megaco	9.8, Appendix B	2
MPλS, GMPLS and OTN support	14.6,14.7	2
OTN and NGN	14.9, Appendix C	2
	Total	**10**

I have taken due care to minimise errors while preparing the book; however, it is possible that some have crept in, inadvertently. I would be grateful for feedback from readers on this and any other suggestions they have for the improvement of the book.

P V Sreekanth
sreekanth27866@yahoo.co.in

Acknowledgements

I started working on *Digital Transmission Hierarchies and Networks* in June 2007 while serving as a professor at IRISET (Indian Railways Institute of Signal Engineering and Telecommunications). During the course of working on the script, I was fortunate to present a paper on *Jitter and BER degradation in OFC networks due to interruptions/impairments* at the IETE National Conference held at, Bengaluru in April 2008 . The session was chaired by Dr Mrutyunjaya H. Kori , Technical Director, Alcatel Lucent Technologies, Bangalore. At the end of the paper, I received warm appreciation from Dr Kori for the paper and later, during personal discussions, several suggestions for improving the script of this book. This helped me to include a unique blend of concepts and practical aspects in the book. The book carries his blessings on the blurb.

I wish to thank A. Seshagiri Rao, Executive Director, Railtel Corporation of India Ltd., Southern Region, for allowing me and the trainees from IRISET to make several visits to Railtel installations, which proved extremely useful in drawing from practical inputs. I thank V Balaram, Director, IRISET, S P Trivedi, Dean, IRISET, and Senior Professors V K Goyal and Ch. Mohan for their encouragement of the book which I undertook to write during my stay at IRISET.

My thanks are due to A P Sukumaran, DGM/Railtel for patiently going through the manuscript and helping me with the pre-submission editorial work. I am grateful to S K Vasishta, MD, Railtel, Rajiv Sinha, Director/POM, Railtel, R. K. Bahuguna, Director/NPM, Railtel and other members of the Railtel family for the warmth shown to me and for making me one of the Railtel family.

I am greatly indebted to Professor P K Vasudevan, Department of Electrical Engineering, IIT Kanpur, for some of his useful observations on the manuscript. I wish to thank my publisher, Universities Press, Hyderabad, for their painstaking effort to make this book a reality.

I thank the participants of the various initial courses, in-service training programs and professional courses with whom I interacted during the two-and-half years of my term at IRISET. The theory and practical sessions I have handled during the period was truly lively and enjoyable, and I owe the creation of this book to them in a large way.

My heart-felt thanks to my parents P Venkateswarlu and Subha; my wife, Kalyani and my children, Koustubha Krishna and Sree Durga, for their wonderful support at home.

Prologue: A Tribute to the Father of IT

Claude Shannon single-handedly laid the mathematical foundations for a technical revolution called the information technology (IT). His contributions to the field of IT are comparable to that of Newton in mechanics and Einstein in modern physics. I would like to put it my own way—he framed the GR (general rules) of modern information theory. Long distance and huge capacity transmission networks like SDH and DWDM networks, countrywide data networks serving the needs of organisations as well as people, facilities like e-mail and the World Wide Web are the fruits of his forays into the difficult and un-explored domains of information theory with exceptional clarity of thought.

Claude Elwood Shannon was born on 30 April 1916 in Petoskey, Michigan, USA. He graduated in 1936 from the University of Michigan in electrical engineering and mathematics. He was inspired by George Boole's great work, *Laws of Thought* and the formalism for logical expressions that George Boole developed in this book, popularly known as *Boolean algebra*. He, like George Boole, invented the mathematics he needed for his wok on 'Information Theory'. In doing so he separated the medium from the message and laid the foundations for all digital communications.

While at MIT, pursuing studies and research for his master's degree, Shannon worked under Vannevar Bush, the inventor of the differential analyser. This device was the building block for all analogue computers at that time. Shannon wanted to take up the task of bringing Boole's precision to the increasingly complex world of telephone networks. In 1940, he published his master's thesis, *A Symbolic Analysis of Relay and Switching Circuits*, in which he showed how Boole's logical symbols could be treated as a series of on or off switches and how binary arithmetic—manipulation of strings of 0s and 1s—could be carried out by electrical circuits. This provided the springboard for the art of computing to jump from analogue domain to digital domain. Shannon's work landed him in Bell Laboratories, which was a turning point in his career.

Shannon's unorthodox progress towards unexpected theoretical insights was not always popular with his colleagues, to whom it appeared that his research methods were lacking in rigour. Soon, however, they realised, 'what rigour could not do, vigour could do'.

In 1948 he published his second landmark paper, *A Mathematical Theory of Communication*, dismissing the universally accepted view that messages were synonymous with the waveforms by which they were sent over lines, and that different lines were needed for the transmission of voice and telegraph messages. Like its predecessor, this paper was instantly recognised as having enormous practical importance. It provided the basis for information theory as well as for showing a method of measuring the efficiency of communications channels, which, adopting the Clausius term from thermodynamics, Shannon called the entropy.

He showed that entropy, a measure of the disorder of a system, can be seen as a lack of information about the system. From this he deduced that disordered systems can be 'cleaned up' using the information they contain, such as their patterns. The paper was highly publicised, and for a decade Shannon was a national figure; he could invite no discouragement from Bell Laboratories.

To telephone engineers who had till then seen the waveform as both the medium and the message, Shannon' work provided a much valued new insight. By treating all messages as strings of binary digits, it was possible to distinguish the message from the medium it travelled on. This allowed engineers to focus on the delivery mechanism without worrying about the source of the message, be it voice, data or telegraph.

You will be surprised to know that Shannon wrote a paper on programming a computer to play chess as early as 1950. In the year 1965, he had an opportunity to discuss the issue with Mikhail Botvinnik, who had been for many years the world chess champion. In Shannon's words, '...the discussion was interesting, but unfortunately it was carried on through a "noisy channel" since the interpreters knew little about either chess or computers.'

Shannon returned to MIT in 1958. In this tenure as a professor at MIT, he continued to develop important ideas in information science, a number of which bear his name. The Shannon capacity, for instance, defines the bandwidth capacity of a local copper loop taking into account the noise levels across the loop; and the Shannon limit, recognising that the capacity of a channel was limited only by the signal-to-noise ratios in the channel, pointed the way to the current, higher-order modulation schemes.

His concept of adding the so-called redundant bits to a message to allow reconstruction of a corrupted message led to the widespread use of error detection and correction codes in data transmission. He was responsible for many developments in cryptography too, harking back to a digital encryption system he had developed for use by Winston Churchill and Franklin Roosevelt.

Let us have a glimpse of the fun-loving nature of this genius. His famous inventions, a two-seater version of the unicycle (nobody was willing to share ride with him because of sudden jugglings he indulged in while riding); unicycle with an off-centre hub (this would bring people out into the corridors to watch him as he rode it, bobbing up and down like a duck); motorised pogo-stick (which, he claimed, would mean that he could abandon the unicycle so feared by his colleagues); an electro-mechanical mouse trapped in a maze (fitted with copper whiskers and a magnet on its wheels, it could find its way through the maze to a 'piece of cheese') and THROBAC-1, which computed in Roman numerals! No one was ever sure whether these activities were part of some new breakthrough or whether Shannon was just amusing himself. Indeed, they all emanated from his fascination for the new and the insight gained through exploring the unknown. For example, the electromechanical mouse was the result of his fascination for artificial intelligence, and it turned out to be one of the earliest attempts to teach a machine to learn.

In 1980, Shannon was on hand when Bell Labs' chess-playing computer, Belle, won the International Computer Chess Championship!

His colleagues at MIT used to wonder how he could remain serious long enough to do any important work. His ability to understand things at lightning speed and the epoch making formulations made possible the convergence of communications and computing. A practical theorist, he was always interested in solving problems, not, as he said, 'creating new ones for someone else to solve'.

A selection of his work appeared in Russian in 1963, but it was not until 1993 that his *Collected Papers*, running to a thousand pages, appeared in English (edited by Neil J. A. Sloane and Aaron D. Wyner). The papers, some of which had previously been classified as secret, fall into three groups: information theory and cryptography; computers, circuits and games; and his doctoral dissertation on population genetics.

There was flow of laurels and honours for Shannon, which included the National Medal of Science in 1966 and the Kyoto Prize in 1985. At University of Michigan, a statue of Shannon was installed in his lifetime—indeed a rare honour. The statue was unveiled by his wife, Mary Elizabeth Shannon.

He was taken ill with Alzheimer's disease and died on 24 February 2001 at the age of 84. He was survived by his wife, son and daughter.

Shannon will forever be remembered as playful genius who invented the bit, separated the medium from the message, laid the foundations for all digital communications and heralded in the IT era. Let us move on to study the three theorems proposed by Shannon, namely the source coding theorem, the channel coding theorem and the information capacity theorem.

Part I
PDH Systems

1

The Three Theorems of IT and Their Outcome

1.1 Overview of Digital Transmission Systems

The establishment of a transmission system, especially medium or long range system covering several hundreds to thousands of kilometres and providing connectivity to large number of intermediate stations calls for a huge investment. The design objective of such a system would be to maximise the utilisation of bandwidth with the least SNR (transmitted signal-power-to-noise ratio) for a prescribed level of performance. This objective is true of analog as well as digital systems.

The digital transmission system, as we know, provides us with the advantages of regeneration of data stream at regular intervals and also integration of all types of traffic—voice, data and video in digitised form. Hence, the design objective of the system, stated more explicitly, would be to pack maximum possible information over the available bandwidth of the medium, transmit the same with the least power (SNR) and ensure that the *bit error rate* (BER) does not exceed a prescribed level. Once the constituents of digital transmission system are understood, the strategy to be followed to achieve the above objective can be outlined.

Fig. 1.1 *Simplified block diagram of a digital transmission system*

Figure 1.1 shows the block diagram of a digital transmission system in its simplest form. The information source can be analog (voice or video) or digital (data file, the data being the digitised form of any source signal). Analog sources require *source encoding*, i.e., digitisation of the output. Sometimes, a digital source may also require source encoding to fit into the scheme of things. The channel is analog, be it RF spectrum on guided or un-guided media or optical spectrum on optical fibre cable, but the transmission is digital. To provide a safeguard against channel noise and to detect errors that may occur due to channel noise, we require *channel encoding.* To match the source traffic (digitised traffic from analog sources or digital traffic from discrete sources) for transmission over an analog channel, *modulation* is required. To maximise the utilisation of bandwidth, *multiplexing* or a systematic combination of traffic from various sources is required. Corresponding to encoding, multiplexing and modulation processes at the transmitting end, we require de-modulation, decoding and de-multiplexing at the receiving end. The digital transmission system, complete in the above respects is represented in Fig. 1.2.

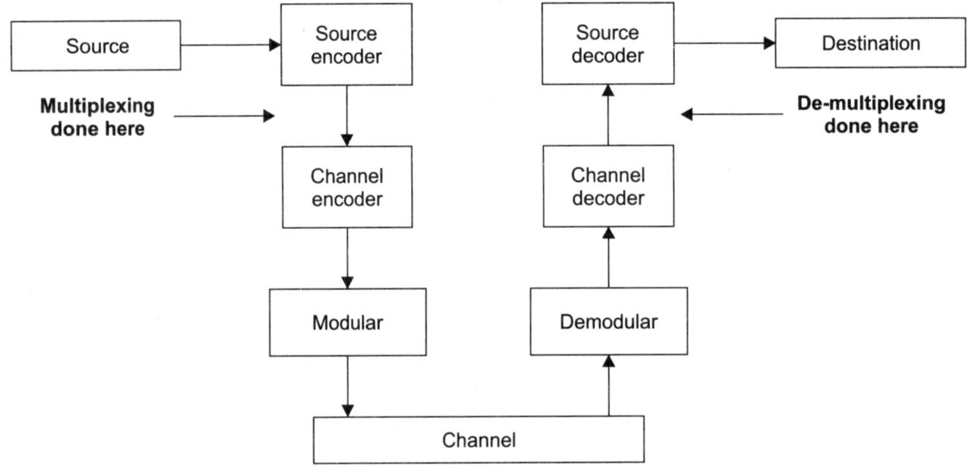

Fig. 1.2 *The block diagram of a complete digital transmission system*

The representation in Fig. 1.2 brings to fore three fundamental issues involved in the design of digital transmission systems.

1. How do we encode outputs of various sources?
2. How to carry such encoded streams over error-prone channels?
3. What is the maximum capacity promised by a channel (medium) when perturbed by noise and how to achieve it?

These three issues are tackled by the three theorems of information theory proposed by Claude E Shannon. These are:

1. The source coding theorem
2. The channel coding theorem
3. The information capacity theorem

1.2 The Three Theorems of IT

1.2.1 Source coding theorem

The source coding theorem gives the limit on the compression and refinement of information generated by a source. According to the theorem, *the number of bits per symbol required to encode the output of a source for a given probability of error [P(e)] cannot be less than the 'entropy' (information content per symbol).*

To fully understand source coding theorem and its implications, we need to know the measure of information first.

Consider a discrete memory-less source whose output is represented by variable S which can take values $S_0, S_1, S_2,\ldots, S_{k-1}$. (*A discrete source is one whose output has discrete values, not a continuous range of values. A memory-less source is one whose instantaneous output depends on instantaneous inputs and not on the past inputs.*)

The amount of information gained after observing event $S = S_k$ occurring with probability p_k is given by

$$I(S_k) = \log_2 (1/p_k). \tag{1.1}$$

The information content or entropy of the source is given by

$$H = \Sigma p_k \log_2 (1/p_k). \tag{1.2}$$

Example: Binary source with output of equal probability

Consider the case of a binary source with output states having 0 and 1 with a equal probability, i.e., ½. The information gained is

$$H = ½ \log_2 [1/(1/2)] + ½ \log_2 [1/(1/2)] = 1 \text{ bit}. \tag{1.3}$$

Thus, one bit is the information we gain when one of the two possible and equally likely events occurs.

Example: Non-binary source with output of equal probability

Consider the case of a binary source whose output states 0 and 1 occur with a non-equal probability of 1/3, 2/3. The information gained is

$$H = 1/3 \log_2 [1/(1/3)] + 2/3 \log_2 [1/(2/3)] = 0.918 \text{ bits}. \tag{1.4}$$

Consider another case of a binary source having outputs 0 and 1 occurring with a non-equal probabilities of 0.2 and 0.8. The information gained is

$$H = 0.2 \log_2 [1/0.2] + 0.8 \log_2 [1/0.8] = 0.722 \text{ bits}. \tag{1.5}$$

Thus, the information content we get when one of two possible events has a greater probability of occurring is less than what we get when both events are equally probable. Also, as one event becomes more and more probable, the information content reduces.

Example: Non-binary source with outputs of equal probability

Consider the variable S taking the values S_0, S_1,…, S_7 with equal probabilities p_0, p_1,…, $p_7 = 1/8$.

$$H = \Sigma[1/8 \log_2 \{1/(1/8)\}]_{\text{taken 8 times}} = 8 \times (1/8) \log_2 8 = 3. \tag{1.6}$$

The symbols required to encode the 8 states = 3 (not less than entropy).

Example: Non-binary source with outputs of non-equal probability

Consider the variable S taking values S_0, S_1, S_2, S_3 with probabilities p_0, p_1, p_2 and p_3 = 0.4, 0.3, 0.2, 0.1 respectively. The entropy in this case would be

$$H = 0.4 \log_2 (1/0.4) + 0.3 \log_2 (1/0.3) + 0.2 \log_2 (1/0.2) + 0.1 \log_2 (1/0.1)$$

$$= 1.8465. \tag{1.7}$$

Let us now check the application of the source coding theorem on all the four examples. This is shown in Table 1.1.

Table 1.1 *Results of entropy case studies*

Ex. No.	Case	No. of states	Symbols required to encode	Entropy
1	Binary source with equal probability output states	1	1	1
2	Binary source with output probability 1/3, 2/3	1	1	0.918
3	Non-binary source with 8 equally probable output states	8	3	3
4	Non-binary source with 4 outputs of probabilities 0.4, 0.3, 0.2 and 0.1	4	2	1.8465

It can be seen that in all the cases the symbols required to encode are greater than the entropy of the source. Thus, the source coding theorem gives the minimum symbols required for encoding or the 'lower bound for encoding', which implies that we cannot carry out compression or refinement below the lower bound of entropy.

The outcomes of the source coding theorem are:

- Concept of information content
- Limit on the compression and refinement of information generated by a source encoder/decoder.

1.2.2 Channel coding theorem

The channel coding theorem gives the limit of transmission rate of information over an error prone channel. According to the theorem, *error-free transmission is possible over a noisy channel if the rate of transmission is less than the channel capacity.*

Let us first understand channel capacity. Refer to Fig. 1.3 showing a discrete memory-less source sending symbols over a discrete memory-less channel. The channel is not ideal, i.e., it is error prone. At this moment we are not probing the causes of error but its effects.

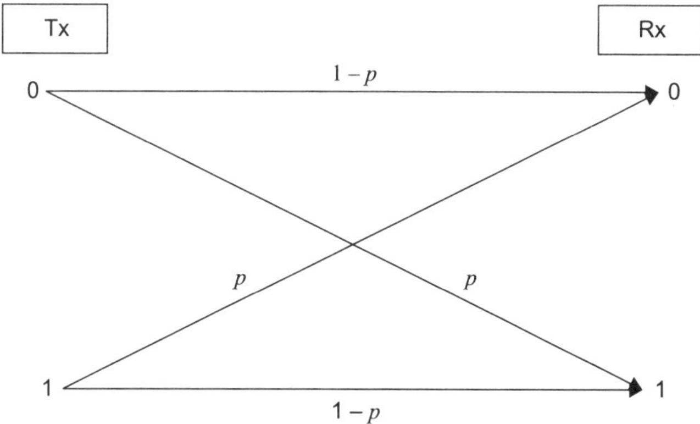

Fig. 1.3 *Understanding channel capacity*

Let p be the transition probability of the channel, i.e. the probability of symbol 1 being sent as symbol 0 or symbol 0 being sent as symbol 1. Then $(p - 1)$ is the probability of correct transmission. In other words, the channel is binary-symmetric. By definition, the channel capacity in bits per use is given by

$$C = 1 + p \log_2 (p) + (1 - p) \log_2 (1 - p). \tag{1.8}$$

Example: Computation of channel capacity for given transition probability

Consider binary symmetric channels with different transition probabilities. The computed capacities are given in Table 1.2.

Table 1.2 *Transition probability and channel capacity*

Nature of channel	Transition probability	Channel capacity in bits per use
Ideal	$p = 0$	$C = 1$
Channel is totally non-ideal (inverting)	$p = 1$	$C = 1$
Channel is not ideal; it is only 50% reliable	$p = 0.5$	$C = 0$
Channel is not ideal but reliability is better	$p = 0.1$	$C = 0.5265$

As can be seen, the channel capacity is minimum for $p = 0.5$ and maximum for $p = 0$ or 1. Therefore, the solution for guarding information on non-ideal channel is to repeat bits, or add redundancy.

If the source has entropy H and emits a symbol every T_s seconds, then the information rate of the source is H/T_s. If each symbol from the source encoder is a k-bit sequence which is encoded into an n-bit sequence by the channel encoder, then the code rate = k/n. Channel coding theorem implies that there exists a coding scheme with code rate $(k/n) \leq C$ for achieving a specified level of performance or the value of $P(e)$.

Channel coding theorem highlights the interplay among C, k/n and $P(e)$. For a given channel whose capacity is known, if we want to improve error performance, i.e. reduce $P(e)$, we have to find suitable channel coding rate (k/n). In other words, for a desired $P(e)$ over a channel of capacity C, a channel encoding scheme of coding rate less than or equal to C can be found out. The lower the value of $P(e)$, the more is the required redundancy.

This is illustrated through an example.

Example: Redundancy required in channel coding for desired *P*(*e*)

Consider a channel with transition probability 0.01. From equation (1.8), the capacity of the channel can be found as 0.9192. By adding redundancy we can achieve desired level of performance or $P(e)$ value. Let n be the length of channel encoded word. If n is odd, by majority decision the channel decoder decides in favour of 0 or 1.

It can be shown that

$$\text{For } n = 3,\ P(e) = 3p^2(1-p) + p^3. \tag{1.9}$$

$$\text{For } n = 5,\ P(e) = 10p^3(1-p)^2 + 5p^4(1-p) + p^5. \tag{1.10}$$

For the general case, if the encoded word has odd number length $n = 2m + 1$,

$$P(e) = \Sigma^n C_i\, p^i (1-p)^{n-i}, \tag{1.11}$$

taking the sum from $i = (m + 1)$ to n.

The effect of redundancy on $P(e)$ of the channel under consideration (which has transition probability of 0.01) is shown in Table 1.3.

Table 1.3 *Effect of redundancy on P(e)*

1 bit encoded as	Code rate	Probability of error $P(e)$
3 bits	1/3	3×10^{-4}
5 bits	1/5	10^{-6}
7 bits	1/7	4×10^{-7}
9 bits	1/9	1×10^{-8}
11 bits	1/11	1×10^{-10}

In all the cases the coding rate is less than C. As $P(e)$ is lowered, more redundancy is required to be added. Also, for any $P(e)$, the coding rate cannot exceed channel capacity of 0.9192. *Thus, channel coding theorem gives an upper bound for channel encoding rate.*

1.2.3 Information capacity theorem

While source coding theorem deals with discrete sources, channel coding theorem deals with discrete channels. But the channels used in communications are analog (continuous). Hence the bit stream's electrical characteristics have to be matched with that of the analog channel and made suitable for transmission over distances. This is achieved through a process called modulation.

In case of RF channels, carrier modulation is employed, whereby a parameter of the carrier (amplitude, frequency or phase) is made to assume distinct values to transmit binary information. Depending on the parameter being varied, these schemes are respectively known as ASK (amplitude shift keying), FSK (frequency shift keying) or PSK (phase shift keying). In some cases more than one parameter of the carrier is varied to transmit a set of bits. For example, in case of QAM (quadrature amplitude modulation), both amplitude and phase are varied. In the case of optical fibre, it is the intensity parameter that undergoes modulation.

Whatever be the channel, it has certain finite bandwidth of operation and is prone to noise. Shannon's information capacity theorem gives the limit for the rate of information transmission over a noisy channel over a bandwidth. According to Shannon's theorem, the information capacity of a continuous channel of bandwidth B Hz, perturbed by additive white Gaussian noise of power spectral density $N_0/2$, is given by

$$C = B \log_2 [1 + (P/N_0 B)]. \tag{1.12}$$

The theorem underlines the power–bandwidth trade-off. Let us explore the same.

Consider a digital communication system transmitting at C. Let E_b be bit energy per symbol. Then the average transmitted power would be equal to $E_b C$. Applying Shannon's information capacity theorem,

$$C/B = \log_2 [1+(E_b C/N_0 B)], \tag{1.13}$$

$$E_b/N_0 = (2^{C/B} - 1)/(C/B), \tag{1.14}$$

where C/B is the information capacity per unit bandwidth, i.e. the bandwidth efficiency, E_b/N_0 is the ratio of the bit energy to noise-power spectral density. Thus, we can get the relationship between bandwidth efficiency of a channel and the symbol-energy to noise ratio. For an infinite bandwidth,

$$E_b/N_0 = \lim_{B\to\infty} [2^{C/B} - 1]/[C/B] = \ln 2 = 0.693 \text{ or } -1.6 \text{ dB},$$

Plot of E_b/N_0 vs. bandwidth efficiency is shown in Fig. 1.4. It highlights the trade-off between E_b/N_0 and bandwidth efficiency (power–bandwidth trade-off).

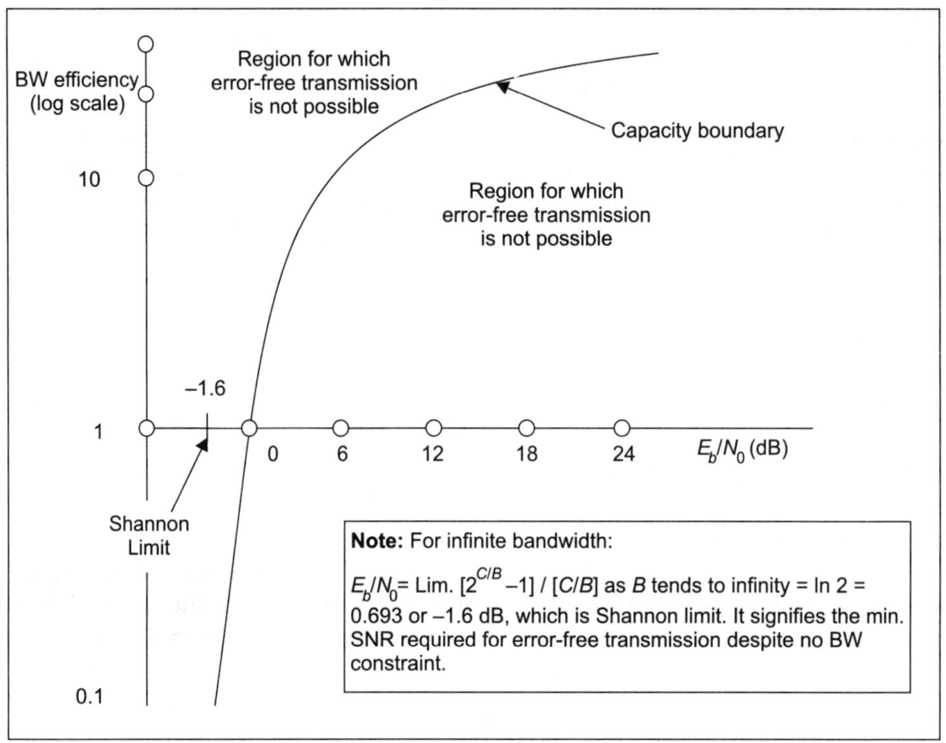

Fig. 1.4 *Plot of E_b/N_0 vs. bandwidth efficiency*

Another way of expressing channel capacity is

$$C = B \log_2 (1 + P/N_0 B) = B \log_2 (1 + \text{SNR}). \tag{1.15}$$

If we raise the 'reliability level', i.e., make BER (bit error rate) specification more stringent, the SNR requirement would increase. Increased SNR is achievable by increasing signal power or reducing noise power. Since noise power is proportional to bandwidth, we have to reduce bandwidth in order to achieve reduction in noise power. This will bring down channel capacity C since C is proportional to bandwidth. Thus, power–bandwidth trade-off is made clearer by this expression.

Information capacity theorem does something more than merely highlighting the power–bandwidth trade-off. What it does more is explained below. We have already discussed that in modulation, either the phase, frequency or amplitude of the carrier can be 'keyed' to assume discrete values for the symbols 0 or 1. If we consider M symbols at a time, we get 2^M possible discrete values. If now the phase of the carrier is varied to assume M discreet values instead of binary, it is M-ary phase shift keying. If we vary both phase and amplitude, it is M-ary QAM (quadrature amplitude modulation). This means that by M-ary PSK or M-ary QAM, we can send M-symbols' information per Hz. But, in any M-ary scheme, we have to safeguard 2^M states against noise and ensure correct decision, i.e., each state received should be clearly discernible with respect to the remaining $2^M - 1$ states in the presence of noise. This task is more demanding in terms of SNR compared to the binary scheme. Such M-ary schemes have practical limitations depending on the media.

Thus, information capacity theorem acts as searchlight to spot transmission schemes for various media under consideration. We will use this searchlight for all media by first computing the maximum data rates possible over various media for normal set of conditions. We will then take examples of working systems and their actual data rates. Later, we will explore the possibilities of M-ary techniques and enhanced data rates.

You will be surprised to know that

1. In some cases we have reached the maximum possible data rates and cannot stretch beyond that with practicable solutions.
2. In some cases, we are yet to reach the maximum rates.

Note: The objective of the book is to explore digital multiplexing hierarchies and their application in networks. Hence, detailed analysis of modulation schemes is not presented. However, equations and computed results which have a bearing on the information capacity and BER are presented.

Let us start our exploration, media-wise.

1.3 Transmission Capacities of Various Media

1.3.1 Copper cable

The copper cable can carry a signal of frequency 1.1 MHz over short distances. For copper cable the noise level is –80 dBm/Hz or 10^{-11}/Hz (considering noise and cross-talk over 1 km length) and the transmitter power level can be 16 dBm or 40 mW. Hence the theoretical limit of channel capacity would be

$$C = B \log_2 (1 + S/N) = (1 \text{ MHz}) \log_2 [1 + (40 \times 10^{-3})/(10^{-11} \times 10^{6})]$$

$$= 12 \text{ Mbps (approximately) over 1 km length.} \qquad (1.16)$$

But the nonlinear characteristic of the copper wire reduces the actual available bandwidth, its characteristic impedance being constant only over 600 Hz to 3000 Hz (2.4 kHz bandwidth). Using modems of QPSK, QAM schemes, the data rate supported steadily increased over the years to 57.6 kbps (V.90 modem). The dial-up connection for internet access on normal telephone wires (pair) using V.90 modems supports data downloads at 57.6 kbps. This was thought to be end of the road for the copper cable in data transmission. With the use of DMT (discrete multi-tone modulation technique), ADSL (asymmetric digital subscriber line) modems are able to support data rates up to 2 Mbps.

Refer to Fig. 1.5 to note that:

- Bandwidth from 0 to 1.1 MHz is divided into 256 sub-carriers (numbered 0 to 255) with a separation of 4.3125 kHz.
- Sub-carrier 0 is for passing 4 kHz analog voice. Sub-carrier 1 to 5 are reserved as guard band.
- Sub-carriers 6 to 37 (a total of 32) are for up-stream traffic and sub-carrier 38 to 255 (a total of 218) are for down-stream traffic.
- Each sub-carrier can support a maximum of 15 bits/Hz, implying 60 kbps per sub-carrier. Hence the limits for up-stream and down-stream data rates are 32 × 60 kbps or 1.8 Mbps and 218 × 60 kbps or 13 Mbps respectively.
- Interference and distortion, which depend on circuit length, limit these rates. ADSL connections are offered over a maximum length of 5.5 km. The operating data-rates are: up-stream, 1.5 Mbps (minimum of 640 kbps) and down-stream, 8 Mbps (minimum of 2 Mbps).

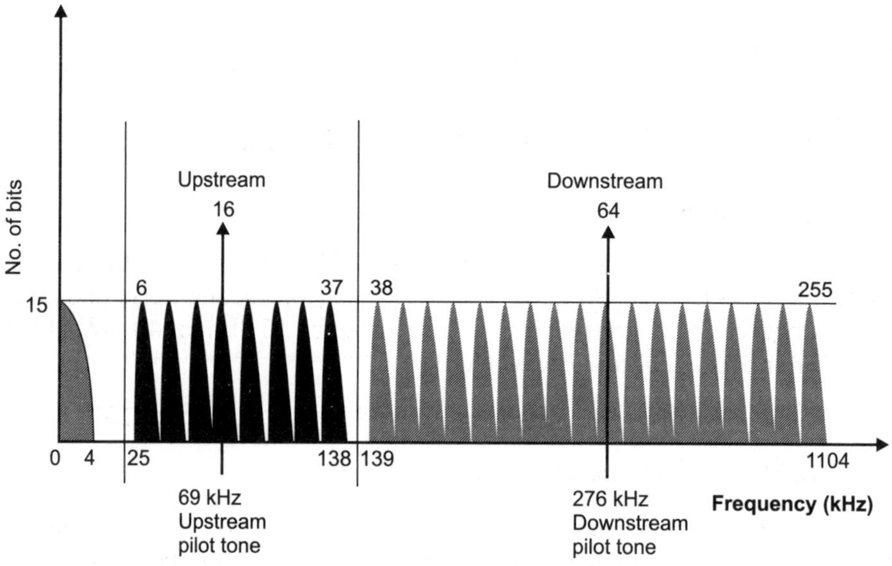

Fig. 1.5 *Discrete multi-tone modulation technique*

Operation up to 5.5 km loop is possible by reducing interference and cutting on data speed. Deployment of ADSL is shown in Fig. 1.6.

Suppose we are able to double the bandwidth. We then get a data rate of 25 Mbps which can be supported. The latest ADSL2+ technology doubles the maximum frequency used for downstream data transmission from 1.1 MHz to 2.2 MHz and offers additional 256 sub-carriers for down-stream and thereby supports a data rate of 25 Mbps downstream. The same is illustrated in Fig. 1.7.

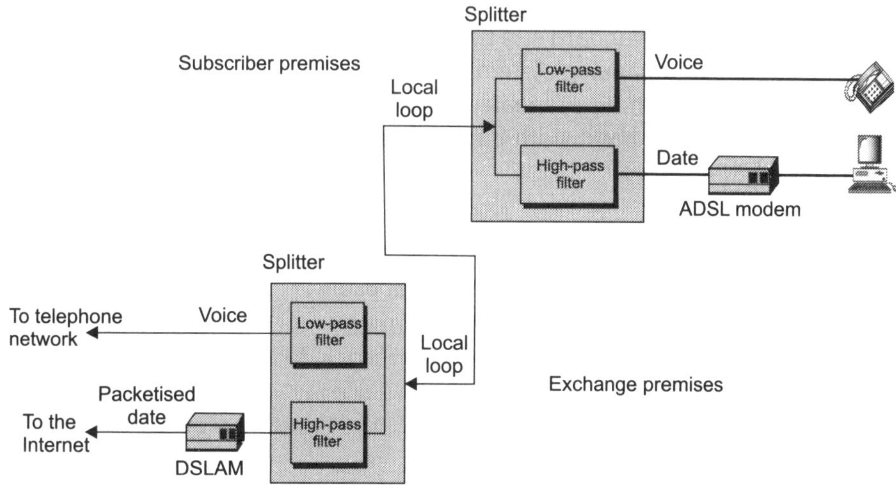

Fig. 1.6 *Deployment of ADSL*

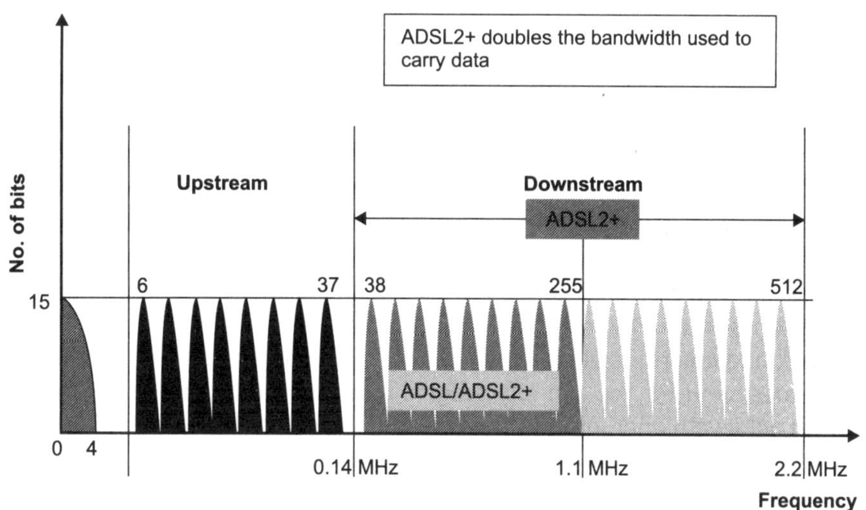

Fig. 1.7 *Bandwidth division in ADSL2+*

To summarise, in case of copper cable the channel capacity limit using conventional modems is far below the theoretical value of 1 Mbps. It is possible to get to a new maximum using DMT modulation technique.

1.3.2 Coaxial cable

Digital transmission systems on coaxial cables were designed primarily to replace existing analog systems using the same types of coaxial pairs and the same interval for repeaters as in the case of analog systems. The cables used are 1.2 mm/4.4 mm (small size pair) and 2.6mm/9.5 mm (standard size pair)[1]. Repeaters in the case of digital coaxial system are regenerative (in analog system, these were amplifiers). The existing location of amplifiers were used to house repeaters (4.7 km for 2.6/9.5 mm cable and 2.1 km for 1.2/4.4 mm cable). The bandwidth budget is presented in Table 1.4 for 2.6/9.5 mm cable and 1.2/4.4 mm cables.

Table 1.4 *Bandwidth budget for coaxial cables*

Parameter	**2.6/9.5 mm cable**	**1.2/4.4 mm cable**
Attenuation (dB/km)	$2.33 f^{1/2} + 0.002 f$ (f is the frequency in MHz)	$5.2875 f^{1/2} + 0.025 f$
Repeater spacing	4.7 km	2.1 km
Maximum attenuation for which equalisation is possible (to correct for amplitude distortion)	84 dB	84 dB
Allowed attenuation per km	17.9 dB	40 dB
Max. allowed frequency (as per attenuation formula)	58.4 MHz	58.4 MHz

It is seen from Fig. 1.8 (showing the attenuation characteristic of three varieties of coaxial cables) that cable 0.7/2.9 mm cable has high attenuation and is not suitable for digital transmission links. Hence, only 2.6/9.5 mm and 1.2/4.4 mm coaxial cables are used.

With the use of line codes, i.e., without the complexity of modulators/demodulators at every regenerator, we can support transmission rate up to 58.4 MHz. The transmitted signal being a square wave, the higher harmonics of the frequency would be snubbed. Nevertheless, the waveform is not greatly distorted over the repeater section, and after the equalisation process, the transmitted signal level can be deciphered.

Transmission rate permissible is within 58 Mbps if binary transitions are used. If other coding schemes are used, we can effect signal transition at the maximum rate permissible and yet achieve higher information rate in the available bandwidth.

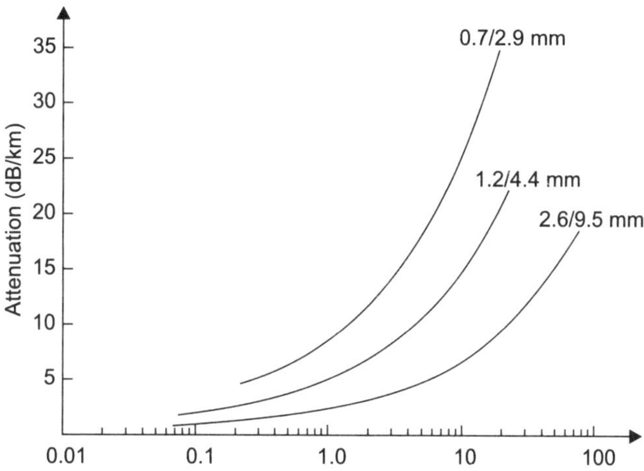

Fig. 1.8 *Attenuation characteristics of three varieties of coaxial cables*

- In case of E4 (139.264 Mbps or 1920 voice channels) transmission, the incoming data from the MUX is at E4 rate (139.264 Mbps). If this is transmitted in CMI form, the baud rate can be 139.264 Mbps/2, i.e., 69.362 Mbps since CMI (coded mark inversion) is 1B2B type line code[1]. Actually, by using FOMOT line code or 4B–3T (4 binary–3 ternary), the baud rate is further reduced to 69.362 × ¾, i.e., 52.224 Mbps. This is an acceptable signal transition rate since it is within 58.4 MHz. Some of the E4 transmission links on 2.6/9.5 mm coaxial cable installed in 1984 in India by the DOT are continuing to be used today.

  Note: Four mode ternary line code (FOMOT) is a ternary coding technique based on trellis coding technique and is applied to baseband digital transmission systems. In ternary coding, for $R = n/n + 1$ $(n = 1,2,3)$ coding rates, a codeword assignment is done to accomplish the basic requirements for line coding in which each n-length binary data sequence is encoded into a $n + 1$ length ternary (+, 0, –) line codeword chosen among the code alphabet with 2^{n+2} elements. Assuming Viterbi decoding, the system error performance is improved by increasing the free Euclidean distance between the coded sequences. The code is $R = 3/4$, i.e. 4 binary–3 ternary. This code has better analytical error performance and better rejection of side lobes than CMI in the power spectral density[2].

- In case of E3 (34.368 Mbps or 480 voice channels) transmission, the incoming data from MUX is at E3 rate (34.368 Mbps). It is within 58.4 Mbps and hence, any line code would be useful. Some of the E3 transmission links on 1.2/4.4 mm coaxial cable installed in 1984 in India by the DOT continue to be in use today.

Why not go beyond E4 on coaxial cable? Let us explore the issue by finding the SNR and then using Shannon's information capacity theorem.

The lower end cut-off frequency from phase distortion considerations is 4.4 MHz on coaxial cable. Hence, the available bandwidth is 54 MHz (58.4 – 4.4 MHz). Using the available bandwidth and ratio of the signal power at the receiver input to the noise power in the available bandwidth, we can find the theoretical limit for maximum supportable data rate.

$$\text{Noise power at the input of the receiver, } N_r = kTBF, \tag{1.17}$$

where k is the Boltzman constant (–228.6 dbW/sec/°K), T the absolute temperature in kelvin, F the receiver noise figure and B the bandwidth of the receiver in Hz.

$kT = -174$ dBm/Hz; $F = 8$ dB for practical receivers and $B = 54 \times 10^6$ Hz. With these values, we get $N_r = -88.7$ dBm.

Signal power = 40 dBm; attenuation and other loss budget = 95 dB. Hence the received signal power = –55 dBm.

$$\text{SNR} = -55 - (-88.7) = 33.7 \text{ dB or } 103.37.$$

Hence, the possible theoretical data rate is given by

$$C_{\text{theoretical}} = (54 \text{ MHz}) \log_2 (1 + \text{SNR}) = 600 \text{ Mbps}.$$

Were there any systems at data rates higher than E4 tried on coaxial? Yes, in Japan, digital coaxial systems were tried at 565 Mbps bit rate (564.992 Mbps by multiplexing four E4 streams, i.e., 7680 voice channels) by re-designing repeater spacing. But, these were not economical solutions and OFC with Gbps transmission capacities came to the centre stage by year 1995.

1.3.3 Radio spectrum for digital microwave links

Long-haul digital transmission networks operate in microwave band as the bandwidth available is large and more channels can be accommodated. Digital microwave terrestrial links for long-haul networks operate on 6 GHz, 7 GHz, 11 GHz and 13 GHz bands as per frequency allotments of different countries.

The case of radio spectrum is different from copper cable or coaxial cable because the available bandwidth is shared by number of operators. Due to path loss or attenuation considerations, the hop length is around 40 km at which the signal needs to be regenerated. We need a pair of carriers on each hop for duplex communication and cannot use the same pair on adjacent hops. Hence the available carriers are shared among the operators and each operator subscribes 6 to 8 carriers. The digital microwave system works with carrier spacing of 30 to 40 MHz. Let us estimate maximum data rate supported on the systems by applying Shannon's information capacity theorem.

Since we have to pack information in the bandwidth available per carrier, we have to find out the maximum data rate modulation scheme-wise and also find optimum solutions. Let us first consider M-ary PSK schemes. The SNR requirement is related to probability of error $P(e)$ by the following (approximate) expression[3]:

$$P(e) = e^{-k\text{SNR}}, \tag{1.18}$$

where $k = \sin^2\pi/M$. $P(e)$ is an estimate while BER is what we actually measure. BER should not exceed the estimated $P(e)$ for a given SNR. Hence, we can use the expression

$$\text{BER} = e^{-k\text{SNR}}. \tag{1.19}$$

$$\log_{10} \text{BER} = -k\text{SNR} \log_{10} e. \tag{1.20}$$

Since BER is expressed as 1×10^{-R}, the order of BER is $R = k.\text{SNR} \log_{10} e$

$$\Rightarrow \text{SNR} = (R/k)/\log_{10} e = R \ln 10/k = 2.3\, R/(\sin^2\pi/M). \tag{1.21}$$

For a BER of 1×10^{-6} end to end on a long-haul link, the BER will be better than 1×10^{-8} (i.e., R, the order of BER = 8) on each hop. We can compute the theoretical SNR requirement of M-ary PSK. Using this SNR value and a bandwidth of 30 MHz, we can compute the theoretical information capacity. The results of the calculation are given in Table 1.5.

Table 1.5 *Theoretical and practical requirements of SNR for M-ary PSK*

***M*-ary PSK**	**SNR requirement as ratio (Theoretical)**	**SNR requirement in dB (Theoretical)**	**Information capacity (Theoretical)**	**SNR requirement in dB (Practical)**
2	18.4	12.65 dB	128 Mbps	18.15
4	36.8	15.65 dB	158 Mbps	21.15
8	125.64	20.5 dB	210 Mbps	26
16	1289.18	31.10 dB	310 Mbps	36.6
32	1915.2	32.82 dB	327 Mbps	38.32

As can be seen from the table, there is a steep rise in the requirement of SNR beyond 8-PSK. Hence, such systems are not practical. BPSK ($M = 2$) is not bandwidth efficient as we are sending 1 bps/Hz. QPSK and 8-PSK are the better choices. An additional 2.5 dB is required towards modem imperfections and 3 dB for RF channel imperfections. Taking this additional requirement into consideration, the practical requirement of SNR is worked out in the last column of Table 1.5.

The noise power at the input of the receiver for practical receivers considering $kT = -174$ dBm/Hz, $F = 8$ dB, and $B = 30$ MHz is

$$N_r = kTBF = -91 \text{ dBm}.$$

For 8-PSK, the minimum input at receiver,

$$Rx_{min} = N_r + SNR = -91 + 26 = -65 \text{ dBm}.$$

For QPSK, the minimum input at receiver,

$$Rx_{min} = N_r + SNR = -91 + 21.15 = -69.85 \text{ dBm}.$$

The transmitter power Tx is 1 W, i.e., 30 dBm, because with higher transmitted power, we have associated problems of cost of power plant, module isolation, over-reach to un-intended hops, etc.

For 8-PSK, system gain = Tx – Rx_{min} = 30 – (–65) = 95 dB.

For QPSK, system gain = Tx – Rx_{min} = 30 – (–69.85) = 99.85 dB.

Considering antenna gains of 43 dB at either side of a hop and path loss of approximately 120 dB, waveguide loss of 5 dB (@ 0.05 dB per metre for a 100 m tower) and other losses (e.g., branching circuit loss = 3 dB) as 8 dB on either side, the system loss (net gain – net loss of antennas) would be

$$(120 + 8 + 8) \text{ dB} - (43 + 43) \text{ dB} = 50 \text{ dB}.$$

This implies that we have a fade margin of 95 – 50 = 45 dB for 8-PSK and 49.85 dB for QPSK scheme.

From Barnett–Vigant criterion[4], this ensures an availability of 99.99%, which is the usually specified availability. Thus, for a BER of 1×10^{-8} on one hop, it is practical to have 8-PSK and QPSK systems. With 8-PSK on 30 MHz bandwidth, we get a data rate of 90 Mbps; the data rate achievable with QPSK is 60 Mbps. The 90 Mbps, 8-PSK system (in USA and Japan, with 56 T1s) and 34 Mbps system (in India, with 16 E1s) are examples of practical working systems.

There is yet another scheme, QAM (quadrature amplitude modulation), which conveys symbols by varying the amplitude as well as the phase of the carrier. 4-QAM and QPSK show the same performance but 8-QAM, 16-QAM, 32-QAM have an SNR advantage[5]. But we do not get the benefits due to nonlinearities in the channel which cannot be equalised beyond 16 QAM. 16 QAM gives 4.2 dB SNR advantage. Hence, for higher rate digital MW systems, 16 QAM is used.

With 16 QAM and 40 MHz bandwidth, we arrive at a theoretical SNR requirement of 32.2 dB; the practical SNR requirement is 37.7 dB, N_r is –92.2 dBm, Rx_{min} = –60 dBm, system gain = 90 dB and fade margin = 40 dB. This scheme can support a bit rate of 140 Mbps.

The parameters of some of the working long haul digital MW links on various bands are summarised in Table 1.6.

As can be seen, the maximum practical data rate is 140 Mbps on long haul digital MW links, though the theoretical limits are around 300 Mbps.

Table 1.6 *Parameters of long haul digital MW links in different bands*

System parameter	6 GHz (USA)	11 GHz (Japan)	7 GHz (Indian Rly)	6 GHz (BSNL)
Frequency band	6 GHz	11 GHz	7 GHz	6 GHz
Modulation	8-PSK	8-PSK	QPSK	16 QAM
Theoretical required SNR for BER of 1×10^{-6}	20.5 dB	20.5 dB	15.65 dB	32.2 dB
Practical required SNR for BER of 1×10^{-6}	26 dB	26 dB	21.15 dB	37.7 dB
System gain	95 dB	95 dB	99.85 dB	90 dB
Fade margin possible	45 dB	45 dB	49.85 dB	40 dB
Bandwidth	30 MHz	30 MHz	30 MHz	40 MHz
BW efficiency (Bit rate/BW)	3	3	2	4
Actual bit rate supported	90 Mbps	90 Mbps	34 Mbps*	140 Mbps
Capacity	56 T1	56 T1	16 E1	64 E1

* Note: In the case of QPSK, though 60 Mbps is possible, we stop at 34 Mbps, as the next hierarchical rate is 140 Mbps.

Q 1.1 *Like in copper medium, why not try higher order QAM and enhance data rates on digital MW links?*

STM radios operating at 155.52 Mbps and 622.08 Mbps are available. Some of them are available in ISM band of 5.8 GHz, which does not require an RF license. The transmission capacities in multiples of Gbps offered by OFC networks made them the work-horse for long-haul digital transmission networks. STM radios are used as the last-mile systems or as the interconnecting systems on OFC networks in situations where the laying of OF cable is problematic.

As an example, the key features of Fiber Logic Communications Inc.'s Opti Q Wave 8310 STM-1 radio (155.52 Mbps) operating in 5.8 GHz ISM band are given in Table 1.7[6].

Table 1.7 *Key features of Opti Q Wave 8310 STM-1 radio*

System capacity	STM-1 (155.52 Mbps)
RF channel bandwidth	32 or 39 MHz
Frequency of operation	5.725 GHz–5.765 GHz 5.811 GHz–5.850 GHz
RF output	+17 dBm
Receiver sensitivity	–74 dBm (for BER of 1×10^{-6})
System gain	>91 dB

(Continues...)

Table 1.7 *Continued*

Max. RF level	–15 dBm
Range	1 to 35 km
Residual BER	1×10^{-10}

Q 1.2 *What are the other bands in which STM radios are available?*

STM radios are available in 6/7/11/13/15/18/23/38 GHz bands, as per ETSI (European Telecom Standards Institute)/FCC (Federal Communicaitons Commission) standards. The key features of WNI Global's STM radios are given in Table 1.8[7].

Table 1.8 *Key features of WNI Global Inc.'s STM radio*

Parameter	6 GHz	7 GHz	11 GHz	13 GHz	15 GHz	18 GHz	23 GHz	38 GHz
Standard	ETSI/ FCC	ETSI	ETSI/ FCC	ETSI	ETSI	ETSI/ FCC	ETSI/ FCC	ETSI/ FCC
Operating freq. (GHz)	5.9–6.4	7.1–8.5	10.7–11.7	12.7–13.3	14.4–15.4	17.7–19.7	21.2–23.6	37–40
Tx power	22 dBm	22 dBm	18 dBm	18 dBm	18 dBm	19 dBm	19 dBm	17 dBm
Rx sensitivity	–68 dBm	–68 dBm	–67 dBm	–67 dBm	–67 dBm	–66 dBm	–66 dBm	–65 dBm
Modulation	128 QAM (ETSI) 64 QAM (FCC)							
Bandwidth	28 MHz (ETSI) 40 MHz (FCC)							

1.3.4 Optical fibre cable (OFC)

Transmission on OFC is at carried out in the 1550 nm – 1650 nm wavelength band, where the attenuation is the lowest (about 0.25 dB/km). While SDH (synchronous digital hierarchy) networks support transmission of 10 Gbps on a single wavelength, WDM (wave division multiplexing) networks support 10 Gbps on each optical carrier. The repeater spacing is decided according to the optical pulse rise time budget (total allowed dispersion of the pulse) and the link power budget (total allowed losses from source to detector over optical fibre, connectors and splices with certain system margin). Let us check-out the maximum data rates possible between adjacent repeaters so spaced.

First, we have to consider quantification of SNR to know its effect on channel capacity. Noise in the case of OFC transmission is given by RIN (relative intensity noise) due to nonlinearity of the fibre's response to varying intensities of source. RIN is expressed in dB/Hz and depends on the source. For a laser diode source, RIN = –160 dB/Hz (10^{-16} /Hz).

For the wavelengths mentioned above, the frequency is in the range of 1900 – 1800 THz. For WDM operation, consider 50 carriers with 100 GHz bandwidth each.

$$\text{RIN power over 100 GHz} = 100 \times 10^{9} \times 10^{-16} = 10^{-5}\ \text{W}. \tag{1.24}$$

Laser source output power = 10 mW.

Hence, $C = B \log_2 (1 + S/N) = (100\ \text{GHz}) \log_2 [1 + (10^{-2}/10^{-5})]$

$= 1$ Tbps/carrier.

In a 128 carrier WDM, the channel capacity would be 128 Tbps!

Operational SDH (synchronous digital hierarchy, which will be presented in detail in Part-II) systems on OFC which operate on 1550 nm wavelength have a maximum of 10 Gbps transmission capacity, and with WDM, the same becomes multiples of 10Gbps. Table 1.9 shows the maximum transmission rates on various media.

Table 1.9 *Data transmission rates supported on various media*

Medium	Frequency band or max. frequency allowed	Maximum data rate possible	Maximum data rate of working systems
Copper	4 kHz (normal) 1.1 MHz (DSL) 2.2 MHz (ADSL –2 +)	57.6 kbps 12 Mbps 25 Mbps	57.6 kbps 12 Mbps 25 Mbps
Coaxial (2.6/9.5 mm)	58.4 MHz	E4 (with 4B – 3T line code)	E4 (139.264 Mbps)
Microwave	6/7/11/13 G Hz	E4 (with 16 QAM)	E4 (139.264 Mbps)
OFC	1550 nm–1650 nm	1 Tbps × Carriers in WDM	10 Gbps per carrier (multiplied by number of carriers in WDM)

The data in Table 1.9 shows that the transmission capacity of OFC is yet to be fully utilised, whereas transmission capacities of the other media have been utilised to the maximum extent possible.

Let us salute Shannon once again for giving us insights on coding.

1.4 The Role of Multiplexing

In the discussion presented so far, we explored the stage-wise processes in the digital transmission system presented in Fig. 1.2, and indeed, we were able to visualise them as the outcomes of the three theorems of Shannon. We were also able to correlate the theorems with working systems on various media and find the maximum data rates possible and note the practical limitations thereof.

Multiplexing is done before modulation. Its purpose is to pack information or digits from as many sources (already in encoded form) as needed so that the composite frame is handed over to the medium through suitable modulation schemes. As different media have different transmission capacities, the packing methodologies or *multiplexing hierarchies* employed also differ.

However, all hierarchies were not formulated at the same time. First, it was just packing of 24 channels/30 channels on copper media into T1/E1 frame respectively. Later, the multiplexing hierarchy named *plesiochronous digital hierarchy* (PDH) was conceived to pack number of T1s/E1s into a hierarchical framework, viz. T2/E2, T3/E3, and T4/E4. This was evolved for meeting the transmission capacities of coaxial cable, digital microwave and 850 nm on OFC. With the longer range and higher data rate transmission made possible by OFC at 1310 nm and 1550 nm, *synchronous digital hierarchy* (SDH) was formulated to map various PDH data rates and also provide superior services across the network. With the advent of wave division multiplexing (WDM) on OFC, more SDH frames could be carried onv different wavelengths. Hence, optical transport hierarchy (OTH) emerged to encapsulate the SDH traffic.

This underlines the need for:

1. A critical study of multiplexing hierarchies followed on coaxial cables and microwave links (plesiochronous digital hierarchy).
2. A detailed study of SDH, its superior features and its practical applications in vogue.
3. A detailed study of OTH, which is essentially the encapsulation of SDH from digital layer onto optical layer for managed light path services.

Part-I of this book presents a critical study of multiplexing concept, PDH system and practical applications. Part-II is devoted to the detailed study of SDH system and its applications. Part-III deals with OTH and its applications.

Summary

In this chapter the three theorems of IT, viz. source coding theorem, channel coding theorem and information capacity theorem are presented. Their outcome, viz. source encoder, channel encoder and modulator with power–bandwidth trade-off for the medium under consideration are discussed. Transmission capacities of various media and how these get enhanced with digital modulation technique are discussed for copper cable, coaxial cable and digital MW transmission. DSL and ADSL on copper cable were presented. Finally, OFC medium's transmission capacity is presented. The role of multiplexing to achieve such capacities is discussed.

References

[1] Sreekanth, P. V., *Digital Microwave Communication Systems* (section 13.8), CMI line code, Universities Press, 2003.

[2] Aygolu, U. and Panayirci, E. New ternary line codes based on trellis structure, *IEEE Transactions on Communications*, 41 (6), pp864–873, Jun 1993.

[3] Kamilo Feher, *Digital Communications: Microwave Applications* (section 3.4), Prentice Hall of India, 1987.

[4] Sreekanth, P. V., *Digital Microwave Communication Systems* (section 15.6.1), Requirements of fade margin for specified system availability, Universities Press, 2003.

[5] John G. Proakis, *Digital Communications*, Fourth Edition (section 5.2.10, Comparison of digital modulation methods), McGraw Hill, 2001.

[6] Opti Q Wave 8310 ISM 5.8 GHz STM-1 Radio, *www.fiberlogic.com*

[7] WNI-Global Inc.'s STM Radios, *www.wnint.com*

2

The Multiplexing Process and Its Evolution

2.1 Fundamentals of Digital Multiplexing Process

In section 1.4, multiplexing was introduced as a means to utilise the available bandwidth by sharing the data to be transmitted among the number of channels. When a number of channels share bandwidth, i.e., the time frame, the digits from the channels are placed one after the other in real time and get transmitted in an order. Hence, digital multiplexing is 'time division multiplexing (TDM)' for 'near simultaneous' transmission. In analog multiplexing on the other hand, each channel's output occupying a specified bandwidth, 4 kHz in case of voice channel, is relocated in the transmission band by frequency division. Hence, analog multiplexing is 'frequency division multiplexing' for 'simultaneous' transmission.

In digital multiplexing or TDM, the utilisation of transmission duration by a number of channels, i.e., the sequential placement of digits of traffic from a number of channels, calls for a disciplined sharing of the available bandwidth or time duration by the associated channels, framing of such time-shared sequences, provision of alignment information in such frames, sequencing transmission of successive frames, provision of error checking capabilities and alarm indications over such transmission/reception processes and consequent supervisory functions.

Digital multiplexing, hence, as a process involves the packing of digits:

- voice/data/video traffic
- their signalling bits
- alignment bits
- error check bits
- alarm bits
- supervisory bits

into regular and hierarchical frame patterns suitable for real time transmission.

Now, the questions that arise immediately are:

1. In the process of digital multiplexing, how does one achieve such packing in real time?
2. How many channels are to be packed for transmission over a medium?
3. What are the data rates?

Packing in real time has to follow a standard pattern of a definite frame size, duration and filling method. The number of channels that can be packed and the data rates that can be supported depend on the medium chosen and the 'power-bandwidth trade-off' settled for. Data rate requirement depends on the traffic to be carried. In case of a mis-match between the requirement and the individual channel rate, suitable aggregation of channels has to be done to meet the requirement.

2.2 Characteristics of Various Traffics

While data rate is the primary characteristic of any traffic or application, the following characteristics of consequence:

1. Is the data rate constant or variable?
2. How 'bursty' is the traffic, i.e., in the allotted bandwidth or time duration, how much is the occupancy?
3. The allowed delay (or delay tolerance) when we fragment the data into a number of units and transmit over successive frames of transmission hierarchy.
4. If some packets are lost, can the system recover or what is the loss tolerance?

While (1) and (2) above depend on the traffic type or the type of application, (3) and (4) depend on the traffic type as well as the unit of fragmentation. It may be noted that multiplexing is associated with the physical layer in the ISO's OSI (seven-layer) model of computer networks. When we use the term 'digital transmission network', the scope of the term 'network' is limited to the physical layer. If it is a voice-only network, the role of digital transmission network is to interconnect switching systems (exchanges) to carry voice traffic of a number of channels between them. All modern telecom networks and switches offer voice + data + video traffic. Hence, digital multiplexing process at the physical layer is required to support the traffic from higher layers, handed over to it in the form of PDUs (packet data units). These units are X.25 packet in the case of packet switching network, ATM cell size in the case of asynchronous transfer mode (ATM), or frame size in the case of frame relay. We have chosen ATM cell for summarising the characteristics of various traffics since it is the mode over ISDN (integrated services digital network).

Let us take a quick look at the characteristics of various voice, data and video traffics (applications) presented in Tables 2.1, 2.2 and 2.3.

Table 2.1 *Characteristics of voice traffic*

Category	Data rate	CBR/ VBR	Burst length	Burst ratio	Cell loss tolerance	Cell Delay tolerance
PCM voice	64 kbps	CBR	1	N/A	10^{-4} to 10^{-6}	10–150 ms
ADPCM	32 kbps	CBR	1	N/A	10^{-4} to 10^{-6}	10–150 ms
Predictive coding	16 kbps	VBR	5–15	2–3 kB	10^{-6} to 10^{-8}	10–150 ms
Voice mail	16–64 kbps	CBR/VBR	1–3	N/A	10^{-6}	500 ms–5 s
High quality voice	192–384 kbps	CBR	1	N/A	10^{-5} to 10^{-6}	10–150 ms
CD quality voice	1.4 Mbps	CBR	1	N/A	10^{-6}	500 ms–25 s
Teleconference	64–192 kbps	CBR	1	N/A	10^{-7} to 10^{-9}	10–150 ms

Table 2.2 *Characteristics of data traffic*

Category	Data rate	CBR/ VBR	Burst length	Burst ratio	Cell loss tolerance	Cell delay tolerance
LAN	1.5–100 Mbps	VBR	Vary	100–1000 B	10^{-12}	10–100 ms
Host–Host File transfer	64 kbps–1.5 Mbps	VBR	1	12 kb–10 MB	10^{-12}	1–500 s
PC file transfer	9.6–64 kbps	VBR	1	1 kb–1 MB	10^{-9}	10–100 s
Client/Server system	10–100 Mbps	VBR	1000	1–500 kB	10^{-9}	10–500 ms
Remote DB access	1–10 Mbps	VBR	1000	100 B–100 kB	10^{-9}	1–10 ms
Remote procedure call	6–60 Mbps	VBR	15–20	60–1000 B	10^{-9}	0.1–100 ms
E-mail	9.6 kbps–1.5Mbps	CBR	1	50–5000 B	10^{-9}	1–10 s
Work-station CAD-CAM	64 kbps–1.5 Mbps	VBR	5	40–100 kB	10^{-9}	1–10 s
Mainframe CAD-CAM	1.5–36 Mbps	VBR	10–100	0.1–1 MB	10^{-9}	10–60 s
Transaction processing	64 kbps–5 Mbps	VBR	40	100–300 B	10^{-9}	1–3 s
Time sharing		VBR	30–100	20–4000 B	10^{-9}	0.1–10 s

As can be seen, the data rate requirement of voice traffic is few kbps and mostly of the constant bit rate type. The allotted time slot or bandwidth is fully utilised by voice traffic, and hence burst ratio is not applicable except in the case of predictive coding. A loss tolerance of 10^{-6} implies that a loss of only one ATM cell in 10^6 is tolerable.

Let us turn our attention to characteristics of data traffic in Table 2.2.

It can be seen that the data rate requirement is in Mbps. Since data traffic is bursty in nature and overheads are more, it has poor burst ratio. The tolerance to loss and delay of ATM cells is more stringent. (In asynchronous transfer mode, the data is carried as cells, with one cell having 53 octets or bytes, of which 48 are information bytes and 5 are overhead bytes)[1].

Let us turn our attention to video traffic where the transmission rate requirement is the highest. The delay tolerance is more relaxed whereas loss tolerance is as stringent as in the case of data traffic. Burst ratio is also poor.

Table 2.3 *Characteristics of video traffic*

Category	Data rate	CBR/ VBR	Burst length	Burst ratio	Cell loss tolerance	Cell delay tolerance
Video telephony	64 kbps–2 Mbps	CBR/ VBR	2–5	2–10 kB	10^{-9}	150–350 ms
Video conference	128 kbps–14 Mbps	CBR/ VBR	2–5	1.6–40 kB	10^{-9}	150–350 ms
Video/ image mail	1–4 Mbps	CBR	1	64 kB–1 MB	10^{-10}	1–5 s
Videotex	64 kbps–10 Mbps	VBR	10	>1 MB	10^{-7}–10^{-10}	0.1–2 s
TV	15–44 Mbps	VBR	2–5	0.5–1.3 MB	10^{-10}	40 ms
HDTV	150 Mbps	VBR	2–5	5–14 MB	10^{-12}	40 ms
Video browsing	2–40 Mbps	CBR	1	0.5–40 MB	10^{-9}	0.1–2 s
Group-4 FAX	64 kbps	CBR	1	256–640 kB	10^{-8}	4–10 s
Medical X-ray	1.5–10 Mbps	CBR/ VBR	25	5–8 MB	10^{-12}	2 s
Medical MRI/ CAT scan	10–200 Mbps	CBR/ VBR	25	250 kB–3 MB	10^{-12}	2 s
High resolution graphics	0.1–10 Gbps	VBR	25	1–100 MB	10^{-12}	10–500 ms

Q 2.1 What is meant by CBR/VBR?

CBR means constant bit-rate and VBR means variable bit rate. Some kinds of traffic can be carried on CBR while others are and some on VBR. Some have the option of CBR/VBR.

Q 2.2 *What is meant by burst length and burst ratio?*

Burst length is the number of PDUs (packet data units) going in a traffic burst. Burst ratio is the ratio of time during which the information is sent to the time the channel is occupied, scaled to bandwidth.

Q 2.3 *What are cell loss and delay tolerances?*

If traffic is carried on ATM cells, the tolerance to loss or delay of one ATM cell in the traffic flow are signified by loss and delay tolerances.

2.3 Time Sharing in *N* Channel Multiplexing

As can be seen from characteristics of various traffics, the data rate requirement varies from a few kbps to several Mbps. Simple digitised transmission of some channels would not serve the ultimate purpose of support for voice + data + video. If we take one channel for digitised transmission of voice as one unit, we would require *N* such units for other traffics (where *N* is varying). Let us study time sharing in *N* channels.

Consider the case of voice transmission using PCM (pulse code modulation). Since the maximum frequency in human voice is below 4 kHz, we sample at Nyquist rate, i.e., 8 kHz. In other words, we have voice samples at the rate 8 k samples per seconds, or a new sample every 125 microseconds.

When we digitise each sample into 8 bits (as in PCM), we are sending traffic of every voice channel at the rate 8×8 k = 64 kbps per seconds. In case of single channel transmission, the eight bits representing any sample occupy the 125 μs duration till the next sample arrives.

Suppose we reduce the width of the encoded pulses representing a sample. We can pack more channels into the 125 μs duration. Let us decide to pack *N* channels in 125 μs duration. If each sample is represented by 8 bits, we have 8*N* bits in 125 μs or a bit rate of 8*N*/125 μs or 64*N* kbps. In such a transmission, to maintain order, we have to pick up samples from each channel cyclically and put them in the nominated slot in the 125 μs duration. Then, every channel gets its nominated slot every 125 μs. In other words, we can have a scheme to multiplex *N* channels provided the medium supports 64*N* kbps data rate.

The transmission capabilities of various media and practical working systems' abilities were presented in section 1.3. The theoretical limit is a fall-out of Shannon's information capacity theorem and the highest rate achieved by a working system is the fall-out of the practical limitations in implementing the technology.

Whatever be the scheme of multiplexing to utilise the available bandwidth or transmission capacity of the medium, it should provide for packing of bits or digits from the channels along with the signalling bits, alignment bits and error check bits, alarm bits and supervisory bits. The way we pack, we have to un-pack as well, and hence we need a hierarchical frame structure. Such a structure requires standardisation, for only then the equipment designed and deployed in various networks in a country or several countries can send/receive such frames with 'full understanding'.

Such standardisation is not a one time job. As higher data rates are made possible on existing media, or newer media with higher data rate transmission capabilities are pressed into service, it becomes necessary to formulate higher order or newer frame structures. Also, such standardised frames are required to provide convenient transmission structures to meet the requirement of switching and networking from the overall perspective of traffic. Figure 2.1 depicts the relationship between transmission, switching and networking.

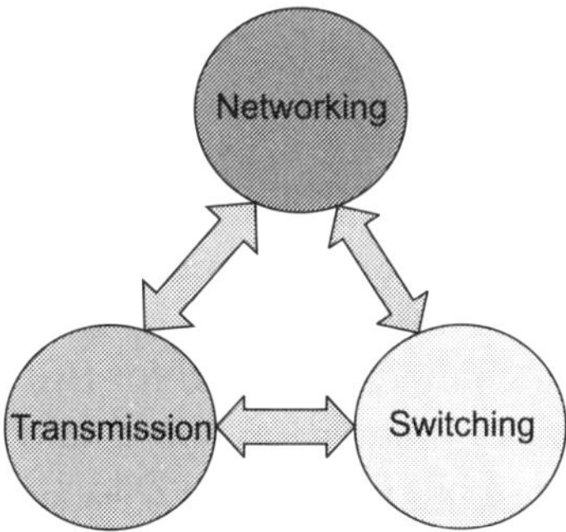

Fig. 2.1 *Interdependence of transmission, switching and networking*

2.4 How CCITT (ITU) Pioneered Evolution of Digital Multiplexing

As discussed in the previous section, the evolution of digital multiplexing (or multiplexing standards) is linked to the switching and networking requirements and the evolution of technology. The evolution of technology cannot be left unsupervised to vendors as it would lead to proliferation of various types of systems and their consequent inter-working problems. The standardisation process is looked after by ITU (known as CCITT till 1992) and its study groups. They come out with recommendations from time to time, some of which are summarised in Table 2.4.

As can be seen, PCM was standardised as 24 channel (T1) in North America by 1972; 30 channel PCM was standardised in Europe, which was later accepted by CCITT. This was just the beginning of digitisation of transmission. As a follow-up, telephone companies in USA and other countries modernised their switching equipment and the existing copper cables (UTP CAT-1) were replaced with UTP CAT-2 cables. (UTP stands for unshielded twisted pair). These cables could carry data up to 64 kbps to user premises over subscriber loops of 5.5 km. This led to the emergence of integrated digital network (IDN), a network capable of transmitting any information, be it voice, data, video or FAX. However, IDN did not have end-to-end digital connectivity of the required bandwidth to offer services to customers through interactive sessions on data terminal equipment.

Table 2.4 *Work done by CCITT study groups till 1992 which influenced the evolution of multiplexing*

Period	Assignment	Contribution
1969–1972	Planning of digital systems (T)	PCM standardisation
1973–1976	Planning of digital systems (X) and integration of services	Specs for digital switching, common channel sigg. (SS7), integrated digital network
1977–1980	Evolution of ISDN	Standardisation of user-network interface; Uniform set of protocols; First ISDN Std.G.705
1981–1984	Network aspects of ISDN	I-Series recommendations
1985–1988	General question on ISDN	Detailed I-Series Rec.
1989–1992	General aspects of ISDN	I.120 Rec.

CCITT's study group issued ISDN recommendations in the year 1980. The connectivity standards were a BRI (basic rate interface) of 192 kbps and a PRI (primary rate interface) of 2.048 Mbps. The channels required for BRI are 2B + D and that for PRI are 30B + D, where B stands for 'bearer (64 kbps)' and D stands for 'delta or signalling (16 kbps)'. Adding overhead bits makes 2B+D to 192 kbps and 30B + D to 2.048 Mbps. ISDN elements in customer premises and connectivity to the switch (exchange) are shown in Fig. 2.2.

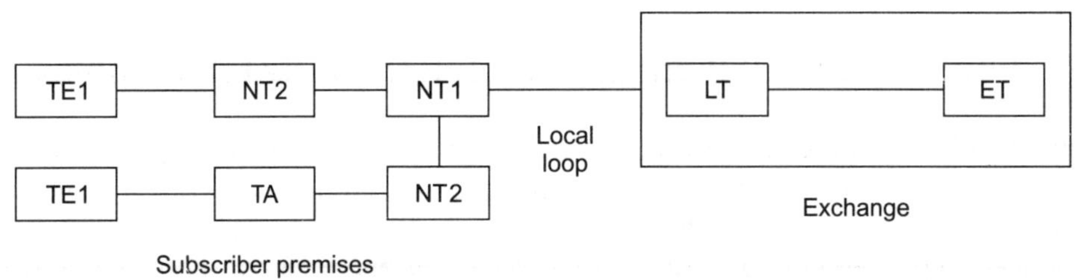

Fig. 2.2 *ISDN entities*

Elements in the above figure are:

NT1: Network termination-1, provides connectivity between the customer's premises/and the exchange. Its interface to the local loop is a U-interface (2 wire) and to NT2 a T-interface (4 wire). It can support up to 8 modules (TEs, TAs, NT2s).

NT2: Network interface-2, provides connectivity to ISDN devices 'TE1 or TE2' via S-interface. It converts T-interface of NT1 to S-interface.

TE1: Terminal equipment-1, is an ISDN compatible device. It can even connect directly to NT1. Examples of TE1 are ISDN phone and ISDN FAX machine.

TE2: Terminal equipment-2, a non- ISDN device. It connects to NT2 via TA (Terminal adapter). Examples of TE2 are analog telephones fax machines and computers without ISDN connection.

TA: Terminal adapter provides ISDN connectivity to non-ISDN devices

LT: Line termination – represents local loop connection at switch

ET: Exchange termination – represents switch to another switch connection.

Let us turn our attention to CCITT's recommendations concerned with ISDN which are presented in Table 2.5. The ISDN recommendations issued till the year 1993 (including the year 1993) are concerned with normal ISDN (or N-ISDN). In line with requirement of N-ISDN, the transmission hierarchy had to support multiples of E1s originating from different sources. The plesiochronous transmission hierarchy (PDH) evolved as a result. 'Plesio' means plural; the bit streams being integrated in PDH are nominally at the same rate, with tolerances prescribed for variations in the data rate due to their origin from different sources.

Table 2.5 *CCITT(ITU-T) recommendations concerned with ISDN (1993–1997)*

Year	Rec. No.	Title
1993	I.414	Overview of layer-1
1993	I.413	User-NW interface
1993	I.211	Service aspects
1993	I.327	Functional architecture
1993	I.356	ATM layer specifications
1997	I.432 Series	Physical layer spec. 155.22/622.08/51.8/25.6 Mbps
1997	I.731,732	ATM equipment related
1997	I.751	ATM NW related

In the year 1997, large bandwidth services, viz. interactive video services (video conferencing, video surveillance) and distribution services (cable TV, HDTV) were contemplated for delivery through ISDN. This ISDN was termed B-ISDN (broadband ISDN).

The user-network interface for N-ISDN and B-ISDN is presented in Fig. 2.3. As can be seen, the link between NT at user premises and the network interface required a data rate of 155.52 Mbps on the up-link (NT to exchange) and 622.08 Mbps on the down-link (exchange to NT). This development also coincided with the availability of a new hierarchy of transmission, viz. synchronous digital hierarchy (SDH), and the OFC with its promise of large bandwidth in the 1550 nm window.

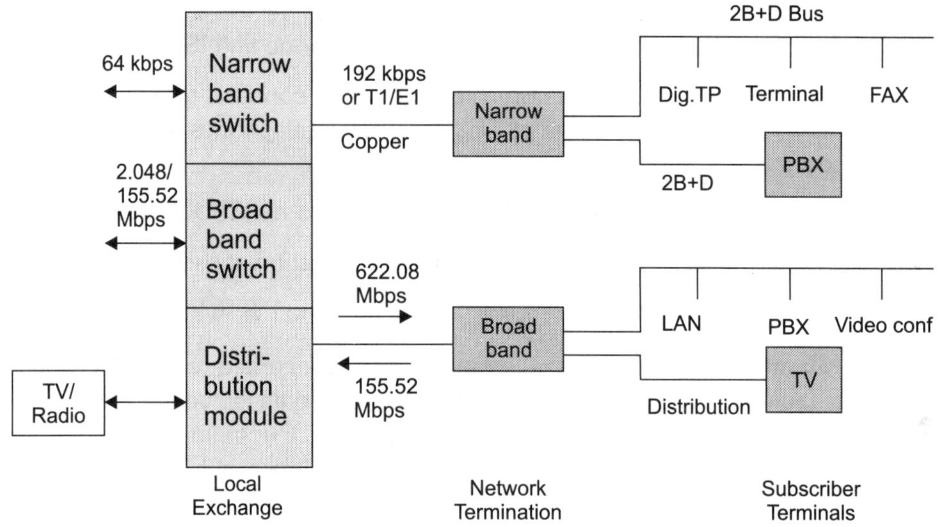

Fig. 2.3 *N-ISDN and B-ISDN user–network interface*

With the advent of WDM technology, the transmission of several SDH multiplexed frames on different wavelengths on the same fibre became a reality. A new hierarchy to encapsulate digital frames of SDH onto optical channels evolved, again governed by ITU recommendations.

2.5 Three Generations of Digital Multiplexing

Plesiochronous or plesio-synchronous digital hierarchy (PDH), synchronous digital hierarchy (SDH) and optical transport hierarchy (OTH) are the three generations of multiplexing pioneered by CCITT (ITU).

PDH is the first generation hierarchy. Since PCM is T1 (24 channel) in North America and E1(30 channel) in Europe, different PDH systems are covered under CCITT (ITU) recommendations. In addition, there is another PDH followed in Japan, which starts with 24 channel PCM but is different at higher levels. These three hierarchies are illustrated in Fig. 2.4. PDH processes and networks are covered in detail in chapters 4 and 5 respectively for the E-hierarchy followed in Europe and India. The hierarchies shown in Fig. 2.4 are for multiplexing 24 channels (Japan, North America) or 30 channels (Europe, India). The number of tributaries multiplexed at the next level are indicated by the number of arrows.

1. E-Hierarchy: E2 from four E1s, E3 from four E2s, E4 from four E3s, E5 from four E4s

2. T-Hierarchy: T2 from four T1s, T3 from six T2s, T4 from six T3s
3. J-Hierarchy: J2 from four J1s, J3 from five J2s, J4 from three J3s

Note: The ITU-T recommendation G.702 on PDH bit rates specifies bit rates up to E4 (139.264 Mbps) only. E5 data rate (564.992 Mbps) is not specified in G.702. However, as few systems were initially tried at this transmission rate by multiplexing 4 E4s, the resultant multiplexed data rate of 564.992 Mbps is referred to as E5.

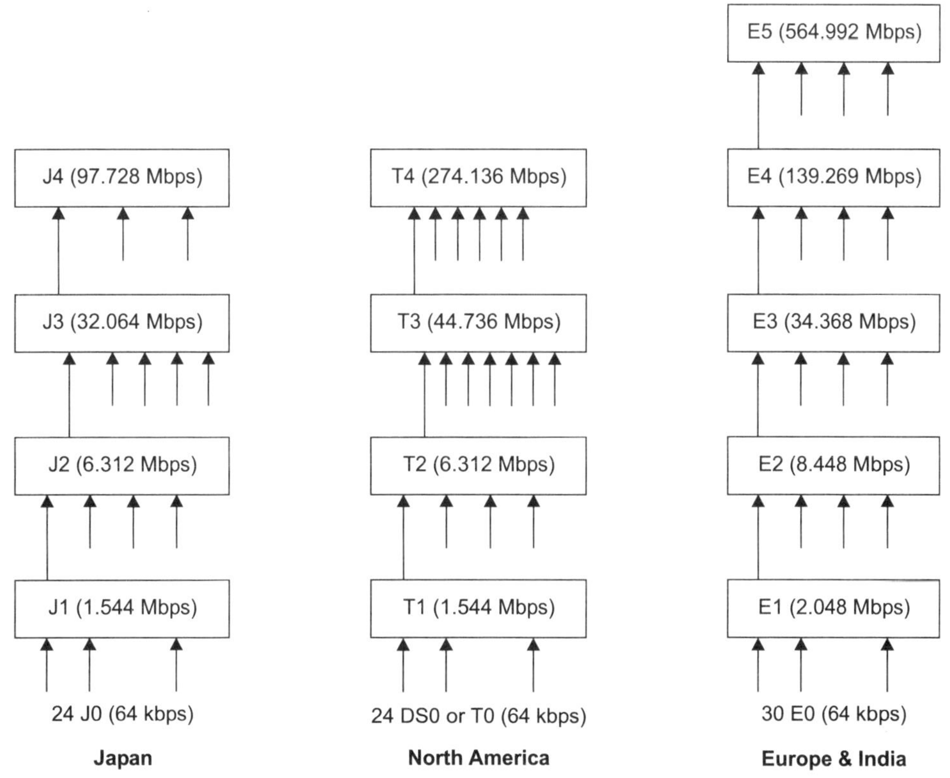

Fig. 2.4 *The three different plesiochronous digital hierarchies*

SDH is the second generation hierarchy. The transport module in SDH is denoted by STM (synchronous transport module). The basic module is STM-1 with a duration of 125 μs, which can carry 63 E1s or 42 E1s + E3 or 21 E1s + 2 E3 s or 3 E3 s or 1 E4 through suitable containers. It can thus be understood that SDH does not eliminate PDH but provides a mapping mechanism for PDH streams. It also supports other transmission modes like ATM and LAN traffic. STM frames are again in a hierarchy that combines four lower frame combinations, i.e., STM-4 provides for mapping of four STM-1 s in the duration of 125 μs by reducing the pulse width or operating at 4 ties the rate of STM-1. Transport modules (frames) and hierarchical rates of SDH defined in ITU-T recommendation G.707 are mentioned below.

STM-1 : 155.52 Mbps
STM-4 : 622.08 Mbps
STM-16 : 2488.32 Mbps (2.5 Gbps approx.)
STM-64 : 9953.28 Mbps (10 Gbps approx.)
STM-256 : 39813.12 Mbps (40 Gbps approx.)

To understand why SDH evolved requires a critical analysis of PDH first. This is covered in chapter 6. SDH process and applications are described in detail in chapters 7 to 10.

OTH, the third generation hierarchy, evolved primarily to manage certain functions like switching and routing at optical level rather than digital (electronic) for handling larger volumes of traffic than SDH on larger networks. OTH provides a digital wrap called OPU (optical payload unit) for STM-N, IP packets, ATM cells and Ethernet frames over optical channels. The hierarchical data rates in OTH are given below.

OPU-1 : 2.5 Gbps
OPU-2 : 10 Gbps
OPU-3 : 40 Gbps
OPU-4 : 160 Gbps

The formulation of OTH, its process description and applications are presented in chapters 11 to 15.

Summary

In this chapter, digital multiplexing is discussed not as a mere process of clubbing a number of channels but as a process for involving interleaving of channel data, alarm bits, framing sequences, error check bits, maintenance functions, etc. The characteristics of voice, data and video traffics are discussed. The evolution of digital multiplexing to meet increasing traffic demands, the work done by the CCITT study groups and later by the ITU-T study groups to standardise the developments in multiplexing are also discussed. PDH, SDH and OTH, the three generations of digital multiplexing are introduced.

References

[1] William Stallings, *ISDN and Broadband ISDN with Frame Relay and ATM*, 4th Edition, (Chapter 16 ,Asynchronous transfer mode), Prentice Hall, 1998.

3

PCM – The Entry Point

3.1 Basics of PCM Process

The basics of PCM process have been described in detail in several books on digital communications. The objective of this chapter is to summarise the basics rather than give a detailed treatment of it and quickly go on to explore the PCM process to present the points hitherto not covered. A detailed treatment of the basics of PCM with supporting mathematical expressions can be found in [1].

Pulse code modulation (PCM) involves 3 steps: sampling, quantization and encoding. Refer to Fig. 3.1 showing the basics of PCM.

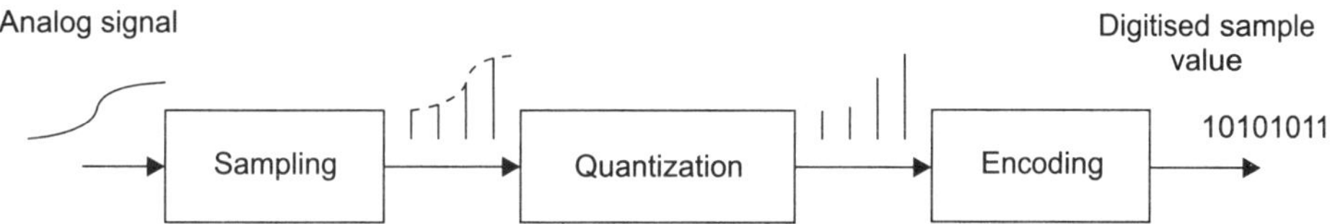

Fig. 3.1 *The Three steps in PCM*

In sampling, the instantaneous values of the analog signal are picked up at regular intervals called the sampling interval. Sampling interval or its inverse, the sampling rate, (number of samples collected per second) depends on the maximum frequency in the spectrum of the signal. According to Nyquist theorem, the sampling rate required for reconstructing a lowpass signal from the samples collected would be twice the maximum frequency in the signal. This criterion is known as 'Nyquist rate'. Undersampling (sampling at less than Nyquist rate) results in 'aliasing' or warping of the spectrum and oversampling (sampling at more than the Nyquist rate) results in excess of samples and more bits of information being carried than is necessary, resulting in waste of bandwidth.

The process of restricting the sample values to a finite set of levels is called quantization. As we know, the resultant samples of the sampling process performed on an analog signal have a continuous range of values. Encoding every possible value into a bit sequence implies an infinite number of bits to encoded at all possible values. In practice, we restrict the sample values to a few quantization levels and round off the exact sample value to the nearest quantization level value. Thus, it is possible to encode the samples with finite number of bits.

The process of representing quantized sample value by finite bit sequence is known as 'encoding'. If the quantization level is chosen as M, we can encode the value in 'n' bits, where

$$2^{n-1} < M \leq 2^n.$$

At the receiving end of the communications system, the received sequence is decoded to the quantization level to reconstruct the analog signal.

3.2 Exploring the PCM Process

The PCM process is simple to understand. Yet, certain aspects of the process need exploration rather than quick glance as they have a bearing on the practical implementation of multiplexing hierarchies. Non-uniform quantization, encoding criteria and framing for 30 channel PCM are some of the aspects that we will look at.

3.2.1 Non-uniform quantization

Why non-uniform quantization?

Non-uniform quantization is resorted to in PCM to take care of the fact that more quantization levels are required on the feeble side than on on the louder side to protect the message. The ITU-T recommendation G.711 for non-uniform quantization standards is as per the

- *A law* followed in India and Europe in 30-channel PCM
- μ *law* followed in North America and Japan in 24-channel PCM

These laws prescribe linear quantization for low input levels and logarithmic quantization for high input levels.

μ-law: μ-law compression is governed by the equations

$$|v| = \log(1 + \mu|s|) / \log(1 + \mu),$$

where s and v are normalised input and output voltages and

$$d|s| / d|v| = [\log(1 + \mu) / \mu] (1 + \mu|s|).$$

For low input levels, $\mu|s| << 1$ and hence the relationship is linear. For high input levels, $\mu|s| >> 1$ and the relationship is logarithmic. The practical value of μ is around 100.

A-law: If s and v are normalised input and output voltages, A-law compression is governed by the equations

$$|v| = A\,|s| \,/\, (1 + \log A) \text{ for } 0 \le |s| \le 1/A$$

$$= (1 + \log A\,|s|) \,/\, (1 + \log A) \text{ for } 1/A \le |s| \le 1,$$

$$d|s| \,/\, d|v| = (1 + \log A) \,/\, A \text{ for } 0 \le |s| \le 1/A$$

$$= (1 + \log A)\,|s| \text{ for } 1/A \le |s| \le 1.$$

The practical value of A used in PCM codec is around 100.

For non-uniform quantization, a compressor is used before the encoder and an expander after the decoder. The combination of compression and expansion as per A-law or μ-law is known as A-law companding or μ-law companding respectively.

Practical implications of A-law and μ-law in PCM

The following implications must be borne in mind while applying A-law or μ-law in PCM:

1. In A-law, all the resulting even bits are inverted before the octet is transmitted. This is to provide plenty of 0/1 transitions to facilitate the clock recovery process in PCM receivers. Thus, a salient A-law encoded PCM channel has the 8-bit samples coded as 0 × 55 instead of 0 × 00 in the octets (or 0 × 05 if the sign bit happens to be set); a salient μ-law encoded PCM has 0 × FF in its 8-bit samples.
2. *Relative performance of A-law and μ-law:* PSQM testing with random voice grade signal under ideal conditions yields MOS (mean opinion scores) of 4.45 for G.711 μ-law and 4.45 for G.711 A-law. PSQM testing with random voice grade signal under network stress yields MOS of 4.13 for G.711 μ-law and 4.11 for G.711 A-law.

Q 3.1 *What is PSQM testing?*

PSQM stands for perceptual speech quality measurement. The PSQM test is based on the ITU-T P.861 and P.50 recommendations.

P.861 defines a method for estimating the subjective quality of voice-band speech codecs. It is designed to measure the quality of voice channels, especially where the traditional transmission impairment measurements (TIMS) cannot be made. The method is based on research carried out in human psycho-acoustic perception. It assigns an objective figure of merit to a voice channel which is equivalent to that which would be assigned by surveying real human listeners.

P.50 defines an artificial voice for use in PSQM tests. It includes both genders (male and female) and is aimed at reproducing the essential characteristics of human speech for transmission testing. The characteristics include long-term average spectrum, short term spectrum, instantaneous amplitude distribution, voiced and unvoiced structure of speech waveform and syllabic envelope.

The PSQM test has been shown to correlate very well with MOS derived from actual listeners. The output of the PSQM algorithm is called the PSQM value. It indicates the degree of quality degradation caused by the whole communication system under test. The PSQM value ranges from 0 to 6.5. A value of 0 means that there is no degradation (perfect quality), while a value of 6.5 indicates the highest possible degradation. An MOS score has a range from 1 to 5 and is inversely proportional to the PSQM value. MOS 5 means excellent speech quality while MOS 1 indicates the other extreme.

3.2.2 Quantization levels for voice channel in PCM: practical considerations

Quantization levels are decided based on quantization noise considerations. Because the quantization process involves the rounding off of a sample value to the nearest quantization level (an approximation being carried out), quantization noise is introduced in the system. Signal-to-quantization-noise ratio is given by

$$SQ_NR\ (\text{dB}) = 1.8 + 6n,$$

where $2^n = M$ (the number of quantization levels).

Increasing the number of quantization levels improves SQ_NR but the bandwidth requirement increases because of the larger number of bits required for encoding the larger number of levels.

Q 3.2 According to ITU-T Rec. G.711, PCM encoding is carried out with 8 bits per sample, i.e., with 256 quantization levels. Why?

The ITU-T Rec. G.703 stipulates that the signal-to-quantization-noise ratio for telephone channel should be better than 45 dB. This can be achieved with $n = 8$ because with $n = 8$, the SQ_NR (in dB) $= 1.8 + 6(8) = 49.8$ dB. It also implies that the quantization levels are $2^8 = 256$ in number.

3.3 Exploring E1 Frame Structure

An E1 frame consists of 32 time slots numbered TS-0 to TS-31. Since the sampling rate for voice channel is 8 kHz, the sampling interval works out to 125 μs. This means that we have a sample from each channel to be uploaded onto E1 frame every 125 μs. Hence, the duration of E1 frame is 125 μs. A simplified way of looking at an E1 frame is to consider it as 30 time slots carrying 30 channels of digitised voice, of which TS-0 is for frame alignment or synchronisation data and TS-16 is for signalling data. Besides these, there are several other details that need to be explored.

The concept of physical and logical channel

TS-0 to TS-15 are for carrying bits (8-bit encoded speech samples or data bits) of the channels numbered 1 to 15. TS-17 to TS-31 are for carrying bits (8-bit encoded speech samples or data bits) of channels numbered 16 to 30. When carrying data traffic, we can aggregate more slots for one channel (for higher data rate requirement) or pack more data circuits into one time-slot (for low data rate requirement), i.e., one time-slot is one physical channel; the logical channel can be more than one time-slot or part of one time slot.

Frame alignment procedure

The TS-0 slot of every alternate frame carries FAS (frame alignment sequence), which is a 7-bit sequence, 0011011. The frame alignment procedure according to ITU-T Rec.G.706 is shown in Fig. 3.2.

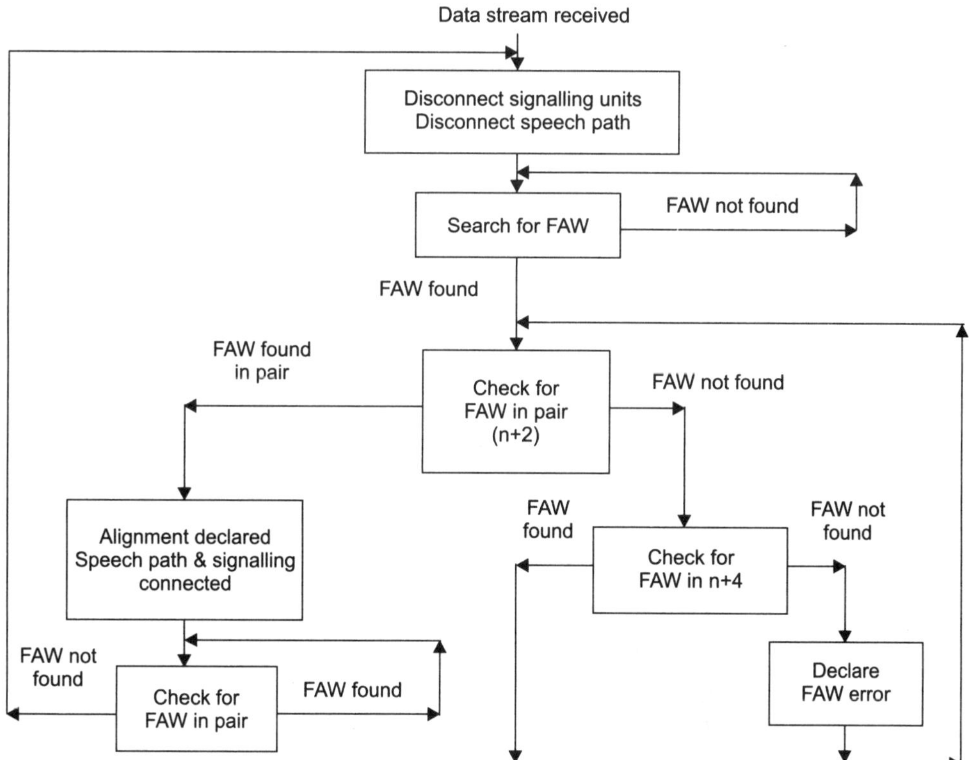

Fig. 3.2 *Frame alignment procedure in PCM*

CRC procedure

16 frames together constitute a multi-frame. The frames which have FAS in TS-0 (alternate frames) are numbered F0, F2, F4, F8, F10, F12 and F14. The leftover first bits of TS-0 of these frames carry the CRC bits, as shown in Fig. 3.3. CRC bits are numbered C1, C2, C3, C4. Since they are four of them carried one bit in alternating frames, 8 frames are required to send all the four CRC bits. These are computed on 8 frames, i.e., on one half of the multi-frame (2048 bits) called the SMF (sub-multi-frame) using the CRC polynomial $x^4 + x + 1$, and are placed as C1, C2, C3, C4 in the next-half of multi-frame or SMF. For the very first multi-frame, in the first SMF, these are set as 0s since there are no preceding multi-frames.

Q 3.3 When the divisor polynomial is $x^4 + x + 1$ (equal to number 19), the reminder can be any value up to 18, which require more than 4 bits. How is it possible to get only 4 CRC bits?

SMF on which CRC is being computed is multiplied first by x^4 and then divided by $x^4 + x + 1$. This gives a reminder of 4 bits only.

Frame No.	Bits 1 to 8 in TS-0							
	1	2	3	4	5	6	7	8
0	C1	FAW (0011011)						
1	0	1	A	S4	S5	S6	S7	S8
2	C2	FAW (0011011)						
3	0	1	A	S4	S5	S6	S7	S8
4	C3	FAW (0011011)						
5	1	1	A	S4	S5	S6	S7	S8
6	C4	FAW (0011011)						
7	0	1	A	S4	S5	S6	S7	S8
8	C1	FAW (0011011)						
9	1	1	A	S4	S5	S6	S7	S8
10	C2	FAW (0011011)						
11	1	1	A	S4	S5	S6	S7	S8
12	C3	FAW (0011011)						
13	E	1	A	S4	S5	S6	S7	S8
14	C4	FAW (0011011)						
15	E	1	A	S4	S5	S6	S7	S8

Fig. 3.3 *TS-0 bit assignment on a multi-frame*

Note: F0 to F15, i.e., 16 frames, make one multi-frame. Of these, the first half (F0 to F7) make first sub-multi frame, SMF-1, while the remaining (F8 to F15) make the second sub-multi frame, SMF-2.

Alarms and supervisory

The TS-0 of frames not carrying the FAS, i.e., F1, F3, F5, F7, F9, F11, F13 and F15 carry alarm and supervisory information. The A bit carries alarm information and S4 to S7 the supervisory bits.

The S bits may be used as a message-based logical channel within the E1 frame. When any intermediate access to data link on E1 between two termination points (e.g., drop–insert of channels at intermediate stations on a thin route) is required, the S bits can be used for reporting error performance to the central control or from peer to peer (e.g., paired MUX). In such situations it is essential to see that the logical path termination role played by CRC-4 be not invalidated or impaired. In other words, CRC-4 bits are to be end-to-end transparent. Hence, any changes to the S bits within an SMF at an intermediate station should not lead to recalculation of CRC-4 bits over the whole SMF; rather, their update as a linear recoding function in respect of specific deterministic binary changes should affect only the S bits [2].

E bits' utility

E bits indicate bit error detection in SMFs in CRC procedure. These E-bits are set to 0 until both basic frame and the CRC-4 multiframe alignment are established. Thereafter, the E-bits are be used to indicate the received errored SMFs by setting the binary state of one E-bit from 1 to 0 for each errored SMF.

According to ITU-T Rec. G.704, any delay between the detection of an errored SMF and the setting of the E-bit that indicates the error state must be less than 1 second.

1. *Utility of national bit and the bit adjacent to it:* The first bit in TS-0 of frames not carrying FAW is termed the national bit. It is set to 1 to link to international networks. In case of operation on exclusive path, this bit can be used as logical channel for NMS (network management system) connectivity. The data rate of such channel would be 1 bit for every 2 frames, i.e., 250 μs or 4 k baud. More details on this are given in section 3.6 in the discussion on NMS.

   The second bit in TS-0 of frames not carrying FAW is set to 1 so that FAS is not falsely identified when S bits assume the same values of respective positions in FAS.

2. *Signalling information:* TS-16 carries the signalling information of the 30 channels on a shared basis, i.e., although it is single physical channel (time-slot), it is logically 30 signalling channels. Since signalling is carried in the same frame, it is channel associated signalling. Bit assignments of TS-16 in a multi-frame are shown in Fig. 3.4. Classification of signalling (subscriber line signalling, inter-exchange signalling), the states thereof, various methods of carrying the same and the details of how PCM converts inter-exchange signalling to TS-16 bits are presented in the details of in [3].

   2 bits or 3 bits can be used to encode signalling states out of the 4 bits allotment of each channel. The remaining bit(s) are set by the manufacturer to

specific value. So, we have to check out the signalling bit map to ensure compatibility of signalling between primary MUXes of different makes.

Multi-frame alignment is required for the correct CRC procedure and any errors appearing there will be reported as alarms. Hence, there is provision for MFAW and multi-frame alarm in TS-16 of F0 in a multi-frame.

Frame No.	Bits 1 to 8 in TS-16							
	1	2	3	4	5	6	7	8
0	MFAW (0110)				Multi-frame alarm			
1	CH-1 Signal				CH-16 Signal			
2	CH-2 Signal				CH-17 Signal			
3	CH-3 Signal				CH-18 Signal			
4	CH-4 Signal				CH-19 Signal			
5	CH-5 Signal				CH-20 Signal			
6	CH-6 Signal				CH-21 Signal			
7	CH-7 Signal				CH-22 Signal			
8	CH-8 Signal				CH-23 Signal			
9	CH-9 Signal				CH-24 Signal			
10	CH-10 Signal				CH-25 Signal			
11	CH-11 Signal				CH-26 Signal			
12	CH-12 Signal				CH-27 Signal			
13	CH-13 Signal				CH-28 Signal			
14	CH-14 Signal				CH-29 Signal			
15	CH-15 Signal				CH-30 Signal			

Fig. 3. 4 *Bit assignment in TS-16 on a multi-frame basis*

3.4 Limitations of Normal Primary MUX

A primary MUX in its simplest form has been discussed in the previous section. Consider a thin-route application link of 10 to 12 stations with primary MUX equipment at every station with the requirement of a few channels at intermediate stations. At the two end stations, we require one primary MUX each. At intermediate stations, we require a pair of primary MUXes if we drop the channels to VF level from an E1 frame. In that case, after dropping channels to voice frequency (VF) level, we connect the required channels to the local exchange of intermediate station or other applications like data circuits and patch the balance channels to the 2nd primary MUX at the VF level. This is illustrated in Fig. 3.5.

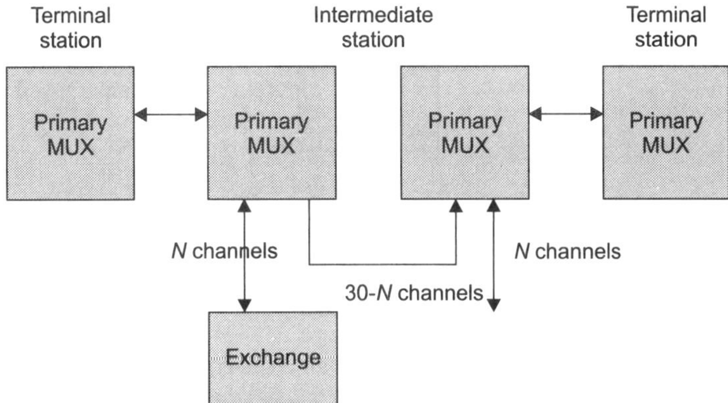

Fig. 3.5 *VF level patching at Intermediate station when not using 'drop-insert' at time-slot level*

The above arrangement increases the hardware requirement, which calls for more effort in wiring. Any alterations in channel distribution plan requires re-wiring at every station. If we require conferencing at the VF level, we have to consume more slots for VF level patching. Also, any interruption at an intermediate station terminates the link at that point.

There are other drawbacks in earlier versions of primary MUX: voice access modules were exclusive cards for each variety (voice access modules were different for E & M signalling, loop signalling etc.,) and there was not much flexibility in using them either. Also, data circuit interface cards were exclusive for various low speeds (below 64 kbps) and for 64 kbps speed. There was no aggregation beyond 64 kbps.

Definitely, we require a primary MUX with add–drop and other superior features. Let us summarise the features required and then examine how they can be achieved technically.

3.5 Add–Drop and Other Superior Features for Thin Routes

The following superior features are desirable in thin route transmission links where provisioning of channels is done through primary MUXes for various applications:

1. It should be possible to drop channels from the corresponding time slots in E1 frame and insert channels in the vacant slots, i.e., facilitate drop–insert at time-slot level. This should be feasible in both directions at an intermediate station, i.e., cross-connection of time-slots in the E1s coming from two different directions for local utilisation. (Refer to Fig. 3.6. Out of 30 slots from E1 of A-direction and 30 slots from E1 of B direction, a total 30 cross-connections can be for use at the intermediate station) This feature is known as 'add–drop with digital cross-connect'.

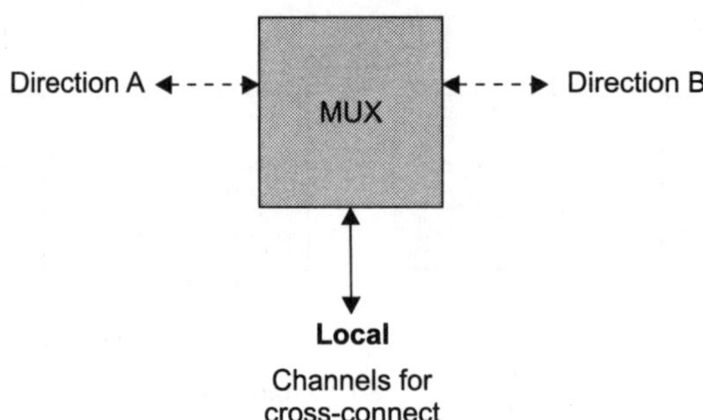

Fig. 3.6 *Add–drop with cross-connect*

2. There should be a common voice channel interface with adaptors for any type of voice circuit access instead of a variety in hardware. Also, it should be possible to adjust transmit and receive levels instead of manual intervention. This feature is known as 'flexibility in voice channel access'.
3. It should be possible to use VF channels in conference mode. This feature is known as 'voice conferencing'.
4. It should be possible to use time-slots for digital conferencing, i.e., we can apply digital logic for the bits coming through a time-slot from every station so that, it is helpful in polling type network where we act upon information uploaded by the network elements on a nominated time-slot. This feature is known as 'digital conferencing'.
5. It should be possible to share one PCM slot (64 kbps) among a number of sub-rate (less than 64 kbps) or low speed channels, to the extent possible. This is known as 'multiple low-speed data interface feature'.
6. It should be possible combine some slots for aggregating bandwidth as $N \times 64$ kbps.
7. It should be possible to create a spur link by branching up to 15 slots from two directions to a third direction and creating an E1. Thus, we need not drop any channel (or slot) at the spur derivation point. This feature is known as 'E1 branching'.
8. It should be possible to provide DSL (digital subscriber loop).
9. It should be possible to program the primary MUX through a local terminal or from a central location by downloading configuration details (channel drop–insert options and type of channels). This feature is known as 'programmability through EMS (element management system) locally or through NMS (network management system) from the central location. Also, it should be possible to display alarms and execute tele-commands for certain tests.

10. It should be possible to automatically change configuration in case of fault in any direction while working in add–drop and cross-connect mode. This is known as 'switch-over to alternate configurations in case of faults'.

11. It should be possible to monitor the performance of the equipment in terms of parameters like error seconds, BER, etc., as per ITU-T norms.

With all the above incorporated, we will have achieved the best possible features and flexibility. Indeed, such primary-MUXes are available and are being used in all new installations (E1 handling will continue in every transmission hierarchy because it is the basic entry point). Such primary MUXes are known as ' add/drop primary digital MUXes. These come with flexible time-slot allotment/aggregation, VF interfaces with conferencing facility, and are programmable through the local terminal or central supervisory terminal.

The general block diagram of such an add–drop MUX possessing superior features is shown in Fig. 3.7.

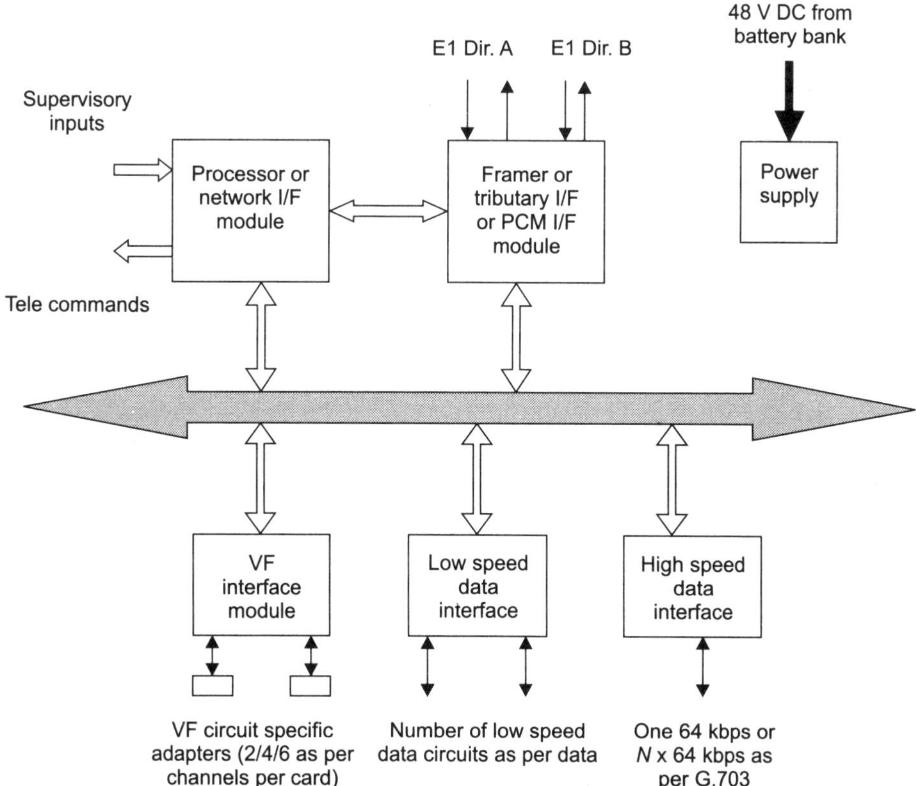

Fig. 3.7 *Block diagram of add–drop MUX with flexible features*

Such a MUX has the following modules in its hardware, besides the power supply and mother board:

a. The network interface module (similar to processor module of normal primary MUX but with superior capabilities)

b. Tributary module (similar to framer module of normal primary MUX, but with superior capabilities)

c. VF modules (common interface with adaptors for any type of voice circuit access)

d. $N \times 64$ data circuit modules (can be used for single 64 kbps interface or aggregate of few 64 kbps time-slots).

Let us see how the above nine superior features are possible technically.

3.6 How Superior Features are Achieved

3.6.1 Add–drop with digital cross-connect

In this section, we shall study how add–drop and cross connect are achieved through time-space interchange process, how local access is provided for channels and how signalling information is cross-connected in case of such channels.

The add–drop feature is achieved through time-space interchanger unit which can handle four time-division multiplexed serial bit stream up to 4 Mbps (Fig. 3.8).

The two 2 Mbps streams received from the network are presented to the two ports (A and B) of the interchanger. The third port (C) is used for interfacing to the local channel access modules and the fourth port (D) is utilised for voice conferencing. The unit performs serial-to-parallel format conversion and stores the data in an internal time-slot memory. For outgoing data, the unit uses a connection memory where the address of data is stored through the microprocessor. This connection memory specifies the interchange by defining when and how the time slots leave the time slot memory. The output section performs the inverse function of the input section and converts the byte-wide data to a corresponding serial stream.

The information of the time slot interchange among the four ports are defined by the users through the use of the element management system (EMS) locally or remotely through network management system (NMS) from a central control. With the help of a semi-graphical interactive screen, the user can define which time slots are to be added and dropped, which channels are to undergo conferencing with local voice and which the channels are to pass through. All the switching and interchange operation is done digitally, and so the voice channels do not suffer any loss of quality. All the mapping and downloading operation of the cross-connect information to the respective nodes is done through software, which opens up logical channel through unused *N*-bits in TS0 of frames not carrying FAW. (1 bit in every alternate frame as discussed in section 3.3 (shown in Fig. 3.3) makes it 1 bit every 250 μs, i.e., a virtual 4 kbps channel through E1 frame connects the nodes for NMS function.)

How local access is provided in add–drop and cross-connect

Port C of the time-space interchanger is used to add and drop local channels. Timing information as obtained from the line interface unit is used to derive respective time-slot pulses. The channel units interface to these synchronising pulses including MCLK and the serial 2 Mb data stream to get access to specific time-slot windows for communication.

How signalling information is cross-connected in add–drop function

The signalling information is sent through time slot 16 of the E1 frame, as seen in section 3.3. The time-space interchanger unit cross-connects complete 8 bits of a time-slot and does not deal with the interleaved signalling information. The signalling information is extracted separately in a TS16 processor which also performs a time-space switching of signalling information in parallel to the time-slot interchanger. The processed signalling information is then presented to two line interface units and local channel modules in the desired format.

Signalling information of conference channels are derived by logically ANDing the corresponding signalling data.

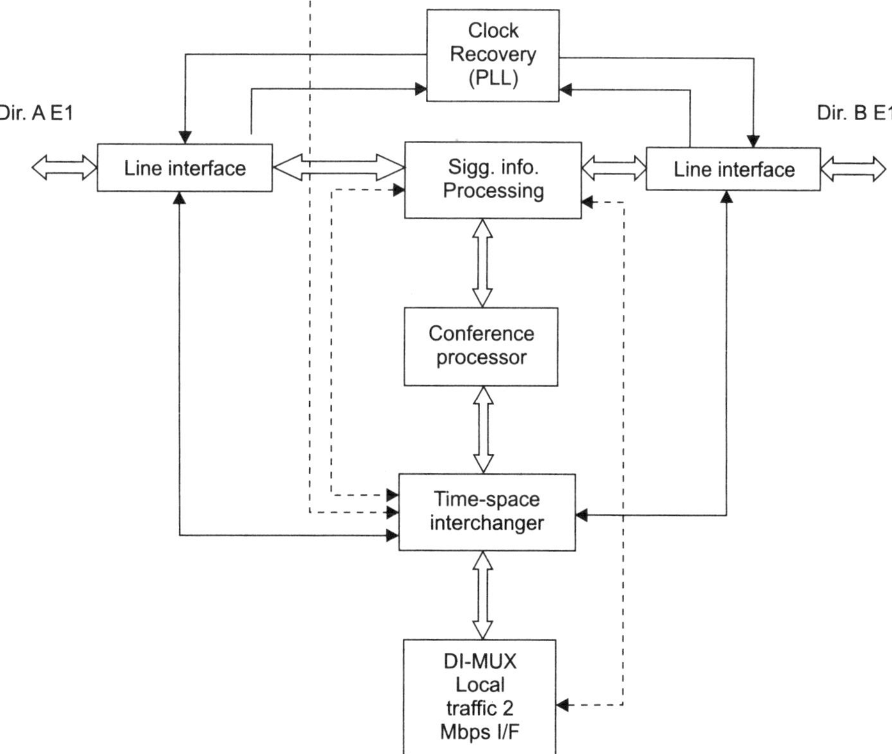

Fig. 3.8 *Time-space interchange function in framer or PCM I/F card*

3.6.2 Flexibility in voice channel access

PCM, the entry point of transmission network, should support any type of voice channel from switch (exchange), viz.,

a. 2W/4W E&M Interface for STD circuits: In case of 2W E&M, two conductors are for transmitting as well as receiving (operating through 2W/4W hybrid) and two are for signalling. They are termed E lead and M lead respectively, i.e., a total of four conductors:

   In case of 4W E&M, two conductors are for transmitting, two for receiving (no need for hybrid) and two for signalling (termed E lead and M lead), i.e., a total of 6 conductors

b. Subscriber interface: A subscriber of an exchange can be at a different station (remote subscriber), in which case the subscriber interface card on PCM provides the subscriber-loop connection (2 conductors which carry speech as well as signalling) for the remote subscriber.

c. Exchange interface: PCM connected to the exchange has an exchange interface corresponding to the subscriber interface. Refer to Fig. 3.9 showing the remote subscriber arrangement.

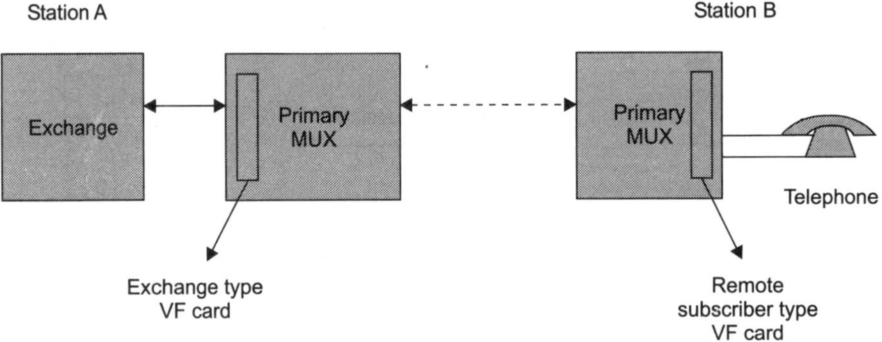

Fig. 3.9 *Remote subscriber of station A at station B*

d. Hot line interface: A hot line is provided between two stations via PCM slot using the hot-line interface card. Two conductors are extended to the telephone from the hot-line interface card (Fig. 3.10).

A detailed description of the circuit operation of each type of card, signalling in telephony (subscriber line signalling, exchange-to-exchange signalling), methods and standards, the encoding of signalling information into bits suitable for carrying in TS-16 of E1 frame, etc., are covered in detail in[2].

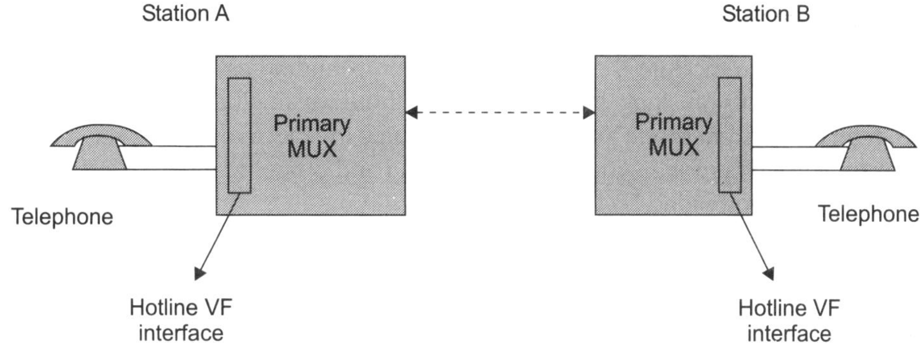

Fig. 3.10 *Hot line between station A and station B*

It may be noted that voice channel cards of different varieties were provided in earlier versions of primary MUX equipment. In the latest versions, to enhance the flexibility of the type of access, we have basic voice access card or module with piggy-back attachments and jumper settings to select variety. The basic voice module provides 4W voice with 3 bit signalling as per ITU-T format. This is then converted to the necessary signalling format in the plug-in module as required for the specific interface.

The level of the speech signal is adjusted by the PCM codec on the access module internally, depending on certain parameters. These parameters are downloaded to the codec through a backplane serial bus by the network interface module. The total range of adjustment is 20 dB in steps of 0.1 dB for both transmit and receive direction.

3.6.3 Voice conferencing

Conferencing of voice is achieved first by normalising the A-law encoded data, adding it to the respective normalised data of the associated channel and then converting back the normalised sum data to the A-law format. This is done in real time in digital format by a separate conferencing unit conforming to G.711 in A-law format.

3.6.4 Digital conferencing

Conferencing of digital data is different from that of voice and is sometimes required in a network to implement polled data. This is achieved by logical ANDing of the respective time slot data. Since the requirement of digital conferencing is less, some of equipment do not offer and some offer this feature as an option on one of the time-slots.

The fourth port of the time-space interchanger presents the individual channel data which require conferencing to the external conferencing unit. These channel data are presented to the digital conferencing processor (implemented using VLSI gates and EPROM), the output of which is then received back through the same port for necessary routing through time-space switch.

3.6.5 Multiple low speed data interface (on single PCM slot) feature

The low speed data module is primarily meant for extending asynchronous low speed RS232C data circuits on single PCM time-slot, without wasting bandwidth. The module works in bit transparent mode and acts as a data pipe irrespective of the data. It can also work in synchronous mode, wherein the incoming and outgoing data are synchronised to the clock signal provided by the module

Each time slot of the 2 Mb frame provides a bit rate of 64 kbps. Low speed data channels can very well be accommodate into such a high bit rate channel. However, for better utilisation of the information bandwidth, a V.110 sub multiplexing format is used to support multiple low speed data on a single time slot.

This module supports baud rate ranging from 600 baud to 19.2 k baud. The number of bits of a time slot required for the same is given below.

Low speed channel data rate	No. of bits required in time-slot per channel	No. of channels which can be supported on one slots
4800 baud	1 bit	8
9600 baud	2 bits	4
19.2 k baud	4 bits	2

(The reader can check that sub-rate supported = 64000/number of bits used and number of such channels through one PCM time slot = 8/number of bits.

Higher speed data circuits (48/57.6/64 kbps), however, consume the complete time slot for transmission. Only one such circuit can be carried on one PCM time slot.

The channel supports all the handshaking lines required for the V.24 protocol. The handshaking signals are status information of the DTE (computer) and DCE (modem) connected at two nodes of the network which are just extended through the PCM link.

To exploit the network bandwidth efficiently for data communications, generally, such interface is allowed on one or two slots only (as we do not have low-speed data circuit requirement in large numbers). Moreover, utilisation of these slots is optional, i.e., if we do not have any low-speed data circuit requirement at all, we can also use these slots for voice or normal speed data circuit.

3.6.6 Aggregating bandwidth through $N \times 64$ data interface

The $N \times 64$ data interface is meant for extending the WAN-based data network or video conference circuits through primary MUX. It works in synchronous mode with V.35 or V.36 interfaces with the clock slaved from data communication equipment (DCE). The module supports configuration from 64 kbps ($N = 1$) up to 1920 kbps ($N = 30$).

3.6.7 E1 branching feature

This module is tested to create a third tributary by cross-connecting the required time slots from port-A or port-B to port-C. It gives out 2 Mbps in line-coded format (HDB3 as per ITU-T recommendations). By this module it is possible to create a spur link by diverting the required time slots without dropping channels at voice/data level. All the NMS functions are available in spur link.

3.6.8 DSL interface

The digital subscriber line interface circuit is designed to provide ISDN basic rate access (2B+D) at U-interface. Full duplex digital transmission at 160 kbits on a single twisted pair in conjunction with the high performance 2B1Q line code allows the IDSL to meet the loop length requirements geater than 4 km of the digital subscriber loop at the U-interface.

It consists of two parts, IDSL-LT (at the MUX side) and IDSL-NT(at the remote side).

IDSL-LT

The two B-channels and the D-channel to be transmitted on the line are input to the LT (on the 2 Mb bus) into the transmit interface. The sync word and the maintenance bits are added to the data which is then formatted, scrambled and digitally encoded into 2B1Q symbols. The transceiver transfers full duplex, time division multiplexed data at 160 kbps. This includes two 64 kbps PCM data channels (B-channels), a 16 kbps signalling channel (D-channel) and a 16 kbps for synchronisation and overhead requirements. The two 64 kbps channels are defined as Bl and B2 channels and carry subscriber information (data). The LT will transfer both B-channels transparently from the 2 BM bus to the line port and vice versa once the device has acquired super frame synchronisation with the NT. The transceiver uses the 2B1Q line code, which is a four level pulse amplitude modulated (PAM) code.

IDSL-NT

The function of NT is to provide customer equipment interface to the U interface and to communicate to the LT synchronously. The NT can be configured through network monitoring system as per requirement. It derives the two B-channels from the IDSL line and directs them to either G.703 or V.35 interface (as per configuration) to provide customer access. The NT consists of two G.703 port, two V.35 port and one U Interface. Power to the NT is to be provided by a separate 5V power supply unit.

3.6.9 About EMS or NMS

How configuration is done through EMS or NMS

The configuration information is downloaded into the equipment through a serial interface connected locally to a PC or laptop on which EMS (element management system) software runs. The downloaded information is stored in an EPROM on the network interface module (also known as processor module).

The configuration information can also be downloaded into the equipment remotely from a central supervisory terminal running the NMS software. The data channel for downloading configuration information is a polled data channel derived through unused national bits (of E1 frame which are in TS-0 of frames not carrying FAW) or through the overhead bit-stream of the transmission equipment (e.g., service channel of digital MW radio system). Bits 4 to 8 of TS-0 not carrying FAW (SV bits S4 to S8) are generally used for carrying peer-to-peer (MUX to MUX) monitoring information like identifier code and error information.

The various information/command that can be usually downloaded from the EMS or NMS are channel cross-connect table, interface configuration table, status response command, diagnostic command, alarm history upload command and global time-set command.

Communication between NMS and MUX for supervision and parameter uploading/downloading is done through a frame format whose structure can be as shown in Fig. 3.11 below, or with some modifications like parity byte, frame type byte, shorter frame for acknowledgement, etc.

Frame Header	Address	Equipment identity	Command	Data	End of frame

Fig. 3.11 *Frame format for communication between NMS and MUX*

Set of functions performed by EMS/NMS

The set of functions performed by EMS/NMS are listed below.

1. Configuration of MUX, card-wise, including parameters like Tx-Rx levels
2. Configuration of timeslots for normal flow, add–drop and cross-connect
3. Execution of supervisory functions (diagnostics and telecommands)
4. Status and alarm history logging
5. Viewing of logged data with specified query
6. Real time monitoring (consequent actions like alternate configurations)
7. Performance monitoring (like error seconds, BER measurement, etc.) of equipment/network.

Functions 1 and 2 are already described. Alarms and supervisory functions (diagnostics, telecommands, i.e., Sl No. 3, 4 and 5 above) are explained now. Sl No. 6 is explained in section 3.6.9 while Sl. No.7 is self explanatory.

Alarms monitoring function of NMS

The various alarms to be uploaded to the NMS (from framer or network interface module) and also to be displayed as per ITU-T recommendations are:

a. Tributary alarms: (both for tributaries A and B)
    i. 2 Mb loss (RXF)

ii. All ones (AIS)

iii. Sync loss (SYNC)

iv. Remote multiframe alarm (RMA)

v. BER

vi. Remote alarm (RA)

b. System related alarms

i. Problem with network interface module or tributary module

ii. Configuration/Set-up error

iii. Power supply unit alarm

c. External alarms (inputs)

Apart from the network alarms, the equipment has provision to monitor (through opto-isolation) some external alarms. These are ON/OFF type in nature, e.g., charger ON/OFF, door open/close, blower or AC ON/OFF, etc. Some equipment also monitor the battery voltage and the temperature inside the equipment.

Supervisory role of NMS

Supervisory functions fall into two categories:

a. Diagnostic commands

i. Tributary-A local loopback: 2 Mb stream coming out of Trib-A is looped back into the input.

ii. Tributary-B local loopback: Same as that of Trib-A but with Trib-B.

iii. Tributary-A remote loopback: 2 Mb stream going into of Trib-A is looped back into the input.

iv. Tributary-B remote loopback: Same as Trib-A but with Trib-B.

v. Digital bypass: This command is primarily meant for isolating the local access. However due to the flexibility of the channel mapping through the cross-connect table, the configuration need not be simple connections of channels of tributary A to tributary B. This may be utilised as switching of a modified table of a particular FlexMUX remotely, temporarily for some diagnostic purpose.

vi. Injection of 1 kHz tone to a particular voice channel. Level of this tone is typically 0 dBm.

vii. Loopback of individual time slot of either tributary A, tributary B or local port.

b. Telecommands

Manufacturers usually provide the feature of remote control of output relays (2 Nos.) on the framer (network interface) module. Through telecommands, these can be switched ON and OFF. After the command is received, the network interface module executes the command and uploads the success or failure of the command execution. The network management system prompts the status of command execution on the screen accordingly.

Using potential free contacts of these relays, we can turn-on or off certain functions like lighting, change-over of power supply, starting generator, etc.

Alarm and supervisory history storage and query

Usually, manufacturers make provision on framer (network interface) module to store history of the alarm status in an internal FIFO buffer with time-stamp. Whenever there is a change of alarm or supervisory input/output status, it stores the complete snapshot of the equipment marked with time. Resolution of this event logging is usually 1 minute and all the changes that have occurred in the previous 1 minute are latched in the snapshot. Thus, more than one event can be acquired in each snapshot. The recommended depth of the FIFO is 32 snapshots. If the number of snapshots is more than 32, then the first snapshot is popped out to make room for the latest one. Past alarms can be queried and retrieved through EMS/NMS.

3.6.10 Switch-over to alternate configurations in case of faults

Faults in provisioning of cross-connects at intermediate station should not lead to interruption in services on the entire link. In such situations, it should be possible to change-over to alternate configurations. Hence, the equipment management supports downloading of the following cross-connect tables to take care of the various conditions of the network:

a. Cross-connect table for normal operation

b. Alternate cross-connect table when E1 from one direction (say direction-A) is having a major alarm (recovery from fault in direction-A).

c. Alternate cross-connect table when E1 from second direction (say direction-B) is having major alarm (recovery from fault in direction-B).

d. Modified cross-connect table when E1 of direction-A receives a remote alarm (recovery from remote fault in direction-A).

e. Modified cross-connect table when E1 of direction-B receives a remote alarm (recovery from remote fault in direction-B).

f. Digital bypass.

The cross-connect tables change-over automatically from one to the other depending on the various alarm conditions of the E1 from the two directions. Digital bypass table is however switched remotely through a command from the network supervisor through the NMS system. The tables are prepared by the user through a menu driven screen on the NMS computer, which are then downloaded. All the tables can be updated online as and when required.

All the above features are achieved by the functional distribution mainly among the processor modules (referred to as network interface module by some manufacturers) and the framer module (referred to as tributary or PCM interface module), as discussed in the next section. We are not presenting any block diagram or design of these modules here. The basic design and circuits are presented in [3].

3.6.11 Performance monitoring

Performance data like error second, severed error second, degraded minutes, etc., of a particular terminal being selected by the user can be observed. In case of any problem in any terminal, the 'Super-Net' can find the faulty area very intelligently. All alarm status can be scanned with occurrence time in a history file in computer and the details of alarms can be retrieved from the logged file on the basis of a query, afterwards. Apart from this, the alarm history of 32 events are stored in the system and can be retrieved using EMS when a central monitoring is not present.

Q 3.4 What are the primary MUXes available which support the superior features discussed so far?

Webfil's Flexi-MUX and Nokia's add–drop MUX support the features discussed so far, with certain variants.

3.7 Functions of Vital Modules of Add–drop Programmable MUX

3.7.1 Functions of processor/network interface module

The processor or network interface module takes care of the EMS/NMS interface and alarm acquisition function from various internal and external sources. In the previous versions, microprocessors were used. Nowadays, microcontrollers are being used.

The network interface module is used for exchange of information among the network manager, the framer (PCM interface or tributary module) and the various access modules in the MUX. It has minimum four communication interfaces for information exchange:

One serial port for EMS (local terminal)
One serial port for NMS (connected to service channel of transmission system)

One serial link to framer
Back-plane bus for access modules

During power on, the microcontroller of the processor (network interface) module downloads the cross-connect tables from the EEPROM to the tributary module, which invokes the desired structure of the drop/insert configuration of the site. The microcontroller also downloads the analog and digital channel parameters to the particular access modules through backplane bus.

The processor (network interface) module analyses all the packets it receives from the network supervisor and responds only to the one meant for the particular equipment. In response to the status response command, it uploads the current status of the equipment.

3.7.2 Functions of framer/PCM interface/tributary module

The framer or PCM interface or tributary module are the different names given to the same module by different manufacturers. If the processor (network interface) module is the brain of the MUX, the tributary module is the heart of the system. It has interfaces to the E1s from two directions and realises the add–drop function of the channels through digital cross-connect. The functions of this module are:

1. Extract the master clock of the network from the incoming E1 and synchronise the node (primary MUX) to it.
2. Generate 2 Mbps internal TDM bus for channel units.
3. Synchronise the incoming E1s and local E1.
4. Cross-connect 64 kbps digital data along with the signalling data.
5. Perform PCM summing of conference channels along with logical summing of corresponding signalling data.
6. Generate 2 Mbps HDB-3 encoded interface signals complying with ITU-T G.703.
7. Collect and analyse equipment and line faults and take necessary actions.
8. Control jitter to a limit well below the ITU-T limit (G.823) (refer to chapter 5).
9. Report to processor module about the alarm status of the E1s and the module itself.
10. Support a communication channel through the use of national bits of TS0 for NMS control.

Summary

In this chapter, the process of PCM is explored in respect of aspects not usually covered in standard books, viz. framing, CRC generation and supervisory functions. The concepts of 'drop–insert', physical channel, logical channels are explained. How the features of programmability and network management can be achieved through E1 frame-borne logical channels are explained. Newer interfaces like conference facility, aggregate data channels, E1 branching and DSL are presented. The role of NMS is explained.

References

[1] Sreekanth, P. V., *Digital Microwave Communication Systems* (chapter 3), Universities Press, 2003.

[2] Alarms and Supervisory, ITU-T Rec. G.706, Annexure-C.

[3] Sreekanth, P. V., *Digital Microwave Communication Systems* (chapter 7), Universities Press, 2003.

4

PDH MUX-DEMUX Process

4.1 Multiplexing Methodology in PDH

4.1.1 Concept of bit interleaving

One of the most challenging tasks faced by a team leader is to get going a team of people who are themselves individuals with strong identities and ideas. In plesio-synchronous digital hierarchy, a similar situation prevails because the tributaries to be integrated at every hierarchical interface are characterised by the following disparate properties:

1. They are not synchronous, i.e., not clocked by the same source.
2. They might have arrived at the hierarchical interface through different sources, have varied trails and transits in the network, and are hence subjected to variations in clock rate.

In other words, the tributaries are not at identical rate but are nominally at same data rate; the variations being, of course, within certain prescribed limits, viz.,

2.048 Mbps ± 50 ppm for E1
8.448 ± 30 ppm for E2
34.368 ± 20 ppm for E3
139.264 ± 15 ppm for E4.

The challenging task of integrating such bit streams or tributaries at every hierarchical level is accomplished in PDH by bit interleaving, i.e., taking one bit at a time from each tributary cyclically, and while doing so, resorting to 'justification process' to tackle the variations in the clock.

Let us examine the process of justification.

4.1.2 Concept of justification

In plesio-synchronous digital multiplexing, four tributaries, nominally at the same data rate but having variations within certain limits are multiplexed to form the next higher rate data stream in the hierarchy. To elaborate,

Four E1s at 2.048 Mbps ± 50 ppm are multiplexed to form E2 at 8.448 Mbps.
Four E2s at 8.448 Mbps ± 30 ppm are multiplexed to form E3 at 34.368 Mbps.
Four E3s at 34.368 Mbps ± 20 ppm are multiplexed to form E4 at 139.264 Mbps.

Since the clocking rates of the four tributaries being multiplexed are not exactly identical, it is not possible to get the next higher clock rate as an exact multiple of 4 times the individual tributary's clock rate. Also, the clock rate of any individual tributary is not constant but varies within tolerance limits. In such a situation, we cannot make any one of the tributary clock as the master clock for the integration process.

Refer to Fig. 4.1(a) showing the problem of write-in and read-out clocks with slightly differing speeds.

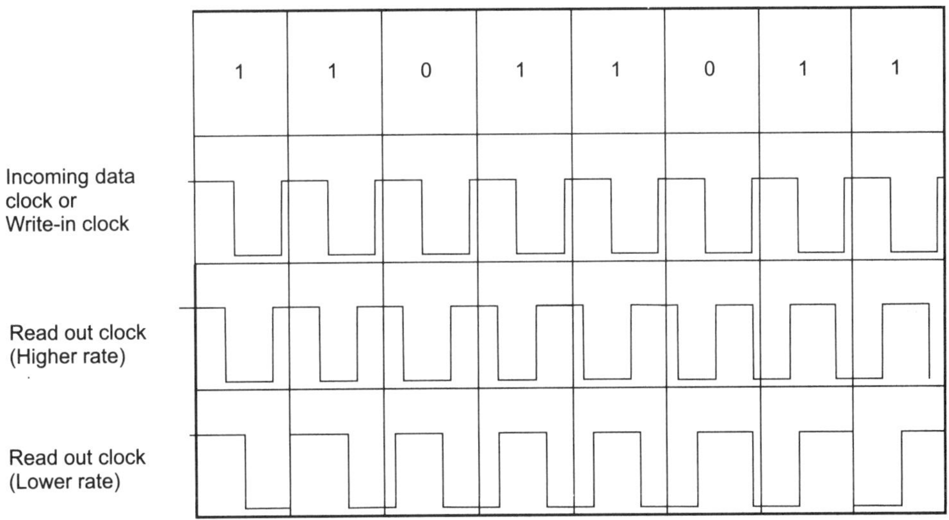

Fig. 4.1(a) *Write-in and read-out clocks with slight differences in their speeds*

Suppose we write the incoming data in the memory at every falling edge of receive clock and read out using a higher speed clock. To begin with, both clocks have no lead/lag with respect to the other, i.e., they are at the same phase. After some clock intervals the time lag between the clock signals may become equal to one full clock interval. At this juncture we are about to read the bit from the memory twice. Hence, whenever the time lead of read-out clock over write-in clock (or phase lead) becomes one full clock interval (or 2π radians of phase lead), we have to suppress the read-out clock in order to avoid re-reading of the same bit. This process of suppressing read-out clock is known as perforation.

However, the outgoing frame is clocked to a higher rate read-out clock. Hence, whenever we perforate read-out clock, we have to insert a dummy bit in the outgoing data frame in the corresponding bit interval. This is not the end of it. We have to intimate the receiving end about the insertion of the dummy bit through certain control bits which are to be inserted in the outgoing data frame itself. This entire process is known as justification. Refer to Fig. 4.1(b) showing justification while multiplexing on the transmit side.

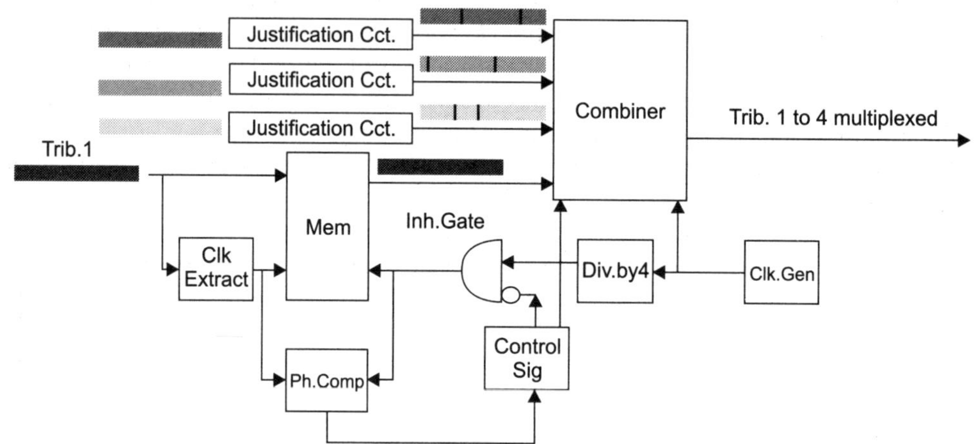

Fig. 4.1(b) *Justification process on the transmit side while multiplexing 4 tributaries*

To summarise, justification is the process involving the following processes:

1. *Elastic memory write-in and read-out:* Storing of the incoming data stream bit by bit in the elastic memory using the write-clock (lower rate clock), which is extracted from the incoming data stream, and reading it out bit by bit by using the read-out clock which is at higher rate.
2. *Perforation of read-out clock:* Suppressing the read-out clock for one interval whenever the time lead of the read-out clock over the write-in clock reaches a full clock period of the read-out clock (or a pre-determined percentage of the read-out clock.)
3. *Insertion of justification information bit:* Insertion of a dummy bit in the outgoing data frame in the bit interval corresponding to a perforated interval.
4. *Insertion of justification control bits:* Insertion of justification control bit / bits in the outgoing data frame to intimate the receiving end about the justification action so that the bit received in the position nominated for justification bit is given the correct interpretation (treated as dummy bit if justification is done; otherwise to be treated as information bit).

De-justification, the converse of justification is illustrated in Fig. 4.2.

The justification process described above is known as positive justification because the read-out clock of the elastic memory is faster than the write-in clock. If the read-out clock is slower than the write-in clock, then, whenever the time lag of the read-out clock over the write-in clock (or phase lag) becomes one full clock interval (or 2π radians of the phase difference), one bit from the elastic memory is missed. To avoid this, the read-out address counter has to be decremented. This process is known as negative justification.

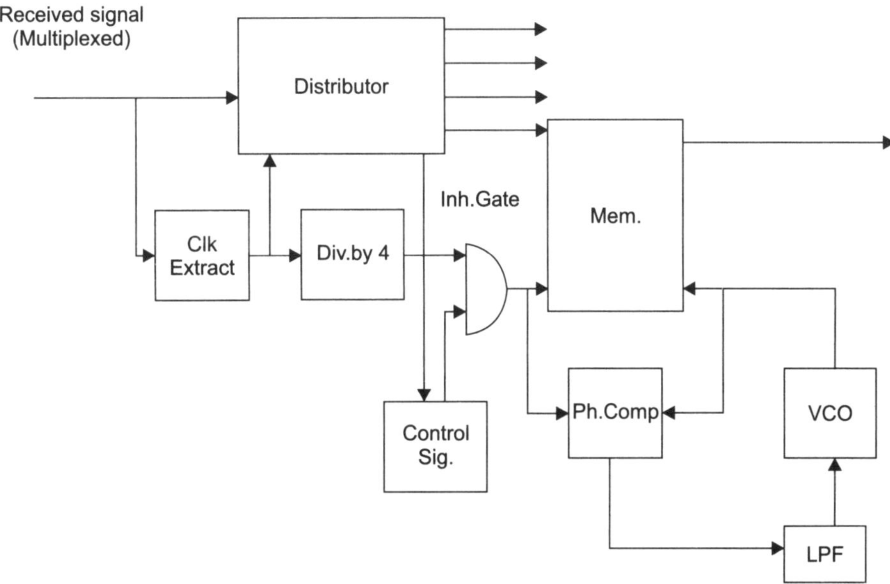

Fig. 4.2 *De-justification on the receive side*

In PDH systems, positive justification is used.

4.2 ITU-T Recommendations Relevant to PDH Systems

4.2.1 Recommendations for E-hierarchy

The ITU-T recommendations of relevance to PDH systems of E-hierarchy are the following.

G.742 G.742 specifies the characteristics of second order digital multiplexing equipment operating at 8448 kbps (E2 frame) using positive justification. The frame length specified is 848 bits. Jitter transfer function, alarms and consequent actions are also defined in G.742.

G. 751 G.751 specifies the characteristics of third order digital multiplexing equipment operating at 34368 kbps (E3 frame) and fourth order digital multiplexing equipment operating at 139264 kbps (E4 frame) using positive justification (frame length is 1536 bits for E3 and 2928 bits for E4). Jitter transfer function, alarms and consequent actions are also defined in G.751.

G.823 G.823 is concerned with the control of jitter and wander in PDH networks of E–hierarchy (E1, E2, E3, E4).

G.742 and G.751 are discussed in this section. Discussion on G.823, which deals with jilter control and jitter transfer function is deferred to chapter 6 on PDH networks and reliability issues.

Q 4.1 What are G.744, G.745, G.753 and G.754 about?

These are not concerned with PDH systems of E-hierarchy.

- G.744 is for 8448 kbps bit rate but the frame length is 1056 bits.
- G.745 is for second order multiplexing with positive/zero/negative justification and frame length of 1056 bits.
- G.753 gives the characteristics of third order digital multiplex equipment operating at 34368 kbps using positive/zero/negative justification. The frame length is 2148 bits for multiplexing four 8448 Mbps streams of G.744 format.
- G.754 gives the characteristics of fourth order digital multiplex equipment operating at 139264 kbps using positive/zero/negative justification. The frame length is 2176 and it is for multiplexing four 34368 Mbps streams of G.753 format.

4.2.2 ITU-T recommendation G.742 for 8.448 Mbps (E2)

As already mentioned, ITU-T recommendation G.742 is for second order digital multiplex equipment operating at 8.448 Mbps using positive justification. The salient points of this recommendation are given below.

1. The nominal bit rate should be 8.448 Mbps. The tolerance on that rate is ±30 ppm.
2. The frame structure should be as per Table 4.1.
3. The loss or recovery of frame alignment is assumed under the following circumstances; the consequent actions are also given below.
   - Loss of frame alignment is assumed to have taken place when four consecutive frame alignment signals have been incorrectly received in their predicted positions.
   - When frame alignment is assumed to be lost, the frame alignment device should decide the alignment to be effectively recovered when it detects the presence of three consecutive frame alignment signals.

- The frame alignment device, having detected the appearance of a single correct frame alignment signal, should begin a new search for the frame alignment signal when it detects the absence of the frame alignment signal in one of the two frames that follow.

Table 4.1 *ITU-T recommendations for frame structure of E2*

Tributary rate	2.048 Mbps
Number of tributaries	4
Frame structure	**Bit number**
	Set-I
Frame alignment signal (1111010000)	Bits 1 to 10
Alarm indication to remote equipment	Bit 11
Bit reserved for national use	Bit 12
Bits from tributaries	Bits 13 to 212
	Set-II
Justification control bits (*Cj*1)	Bits 1 to 4
Bits from tributaries	Bits 5 to 212
	Set-III
Justification control bits (*Cj*2)	Bits 1 to 4
Bits from tributaries	Bits 5 to 212
	Set-IV
Justification control bits (*Cj*2)	Bits 1 to 4
Justification information bits	Bits 5 to 8
Bits from tributaries	Bits 9 to 212
Frame length	848 bits
Bits per tributary	206 bits
Maximum justification rate per tributary	10 kbps
Nominal justification ratio	0.424

4. The method of multiplexing is cyclic bit interleaving in the order of tributary numbering, with positive justification
5. The justification control signal should be distributed and should use the *Cjn*-bits mentioned in the frame structure. Positive justification should be indicated by the signal 111 and a no justification by the signal 000. Majority decision is recommended.
6. Bit 11 of Set-I is used to transmit alarm indication to the remote (pairing MUX) equipment when specific fault conditions are detected in the MUX equipment.
7. Bit 12 of Set-I is reserved for national use. When the digital path crosses the border, this bit is fixed at 1.

8. The digital multiplex equipment should detect the following fault conditions and alarms should to be generated for alerting to the equipment at the other end.
   - Failure of power supply
   - Loss of an incoming 2048 kbps signals at the inputs of the multiplexer
   - Loss of the incoming signal at 8448 kbps at the input of the DEMUX (AIS)
   - Loss of frame alignment
9. After the detection of a fault condition, appropriate actions should be taken as specified below.
   - Prompt maintenance alarm indication to be generated to signify that the performance is below acceptable standards and that maintenance attention is required locally.
   - When an alarm indication signal (AIS) is detected at the input of the demultiplexer, the maintenance alarm indication associated with the loss of frame alignment should be inhibited promptly.
   - The location and provision of any visual and/or audible alarms activated by this maintenance alarm indication is left to the discretion of individual manufacturer.
   - Alarm indication to a remote multiplex equipment is generated by changing the state from 0 to 1 for bit 11 of Set-I of the outgoing frame of the multiplexer.
   - In case of an alarm indication signal (AIS) at the input of the demultiplexer, an 'all 1's signal (AIS)' is applied to all the four tributary outputs from the demultiplexer.
   - In case of the absence of input of a tributary to the multiplexer, AIS is applied to the time slots of the 8448 kbps signal at the output of the multiplexer corresponding to the relevant 2048 kbps tributary.
   - The method of transmitting the AIS at the output port of the multiplexer in time slots corresponding to a faulty input tributary should be such that the status of the justification control digits is controlled to ensure that the AIS is within the tolerance specified for that tributary.
   - The bit rate of the AIS at the output of the multiplexer equipment or at the output of the demultiplexer equipment should be in accordance with the interface specifications.
   - The equivalent binary content of the AIS at the output of the DEMUX or at the output of the MUX is nominally a continuous stream of 1s. The strategy for detecting the presence of the AIS should be such that the AIS is detectable

even in the presence of an error ratio of 1×10^{-3}. However, a signal with all the bits except the frame alignment signal in the 1's state should not be mistaken for an AIS.

- The fault detection and generation of alarms for passing information to the other end should be completed within a time limit of 1 ms.

4.2.3 ITU-T recommendation G.751 for 34.368 Mbps (E3)

As already mentioned, ITU-T recommendation G.751 is for third order digital multiplex equipment operating at 34.368 Mbps and fourth order digital multiplex equipment operating at 139.264 Mbps using positive justification. The salient points in this recommendation for 34.368 Mbps are given below.

1. The nominal bit rate should be 34.368 Mbps. The tolerance on that rate is ±20 ppm.
2. The frame structure should be as per Table 4.2. Rest of the points are similar to those in the case of E2.

Table 4.2 *ITU-T recommended frame structure for E3*

Tributary rate	8.448 Mbps
Number of tributaries	4
Frame structure	**Bit number**
	Set-I
Frame alignment signal (1111010000)	Bits 1 to 10
Alarm indication to remote equipment	Bit 11
Bit reserved for national use	Bit 12
Bits from tributaries	Bits 13 to 384
	Set-II
Justification control bits ($Cj1$)	Bits 1 to 4
Bits from tributaries	Bits 5 to 384
	Set-III
Justification control bits ($Cj2$)	Bits 1 to 4
Bits from tributaries	Bits 5 to 384
	Set-IV
Justification control bits ($Cj2$)	Bits 1 to 4
Justification information bits	Bits 5 to 8
Bits from tributaries	Bits 9 to 384
Frame length	1536 bits
Bits per tributary	378 bits
Maximum justification rate per tributary	22.375 kbps
Nominal justification ratio	0.436

4.2.4 ITU-T recommendation G.751 for 139.364 Mbps (E4)

The salient points of ITU-T recommendation for 139.364 Mbps are given below.

1. The nominal bit rate should be 139.264 Mbps. The tolerance on that rate is ± 15 ppm.
2. The frame structure should be as per the Table 4.3. Rest of the points are similar to those in case of E2 and E3.

Table 4.3 *ITU-T recommended frame structure for E4*

Tributary rate	34.368 Mbps
Number of tributaries	4
Frame structure	**Bit number**
	Set-I
Frame alignment signal (1111010000)	Bits 1 to 12
Alarm indication to remote equipment	Bit 13
Bit reserved for national use	Bits 14 to 16
Bits from tributaries	Bits 17 to 488
	Set-II
Justification control bits ($Cj1$)	Bits 1 to 4
Bits from tributaries	Bits 5 to 488
	Set-III
Justification control bits ($Cj2$)	Bits 1 to 4
Bits from tributaries	Bits 5 to 488
	Set-IV
Justification control bits ($Cj3$)	Bits 1 to 4
Bits from tributaries	Bits 5 to 488
	Set-V
Justification control bits ($Cj4$)	Bits 1 to 4
Bits from tributaries	Bits 5 to 488
	Set-VI
Justification control bits ($Cj5$)	Bits 1 to 4
Justification information bits	Bits 5 to 8
Bits from tributaries	Bits 9 to 488
Frame length	2928 bits
Bits per tributary	723 bits
Maximum justification rate per tributary	47.563 kbps
Nominal justification ratio	0.419

Mux and demux processes at various hierarchical levels are described in detail in the following sections.

Note: Though E5 is defined in ITU-T Rec. G., we are not discussing the same here as there are very few E5 PDH links operational in the world and none in India. We have SDH links on OFC offering better transmission speeds over long-haul links.

4.3 Multiplexing E1s to form E2

Integration of 4 E1s (2.048 Mbps tributaries) to form one E2 (8.448 Mbps) stream as per ITU-T recommendation G.742 is described in this section.

Bit stream integration is to be done by bit interleaving and positive justification, tributary wise. Hence, we need elastic memory with justification control circuits associated with each E1 tributary and a multiplexed memory (buffer) to hold the bit-interleaved multiplexed frames of E2 format so as to be sent out frame by frame.

Refer to Fig. 4.3(a). It shows the elastic memory with justification control circuits associated with one E1 tributary. Similar setup exists for the other three tributaries also. This figure is to be read in conjunction with the frame structure of E2 stream given in Fig. 4.3(b).

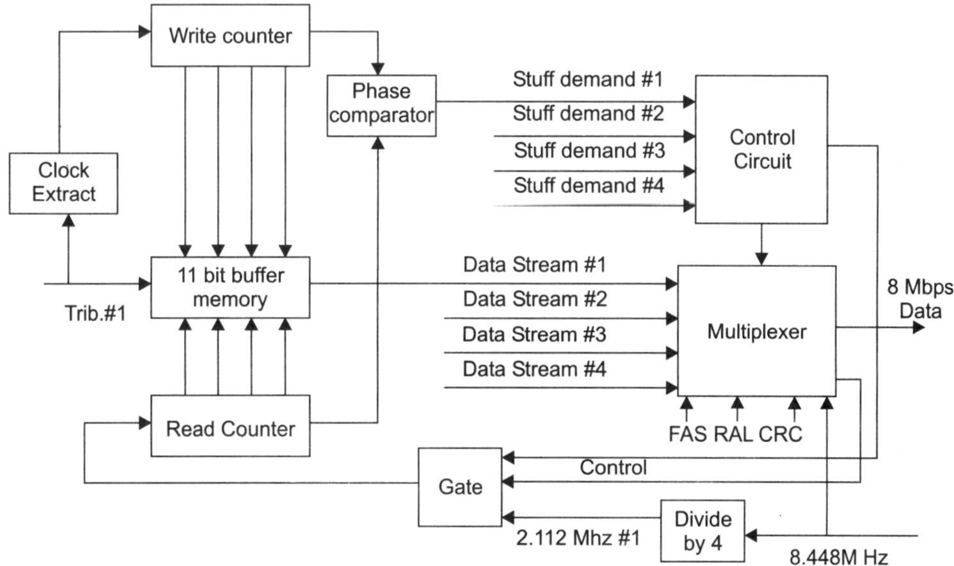

Fig. 4.3(a) *Multiplexing 4 E1s into E2*

Let us examine the E2 frame structure given in Fig. 4.3(b). Each E2 frame has 848 bits, starting with 12 bit pattern (consisting of 10 bit frame alignment sequence or FAS,

1 alarm bit and 1 national bit). Bits from the incoming E1 tributaries are organised as sets, with each set having a bit from each tributary by bit-interleaving process. Each set of 4 bits is called bit-set. There are 206 such bit-sets plus 3 sets with nomenclatures C1, C2 and C3 (which carry 'justification control bits' or JCBs). C1.1, C2.1, C3.1 are the bits used to keep a note in the frame that justification has been done in respect of tributary-1. Similarly, C1.2, C2.2, C3.2 convey that justification has been done in respect of tributary-2 and so on. All 206 bit-sets except bit-set 155 carry information bits of individual tributaries. Bit-set 155 may carry information bit or justification bit as the case may be.

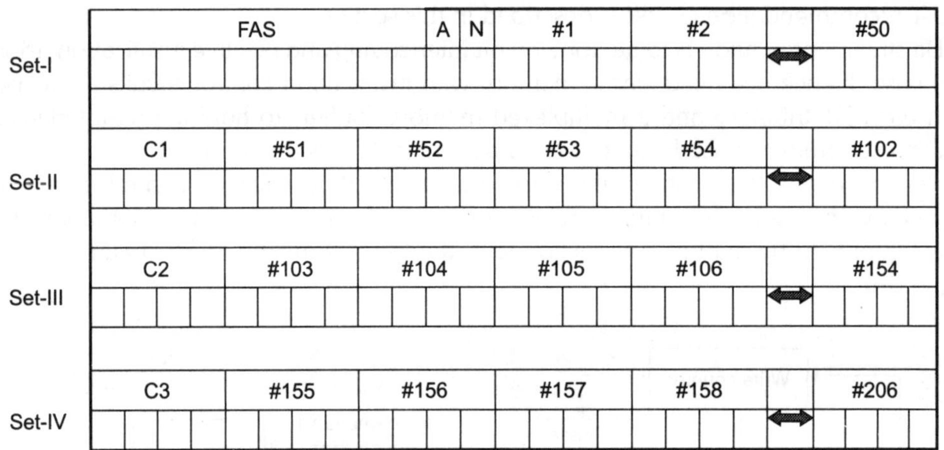

Frame size: 212 bits × 4 sets = 848 bits
FAS: 1111010000. A: remote alarm (1-alarm) N: Resvd. #1 to #206 (excl.#155): Each a set of 4 bits; 1 bit from each tributary
C1 to C3: Justification control bits (1st bit of each for Trib.1, etc. – 111 ⇒ Justification)

Fig. 4.3(b) *Frame structure of E2*

Let us refer to Fig. 4.3(a) again. The 8.448 MHz reference clock is divided by 4 to get 2.112 MHz clock, which is applied as read-out clock to the elastic memory associated with each of the four 2.048 Mbps tributaries. The elastic memory's write-in clock is 2.048 MHz recovered from the incoming stream (tributary).

The phase difference between 2.048 MHz write-in clock and 2.112 MHz read-out clock is monitored. When the phase difference exceeds a threshold, there is need for justification. Then the read-out clock is perforated, i.e., the data bit already written in elastic memory is not read out to E2 frame (under construction in the multiplexed memory) but a dummy bit is stuffed in the corresponding bit position of bit-set 155 in the E2 frame.

The above arrangement gives rise to certain questions. If these questions are answered, the PDH multiplexing concept is in principle de-mystified.

Q 4.2 *Why is positive justification done while multiplexing E1s to form E2 when negative justification scheme is also available?*

When we multiplex E1s to form E2s (or other hierarchical levels of PDH), we are executing bit interleaving and creating a frame. We require frame alignment sequence and also exchange of alarm information. Hence extra bits are needed. So, the multiplexed stream data rate is to be greater than 4 times E1. In other words, the per tributary read-out rate is to be greater than E1. This also helps in tackling variations in E1 clock speed (within ±50 ppm).

Q 4.3 *We have reserved bit-set 155 for justification bits. What is the guarantee that justification is done in respect of all tributaries in this set when we do not have control over their clocking rates?*

We have to judge the need for justification, tributary-wise, in the same bit-set instant. Only then can we locate JCBs together in C1, C2 and C3 (else, we have no means of knowing in which bit-set justification is done for which tributary). The nominated bit-set for judging the need for justification is bit-set 155.

There is no guarantee that the lead of read-out clock over write-in clock of elastic memory reaches full clock period at the same instant for all tributaries. Hence, instead of the criterion of time lead reaching full clock period for carrying out justification, we have to follow the criterion of time lead exceeding certain threshold level of clock period to judge the need for justification in the same bit-set for all the tributaries.

Q 4.4 *What is the criterion to decide justification? Is it the accumulation of time lead (of 2.112 MHz over 2.048 MHz) being greater than one clock period of 2.112 MHz?*

During the picking up of bits to fill-up set-155, judging of the need for justification is done on each tributary at the respective bit instant in set-155. If the accumulated time lead of 2.112 MHz clock over 2.048 MHz clock is within 50% of one clock period of 2.112 MHz, by the time next set-155 is filled-in (in next frame), the accumulated time lead does not exceed one clock period of 2.112 MHz. Thus, we are sure of avoiding re-reading of the same bit from the elastic memory. This is the reason why 50% is chosen as the threshold rather than a full clock interval.

Q 4.5 *We are sending JCBs of all tributaries together as a set. If so, why are 3 bits per tributary used to convey that justification has been done?*

The vital information whether justification is done or not is conveyed with some redundancy so that a majority voting criterion can be followed at the receiving end. This reduces the scope for errors.

Q 4.6 Why are the 3 bits used scattered in the frame?

Since data traffic errors are bursty in nature, it is better to keep the justification control bits scattered in a frame.

Q 4.7 Why is the frame size chosen as 848 bits?

Frame size is chosen in such a way that the rate of information in-flow from the incoming tributary is equal to the average rate of information out-flow per tributary through the multiplexed stream. Let us assume a frame size of N bits. We will prove N to be 848 in the case of E2.

In a frame of N bits of E2, the actual information bit flow per tributary is $(N - k)/4$, where k depends on the number of FAS, alarm and national bits, justification bits and justification control bits (JCBs) in each frame. This outflow should match the inflow through the incoming E1 on an average.

In a frame of N bits of E2, the information bits are

$N - 24$ (12 bits of FAS, A and N and 12 JCBs, i.e., a total of 24 are subtracted) if no justification is done, or $N - 28$ (12 bits of FAS, A and N; 12 JCBs and 4 justification bits, i.e., total 28 bits subtracted) if justification is done.

Number of frames per second = Bit rate/Number of bits per frame = $8.448 \times 10^6/N$.

Outflow of information bits = Frames per second × Number of information bits per frame
$= (8.448 \times 10^6 / N)\ (N - 24)$ if justification is not done
$= (8.448 \times 10^6 / N)\ (N - 28)$ if justification is done.

Therefore, outflow per tributary
$= (8.448 \times 10^6 / N)\ (N - 24) / 4$ if justification is not done
$= (8.448 \times 10^6 / N)\ (N - 28) / 4$ if justification is done.

This gets simplified as
$(2.112 \times 10^6 / N)\ (N - 24)$ if justification is not done,
or
$(2.112 \times 10^6 / N)\ (N - 28)$ if justification is done.

The inflow is at 2.048 Mbps rate per tributary.

Equating the outflow to inflow, we get:
$(2.112 \times 10^6 / N)\ (N - 24) = (2.048 \times 10^6)$ if justification is not done
or
$(2.112 \times 10^6 / N)\ (N - 28) = (2.048 \times 10^6)$ if justification is done.

This gets simplified as
$2.112\ (N - 24) = 2.048\ N$ if justification is not done

or

2.112 $(N - 28) = 2.048\, N$ if justification is done.

From the above we get $N = 792$ if justification is not done and 924 if justification is done. We cannot take an average of these two values to decide N. Rather, we should take the weighted average.

If the nominal justification or the probability that the justification done is p, then the probability that justification is not done is '$1 - p$'. Then, $N = (1 - p)\, 792 + p\, 924$.

According to ITU-T Rec. G.742, the nominal justification ratio while integrating E1 streams to form E2 is 0.424, i.e., $p = 0.424$. Using this value we get $N = 748$.

Q 4.8 Why is the elastic memory size chosen as 11 bits?

FAS is a 10-bit sequence, hence elastic memory is chosen as 11 bits.

4.4 Demultiplexing E2

Refer to Fig. 4.3(c). In the process of demultiplexing E2 for retrieval of 4 E1s, the bits from the incoming E2 are segregated tributary-wise by the demultiplexer and are written in the elastic memory using 2.112 MHz write-in clock. This 2.112 M Hz write-in clock is derived from the incoming E2 (8.448 MHz clock extracted and then 'divided by 4'). The data bits from the elastic memory are readout using 2.048 MHz clock.

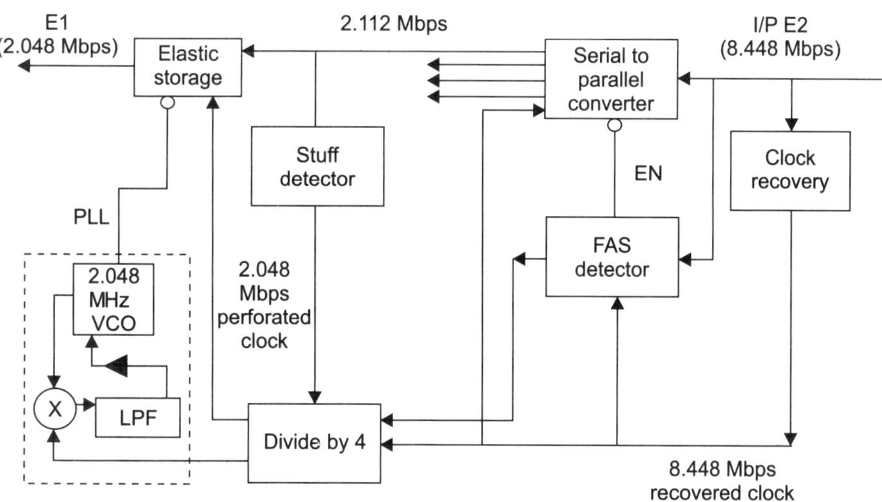

Fig. 4.3(c) *Demultiplexing E2 to get E1*

FAS, A & N bits and JCBs need not be written in the tributary elastic memory. Hence, during these intervals, the 2.112 MHz write-in clock is perforated. Also, when majority of JCBs pertaining to a tributary are 1's, we know that justification was done in bit-set 155 and that the write-in clock needs to be perforated. Because of these perforations, the effective rate of writing in the elastic memory is 2.048 Mbps and hence can be read out by a 2.048 MHz clock.

We require one full frame size of E2 as buffer memory before we transfer the data to the demultiplexer because the JCBs in current frame tell about justification action in bit-set 155 of the previous frame. The JCB monitoring unit applies the perforation command with the necessary timing delay.

4.5 Multiplexing E2s to Form E3

Integration of four E2s (8.448 Mbps tributaries) to form one E3 (34.368 Mbps) tributary is described in this section. The bit stream integration is achieved by bit interleaving and positive justification, tributary-wise. Therefore, we need an elastic memory that can hold justification control circuits associated with each E2 tributary and multiplexed memory (buffer) for holding the bit-interleaved multiplexed frames of the E3 format that are to be sent out frame by frame.

Refer to Fig. 4.4(a). It shows the elastic memory with justification control circuits associated with one E2 tributary. Similar setups exist for the other three tributaries as well. This figure is to be read in conjunction with the frame structure of E3 stream given in Fig. 4.4(b).

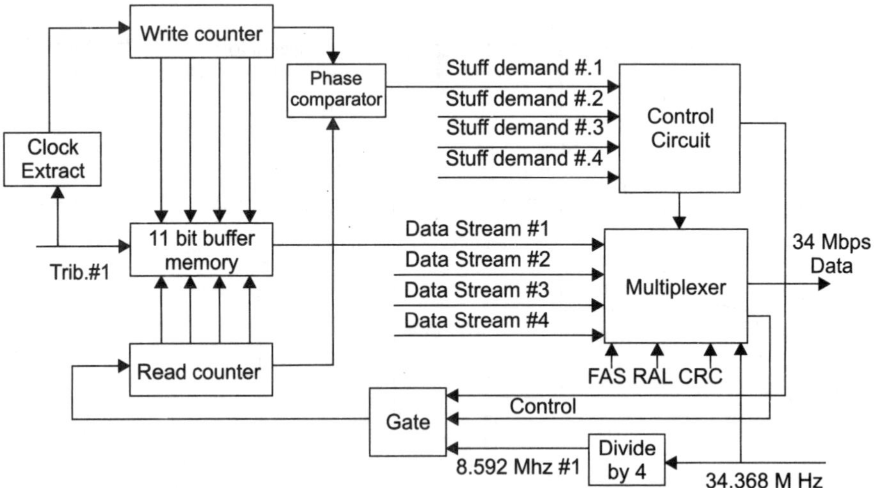

Fig. 4.4(a) *Multiplexing 4 E2s to form E3*

Let us examine the E3 frame structure given in Fig. 4.4(b). Each E3 frame has 1536 bits, starting with 12 bit pattern (consisting of 10 bit frame alignment sequence or FAS, 1 alarm bit and 1 national bit). Bits from the incoming E2 tributaries are organised as sets, with each set containing a bit from each tributary inserted by bit-interleaving process. A set of 4 bits is called a bit-set. There are 378 such bit-sets and 3 sets with nomenclatures C1, C2, C3 which carry the justification control bits or JCBs; bits C1.1, C2.1, C3.1 used indicate that justification has been done in respect of tributary 1. Similarly, C1.2, C2.2, C3.2 convey information whether justification has been done in respect of tributary-2 and so on. All 378 bit-sets except bit-set 284 carry information bits of individual tributaries. Bit-set 284 may carry information bit or justification bit as the case may be.

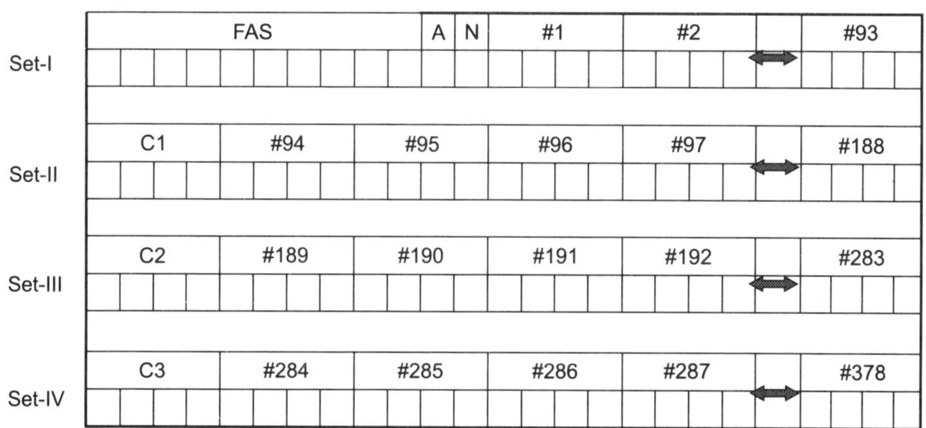

Frame size: 384 bits × 4 sets = 1536 bits
FAS: Frame alignment signal (1111010000) A: Remote alarm (1-Alarm) N: Resvd. #1 to #378 (excl.#284): Each a set of 4 bits, one bit from each tributary
C1 to C3: Justification control bits (1st bit of each for trib.1, etc. – 111⇒ Justification)

Fig. 4.4(b) *Frame structure of E3*

Let us refer to Fig. 4.4(a) again. The 34.368 MHz reference clock is divided by 4 to get 8.592 MHz clock which is applied as read-out clock to the elastic memory associated with each of the four 8.448 Mbps tributaries. The elastic memory's write-in clock (8.448 MHz) is recovered from the incoming stream (tributary).

The phase difference between 8.448 MHz write-in clock and 8.592 MHz read-out clock is monitored. When it exceeds a threshold, there is a need for justification. Hence, the read-out clock is perforated, i.e., the data bit already written in elastic memory is not read out to the E3 frame (under construction in multiplexed memory) but a dummy bit is stuffed in the corresponding bit position of the bit-set 284 in the E3 frame.

All the exploring questions related to multiplexing have been tackled in section 4.3 and need not be repeated here except for one pertaining to frame size.

Q 4.9 Why is frame size 1536 bits for E2 to E3 multiplexing?

The frame size is chosen in such a way that the rate of information inflow from the incoming tributary is equal to the average rate of information outflow per tributary through the multiplexed stream. Let us assume a frame size of N bits. We will prove N to be 1536 in the case of E3.

In a frame of N bits of E3, the actual information bit flow per tributary is $(N - k) / 4$, where k depends on the number of FAS, alarm and national bits, justification bits and justification control bits (JCBs) in each frame. This outflow, on the average, should match the inflow through incoming E2.

In a frame of N bits of E3, the information bits are $N - 24$ (12 bits of FAS, A and N; 12 JCBs, i.e., a total of 24 to be subtracted) if no justification done.
or
$N - 28$ (12 bits of FAS, A and N; 12 JCBs; and 4 justification bits, i.e., a total of 28 bits to be subtracted) if justification is done.

Number of frames per second = Bit rate/Number of bits per frame = $34.368 \times 10^6/N$.

Outflow of information bits = Frames per second × Number of information bits per frame
$= (34.368 \times 10^6 / N)\ (N - 24)$ if justification is not done
$= (34.368 \times 10^6 / N)\ (N - 28)$ if justification is done.

Therefore,
Outflow per tributary
$= (34.368 \times 10^6 / N)\ (N - 24) / 4$ if justification is not done
$= (34.368 \times 10^6 / N)\ (N - 28) / 4$ if justification is done.

This gets simplified as –
$= (8.592 \times 10^6 / N)\ (N - 24)$ if justification is not done
$= (8.592 \times 10^6 / N)\ (N - 28)$ if justification is done.

Inflow is at the rate of 8.448 Mbps per tributary.

Equating the outflow to the inflow, we get
$(8.592 \times 10^6 / N)\ (N - 24) = (8.448 \times 10^6)$ if justification is not done
$= (8.448 \times 10^6)$ if justification is done.

This gets simplified as
$8.592\ (N - 24) = 8.448\ N$ if justification is not done
$= 8.448\ N$ if justification is done.

From the above we get $N = 1432$ if justification is not done and 1670 if justification is done. We cannot take an average of these two values to decide N. Rather, we should take their weighted average.

If the nominal justification is 'p', i.e., the probability that justification done is p, then the probability that justification is not done is $1 - p$. Then, $N = (1 - p).1432 + p.1670$. According to ITU-T Rec. G.751, nominal justification ratio while integrating E2 streams to form E3 is 0.436, i.e., $p = 0.436$. Using this value we get $N = 1536$.

4.6 Demultiplexing E3

Refer to Fig. 4.4(c) showing demultiplexing of an E3 into four E2s.

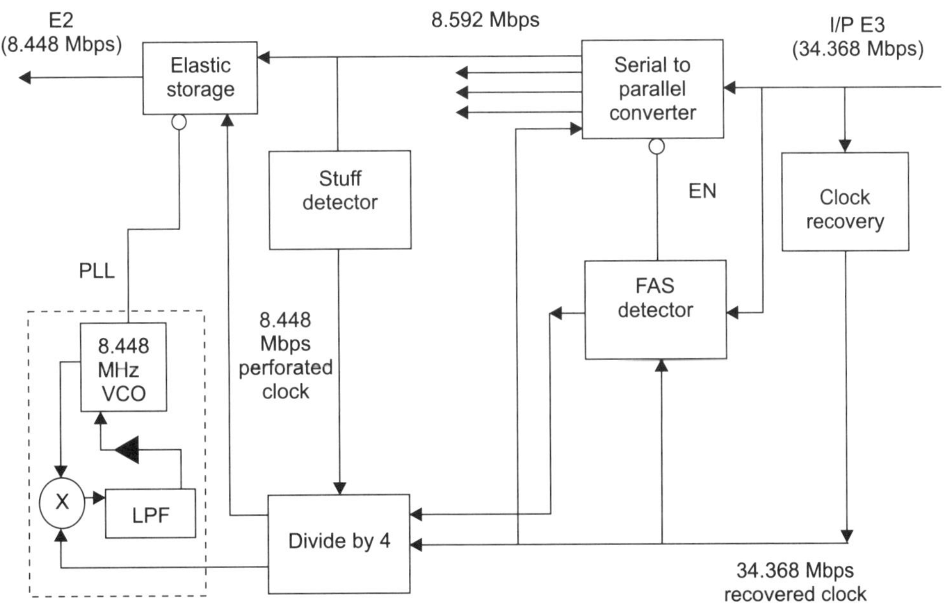

Fig. 4.4(c) *Demultiplexing E3 into E2s*

In the process of demultiplexing E3 for retrieval of four E2s, the bits from incoming E3 are segregated tributary-wise by the demultiplexer and written in the elastic memory using the 8.592 MHz write-in clock. This 8.592 MHz write-in clock is derived from the incoming E3 (34.368 MHz clock extraction and then 'divide by 4' action). Data bits from elastic memory are read out using 8.448 MHz clock.

FAS, A & N bits and JCBs need not be written in the tributary elastic memory. Hence, during these intervals the 8.592 MHz write-in clock is perforated. Also, when the majority of JCBs pertaining to a tributary are 1s, we know that justification was done in bit-set 284 and write-in clock needs to be perforated. Because of these perforations, the effective rate of writing in the elastic memory is 8.448 Mbps and hence can be read out by the 8.448 MHz clock.

We require one full frame size (of E3) as buffer memory before we transfer the data to the demultiplexer because the JCBs in current frame tell about the justification action in bit-set 284 of the previous frame. The JCB monitoring unit applies perforation command with necessary timing delay.

4.7 Multiplexing E3s to Form E4

Integration of four E3s (34.368 Mbps tributaries) to form one E4 (139.264 Mbps tributary) is described in this section.

Bit stream integration is to be achieved via bit interleaving tributary-wise with positive justification. So, we need elastic memory with justification control circuits associated with each E3 tributary and a multiplexed memory (buffer) to hold the bit-interleaved multiplexed frames of E4 format so that they can be sent out frame by frame.

Refer to Fig. 4.5(a) illustrating multiplexing of 4 E3s into an E4.

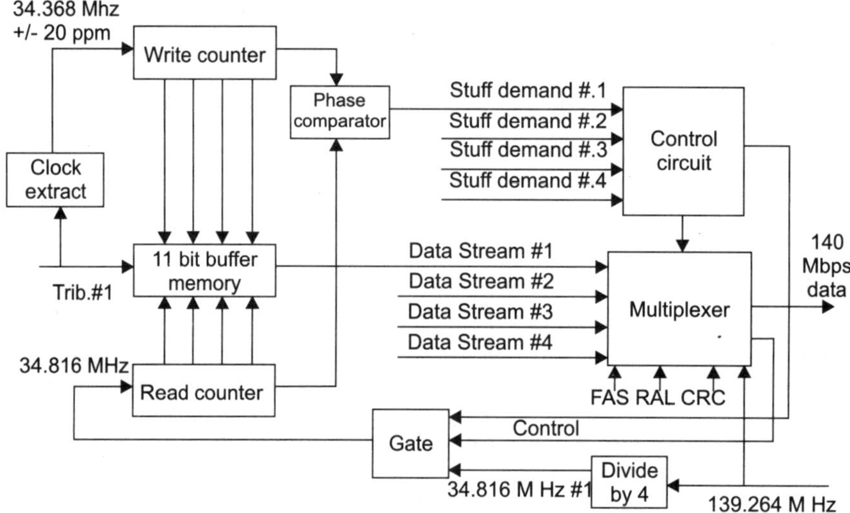

Fig. 4.5(a) *Multiplexing 4 E3s into E4*

Elastic memory and justification control circuits associated with one E3 tributary are shown in Fig. 4.5(a). Similar setups exist for other three tributaries as well. This figure is to be read in conjunction with frame structure of E4 stream given in Fig. 4.5(b).

Let us examine the E4 frame structure given in Fig. 4.5(b). Each E4 frame has 2928 bits, starting with a 16-bit pattern (consisting of a 12 bit frame alignment sequence or FAS, 1 alarm bit and 3 N bits). Bits from incoming E3 tributaries are organised as sets, with each set having a bit from each tributary by bit-interleaving process. Each set of 4 bits is called a bit-set. There are 723 such bit-sets plus 5 sets with nomenclatures C1, C2, C3, C4 and C5 (which carry JCBs). C1.1, C2.1, C3.1, C4.1 and C5.1 are the bits used to keep a note in the frame that justification has been done in respect of tributary-1. Similarly, C1.2, C2.2, C3.2, C4.2, C5.2 convey that justification has been done in respect of tributary-2 and so on. All 723 bit-sets except bit-set 603 carry information bits of individual tributaries. Bit-set 603 may carry an information bit or justification bit as the case may be.

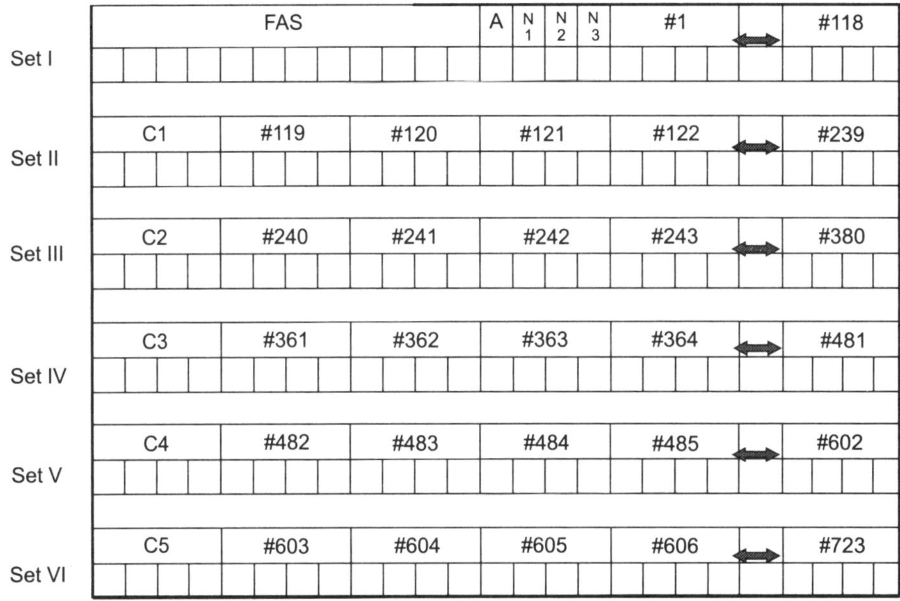

Fig. 4.5(b) *Frame structure of E4*

Let us refer to Fig. 4.5(a) again. The 139.264 MHz reference clock is divided by 4 to get 34.816 MHz clock which is applied as read-out clock to the elastic memory associated with each of the four 34.368 Mbps tributaries. The elastic memory's write-in clock is the 34.368 MHz recovered from the incoming stream (tributary).

The phase difference between the 34.368 MHz write-in clock and 34.816 MHz read-out clock is monitored. When the phase difference exceeds a threshold, there is a need for justification. Hence the readout clock is perforated, i.e., the data bit already written in elastic memory is not read out to E4 frame (under construction in multiplexed memory) but a dummy bit is stuffed in the corresponding bit position of bit-set 603 in the E4 frame.

Q 4.10 *Why is the frame size 2928 bits for E3 to E4 multiplexing?*

Frame size is chosen in such a way that the rate of information inflow from the incoming tributary is equal to average rate of information outflow per tributary through the multiplexed stream. Let us assume a frame size of N bits. We will prove N to be 2928 in the case of E4.

In a frame of N bits of E4, the actual information bit flow per tributary is $(N - k) / 4$, where k depends on the number of FAS, alarm and national bits, justification bits and justification control bits (JCBs) in each frame. This outflow, on an average, should match the inflow through the incoming E3.

In a frame of N bits of E4, the information bits are

$N - 36$ (16 bits of FAS, A & N and 20 JCBs, i.e., a total 36 bits are to be subtracted) if no justification done;

$N - 40$ (16 bits of FAS, A & N; 20 JCBs and 4 justification bits, i.e., a total of 40 bits are subtracted) if justification is done.

Number of frames per second = Bit rate/Number of bits per frame = $139.264 \times 10^6 N$.

Outflow of information bits = Frames per second × Number of information bits per frame
= $(139.264 \times 10^6 / N)\ (N - 36)$ if justification is not done
= $(139.264 \times 10^6 / N)\ (N - 40)$ if justification is done.

Therefore,
Outflow per tributary = $(139.264 \times 10^6 / N)\ (N - 36) / 4$ if justification is not done
= $(139.264 \times 10^6 / N)\ (N - 40) / 4$ if justification is done.

This gets simplified as $(34.816 \times 10^6 / N)\ (N - 36)$ if justification is not done
or
$(34.816 \times 10^6 / N)\ (N - 40)$ if justification is done.

The inflow is at the rate of 8.448 Mbps per tributary.

Equating outflow to inflow, we get
$(34.816 \times 10^6 / N)\ (N - 36) = (34.368 \times 10^6)$ if justification is not done
$(34.816 \times 10^6 / N)\ (N - 40) = (34.368 \times 10^6)$ if justification is done.

This gets simplified as
$34.816\ (N - 36) = 34.368\ N$ if justification is not done.
$34.816\ (N - 40) = 34.368\ N$ if justification is done.

From the above, we get $N = 2798$ if justification is not done and 3108 if justification is done. We cannot take an average of these two values to decide N. Rather, we should take their weighted average.

If the nominal justification is p, i.e., the probability that justification done is p, then the probability that justification is not done is $1 - p$. Then, $N = (1 - p).2798 + p.3108$.

According to ITU-T Rec. G.751, nominal justification ratio while integrating E2 streams to form E3 is 0.419, i.e., $p = 0.419$. Using this value, we get $N = 2928$.

4.8 Demultiplexing E4

Refer to Fig. 4.5(c), illustrating demultiplexing of E4 into 4 E3s. In the process of de-multiplexing E4 for the retrieval of 4 E3s, the bits from the incoming E4 are segregated tributary-wise by the demultiplexer and the written in the elastic memory using write-in clock of 34.816 MHz. This 34.816 MHz write-in clock is derived from the incoming E4 (139.264 MHz clock extraction and then divide by 4). Data bits from elastic memory are read-out using 34.368 MHz clock.

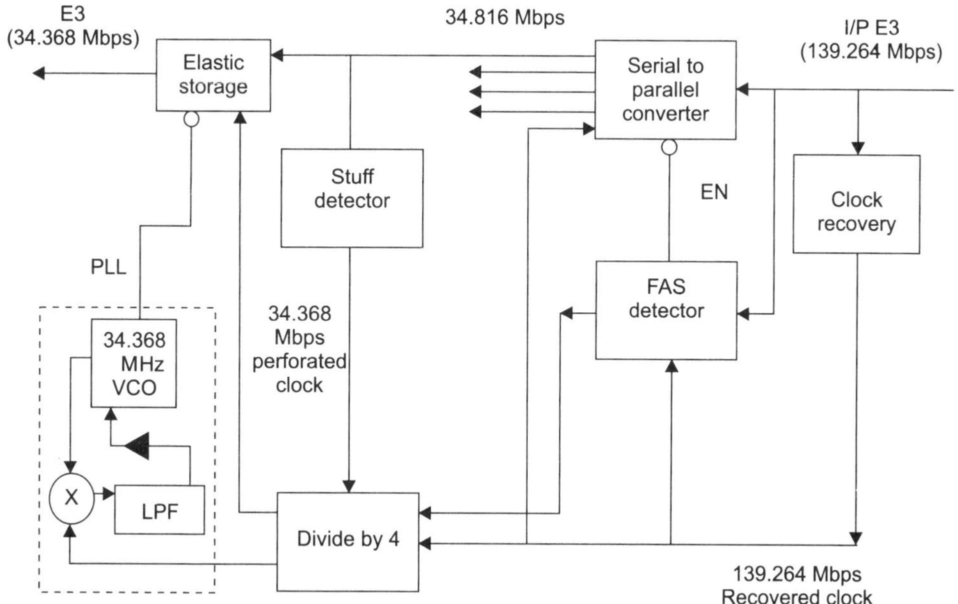

Fig. 4.5(c) *Demultiplexing E3 into E2s*

FAS, A & N bits and JCBs need not be written in the tributary's elastic memory. Hence, during these intervals, 34.816 MHz write-in clock is perforated. Also, when majority of JCBs pertaining to a tributary are 1s, we know that justification was done in bit-set 603 and that the write-in clock needs to be perforated. Because of these perforations, the effective rate of writing in the elastic memory is 34.368 Mbps and hence we can read out using a 34.368 MHz clock.

We require one full frame size of E4 as buffer memory before we can transfer the data to the demultiplexer because the JCBs in the current frame tell about justification action in bit-set 603 of the previous frame. The JCB monitoring unit applies perforation command with necessary timing delay.

Summary

In this chapter the basics of PDH multiplexing, viz. bit-interleaving and justification are explained. PDH multiplexing on E-hierarchy (E2, E3, E4) is explained, and the frame sizes and structures, justification and de-justification provisions for all the 4 tributaries being integrated are explored. De-multiplexing is also explained.

References

[1] Frame structure for E2, ITU-T Rec. G.742, 1993.

[2] Frame structure of E3 and E4, ITU-T Rec. G.751, 1993.

5

PDH Network Reliability Issues

5.1 Introduction

Noise, inter-symbol interference (ISI) and jitter are three commonly encountered problems in digital communication networks. All the three manifest as bit errors, i.e., they contribute to an increase in the probability of error or BER in their own ways. Signals corrupted by noise can be wrongly interpreted at the receiving end, hence noise contributes to BER. ISI imposes a limitation on the rate of transmission as the spread of pulse to next bit interval causes errors in detection. Consequently, we limit the transmission rate to one which produces a comfortable level of ISI. Even after accepting the limitation and operating within that, any additional ISI can cause errors. Noise and ISI can therefore be considered as attacks on the amplitude domain of the signal.

Jitter has an altogether different attack mechanism. Jitter is essentially the shift of recovered clock from the ideal position, i.e., the phase error of the recovered clock with respect to the transmitted clock. A signal reconstructed using recovered clock which is jittery has pulses shifted from their ideal position, giving scope for wrong interpretation. So, jitter attacks the phase domain. While jitter is a short term variation, there is another form of jitter, identified as 'wander', which is long term variation in significant instances of the transmitted digital signal.

System design considerations emanate from the following performance requirements:

- Required rate of transmission
- Tolerable BER
- Required availability of link
- Permitted levels of jitter and slips

In this chapter we will first discuss the system design considerations in tackling noise and ISI. Later, we will discuss the sources of jitter (and wander) in PDH networks, the ITU-T recommendations on their limits and techniques to reduce jitter.

5.2 System Design Considerations in Tackling Noise

5.2.1 Digital MW radio links

Noise is essentially taken care of by modulation technique on radio links. The modulation on technique adopted depends on the power-bandwidth tradeoff considerations for a required rate of transmission and specified BER tolerance on a digital MW radio link. These trade-offs were presented for various modulation schemes on radio links in section 1.3.3, where we established the suitability of 8-PSK, QPSK, 16 QAM as the modulation schemes for various rates of data transmission. This is not the end of the story. We have to decide system margin based on the availability requirement. System margin is the margin we have between the transmitted power and the receiver sensitivity, considering all the gains and all the losses. The system margin of digital MW radio link is given by

$$\text{System margin} = \text{Transmitted power} + \Sigma\,\text{Gains} - \Sigma\,\text{Losses} - \text{Receiver sensitivity (in dB)}. \tag{5.1}$$

After deciding the modulation scheme, transmitted power and receiver sensitivity, we cannot compute and declare the available system margin. Rather, we have to work towards a system margin which ensures certain availability of the link. In the case of digital MW links, the system margin is the fade margin (FM), given by Barnett–Vigant reliability formula[1].

$$\text{FM} = 30 \log d + 10 \log (ABf) - 10 \log (1 - R) - 70, \tag{5.2}$$

where d is the hop length, f the carrier frequency, R the reliability or availability objective (for example 99.99% availability implies $R = 0.9999$) and A = roughness factor (4 for smooth terrain, 1 for average terrain, 1/4 for rough terrain), B = factor to convert worst month probability of fading to annual probability (1/2 for lakes, 1/4 for inland area, 1/8 for mountains and rough areas).

For any given hop all parameters are known except R. Reliability objective R is to be decided as per services offered. So, for a given hop, we can state

$$\text{FM} = k - \log (1 - R), \tag{5.3}$$

where k is the specific value we get for a hop considering parameters other than R on RHS of equation (5.2).

If R is 0.9999, FM $= k + 40$ dB,
If R is 0.99999, FM $= k + 50$ dB.

This implies an increase by an order in R requires additional 10 dB of FM.

For utilities carried on a digital MW link, the availability required is 99.9999%, i.e., R is 0.999999. Hence, fade margin or system margin for each hop is high.

Referring to equation (5.1), we can achieve the required FM by increasing the transmitted power or by an increase in gain elements (using better directivity antennas) or through the use of a better designed receiver which can work with low sensitivity or even resort to diversity techniques (space diversity or frequency diversity to forbid losses beyond a particular value). Increasing the transmitted power beyond 1 W (30 dBm) increases the cost of power plant; it poses over-reach problem and the size of equipment may also increase. Similarly, we have to evaluate the pros and cons of other options. A detailed treatment of the same is given in [2]. Increase of antenna size may increase the tower cost, because of which diversity, especially frequency diversity, may be a better option.

5.2.2 OFC links

Noise due to EMI is not a problem in OFC. Noise is the 'relative intensity noise' (RIN) as mentioned in section 1.3.4 However, we have to provide sufficient margin against losses (fibre attenuation, splice loss (loss at joints), connector losses) on the link. Moreover, receiver sensitivity is a function of BER and data rate.

One such analysis or link power budgeting is presented below.

Consider an OFC link with a laser diode source (+3dBm) and an APD detector (sensitivity of –32 dBm for BER of 1×10^{-12} at the E4 data rate under consideration). Considering a connector loss of 1 dB, fibre attenuation of 0.25 dB per km, patch cord loss of 3 dB and splice loss of 0.1 dB, find the system margin over a 60 km link.

Available power budget = source power – receiver sensitivity = +3 – (–32) = 35 dB.
Attenuation in cable = 0.25 × 60 = 15 dB.
Splice loss (assuming 3 km drum, which is the usual case; we require 20 drums i.e., 20 splice joints on the link) = 20 × 0.1 = 2 dB.
Patch cord losses = 3 dB on either side = 6 dB total.
Connector losses = 1 dB on either side = 2 dB total.
Net losses = 15 + 2 + 6 + 2 = 25 dB.

Available system margin = 35 dB – 25 dB = 10 dB.

This margin can take care of any extra RIN or splice loss or fibre losses.

So, the key to tackling RIN related problems in OFC is through the choice of source, detector and proper system margin according to power budget parameters.

5.3 Tackling ISI

5.3.1 In digital radio links

Any transmitted pulse can be considered as the weighted summation of a basic frequency component and its harmonics, i.e., it is essentially a wideband signal with different spectral components of different amplitudes. When such a stream of pulses is transmitted, it travels through filters, interconnection cables, transmission media and are subjected to their nonlinear effects. The filters, interconnection cables and media do not have ideal transfer characteristics over the entire band of frequencies under consideration and hence cause distortion or spread of pulses which manifests themselves as ISI. The problem of ISI is more pronounced in digital radios. If we apply equalisers which compensate for the distortions introduced by the filters, interconnection cables and media, the effect of ISI can be minimised.

In digital radio equipment, the following equalisers are used.

1. **Delay equaliser:** Delay equaliser equalises the group delay developed in RF branching filter of the radio equipment. If the bandwidth of the radio link is *B* MHz, it is centred around the RF carrier in the transmitted spectrum. The RF filter does not have a flat transfer function in the band under consideration, (*B* MHz), but has nonlinearities. Delay time is more at the edge of the band compared to the centre of the band. Hence, we need to compensate for the same.

   After superheterodyning, the modulating envelope of the bandwidth of *B* MHz gets translated to IF band. Now, we apply delay equalisation to achieve constant delay at the centre of the band and less delay towards the edges of the band (as shown in Fig. 5.1) to compensate for the nonlinearities caused by RF filters.

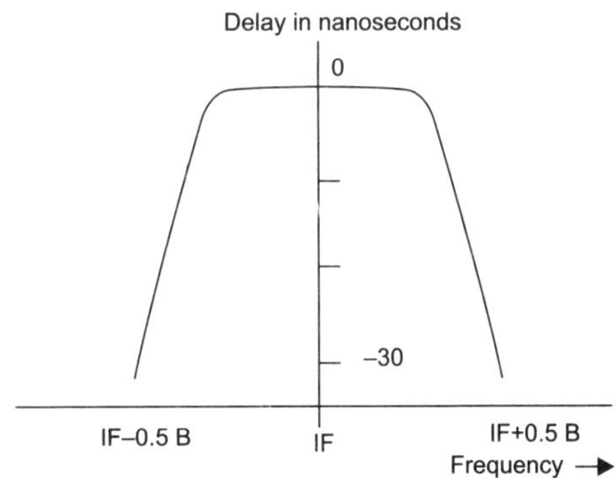

Fig. 5.1 *Delay equaliser characteristic*

2. **Transversal equaliser:** Transversal equaliser compensates for both amplitude and delay distortions caused by selective multipath fading in the transmission band. It consists of a transversal filter, control signal generator and an IF amplifier. Refer to Fig. 5.2(a).

   The transversal filter takes the IF signal and applies it to 4 delay line units. Different weights are applied to the delay lines' outputs which are added up to get the in-phase and quadrature components of the IF signal. In normal conditions, the IF signal at the centre line is taken as the signal and tap weights for other lines are almost negligible. In case of distortion caused by selective fading, tap weights on delay lines are increased dynamically by a control circuit and we get a compensated in-phase and quadrature components at the output of the transversal filter. The control signal works under the control of phase demodulator of the receiver to decide the tap weights.

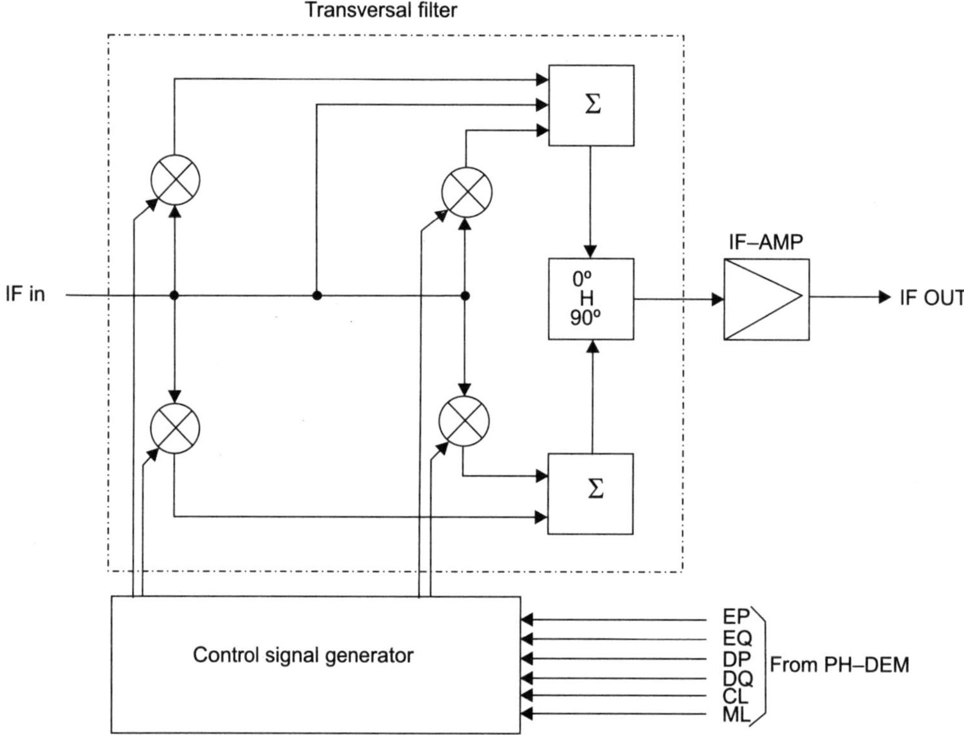

Fig. 5.2(a) *Functional block diagram of transversal equaliser*

3. **Line equaliser:** While delay equaliser and transversal equaliser are placed after the IF signal, line equaliser is placed at the input stage. There is a fixed equalisation circuit which pre-compensates the distortion that would occur in the interconnecting cables.

   Refer to Fig. 5.2(b). The incoming line coded data signal is applied to the equalisation circuit with automatic feedback control circuit. The DET circuit detects the current at the output and applies a signal to control the frequency characteristics of the equaliser.

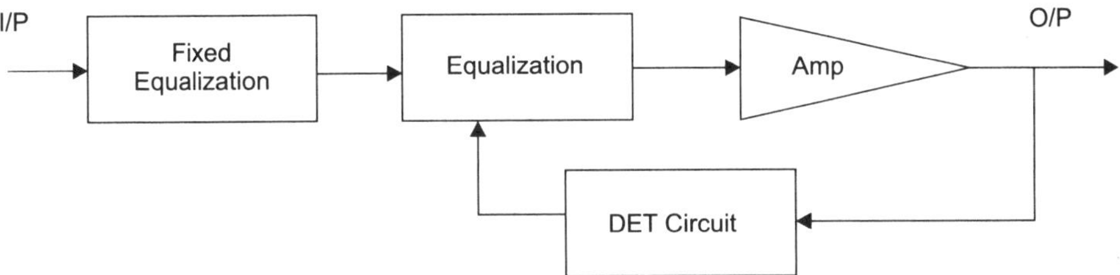

Fig. 5.2(b) *Functional block diagram of line equaliser*

5.3.2 In OFC links

In OFC links, dispersion of the transmitted pulse takes place due to ISI. The more the dispersion, the more is the rise time (or fall time) of the system. The following are three commonly encountered rise time elements of an OFC link.

- Transmitter rise time.
- Dispersion rise time of the fibre.
- Receiver rise time

The total system rise time is given by the equation

$$t_{sys} = (t^2_{Tx} + t^2_{Rx} + t^2_{\text{fibre}})^{1/2} \tag{5.4}$$

Rise time analysis is a means for ensuring the minimum permissible data rate over an optical fibre link. Conventionally, the total transition time degradation of a digital link should not exceed 70% of an NRZ bit period or 35 percent of a bit period for RZ bit rate.

Let us find out the rise times quantitatively.

Transmitter rise time

Transmitter rise time is primarily due to light source and its drive circuitry. It is of the order of few nano seconds for an LED source and few picoseconds (25 picoseconds) for a laser source. High speed links invariably use laser sources.

Receiver rise time

Receiver rise time results from the bandwidth of the receiver. It is given by

$$t_{Rx} = 0.35 \,/\, (\text{Receiver BW}) \tag{5.5}$$

Fibre rise time or dispersion

In long distance transmission links, only single mode fibres are used. In single mode fibre, there is no multimodal dispersion (dispersion due to multiple propagation modes) and dispersion is due to chromatic dispersion alone. Chromatic dispersion has two components:

- *Material dispersion:* Different wavelength components in the pulse have different refractive indices in the core of the fibre and hence have different travel times through the fibre from the Tx end to the Rx end. This causes pulse spread known as material dispersion.
- *Waveguide dispersion:* Though propagation of light in an OFC is only through the core and by total internal reflection, a minute portion of pulse energy goes through the cladding and arrives at the receiving end adding up as spread of the pulse in time domain at the receiving end.

The resultant of material dispersion and waveguide dispersion is chromatic dispersion. It is expressed in pico seconds/nm-km. For example, a dispersion factor of 2 ps/nm-km means a spread of 2 ps in the pulse for every nanometre of spectral width of the light source over a 1 km link. The transmission rate supported by the fibre for a given link should be so chosen that the pulse spread does not stray into the adjacent bit's duration, i.e., there is be no ISI or very little ISI, which makes the pulses discernible. Dispersion compensated fibres are used on long distance transmission links to keep the dispersion to a low value so that ISI is under check.

Rise-time analysis is made clear in the following example.

Consider an OFC link operating at 1310 nm. The laser source rise time is 25 ps, spectral width of the laser is 2 nm and fibre dispersion is 2 ps/nm-km. Find the feasibility of a 100 km E4 link.

Transmitter rise time = 25 ps.
Receiver rise time = $0.35 / (140 \times 10^6) = 2.5$ ns = 2500 ps.
Dispersion rise time of fibre = Dispersion factor × Spectral width of the source × length of link = 400 ps.

System rise time = $(25^2 + 2500^2 + 400^2)^{1/2} = 2532$ ps = 2.532 ns.

70% of the bit interval time for 140 Mbps NRZ data string = $0.7 / (140 \times 10^6) = 50$ ns.

Since the system rise time is less than 70% of the bit interval time for NRZ coding, we can finalise the link. If system rise time is more than the prescribed limit, we have to use regenerative repeaters after the permissible length.

5.4 Jitter in PDH Networks

5.4.1 What is jitter?

Jitter is the time-shift of the recovered clock from the original symbol clock. Since jitter is a time-shift, it is measured as a fraction of the symbol clock period. Hence, the unit for jitter measurement is UI (unit interval). Jitter is illustrated in Fig. 5.3.

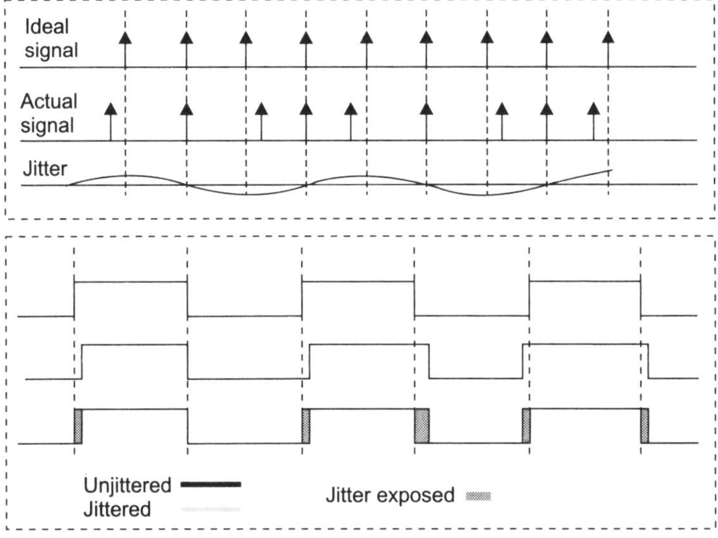

Fig. 5.3 Illustration of jitter

Q 5.1 What do you understand by a jitter of 0.05 UI on a clock of 2.048 Mbps?

A 2.048 Mbps clock has a clock period of 488 ns. Hence a 0.05 UI would mean a time shift of 2.44 ns.

5.4.2 Causes of jitter in PDH networks

The causes of jitter are summarised in Table 5.1.

Table 5.1 *Types of jitter and their causes*

Jitter variety	Causes
Very low frequency jitter	1. Short term variations in propagation delays 2. Temperature variation delays
Low frequency jitter	1. Inherent instability of clock sources
Noise induced jitter	1. Phase noise in crystal controlled oscillator 2. Noise in logic circuits
Multiplexed induced jitter	1. Insertion / Removal of justification bits, justification control bits, FAS 2. Waiting time jitter in de-justification circuit (clock perforated when justification bits and control bits are received in the incoming frame, waits for valid information bit in the received frame)
Regenerator jitter	1. Imperfect timing recovery 2. Imperfections in timing of outgoing frame

5.4.3 Wander and its causes

While jitter is short term variation, long term variation in a significant instrument of the digital signal transmission system is known as wander. It is caused by

- Long term changes in propagation delay
- Ageing of clock circuitry.

A comparison of jitter and wander is presented in Table 5.2.

Table 5.2 *Jitter and wander compared*

Jitter	Wander
Jitter is time shift in the clock recovered from the symbol clock	Wander is long-term variation in significant instants of the digital signal
Extraction of timing information and insertion of bits are the main causes of jitter	Changes in propagation delay, ageing of clock circuitry cause wander
Jitter is unpredictable	Wander is predictable

5.4.4 Effects of jitter and wander

If the accumulation of jitter is unchecked over a number of hops, it can result in the following.

1. Introduction of uncontrolled slips in the digital signals through store spillage and depletion in the MUX equipment which have buffer store and phase comparators to take care of the justification process.
2. Degradation of digitally encoded analog information as a result of phase modulation of reconstructed samples in the digital-to-analog conversion device at the end of the link.

Wander, if not accommodated properly at the input ports, may cause slipping of bits. For lines carrying utilities, such slips should not be more than 1 in 70 days.

Let us go on to study ITU-T recommendations regarding control of jitter and wander.

5.5 ITU-T Recommendations Regarding Control of Jitter and Wander

The ITU-T recommendations cover the following aspects regarding the control of jitter and wander:

1. Network limits for jitter at hierarchical interfaces
2. Network limits for wander at hierarchical interfaces

3. Input jitter and wander tolerance at hierarchical levels
4. Limits for intrinsic jitter (output jitter in the absence of input jitter)
5. Limits for manifestation of input jitter at the output, i.e., allowed jitter transfer
6. Guidelines to measure and control jitter

Let us examine them one by one.

5.5.1 Network limits for output jitter at hierarchical interfaces

ITU-T Rec. G.823 prescribes the maximum permissible levels of jitter at hierarchical interfaces in a PDH network. Jitter, as measured over a 60 second interval using specified measurement filters, should not exceed the limits specified in Table 5.3.

Table 5.3 *Limits for output jitter at hierarchical interfaces in PDH network*

Data rate (kbps)	1 UI (ns)	NW limits of peak-to-peak jitter		Measured BPF lower cut-off f_1 or f_3 and upper cut-off f_4		
		B1 UI Measured f_1–f_4	B2 UI Measured f_3–f_4	f_1 Hz	f_3 kHz	f_4 kHZ
64	15600	0.05	0.01	20	3	20
2048	488	0.25	0.06	20	18	100
8448	118	0.25	0.06	20	3	400
34368	29.1	0.25	0.05	100	10	800
139264	7.18	0.25	0.025	200	10	3500

It may be noted that with regard to the values given in Table 5.3,

a. The limits given in the table should be met for all operating conditions regardless of the amount of equipment preceding the interface. In other words, these limits are compatible with the input tolerance limits of the concerned hierarchical equipment where the output under consideration is applied.
b. The high-pass measurement filters has first-order characteristic and a roll-off of 20 dB/decade.
c. The low-pass measurement filters have a maximally-flat, Butterworth characteristic and a roll-off of –60 dB/decade.

5.5.2 Network limits for output wander at various PDH interfaces

PDH output interfaces are E1, E2, E3 and E4. These outputs are formed by the multiplexing of lower hierarchical input streams. Hence, wander limit is expressed in MRTIE (maximum relative time interval error), i.e., the maximum time interval error accumulated by the network relative to the input clock.

ITU-T Rec. G.823 specifies network limit for wander at 2.048 Mbps, 34.368 Mbps and 139.264 Mbps interface. These are mentioned in Table 5.4.

2.048 Mbps interface output wander limit: The maximum level of wander that may exist at a 2.048 Mbps network interface, expressed in MRTIE should not exceed the limit given in Table 5.4. The same is shown graphically in Fig. 5.4.

Table 5.4 *2.048 Mbps interface output wander limit*

Observation interval τ (seconds)	MRTIE requirement (μs)
0.05 < τ ≤ 0.2	46 τ
0.2 < τ ≤ 32	9
32 < τ ≤ 64	0.28 τ
64 < τ ≤ 80	18

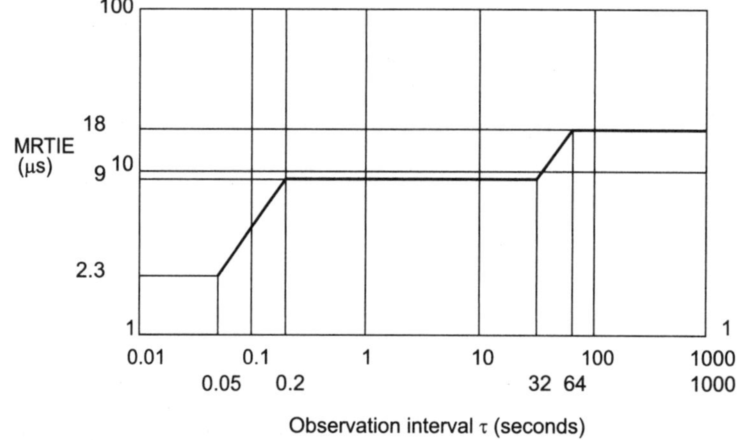

Fig. 5.4 *Graphical representation of 2.048 Mbps interface output wander limit*

34.368 Mbps interface output wander limit: The maximum level of wander that may exist at a 34.368 Mbps network interface, expressed in MRTIE, should not exceed the limit given in Table 5.5. The same is shown graphically in Fig. 5.5.

Table 5.5 *34.368 Mbps interface output wander limit*

Observation interval, τ (seconds)	MRTIE requirement (μs)
0.05 < τ ≤ 0.073	14 τ
0.073 < τ ≤ 2.5	1
2.5 < τ ≤ 10	0.4 τ
10 < τ ≤ 80	4

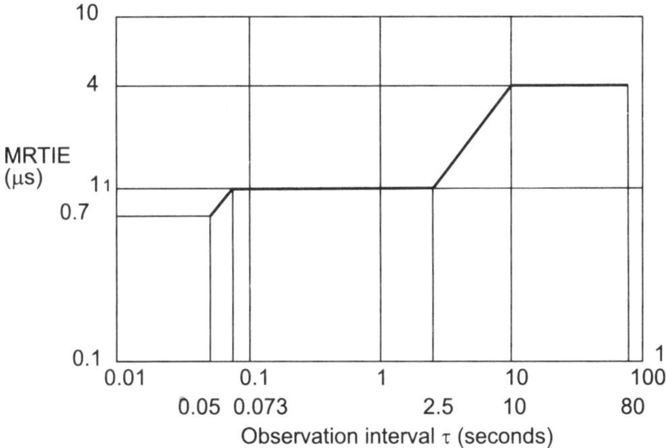

Fig. 5.5 *Graphical representation of 34.368 Mbps interface output wander limit*

139.264 Mbps interface output wander limit: The maximum level of wander that may exist at a 139.264 Mbps network interface, expressed in MRTIE, should not exceed the limit given in Table 5.6. The same is shown graphically in Fig. 5.6.

Table 5.6 *139.264 Mbps interface output wander limit*

Observation interval, τ (seconds)	**MRTIE requirement (μs)**
$0.05 < \tau \leq 0.15$	$6.8\ \tau$
$0.15 < \tau \leq 2.5$	1
$2.5 < \tau \leq 10$	$0.4\ \tau$
$10 < \tau \leq 80$	4

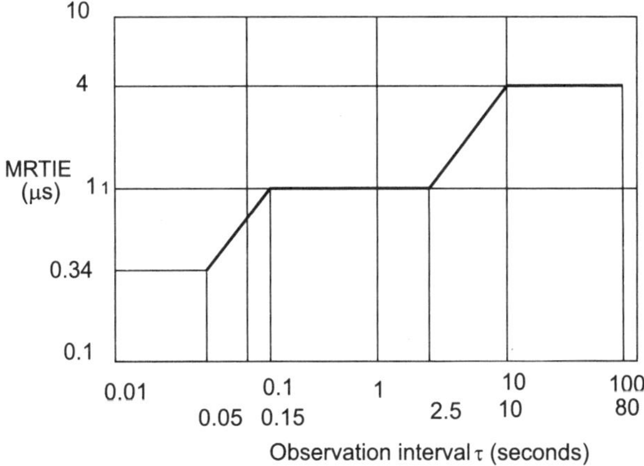

Fig. 5.6 *Graphical representation of 139.264 Mbps interface output wander limit*

5.5.3 Input jitter and wander tolerances at network interfaces

Since multiplex equipment are connected interface-wise to the transmission equipment in PDH networks (E3 or 2/34 MUX to 34 Mbps radio, E4 or 140 Mbps equipment to 140 Mbps OLTE, E1 to 2 Mbps UHF radio or WS port of MW radio, etc.), it is necessary to arrange the input ports of all equipment types that are capable of accommodating levels of jitter and wander up to at least the minimum limits as defined in ITU-T Rec. G.823. These limits are termed 'input jitter and wander tolerances'.

Definition of input jitter and wander tolerance (ITU-T Rec. G.823)

The jitter and wander tolerance of a PDH interface indicates the minimum level of phase noise that the input port should accommodate whilst

a. not causing any alarms
b. not causing any slips
c. not causing any bit errors.

All digital input ports of equipment should be able to tolerate a digital signal that has:

a. electrical characteristics in accordance with the requirements of ITU-T Rec. G.703.
b. clock variations of ±1000 ppm on 64 kbps, ±50 ppm on 2.048 Mbps, ±30 ppm on 8 Mbps, ±20 ppm on 34.368 Mbps and ±15 ppm on 139.264 Mbps.
c. digital input ports of equipment should be able to tolerate input jitter and wander of at least the values (lower limits) given (hierarchical data, rate-wise) below.

For 64 kbps

The level of jitter and wander that can be accommodated by a 64 kbps co-directional network interface, expressed in peak-to-peak sinusoidal phase amplitude, should not be less than the values given in Table 5.7. The same is shown graphically in Fig. 5.7. The test sequence to be used is a PRBS of length $2^{11} - 1$, defined in ITU-T recommendation O.150.

Table 5.7 *64 kbps input jitter and wander tolerance*

Frequency f(Hz)	Requirement (peak-to-peak phase amplitude)
$12\ \mu < f \leq 4.3$	18 μs
$4.3 < f \leq 20$	$77 f^{-1}$ μs
$20 < f \leq 600$	0.25 UI
$600 < f \leq 3$ k	$150 f^{-1}$ UI
3 k $< f \leq 20$ k	0.05 UI

Note: 1 UI = 15.6 μs

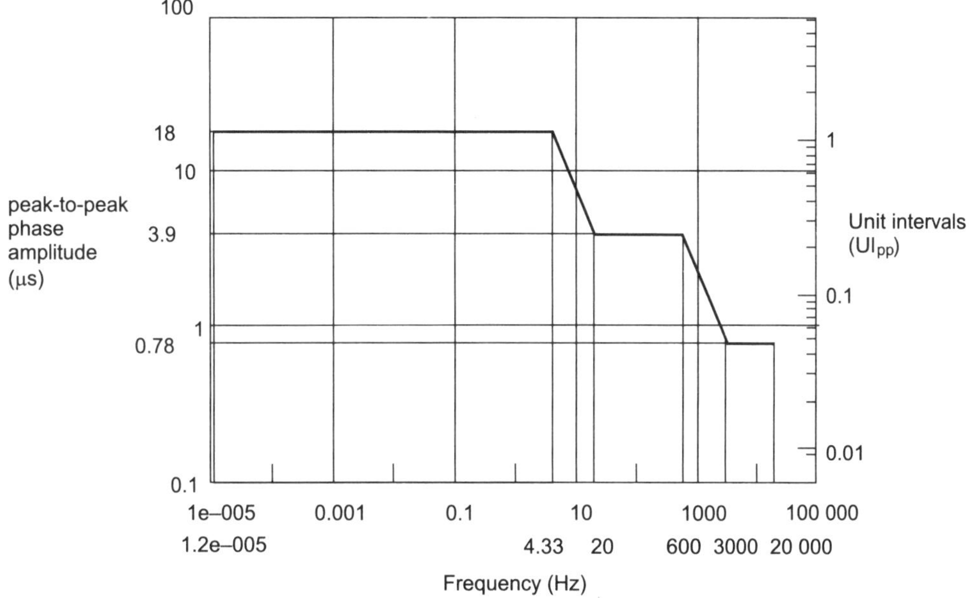

Fig. 5.7 *Graphical representation of 64 kbps input jitter and wander tolerance*

For 2.048 Mbps

The level of jitter and wander that can be accommodated by a 2.048 Mbps network interface, expressed in peak-to-peak sinusoidal phase amplitude, should not be less than that of the values given in Table 5.8. The same is shown graphically in Fig. 5. 8. The test sequence to be used is a PRBS of length $2^{15} - 1$, defined in ITU-T Rec. O.150.

Table 5.8 *2.048 Mbps input jitter and wander tolerance*

Frequency f(Hz)	Requirement (peak-to-peak phase amplitude)
$12\ \mu < f \leq 4.88\ \mu$	18 μs
$4.88\ \mu < f \leq 10\ \mu$	$0.088 f^{-1}$ μs
$10\ \mu < f \leq 1.67$	8.8 μs
$1.67 < f \leq 20$	$15 f^{-1}$ μs
$20 < f \leq 2.4$ k	1.5 UI
2.4 k $< f \leq 18$ k	$3600 f^{-1}$ UI
18 k $< f \leq 100$ k	0.2 UI

Note: 1 UI = 488 ns

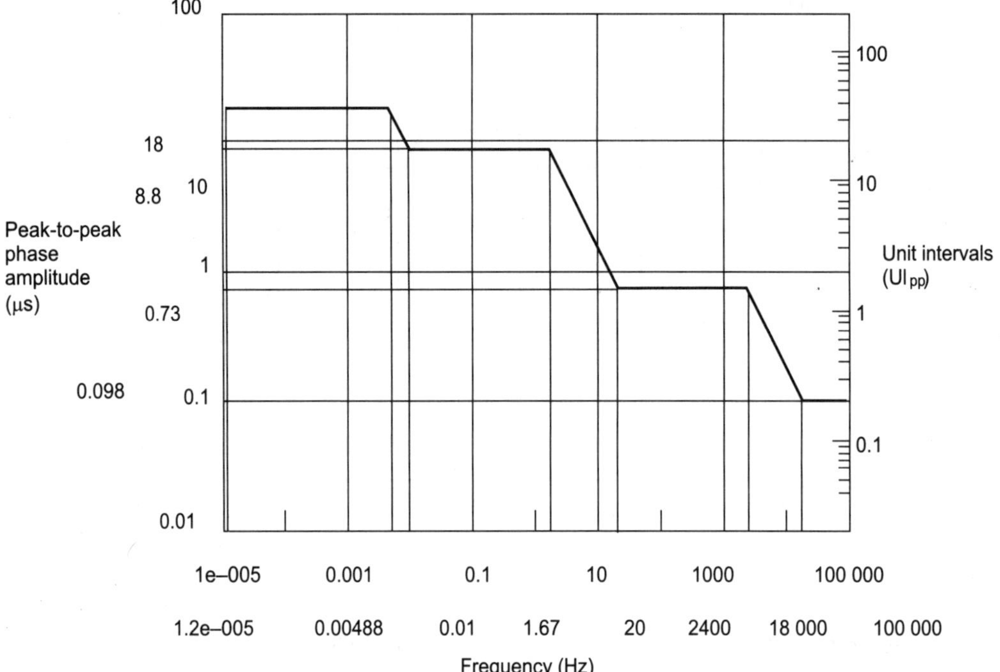

Fig. 5.8 *Graphical representation of 2.048 Mbps input jitter and wander tolerance*

For 8.448 Mbps

The level of jitter and wander that can be accommodated by a 8448 kbps network interface, expressed in peak-to-peak sinusoidal phase amplitude, should not be less than the values given in Table 5.9. The same is shown graphically in Fig. 5.9. The test sequence to be used is a PRBS of length $2^{15} - 1$, defined in ITU-T Rec. O.150.

Table 5.9 *8.448 Mbps input jitter and wander tolerance*

Frequency *f*(Hz)	**Requirement (peak-to-peak phase amplitude)**
$20 < f \leq 400$	1.5 UI
$400 < f \leq 3$ k	$600 f^{-1}$ UI
3 k $< f \leq 400$ k	0.2 UI

Note: 1 UI = 118 ns

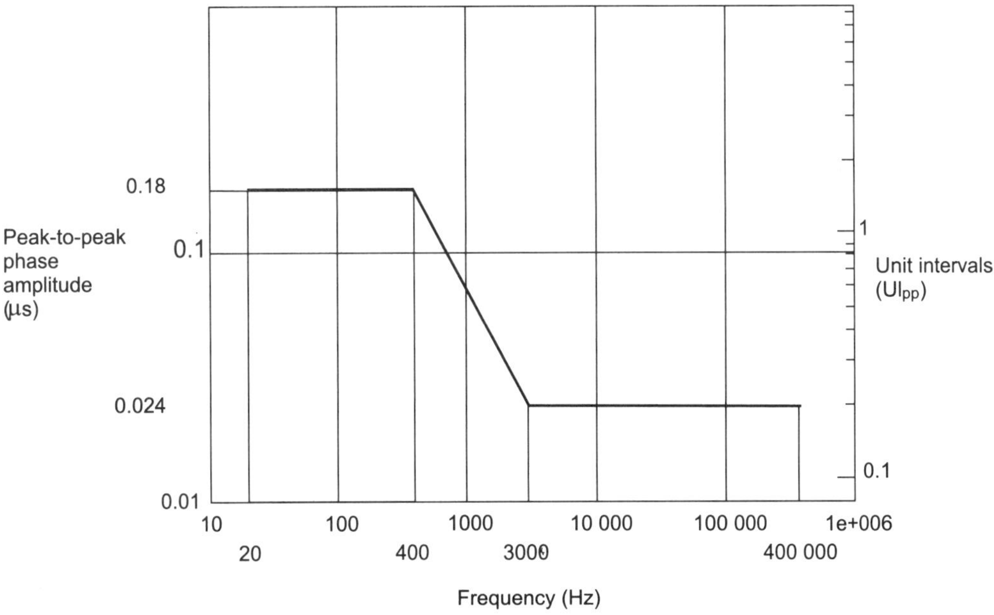

Fig. 5.9 *Graphical representation of 8.448 Mbps input jitter and wander tolerance*

For 34.368 Mbps

The level of jitter and wander that can be accommodated by a 34.368 Mbps network interface, expressed in peak-to-peak sinusoidal phase amplitude, should exceed the values given in Table 5.10. The same is shown graphically in Fig. 5.10. The test sequence to be used is a PRBS of length $2^{23} - 1$, defined in ITU-T Rec. O.150.

Table 5.10 *34.368 Mbps input jitter and wander tolerance*

Frequency f(Hz)	Requirement (peak-to-peak phase amplitude)
10 µ < $f \leq$ 32 µ	4 µs
32 µ < $f \leq$ 130 µ	0.13 f^{-1} µs
130 µ < $f \leq$ 4.4	1 µs
4.4 < $f \leq$ 100	4.4 f^{-1} µs
100 < $f \leq$ 1 k	1.5 UI
1 k < $f \leq$ 10 k	1500 f^{-1} UI
10 k < $f \leq$ 800 k	0.15 UI

Note: 1 UI = 29.1 ns

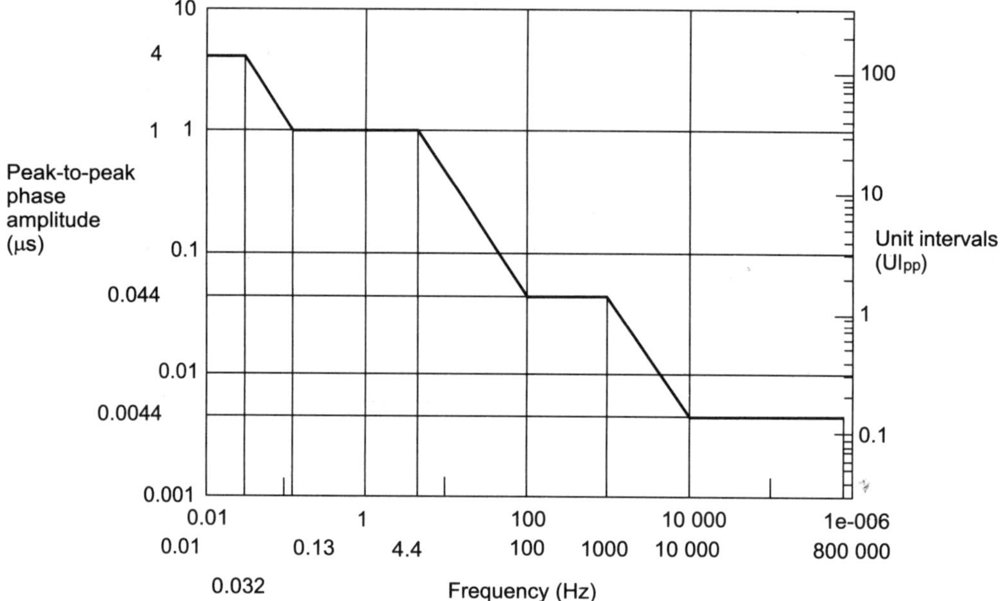

Fig. 5.10 *Graphical representation of 34.368 Mbps input jitter and wander tolerance*

For 139.264 Mbps

The level of jitter and wander that can be accommodated by a 139 264 kbit/s network interface, expressed in peak-to-peak sinusoidal phase amplitude, should not be less than the values given in Table 5.11. The same is graphically represented in Fig. 5.11. The test sequence to be used is a PRBS of length $2^{23} - 1$, defined in ITU-T Rec. O.150.

Table 5.11 *139.264 Mbps input jitter and wander tolerance*

Frequency *f*(Hz)	**Requirement (peak-to-peak phase amplitude)**
10 μ < $f \leq$ 32 μ	4 μs
32 μ < $f \leq$ 130 μ	0.13 f^{-1} μs
130 μ < $f \leq$ 2.2	1 μs
2.2 < $f \leq$ 200	2.2 f^{-1} μs
200 < $f \leq$ 500	1.5 UI
500 < $f \leq$ 10 k	750 f^{-1} UI
10 k < $f \leq$ 3.5 M	0.075 UI

Note: 1 UI = 7,18 ns

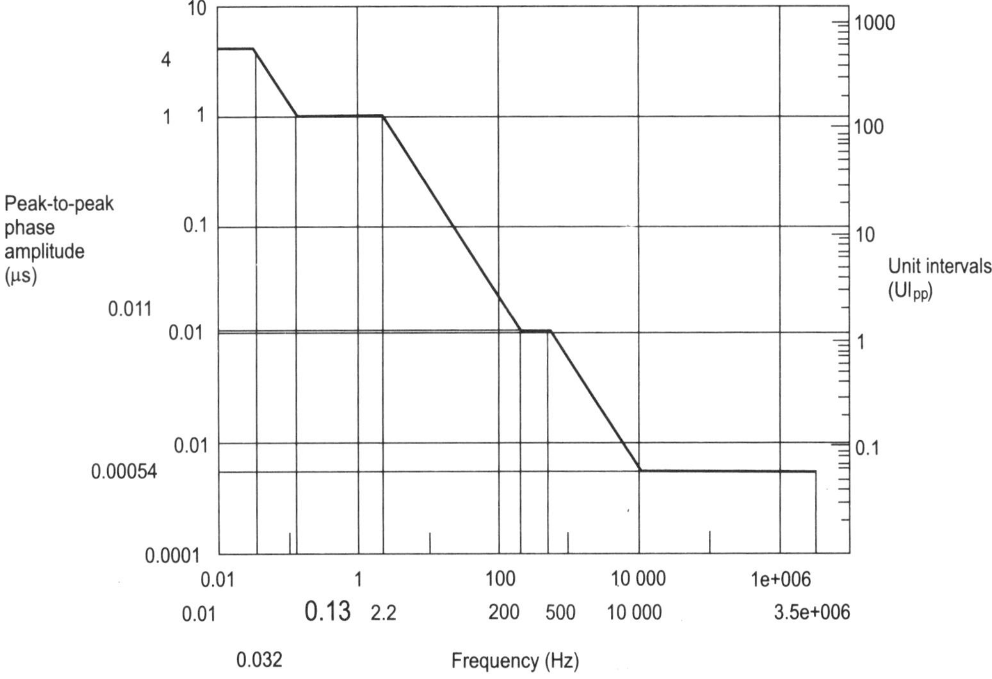

Fig. 5.11 *Graphical representation of 139.264 Mbps input jitter and wander tolerance*

When assessing interface tolerance, two different operating conditions of equipment are to be distinguished:

1. Non-synchronised operation, where the receiving equipment is not being timed from a source that is synchronous with the interface under consideration. In this case, it is the equipment's capability to accommodate phase variation on the incoming signal (in terms of clock recovery circuitry and synchroniser/de-synchroniser buffers) that is of interest.
2. Synchronised operation, where the receiving equipment is being timed from a source that is synchronous with the interface under consideration. In this case the slip buffer dimension and operation is also of interest.

5.5.4 Limits for intrinsic jitter

Intrinsic jitter is the jitter generated in the equipment itself, i.e., output jitter measured in the absence of input jitter. ITU-T Rec. G.732, G.742 and G.751 specify maximum levels of output jitter in the absence of input jitter (i.e., intrinsic jitter) for various equipment. These are given below.

At 2.048 Mbps output (ITU-T Rec. G.732)

In the absence of input jitter in the 64 kbps signals being multiplexed to form 2.048 Mbps signal, the peak-to-peak jitter at the 2.048 Mbps output should not exceed 0.05 UI when measured in the frequency range of 20 Hz to 100 kHz.

At 8.448 Mbps output (ITU-T Rec. G.742)

In the absence of input jitter in the 2.048 Mbps signals being multiplexed to form 8.448 Mbps signal, the peak-to-peak jitter at 8.448 Mbps output should not exceed 0.25 UI when measured in the frequency range up to 100 kHz.

When measured with an instrument incorporating a bandpass filter having a lower cutoff frequency of 18 kHz, a roll-off of 20 dB/decade and an upper limit of 100 kHz, the peak-to-peak output jitter should not exceed 0.05 UI with a probability of 99.9% during a measurement period of 10 s.

At 34.368 Mbps output (ITU-T Rec. G.751)

In the absence of input jitter in 8.448 Mbps signals being multiplexed to form 34.368 Mbps signal, the peak-to-peak jitter at 34,368 Mbps output should not exceed 0.25 UI when measured in the frequency range up to 400 kHz.

When measured with an instrument incorporating a bandpass filter having a lower cutoff frequency of 3 kHz, a roll-off of 20 dB/decade and an upper limit of 400 kHz, the peak-to-peak output jitter should not exceed 0.05 UI with a probability of 99.9% during a measurement period of 10 s.

At 139.264 Mbps output (ITU-T Rec. G.751)

In the absence of input jitter in 34.368 Mbps signals being multiplexed to form 139.264 Mbps signal, the peak-to-peak jitter at 139.264 Mbps output should not exceed 0.3 UI when measured in the frequency range up to 800 kHz.

When measured with an instrument incorporating a bandpass filter having a lower cutoff frequency of 10 kHz, a roll-off of 20 dB/decade and an upper limit of 800 kHz, the peak-to-peak output jitter should not exceed 0.05 UI with a probability of 99.9% during a measurement period of 10 s.

5.5.5 Jitter transfer function

The jitter transfer characteristics define the ratio of output jitter amplitude to input jitter amplitude for various jitter frequencies for a given bit rate. When jitter is present at the digital input port of an equipment, some portion is transmitted to the output port (only the higher frequency jitter components get filtered). To control jitter in cascaded homogeneous digital equipment, it is important to restrict the value of jitter gain.

Figure 5.12 indicates the general shape of jitter transfer characteristic. The values of parameters x, y, f_0, f_5, f_6 and f_7 are tabulated in Table 5.12.

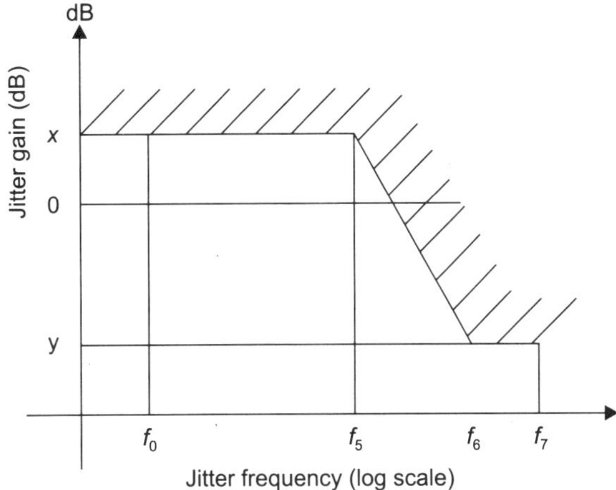

Fig. 5.12 *General shape of jitter transfer function*

Table 5.12 *Interpretation of jitter transfer function parameters*

Data interface	f_0	f_5	f_6	f_7	x	y
2.048 Mbps	10 Hz	40 Hz	100 Hz	10 kHz	0.5	–29.1
8.448 Mbps	10 Hz	40 Hz	400 Hz	100 kHz	0.5	–19.5
34.468 Mbps	10 Hz	100 Hz	1 kHz	400 kHz	0.5	–19.5
139.264 Mbps	10 Hz	300 Hz	3 kHz	800 kHz	0.5	–19.5

5.6 ITU-T Recommendations Regarding Jitter Measurement

The general setup to measure jitter and wander tolerance is shown in Fig. 5.13. It may be noted that:

1. The equipment under test clock can be externally synchronised (if a reference input is available) or synchronised from the interface under test.
2. The clock generator can be used to generate a fixed frequency offset, Δf, upon which the jitter and wander is modulated. The frequency offset must be limited to the values applicable to the interface or equipment under test. The frequency offset should be held constant during the stabilisation period and the subsequent measurement. The allowed frequency offset can be dependent on the path that the measurement signal takes through the system and the way that the equipment under test clock is synchronised.
3. A modulation source is used to superpose a jitter or wander effect, $\Delta\Phi$, on top of the clock signal, which can also have a fixed frequency offset Δf. These jitter and wander phase perturbations usually have a sinusoidal, triangular, or noisy

(PRBS-generated) characteristic. The exact perturbations are prescribed in the applicable jitter and wander tolerance requirements.

4. The test-pattern *P* must match the bit rate of the particular interface that is being subjected to jitter and wander tolerance test. Pattern *P*´ is not necessarily the same as pattern *P*, but is of importance in that a part of pattern *P* is present in *P*´. This part, let us call it *Q*, is passed transparently through the equipment under test. The bit error detector can only search for errors in this common part *Q*.

5. The attenuation function is needed for optical interfaces to be able to determine the 1 dB sensitivity penalty (in terms of optical power) at a certain bit-error ratio. For electrical interfaces, the attenuation (frequency dependent) should represent the worst-case cable length.

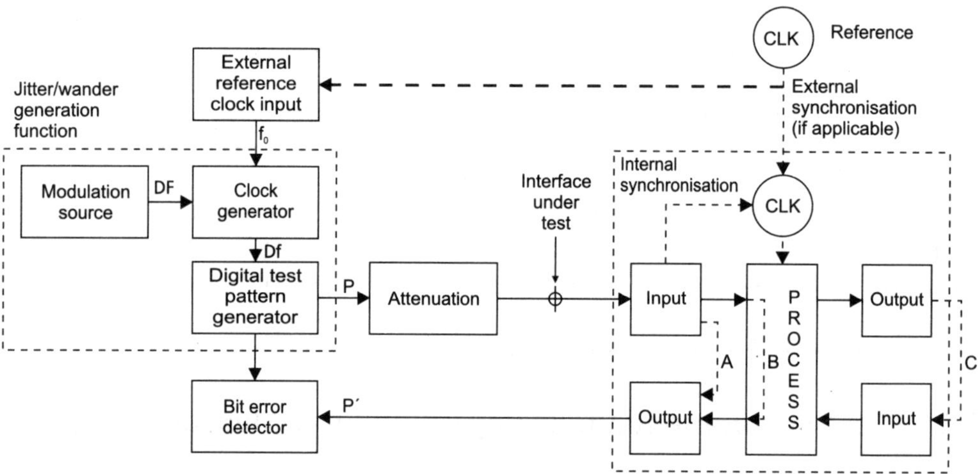

Fig. 5.13 *Measurement setup to measure jitter and wander tolerance*

5.7 Jitter Accumulation

In the previous section we have seen the limits to be observed for jitter and wander at hierarchical outputs and tolerances to be exhibited by the equipment at hierarchical inputs. We have also seen how jittery an equipment can intrinsically be, i.e., limits for its output jitter in the absence of input jitter. We have also seen the limits of how much input jitter can be transferred to output jitter, i.e., the jitter transfer function.

When we operate a single hop of say 34 Mbps link or 140 Mbps link, we have 2 or 3 levels of hierarchy with all equipment obeying the limits. We do not face any problem due to jitter in this case. When more hops are operational on the link, we have jitter accumulation of all the hops after some hops. This accumulated jitter may cross the prescribed limits.

If jitter on a single hop is J_0 UI, on N-hops (repeaters) the jitter accumulation would be [2]:

1. Truly random case: $J_N = J_0 \times N^{1/4}$,
   Pattern dependent case (there is single repetitive pattern): $J_N = J_0 \times (2NA)^{1/2}$, where A is a factor dependent on the number of hops (repeaters) and damping factor of PLL of clock recovery circuitry.
2. Alternate repetitive patterns transmitted: $J_N = J_0 \times N$.

From the above, we can draw the following inferences.

1. Pattern dependent jitter accumulates more rapidly than truly random jitter. So, we have to scramble patterns, if any, in the transmitted sequences and unscramble them at the receiving end.
2. As the number of regenerators increases, accumulation of jitter increases. Since we cannot cut down the number of hops for the sake of jitter, we can recover clock with high Q PLL.

The two techniques mentioned in these inferences are indeed the jitter reduction techniques. These are explained in the next section.

5.8 Jitter Reduction Techniques

5.8.1 Use of scrambler–descrambler

A scrambler breaks up the repetitive patterns in the transmitted sequence and is used in transmission equipment for this purpose. To get back the actual transmitted signal, a de-scrambler is used at the receiving end. Digital radios, OLTEs or optical interfaces make use of a scrambler–descrambler setup.

Refer to Fig. 5.14 illustrating the scrambler–descrambler operation. A scrambler consists of a shift register. An EX-OR operation is carried out on the tappings at the two stages, (j, k), and the resultant sequence is EX-OR ed with the incoming sequence. Thus, patterns in the incoming sequence are broken and a random sequence is transmitted. Converse operations are likewise performed at the de-scrambler.

In Tables 5.13(a–d), the scrambler–descrambler operations are illustrated for an incoming sequence, 100010001000 (repetitive pattern of 1000).

Table 5.13(a) shows the register contents as the incoming sequence 'shifts-in'. It also shows the sequences tapped at the 4th stage, 6th stage and stages $n–j$, $n–k$ ($j = 4$, $k = 6$) and the result of their X-OR operation. Table 5.13(b) shows the further X-OR output with the incoming sequence and the resultant sequence online (i.e., sent to receiver). Table 5.13(c) illustrates sequence shifting inside the shift register of the descrambler and the X-OR output with tappings from the shift register. Table 5.13(d) illustrates the descrambled receive sequence *Rn*, which is a replica of *In*.

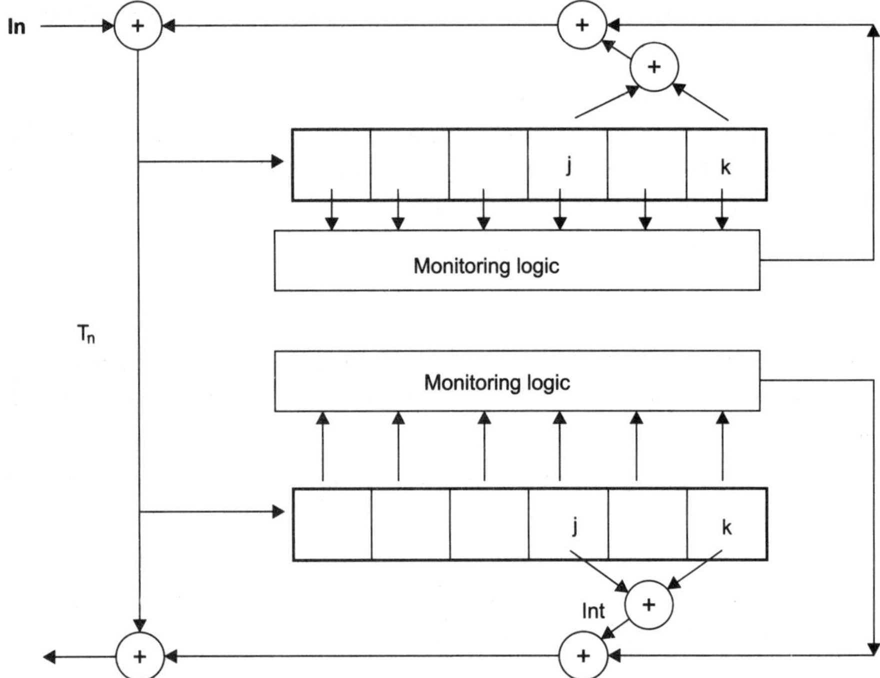

Fig. 5.14 *Operation of scrambler–descrambler*

Table 5.13(a) *Incoming sequence and X-OR of tap (n–j) and tap (n–k)*

In	Register contents						Tap *n–j*	Tap *n–k*	X-OR of tap(*n–j*), tap(*n–k*)
1	1	0	0	0	1	1	0	0	0
0	0	1	0	0	0	0	1	0	0
0	0	0	1	0	0	0	0	1	0
0	0	0	0	1	0	0	0	0	1
1	1	0	0	0	1	1	0	0	0
0	0	1	0	0	0	0	1	0	0
0	0	0	1	0	0	0	0	1	0
0	0	0	0	1	0	0	0	0	1
1	1	0	0	0	1	1	0	0	0
0	0	1	0	0	0	0	1	0	0
0	0	0	1	0	0	0	0	1	0
0	0	0	0	1	0	0	0	0	1

Table 5.13(b) *Resultant online sequence (sequence sent to receiver)*

Tn **(*Int*)**	**Register contents**						**Tap** ***n–j***	**Tap** ***n–k***	**X-OR of Tap(*n–j*), Tap(*n–k*)**
1	1	0	0	0	1	1	0	0	0
0	0	1	0	0	0	0	1	0	0
0	0	0	1	0	0	0	0	1	0
1	1	0	0	1	1	1	0	0	1
1	1	1	0	0	1	1	1	0	0
1	1	1	1	0	1	1	1	1	0
0	0	1	1	1	0	0	1	1	1
1	1	0	1	1	1	1	0	1	1
1	1	1	0	1	1	1	1	0	1
1	1	1	1	0	1	1	1	1	0
0	0	1	1	1	0	0	1	1	1
1	1	0	1	1	1	1	0	1	1

Table 5.13(c) *Sequence inside shift register of descrambler and X-OR with tappings*

Tn **(line)**	**Register contents**						**Tap** ***n–j***	**Tap** ***n–k***	***Int***
1	1	0	0	0	0	0	0	0	1
0	0	1	0	0	0	0	0	0	0
0	0	0	1	0	0	0	0	0	0
0	0	0	0	1	0	0	1	0	1
1	1	0	0	0	1	0	0	0	1
0	0	1	0	0	0	1	0	1	1
1	1	0	1	0	0	0	0	0	1
0	0	1	0	1	0	0	1	0	1
1	1	0	1	0	1	0	0	0	1
0	0	1	0	1	0	1	1	1	0
0	0	0	1	0	1	0	0	0	0
0	0	0	0	1	0	1	1	1	0

Table 5.13(d) *Descrambled sequence Rn (replica of In)*

Int	Register contents						Tap n–j	Tap n–k	*Rn*
1	1	0	0	0	0	0	0	0	1
0	0	1	0	0	0	0	0	0	0
0	0	0	1	0	0	0	0	0	0
1	1	0	0	1	0	0	1	0	0
1	1	1	0	0	1	0	0	0	1
1	1	1	1	0	0	1	0	1	0
1	1	1	1	1	0	0	1	0	0
1	1	1	1	1	1	0	1	0	0
1	1	1	1	1	1	1	1	1	1
0	0	1	1	1	1	1	1	1	0
0	0	0	1	1	1	1	1	1	0
0	0	0	0	1	1	1	1	1	0

5.8.2 De-jitter circuit

A de-jitter circuit is used in a de-MUX. Refer to Fig. 5.15 showing one such circuit.

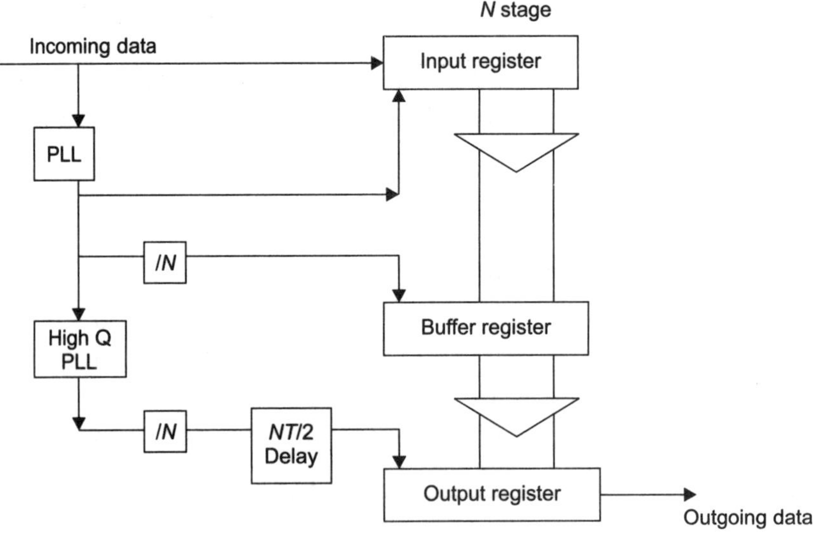

Fig. 5.15 *De-jitter circuit*

- Clock recovered from the incoming digital sequence (say frequency *f*) is used to clock the incoming sequence into a shift register A. Another clock of rate *f*/*N* is used to transfer the contents of shift register A to an intermediate buffer B (since the register length is *N* cells).
- The *N* bits in B are transferred to the output register C in the middle of the frame sequence under the control of *f*/*N* clock. The recovered clock of rate *f* is used to clock the bits out of register C. The clock recovery circuit operating on the incoming sequence has a high *Q*.
- This two-stage clocking in and out with normal-pressure recovered clock and high pressure recovered clock eliminates jitter.

Summary

In this chapter the system design considerations of PDH networks on digital radio as well as OFC in tackling noise and ISI are presented. ITU-T recommendations on control of jitter, wander and measurement of jitter are presented. The problem of jitter accumulation on multiple hops and their solutions are discussed.

References

[1] Barnett-Vigant, FM reliability relationship on digital radio transmission systems, *Proc. IEEE*, ICC-70, 1970.

[2] Sreekanth P V, Digital Microwave Communications System, Chapter 5 Universities Press (2003).

PART II
SDH Systems

6

Evolution of SDH

6.1 What Led to the Evolution of SDH

The inadequacies of PDH, viz. its inability to directly access tributaries from higher hierarchical streams, justification and associated problems and fixed payload structure were always uncomfortable for transmission system designers to live with. Nevertheless, they had to do so till the advances in OFC technology promised higher transmission rates and better transmission frame structures. It was primarily the efforts by the Bell Core Group that paved the way for the evolution of the synchronous digital hierarchy (SDH), the dominant standard for optical transmission and multiplexing for high speed networks. In this chapter, we will examine the inadequacies of PDH and see how the advances in OFC technology heralded SDH.

6.2 Inadequacies of PDH

No direct access – demultiplexing down the hierarchy

In PDH, if an E1 is to be extracted from an E4, it cannot be done directly. First the E4 has to be demultiplexed to four E3s and then the E3 of interest further demultiplexed to four E2s from which the E2 of interest is demultiplexed to four E1s before finally extracting the desired E1. This is because bit-interleaving (along with justification) is employed at every hierarchical stage of multiplexing in PDH.

Justification while integrating

In addition to bit-interleaving while multiplexing at every hierarchical level, justification is resorted to while integrating plesiosynchronous streams. As already discussed in chapter 4, variations in clock speeds of bit streams being integrated (which are almost at the same rate but not exactly identical) require the reading of the bit stream into an elastic storage at the tributary rate and reading out of it from the elastic storage at a rate higher than the tributary rate, thereby creating the need for insertion of justification bits and justification control bits in the multiplexed frame. This process however introduces jitter at every hierarchical level of PDH multiplexing, which, over long-haul networks can accumulate to menacing proportions.

Static payload

Suppose we have the need to allocate certain bandwidth between two points on a PDH network. We will try to find the multiple of E1 closest to the required bandwidth and allocate that many E1s for the purpose. Further, we have to combine the allocated E1s through Stat-MUX process. If the bandwidth plan changes, we do not have the flexibility to re-allocate the bandwidth. The network will have to be reconfigured and the Stat-MUX support systems rewired. Thus the bandwidth division of payload is static.

Multiple standards

PDH is based on T1 (24 channel PCM) in North America and E1 (30 channel PCM) in Europe and India. Integrating services across such networks require elaborate buffer storage and mapping arrangements.

Mapping other data rates

HDLC, frame relay and asynchronous transfer mode (ATM) cannot be mapped easily into PDH streams. ATM's omission was definitely a great drawback, as ATM was devised for encapsulating layer-3 packets into number of equal-sized cells in layer-2 (53 octect in each ATM cell) so as to minimise overheads and add flexibility to handling of ISDN traffic. 53 bytes cannot smugly fit in either E1 or multiples of E1. Higher frames of PDH, like E2, E3 and E4, are not suitable as they are byte-interleaved, whereas ATM traffic is octet (byte)-based.

No support for B-ISDN requirements

B-ISDN requires a minimum 622.08 Mbps for downlink (from exchange to network termination) and 155.22 Mbps for uplink (from network to exchange). A PDH network cannot support such data rates; it calls for multiple PDH links with aggregation and segregation of bandwidth at the end points.

No support for OAM&P functions

For management of network element of any hierarchical level, we require certain bit or byte (preferably byte) allocation in the data frame at that hierarchical level. It is through this bit/byte that the network element (NE) in question communicates with the central monitoring system. If n bits are allocated for this purpose and the frame repeats at rate F per second, the NMS channel data rate would be nF bits/s.

Unfortunately, in PDH systems, E2 or higher level frames do not have a bit or byte allocation for NMS. Only E1 has supervisory bits (5 SV bits in TS-0 of every alternate frame and unused *N* bits (when the link is on exclusive path) not connected to international network. Therefore the PDH MUX equipment takes the support of transmission equipment. For example, digital microwave radio systems and the associated transmission equipment on microwave links, and optical line terminating equipment (OLTEs) and the associated transmission equipment on optical fibre have the provision to integrate SV and NMS traffic with data frames taken from PDH MUXes and transmit these integrated frames. This makes the process complicated as multi-rate multiplexing is resorted to. Even with this process the data rate available is small (few kbps). Hence, we can only have configuration, alarms monitoring, supervisory functions and certain performance monitoring functions through NMS on PDH networks[1].

In large networks handling data rates in hundreds of Mbps to Gbps, we require enhanced functions of operations, administration, maintenance and provisioning (OAM&P), which require hundreds of kbps as the channel data rate. These have to be supported on the transmission frames as logical channels by way of allocating bits commensurate with the required data rate, which is not possible in PDH.

6.3 Guided Light Guides SDH Evolution

The evolution of SDH was guided by guided light, i.e., the evolution of optical fibre communications systems or OFC technology. Figure 6.1 shows the absorption loss and scattering loss, which together constitute the attenuation of transmitted signal in

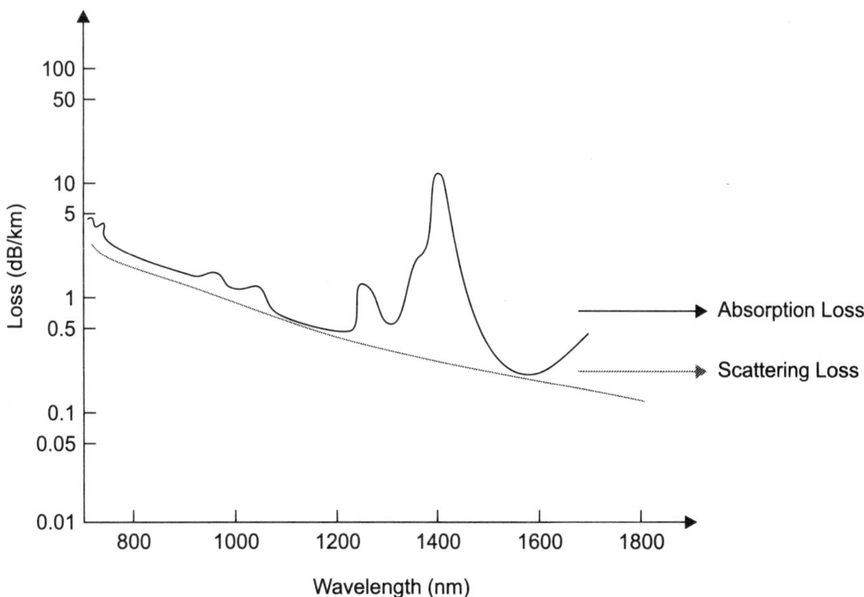

Fig. 6.1 *Attenuation (absorption loss + scattering loss) in OFC as function of wavelength*

OFC, plotted against wavelength. It can be seen that the attenuation is low in three bands around 850 nm, 1310 nm and 1550 nm, but viable technology in the latter bands became available only progressively. In the late 1970s and early 1980s, OFC technology employed 850 nm band (attenuation about 3 dB per km). However, the dispersion phenomena, which causes spreading of the pulse and thereby inter-symbol-interference, put a cap on the transmission rates. The existing PDH rates could be supported up to E2.

By 1985, OFC technology had moved to 1310 nm band where the attenuation in the fibre is about 0.35 dB/km. Inter-modal dispersion was controlled using single mode fibre, which, with its core diameter of about 10 microns and cladding diameter of 125 microns allowed only one mode beyond certain cut-off wavelength. The zero-dispersion phenomenon at 1310 nm, where the constituents of chromatic dispersion viz. material dispersion and waveguide dispersion nullify each other, was an additional advantage. This allowed transmission rates up to 2.5 Gbps over 40 km and lower rates over longer distances. The PDH rates of E4 could easily be supported and there arose the need for higher rate transmission hierarchies.

By year 1990 OFC technology had moved to 1550 nm window with the help of dispersion shifted fibre (DSF). By changing the composition of the fibre, zero-dispersion point was shifted to 1550 nm. Also, the lowest attenuation was at 1550 nm, which is about 0.2 dB/km. This promised data rate support up to 10 Gbps over 80 km. Also, the cost of fibre and electronics had come down and it became economically viable to deploy regenerators for transmission over longer distances.

The promise shown by OFC for higher data rates and the limitations of PDH propelled the evolution of SDH. In fact, the SONET transmission hierarchy of 1987 (STS-1 to STS-48 with data rates of 51.84 Mbps to 2.488 Gbps) evolved into SDH in the 1990s. There was another contributing factor to this evolution, which we shall look at in the following section.

6.4 Bell and the Baby Bells Story

The American Telephone and Telegraph company (AT&T), formed in 1885 as s subsidiary of the American Bell Company, had enjoyed monopoly in its field of operation for over of 100 years. The divestiture of AT&T in 1984 left Bellcore as long distance provider and seven Baby Bells were formed as regional service providers. The 1984 judgement by US district court judge, Harold Green, to open long-distance services to other competitors brewed severe competition and proliferation of service providers.

To retain its hold over the market, the Bell Core Group used the newly evolved OFC technology with its higher-rate transmissions to find a solution to the problems of PDH. They submited the synchronous optical networks (SONET) proposal to ANSI in 1987 based on the synchronous transmission concept, the starting point of which was STS-1 at 50.688 Mbps rate. Later, certain innovations by Metrobus were added to map the T-hierarchy PDH streams onto STS frame, which slightly changed the size and structure of the frame. The ITU-T was however considerate towards the E-hierarchy of PDH and refused to consider STS-1 as the international standard. SDH with STM-1 was declared the international standard. It was done to conveniently map E-hierarchy streams as well as T-hierarchy streams onto the new standard frame format.

This prompted Bell Core to revise SONET frame and syntax and re-fix the starting rate of STS (i.e. STS-1) as 51.84 Mbps. STS-3(OC-3) of SONET was equivalent to STM-1 of SDH (155.22 Mbps). Thus, the ending was happy for all.

6.5 How SDH Was Lapped Up

The SDH equivalence of SONET, after the adjustments by Bell Core, is shown in the Table 6.1.

Table 6.1 *SONET/SDH equivalence*

SONET level	SONET frame	Line rate (Mbps)	SDH equivalent
OC-1	STS-1	51.84	----
OC-3	STS-3	155.52	STM-1
OC-12	STS-12	622.08	STM-4
OC-24	STS-24	1244.16	STM-8
OC-48	STS-48	2488.32	STM-16
OC-96	STS-96	4976.64	STM-32
OC-192	STS-192	9953.28	STM-64

The equivalence, as shown in Table 6.1, and the ability of STM frame of SDH to take-in T-family streams, E-family streams and also support HDLC frames, MAC frames and ATM made it a one point support for all networks, and the standard became extremely popular. By year 2000, i.e., within a few years from the proclamation of standards, 75 million fibre km network based on SDH (including STS of SONET) was functional over the entire world. The power of SDH can be imagined from this.

Since SONET was the predecessor of SDH network, it is worthwhile to take a quick look at basic SONET frame. We shall look at STS-1, the synchronous transport signal of first order.

6.6 STS-1 Frame of SONET

Please refer to Fig. 6.2 showing STS-1 frame. It has 810 bytes organised in 9 rows and 90 columns. The frame duration is 125 μs. Transmission is row-wise, left to right; for each byte also the transmission is from left to right, the MSB being to the left-most.

Columns 4 to 90 (87 columns × 9 rows = 783 bytes) is the synchronous payload area and is known as synchronous payload envelope (SPE) . Of this, column 4 (9 bytes) carries the path overhead (POH). The rest, i.e. columns 5 to 90, is the actual payload area marked to carry user traffic.

Column 1 to 3 in the first three rows, i.e., 9 bytes, carry section overhead (SOH) meant for interpretation by the regenerator equipment or regenerator module of the optical interface. Columns 1 to 3 in rows 5 to 9, i.e., 15 bytes, carry line overhead (LOH), meant for interpretation by the multiplexer equipment. The leftover three bytes in row 4 (column 1 to 3 of row 4) carry the pointer which points to the start of SPE.

Identification of bytes in SOH and LOH are marked in Fig. 6.2. Details of the bytes of SOH and LOH (nomenclature and purpose) are shown in Table 6.2.

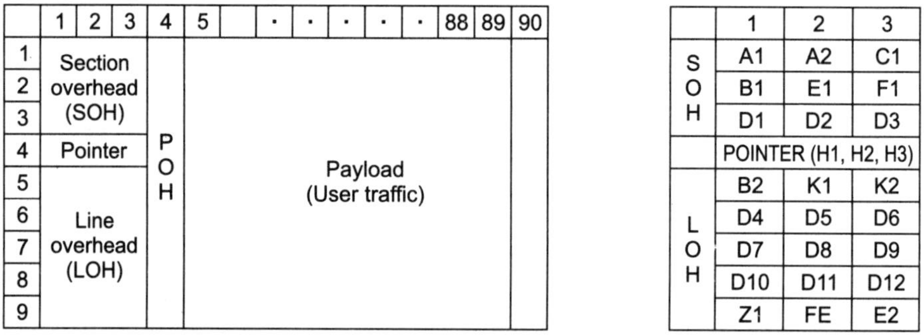

	1 2 3	4	5 · · · · · 88 89	90
1	Section overhead (SOH)	P O H	Payload (User traffic)	
2				
3				
4	Pointer			
5	Line overhead (LOH)			
6				
7				
8				
9				

	1	2	3
SOH	A1	A2	C1
	B1	E1	F1
	D1	D2	D3
	POINTER (H1, H2, H3)		
LOH	B2	K1	K2
	D4	D5	D6
	D7	D8	D9
	D10	D11	D12
	Z1	FE	E2

Fig. 6.2 *SONET's STS-1 frame and identification of bytes in SOH, LOH*

Table 6.2

Section overhead	A1 (Frame sync for frame synchronisation)	A2 (Frame sync for frame synchronisation)	C1 (section trace, for correct pairing of regenerators)
	B1 (Parity check)	E1 (User order wire, i.e., communication channel for maintenance instructions)	F1 (Service provider order wire, i.e., communication channel for maintenance instructions)
	D1 (for NMS)	D2 (for NMS)	D3 (for NMS)
Pointer	H1	H2	H3
Line overhead	B2 (Parity check)	K1 (for automatic protection switching message)	K2 (for automatic protection switching message)
	D4 (for NMS)	D5 (for NMS)	D6 (for NMS)
	D7 (for NMS)	D8 (for NMS)	D9 (for NMS)
	D10 (for NMS)	D11 (for NMS)	D12 (for NMS)
	Z1 (to indicate synchronisation status)	FEBE (to give count of blocks in error)	E2 (User order wire on MUX)

We will not dwell further on these bytes as similar bytes with some more additions and features are used in STM frames of SDH which will be explained in greater detail in chapter 7. The point to note here is that STS frame of SONET was the concept behind the STM frame of SDH.

Summary

In this chapter the evolution of SDH, spurred by the inadequacies of PDH and the advances in OFC technology are discussed. SONET and SDH frame structures are presented.

References

[1] P. V. Sreekanth, *Digital Microwave Communication System*, section 6.7, Bit stream integration in Digital MW Radio, Universities Press, 2003.

7

Multiplexing Process in SDH

7.1 Requirements of SDH Frame

A digital multiplexing process takes in digital streams from various sources and forms a multiplexed stream which is structured as frames. The frame capacity in terms of bytes/ bits would meet the intended traffic requirement. In this process, certain overhead bits/ bytes are added to facilitate performance monitoring of paired equipment by the counterpart, network management functions over the entire network, protection switching mechanisms to diversity paths, and order wire or dedicated channels for users or service providers.

In this section, the requirements of SDH frame to support the intended traffic and the associated functions are discussed.

Frame duration

The frame repetition rate should be such that the source streams maintain continuity in real time after demultiplexing. Since the constituent channels are rated at 64 kbps (1 sample every 125 μs), SDH frame duration is 125 μs, be it STM-1, STM-4, STM-16 or STM-64.

With frame duration remaining the same, higher transmission rates are possible (622.08 Mbps for STM-4, 2.5Gbps for STM-16 and 10 Gbps for STM-64) because the frame takes in more bits in the same duration of 125 μs.

Frame alignment sequence

A multiplexed frame requires start-up identity so that at the receiving end frame boundaries are identified in the received stream and the constituents of the frames are properly located and processed. STM-1 frame has 3 A1 bytes (A1 = 11110110) and 3 A2 bytes (A2 = 00101000) as FAS (frame alignment signal).

Check-sum

Parity check-sum helps in error detection through verification of parity check-sum of the transmitted frame.

Support for OAM&P through NMS

In PDH systems, except for E1 frames, no other frame has the provision for supervisory bytes. These are supported by the transmission equipment in the integrated transmission. For example, 34 Mbps digital radio meant for transmission of E3 (34.368 Mbps) supports DSC (digital service channel) together with SV (supervisory) in the integrated 38.664 Mbps data stream actually transmitted. This was discussed in detail in [1].

SDH networks handle data rates in hundreds of Mbps to Gbps and also have tens of NEs on a trail or path. Several such trails are meshed and protection rings are also formed. Normal NMS functions of PDH networks, viz., alarms, monitoring, supervisory and configuration of primary MUXes would not suffice for SDH networks. It requires enhanced capabilities of operations, administration, maintenance and provisioning (OAM&P) of each NE irrespective of its order in the hierarchy for NMS. OAM&P capabilities of NMS can be put to use only if we have logical channels through STM frames which support hundreds of kbps as channel data rate. For example, each byte position in STM frame repeats every 125 μs, i.e., if we reserve one byte for the purpose in the STM frame, we get a logical channel rate of 64 kbps supported through that byte on the STM frame. If we have N bytes, the data rate would be $N \times 64$ kbps.

One of the objectives of SDH is that the frame should be self sufficient, i.e., it should not depend upon the transmission equipment for carrying any information regarding OAM&P functions. Since OAM&P functions require few hundreds of kbps channel rate as discussed above, around 10 to 15 bytes space in STM frame are required to carry information related to OAM&P functions of NMS.

Support for APS

It should be possible to issue APS (automatic protection switching) command to any equipment in the network for switching over to a redundant fibre or transmission equipment/interface in case of performance degradation/failure of fibre or equipment/interface.

Clock status information

In SDH network, a network element may be connected to more than one neighbour. Any network element should know the clock beat of its neighbours as also make its own clock beat known to its neighbours. So, a clock status byte (known as sync status byte) is required in the frame structure for carrying the information regarding the reference clock being used by the network element sending the frame. This will help in establishing synchronisation with respect to the primary reference clock or subordinate synchronisation units which are strategically placed in the network. In case of interruptions in the network and subsequent recovery through alternative paths, the network elements re-establish synchronisation for themselves for the changed situation.

Trace bytes

While transmitting payloads across the network from one NE to another through intermediate NEs, the delivery should be made where it is intended. So, there has to be a provision in the SDH frame to carry 'trace' information to ensure correct pairing of transmission and reception equipment.

Structured payload and overheads

An NE of SDH network, which is basically a data transmission network, when considered functionally, can be any of the following:

a. Terminal multiplexer (TM): This equipment is the originator/terminator of the traffic on a path, i.e., the end element.

b. Regenerator (RG): This equipment regenerates received data frame.

c. Add drop multiplexer (ADM): This equipment drops part of the payload and facilitates addition to the payload in the vacant portion.

d. Digital cross connect (DXC): This equipment facilitates cross-connection of part payloads across different directions.

Refer to Fig. 7.1(a) showing SDH network. The network originates and terminates on TM (terminal MUX) with RGs, ADMs and DXCs placed in between. From the above functional classification and configuration, it is clear that information needs to be exchanged between like-minded NEs, i.e., between regenerator pairs or MUX pairs or end-to-end path terminators. This implies a proper structural arrangement for SDH frame with clearly marked chunks of overhead bytes for regenerators, MUXes and path terminators. We explore this further in the next section.

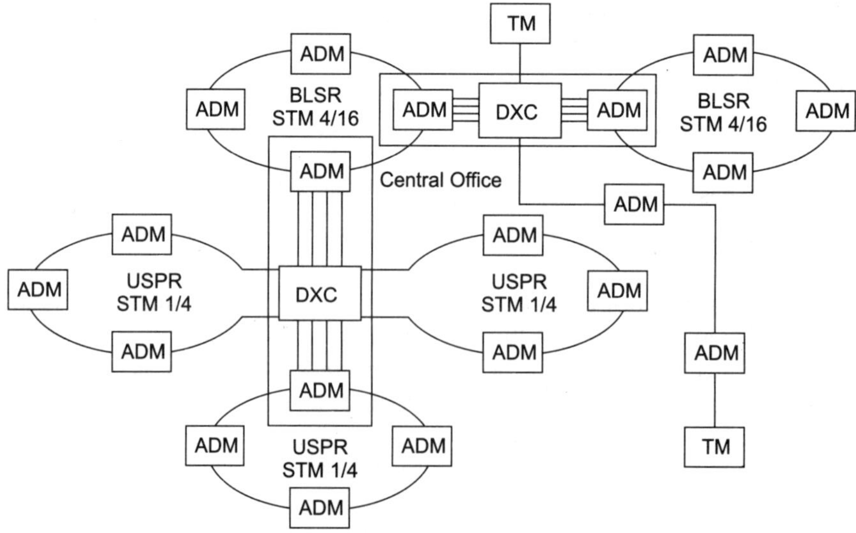

Fig. 7.1(a) *SDH network*

7.2 Concept of Regenerator Section, Multiplexer Section and Path

In the previous section, types of SDH network elements, viz., TM (terminal MUX), RG (regenerator), ADM (add–drop MUX) and DXC (digital cross-connect) have been introduced. The SDH network (NW) comprises several segments interconnecting RGs and MUXes with paths set up between TMs.

If the segment is between two adjacent regenerators, it is a 'regenerator section'. If the segment is between two adjacent MUXes, it is 'multiplexer section'. End-to-end connection between TMs is called a path.

The above segmentation in the physical sense has more than mere organisation of the network to achieve. It serves certain link-engineering as well as logical purposes. The concept of regenerator section, multiplexer section and path in the physical and logical sense are illustrated in Fig. 7.1(b).

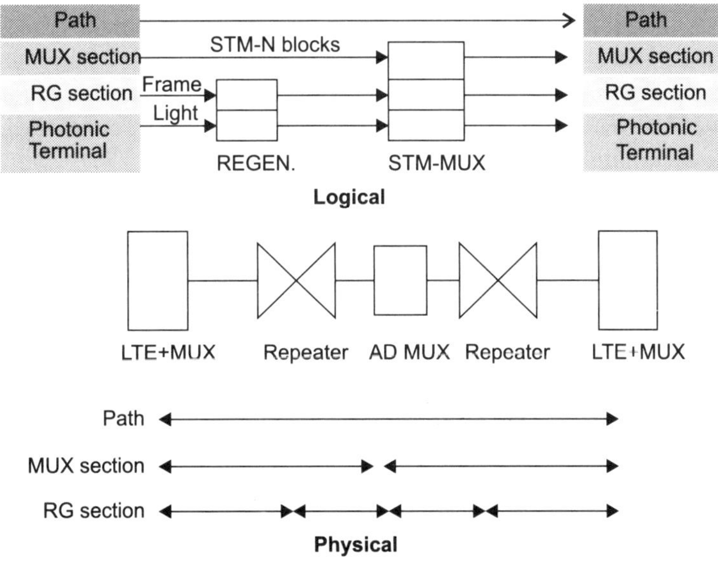

Fig. 7.1(b) *Concept of RG section, MUX section and path in SDH network*

Equipment manufacturers offer NE and MUX together with optical interface. In fact, 'MUX-DEMUX' or add–drop and cross-connect operations on payloads (arriving at/ leaving the NE as STM streams) are done by cross-connect module and the aggregation of traffic into STM-N stream to be sent on fibres is done by aggregate card equipped with suitable optical interface. Aggregation can be for STM-1, STM-4, STM-16 or STM-64. Furthermore, it can be for 1310 nm or 1550 nm and for intra station/short-haul/

long-haul. The combinations offered depend on sustainable transmission rates at the wavelength of operation of the fibre. Details of this are given in chapter 8.

The MUX portion has the interfaces to map/derive various traffics with their respective allowed bandwidths. By changing the aggregate card and cross-connect card, the NE can be upgraded to higher order, i.e., STM-1 can be upgraded to STM-4 and STM-4 can be upgraded to STM-16. Also, by programming the configuration of the equipment, it can be made RG or TM or ADM or DXC. Thus, total flexibility in link engineering is possible by viewing the NW in segments and providing appropriate optical interface and configuration of equipment.

Allocation of overhead bytes for the regenerator section, the MUX section and the path help in logical coupling between ends of such segments. This in turn helps the end NEs of each segment (RGs in case of regenerator segment, MUXes in case of MUX section and TMs in case of path) to exchange information through overhead bytes, monitor paired NEs and also switch over to stand-by facility in case of need.

The above logical coupling implies organisation of the set of overhead bytes pertaining to RGs, MUXes and TMs into exclusive groups so that any NE will process only the overhead bytes meant for it specifically and be transparent to others.

The overhead bytes meant for RGs are organised as RSOH (regenerator section overhead), those for MUXes are organised as MSOH (multiplexer section overhead) and those meant for TMs of a path are organised as POH (path overhead).

Let us now summarise the SDH frame requirements.

Frame duration should be 125 μs.
Frame should have bytes for FAS, parity, NMS, APS and path trace
Frame should be structured as RSOH, MSOH, POH and payload areas.

We are now ready to present STM-1 frame of SDH.

7.3 Description of SDH Frames

7.3.1 Structure of STM-1 frame

As discussed in the previous section, we can visualise the network portions physically as:

Regenerator section: Between regenerators (between any two adjacent optical interfaces)

Multiplexer section: Between multiplexers (between any two adjacent SDH MUX functions)

Path: End-to-end.

Logically, SDH data frame provides for exchange of network related information between NEs in the form of the following.

RSOH (regenerator section overhead): It is accessed at every regenerative element, i.e., at every NE having regenerative function. It may be noted that every NE, whether RG, TM, ADM or DXC, has an optical interface which does optical-to-electrical and electrical-to-optical' conversions. Hence, RSOH is accessed at every NE.

MSOH (multiplexer section overhead): It is accessed at every NE which has MUX, i.e., all NEs except the RG access the MSOH.

POH (path overhead): It is accessed by the end elements of a path, i.e., the TMs.

Refer to Fig. 7.2 showing the STM-1 frame.

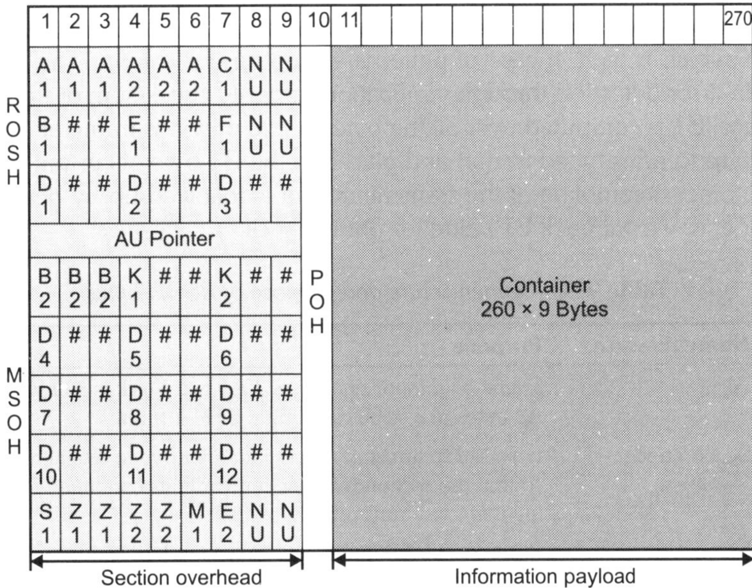

Fig. 7.2 *STM-1 frame structure*

7.3.2 Details of RSOH

Nomenclature of RSOH bytes and their role is given in Table 7.1.

Exploring RSOH

Order-wires E1 and F1

In PDH systems we have two order wires: (i) normal order wire for voice communication in an omnibus fashion between all stations and (ii) express order wire through DSC at selected stations. In SDH systems we have order wires on the section, line and path, separately. Again, over a section we have both normal order wire and special order wire. Normal order wire (E1 channel) is for the user while special order wire (F1 channel) is for the network provider or for special maintenance purpose by the user itself.

Since STM frame repeats every 125 μs, one byte allotment for a purpose implies a bit rate of 64 kbps. Thus, E1 and F1 support 64 kbps channel.

Data communication channel (DCC)

DCC is used for NMS communication to communicate under the network monitoring system, with a centralised software system, graphic user interface for alarms, control, monitoring and administration purposes. Since DCC has a 3-byte allocation, NMS channel over the regenerator section has 3 × 64 kbps, i.e., a data rate of 192 kbps.

Parity byte B1

A frame is scrambled before transmission so that patterns are broken and jitter accumulation, which is high in case of patterns, is kept under control. The parity checksum helps in error detection through verification of parity check-sum of the transmitted frame. Hence, B1 is computed over all the bytes of the previous frame after scrambling (i.e., it pertains to transmitted frame) and placed in the B1-byte of current frame before scrambling. Since scrambling at the transmitter end is followed-up by descrambling at the receiver end, we get back B1 value for parity check.

Table 7.1 *Nomenclature and purpose of RSOH bytes*

Byte	**Nomenclature**	**Purpose**
A1, A2	FAS	Frame alignment signal has 3 A1 bytes (A1 = 11110110) and 3 A2 bytes (A2 = 00101000) in STM-1 frame
J0	Section trace	It is used to transmit access identifier over a regenerator section so that the section receiver can verify its continued connection to the intended transmitter.
Z0	Spare	Reserved for future international standardisation
B1	BIP-8	Parity check-sum is computed over all bytes of previous frame after scrambling and placed in the B1 byte of the current frame before scrambling
E1	Normal order wire	Voice channel between regenerators
F1	Special order wire	Voice/data channel for special maintenance purpose
D1-D3	DCC	Data communication channel for NMS functions

Q 7.1 Why are there 3 bytes of A1 and A2 as FAS?

STM-1 frame *structure* is designed by merging 3 STS-1 frames for compatibility with SONET, commissioned by Bell Core before SDH was set as standard by ITU-T. STS-1 frame has one byte each of A1 and A2 as its FAS, which has been carried over to STM-1 as its FAS.

7.3.3 Details of MSOH

Nomenclature of MSOH bytes and their purposes is given in Table 7.2.

Table 7.2 *Nomenclature and purpose of MSOH bytes*

Byte	Nomenclature	Purpose
B2	Line error monitoring	The three bytes help to determine if transmission error has occurred over a line/ MUX section.
K1, K2	APS bytes	Bytes to carry APS commands
D4–D12	DCC over line	Data communication channel over a line between MUXes for NMS
S1	Synchronisation status byte	It is the 'life byte' of SDH network. It gives the rank of clock under use by the network element so that synchronisation with the best ranked clock source is attempted by any network element in the trail.
M1	REI byte	Conveys count of interleaved bit blocks detected in error over the MUX section
E2	Order-wire	Voice channel between MUXes

7.3.4 Payload area in STM-1 frame

The portion of the STM-1 frame other than overheads and pointer has 261 columns and 9 rows. This is the payload of the frame. In this the first column, i.e.,10th column of STM-1 frame, is reserved for POH (path overhead). The POH field is reserved for information exchange between TMs of a path in a network. The leftover portion, i.e., columns 11 to 270 (a total 260 × columns 9 rows), is designated as SPE (synchronous payload envelope).

This is the area which can be filled with user traffic. This is also known as C-4 or container-4 and is the maximum size of a container used for carrying user traffic. Depending on the traffic type, mapping methods are prescribed, which are dealt with in detail in section 14.5. C-4 container plus POH is known as VC-4 (virtual container-4). Let us first look at the details of VC-4 POH.

7.3.5 VC-4 POH bytes

Nomenclature and purpose of VC-4 POH bytes are given in Table 7.3.

Exploring VC-4 POH

The following questions leave the VC-4 POH fully explored. The readers would have had many of these queries while going through the earlier discussions.

Table 7.3 *Nomenclature and purpose of VC-4 POH bytes*

Byte	Nomenclature	Purpose
J1	Path trace	Enables receiver to verify connection with TX
B3	Path check-sum	Check-sum of previous VC4
C2	Signal label	E4/T6 (European/American)
G1	Path status	Path status and performance conveyed back to VC-4 source terminal. Complete duplex trail can be monitored
F2	Path user channel	Data channel for path user
H4	Position indicator	Generalised position indicator for payloads. Can be payload specific
F3	Path user channel	Data channel for path user
K3	APS	Automatic protection switching at VC-4 path level
N1	NW operator byte	Monitoring NW

Q 7.2 *What is the difference between path trace and path status?*

The receiving terminal on a path verifies its continued connection to the intended transmitter through path trace byte.

Path status conveys the performance of the path back (e.g., error performance, any far-end defect) to the source from the sink.

We have number of paths on a nation-wide network, i.e., large number of TX–RX pairs. Still, we distinguish them with only 8 bits of identification.

The trick is simple: Instead of just one byte code, path trace can carry code of length 16 bytes or 64 bytes as per size of the network. The code bytes are carried through path trace bytes of successive frames. It means, path trace is a multi frame signal.

Q 7.3 *Why do we need two user channels on path?*

The two different user channels can be payload specific and hence provide additional scope for communication by user (maintainer/monitoring equipment) situated at the path ends.

Q 7.4 *We have no APS byte in RSOH; there is one in VC-4 POH and two in MSOH. Why such distinction?*

APS (automatic protection switching) operation is for the MUX section while the path provides alternate 'data pipe or connection' for the interfaced application. At the re-generator, we do not derive any connection for interfacing to any application. Hence, we do not have APS byte in RSOH.

APS over path is at higher order, hence one byte would suffice (in fact 4 bits of one byte would suffice) to convey APS signals. APS over MUX section is at lower order, hence, we need elaborate arrangement of 2 bytes to convey the required signalling for APS.

Q 7.5 *Are there any formatted messages for APS signalling sent through K1 and K2 bytes?*

Yes, for the APS the signalling being sent through K1 and K2 bytes of MSOH for protection of MUX section has the format specified by ITU-T G.841 recommendation. In fact, two formats are given – one for linear protection and the other for ring protection. Linear protection is point-to-point on a path or is mesh based whereas ring protection is on a ring topology over span or ring.

Linear protection messages are shown in Table 7.4. As can be seen, switching option can be locked-out or made manual (i.e., by command from administrator rather than automatic) or prioritised. These different options help in situations of different complexities when there are several cuts/interruptions in the network and we want to keep switching the options to minimal or null and only concentrate on high priority traffic.

Table 7.4 *Linear APS messages*

K1 byte	Condition	K1 byte	Condition
Bits b1 to b4		0100	Exercise
1111	Lock-out of protection	0011	Unused
1110	Forced switch	0010	Reserve request
1101	Signal fail (high priority)	0001	Do not revert
1100	Signal fail (low priority)	0000	No request
1011	Signal degrade (high priority)	Bits b5 to b8	Selects channels used by APS messages
1010	Signal degrade (low priority)	**K2 byte**	
1001	Unused	Bits b1 to b4	Selects bridged channel used
1000	Manual switch	Bit b5	Determines APS architecture
0111	Unused	Bits b6 to b8	000 to 101 are reserved for future use
0110	Wait to restore		110 is for MS-RDI
0101	Unused		111 is for MS-AIS

Ring protection messages are shown in Table 7.5. We have all the options as in linear protection and these can be on span or on ring.

Table 7.5 *Ring APS messages*

K1 byte	**Condition**
Bits b1 to b4	
1111	Lock-out of protection (span) or signal fail (protection)
1110	Forced switch (span)
1101	Forced switch (ring)
1100	Signal fail (span)
1011	Signal fail (ring)
1010	Signal degrade (protection)
1001	Signal degrade (span)
1000	Signal degrade (ring)
0111	Manual switch (span)
0110	Manual switch (ring)
0101	Wait-to-restore
0100	Exerciser (span)
0011	Exerciser (ring)
0010	Reserve request (span)
0001	Reserve request (ring)
0000	No request
Bits b5 to b8	Destination node ID
K2 byte	
Bits b1 to b4	Source node ID
Bit b5	Path code: 0-Short path, 1-Long path
Bits b6 to b8	
000	Idle
001	Bridged
010	Bridged and switched
011	Reserved for future use
100	Reserved for future use
101	Reserved for future use
110	MS-RDI

7.4 Concept of Containers

The term container designated by C describes a defined network-synchronised transmission capacity. Container facilitates interleaving of tributary information inside it, i.e., it acts as a box to upload certain volume of data and download the same at the designated destination.

In SDH frame, the synchronous payload envelope, viz. column 11 to 270 is viewed as one container with sub-containers as illustrated in Fig. 7.3(b). Figure 7.3(a) shows the mapping of traffic into containers in the path layer which are carried by other layers.

The following containers are distinguished from others:

Designation	**Signal to be transmitted**
C-11	1,544 kbps
C-12	2048 kbps
C-2	6312 kbps
C-3	44,736 kbps or 34,368 kbps
C-4	139264 kbps

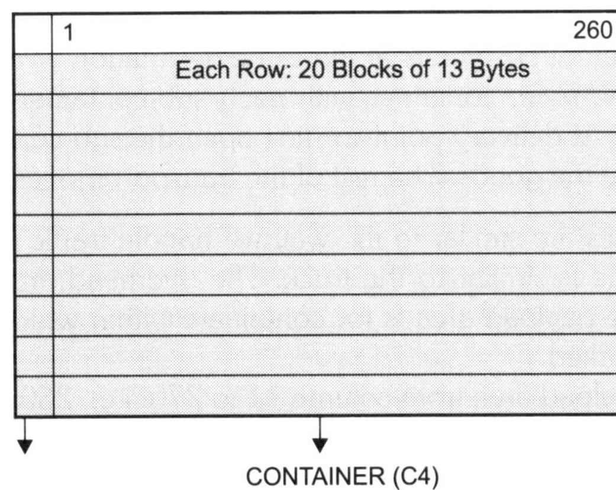

Fig. 7.3(a) *C4 container in STM-1 frame*

Fig. 7.3(b) *C3, C12 containers in STM-1 frame*

(As an analogy, consider a goods transportation company operating trucks between two cities A and F with halts in-between at cities B, C, D and E.)

Case 1: If the goods meant for different destinations are put in one container and kept on a truck and sent from A, at every intermediate station we unload the container, take-out goods meant for that station, re-load the rest and the truck moves on.

Case 2: There are several customers at the same destination (say F) and all their goods are kept in one container. We unload the container, sort-out the goods and hand them over.

These two cases are similar to the way we handle traffic in PDH system. PDH data frame is the container. The E1s interleaved in it are the 'goods' being carried to different destinations or to different customers at the same destination.

Now, we change our strategy slightly.

Case 3: If we now have different compartments in a container which can be individually opened without disturbing the rest, we can open these at the respective intermediate stations for unloading the goods. The rest moves on without getting disturbed.

Case 4: We have different customers at the same destination. We keep sub-containers destination-wise in the main container with each sub-container containing compartments customer-wise. At delivery point we first open the sub-container, then the compartment and take out the goods. The rest of the transport moves on.

These above two cases are similar to the way we handle traffic in SDH system. SDH data frame/STM frame is similar to the truck. The overhead area in the frame is the driver's cabin and the payload area is for containers within which sub-containers and compartments are marked.

The entire STM payload area from column 11 to 270, i.e., 260 columns × 9 rows or 60 × 9 bytes = 2340 bytes can accommodate one C 4 or three C3s or sixty three C-12s or two C3 + twenty-one C12 or one C3 + forty-two C12s. Refer to Fig. 7.4 illustrating the process of formation of a container, a virtual container and so on.

PDH signal or any payload stream supported by SDH is mapped into the container. A path overhead (POH) is added to each container. This unit is called a virtual container (VC).

POH serves the following purposes:

- POH carries information on supervision and maintenance of a path switched in the network.
- POH ensures reliable transport of the container from signal source to destination.

The following VCs are defined in SDH:

VC-11, VC-12, VC-2	LO (lower order) containers
VC-3, VC-4	HO (higher order) containers

All containers transmitted in one large container are termed as LO containers. Those containers that are transmitted directly are termed as HO containers.

The tributary unit, TU, is termed as a component of the higher-order (HO) container inside which the embedded lower-order (LO) container can move, located by the corresponding pointer.

The following TU are distinguished: TU – 11, TU – 12, TU – 2 , TU – 3.

Before getting interleaved in the higher-order container, the TUs are combined into a group or byte interleaved. Such a group is called a TUG (tributary unit group).

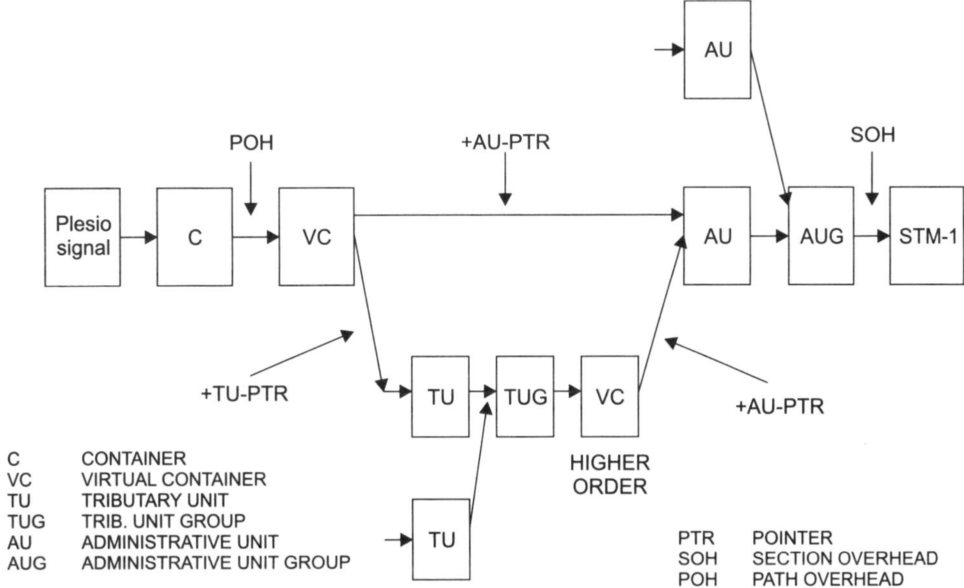

Fig. 7.4 *Formation of containers, virtual containers and AUs in SDH*

The component of the STM-1 frame within which the VC is able to float is termed as administrative unit. The corresponding pointer is called the AU pointer. In other words, 'VC + Pointer' is AU.

It is possible to transmit the following AU in STM-1 frame:

1 × AU - 4
3 × AU - 3

Several AUs are byte-interleaved to one AU group (AUG). The pointer is not 'overhead' like RSOH, MSOH, POH. Attachment of pointer to VC to form AU is for a different purpose, as explained in the following section.

Q 7.6 How are the 63 VC12s in STM-1 frame or VC-4 identified?

VC12s are identified by KLMs. K stands for TUG3 number, L stands for the TUG2 number and M stands for VC-12 number in the TUG2. K can be 1 to 3, L can be 1 to 7 and M can be 1 to 3.

7.5 Pointers

7.5.1 Need for pointers in STM frames

In SDH network, each NE has two or more neighbours with the exception of the end NEs. The SDH network has a vast span connecting several hundreds of NEs. Each NE cannot be however supplied with a separate synchronisation clock. The SDH network provider in fact, takes the primary reference clock (PRC, which is an atomic clock) at one or two places and the NE connected to it starts its operation with respect to the PRC. As the data frames go on the network and reach other NEs, there is degradation due to phase error and jitter accumulation. Hence, at intermediate locations in the trail, SSUs (secondary synchronisation units) are provided which clean up the clock. The rules governing the spacing of PRC, SSU, the number of NEs on each trail and the synchronisation process are discussed in detail in section 7.10.

Let us closely examine STM frame processing in the NE; this is depicted in Fig. 7.5(a).

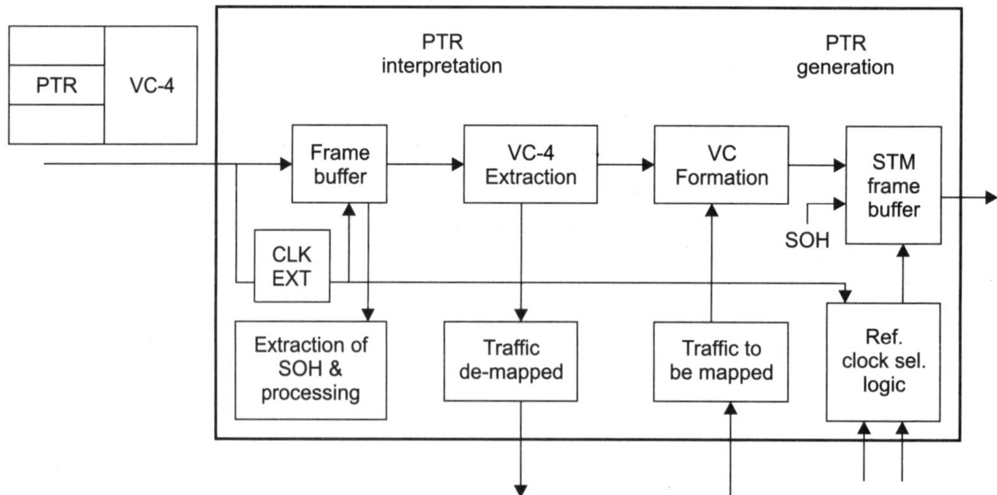

Fig. 7.5(a) *STM frame processing in NE*

Each NE gets data frames from its neighbour as per the transmit clock of its neighbour. After extraction of the clock from the incoming frame, the received data is stored in a buffer by the NE, i.e., the incoming data is clocked into the buffer by the NE as per the received clock. The NE, after examining overhead bytes, deals with the data in the buffer (de-mapping from containers for local use and sending it to neighbouring NEs). However, to send data to the next NE, an STM frame is to be formed. Therefore, the data is first transferred to the output buffer to form a container, POH is added to form VC and section overheads are added to form the STM frame. STM frames are sent out as per the NE clock. The delay involved from input to output is that of two buffers. In a perfect scenario, each NE (the receiver in each NE) gets a clean, uninterrupted clock reference down the trail from its neighbour. So processing delay is constantly maintained and there are no problems of underflow/overflow and online continuity is maintained in the transmission of STM frames.

But, the real operational scenario is different. No NE (receiver) can be wedded to the perfect clock reference forever. There will be degradations in the reference clock or temporary interruptions or link failures, and NEs need to be provided with capability to recover from them.

The receiver clock performance can be characterised by its operation in three scenarios:

- Ideal operation
- Stressed operation
- Holdover operation

In ideal operation there are only short term random phase variations of the received clock. But these can lead to accumulation of phase errors and underflow/overflow of information bits. Stressed operation is the typical mode of operation of a receiver clock, where a receiver clock is expected to receive timing from a source clock over a facility that has short-term impairments and interruptions. The number of interruptions can range from 1 to 100 per day. All interruptions will affect the receiver clock. During an interruption, the timing reference cannot be used. When the reference is restored or if the interruption persists and the NE switches reference, there is a time difference between the local receiver clock and the newly restored reference. This leads to underflow/overflow of information bits.

When there is a cut in the fibre or the equipment fails in the trail, the NE next to it loses clock reference and enters holdover mode of operation. This mode characterises the clock's performance in case all the timing references to the clock are lost. There will be a steady accumulation of phase errors and consequent underflow/overflow of information bits periodically.

In all the three modes of operation discussed above, there is a requirement for tackling clock rate differences as well as phase differences between the incoming frame clock and the NE clock (outgoing frame clock). This difference manifests itself as real-time difference between the incoming frame and the outgoing frame as shown in Fig. 7.5(b) (parts- I and II), which needs to be adjusted. If the overflow or underflow bits are not accommodated, there would be interruption in services. So we require a mechanism within the STM frame to adjust such an overflow/underflow of information bits perpetuated by either degradations in the reference clock or temporary interruptions in the clock or link failure related holdover operation.

Pointers provide a mechanism to adjust the real-time difference between the incoming STM frame and the outgoing STM frame so that the overflow/underflow bits get accommodated.

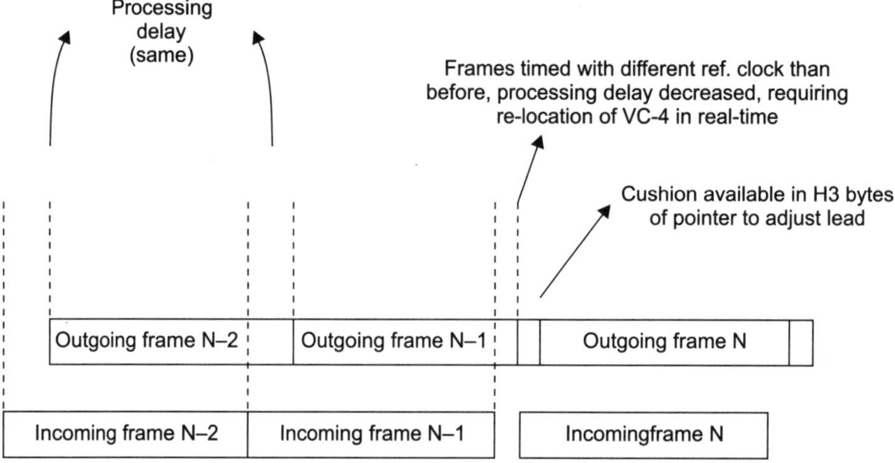

Fig. 7.5(b) *Part-I: New reference clock slower – lag in real time*

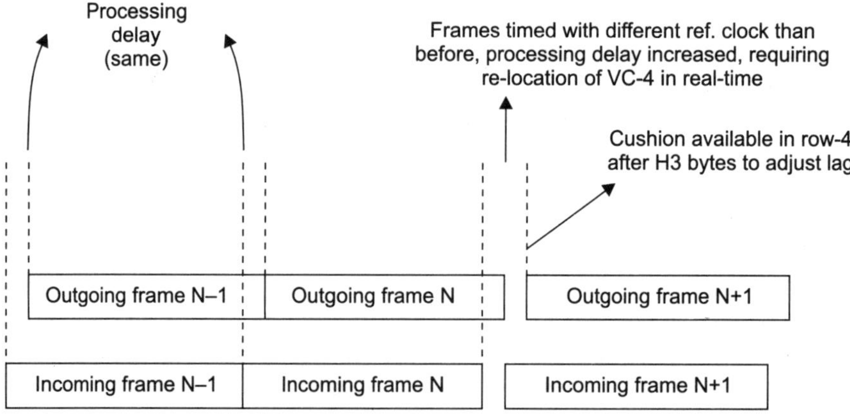

Fig. 7.5(b) *Part-II: New reference clock faster – lead in real time*

A pointer essentially has VC locating bytes (H1 and H2) and cushion bytes (H3) to accommodate lead; bytes after H3 in row 4 of the STM frame accommodates lag, if required. The details of pointer byte-fields and bit-fields are explained in section 7.5.2

and explanation of how pointer adjustments take care of real-time lead or lag is included in section 7.5.3.

7.5.2 Pointer interpretation

Figure 7.5(c) shows the byte structure of AU-4/AU-3/TU-3 pointer. The meaning of various bit fields is presented now.

SS bits (section supporting bits or bits indicating the type of pointer)

These are bits 5 and 6 in the 16-bit word of H1-H2 taken together. SS indicates the AU type. In fact, the pointer structure is similar for the AU-4 (administrative unit for VC-4) and AU-3 (administrative unit for VC-3). SS assumes values 10 for AU-4, AU-3 and TU-3.

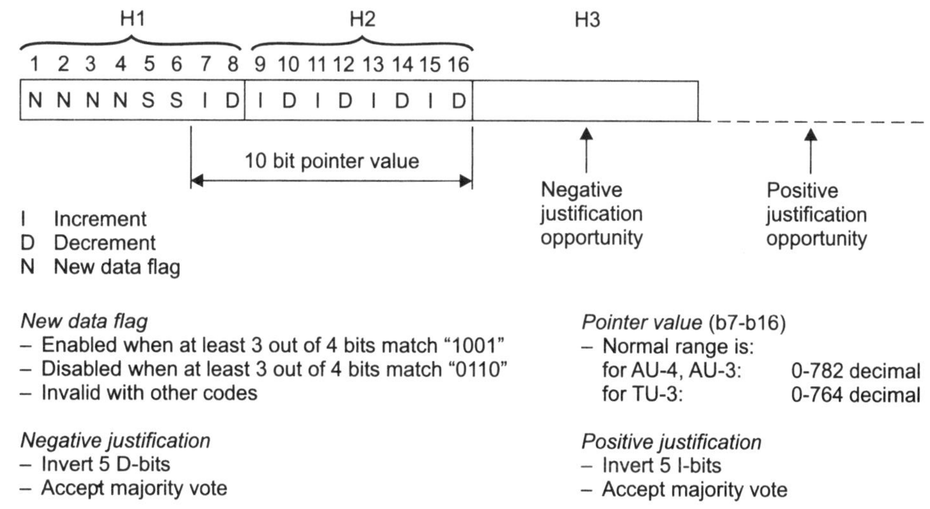

Fig. 7.5(c) *Pointer interpretation*

NDF (New data flag)

When H1 and H2 bytes are viewed together, the first 4 bits of the 16-bit word so formed are termed NDF (new data flag) bits. Normally, these bits are '0110'. If the pointer adjustments go beyond this range, the NDF is set as 1001. (This happens when containers in higher-order STM frames are concatenated, i.e., considered as a combined transport entity. Concatenation is dealt with in section 7.7. The indication of concatenation is given by the following: 1001 in bits 1-4; bits 5-6 unspecified; ten '1's in bits 7-16.)

H1, H2 bytes

These are pointer value bytes. Considering H1-H2 as a 16 bit sequence, the last 10 bits, i.e. bits 7 to 16, give the value of the pointer. This value indicates the offset in three-byte increments between the pointer and the first byte of the VC-4 as shown in Fig. 7.5(d).

Bits 7 to 16 in H1-H2 field serve another purpose in addition to pointer value indication. They indicate positive justification if the I-bits (bits 7, 9, 11, 13 and 15) are inverted from their previous value by majority vote, or negative justification if the D-bits (bits 8, 10, 12, 14 and16) are inverted from their previous value by majority vote in the frame where justification opportunity is presented to clear the accumulated lead/lag bits of the incoming frame with respect to the outgoing frame timing.

H3 bytes

The three bytes marked H3 are pointer action bytes. These are used for frequency justification. The three H3 bytes are called negative justification opportunity bytes. The three bytes after the three H3 bytes in the same row are termed positive justification opportunity bytes. Figure 7.5(d) illustrates negative and positive justification opportunity bytes. Justification process is explained in the next section.

As mentioned already, the pointer value given by the 10 bits are in positions 7 to 16 in the combined 16-bit word H1-H2. Since 10 bits are available, the values possible are 0 to 1023 or a total of 1024 values. However, the range of permissible values is 0–782, i.e., a total of 783 values in case of AU-4 pointer and 0–763 or 764 total values in the case of AU-3 pointer. This will be explained further in the next section.

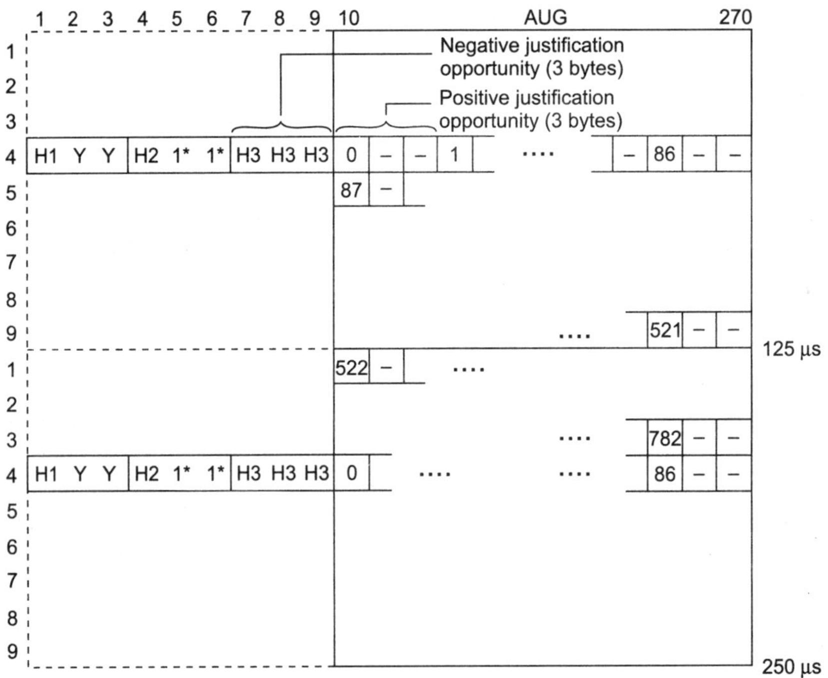

Fig. 7.5(d) *Locating VC-4 using offset value indicated by H1-H2 bytes of the pointer*

7.5.3 Exploring pointer operation

Differences in phase and frequency between two SDH NEs can be handled by the use of payload pointers. If the sending SDH NE is faster than the receiving NE, the receiving NE will introduce a negative pointer adjustment and shift the payload ahead by three bytes (using the three H3 bytes, i.e., the negative justification bytes). Similarly, if the sending NE is slower than the receiving NE, the receiving NE will introduce a positive pointer adjustment and delay the payload by three bytes (using the three bytes after H3, i.e., positive justification bytes). In this manner the receiving NE can keep up with the sending NE without loss of information. Let us look at these two situations in some detail.

When the sending NE is faster than the receiving NE

When the sending NE is faster than the receiving NE, the NE reference clock being slower than the incoming STM frame clock, the 125 μs reckoned by the receiving NE is more than the 125 μs reckoned by the sending NE. As the time-lead develops, the additional bits become 24 in number or equal to 3 bytes. At this juncture 3 negative justification bytes of H3 are to be offered to clear the additional bits. This implies that the alignment of the VC-*n* is to be advanced in time and the pointer value to be decremented by one. This operation is indicated by inverting bits 8, 10, 12, 14 and 16 (D-bits) of the pointer word to allow 5-bit majority voting at the receiver. Three negative justification bytes appear in the H3 bytes in the AU-4 frame containing the inverted D-bits. Subsequent pointers will contain the new offset. The entire process is shown in Fig. 7.5(e).

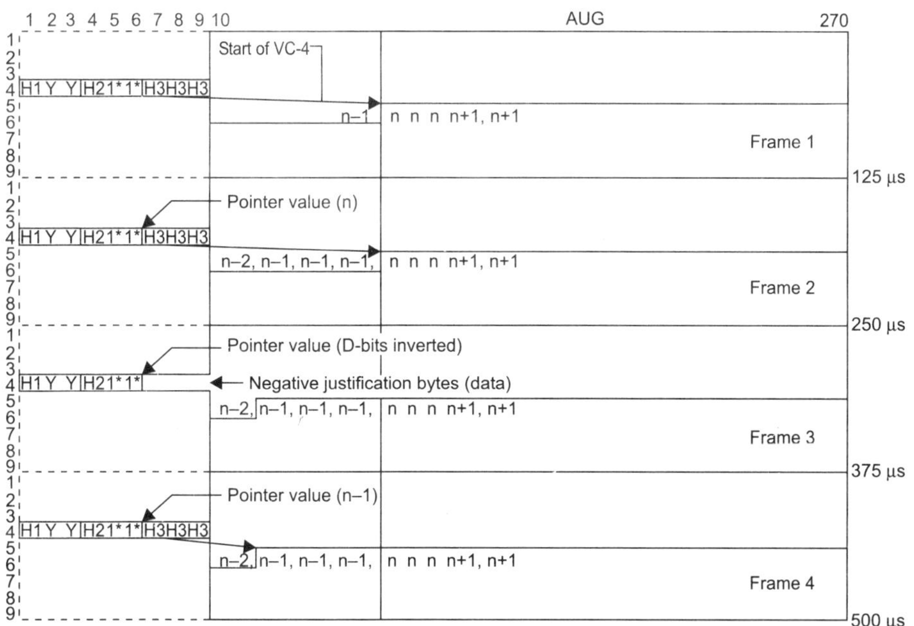

Fig. 7.5(e) *Negative justification and pointer decrement operation*

The above process is not one time process but happens periodically as and when the differences between clocks continue to arise. The frequency of the above justification process (with decrement of pointer) depends on the difference between the sending NE clock and the receiving NE clock.

When the sending NE is slower than the receiving NE

When the sending NE is slower than the receiving NE, the NE reference clock is faster compared to the incoming STM frame clock. In this situation the 125 μs reckoned by receiving NE is less than the 125 μs reckoned by the sending NE. As the time-lag develops, the leftover bits become 24 in number or equal to 3 bytes. At this juncture 3 positive justification bytes (the three bytes next to H3 in the fourth row) are to be offered to clear the leftover bits. This implies that the alignment of the VC-*n* is to be delayed in time and the pointer value is to be incremented by one. This operation is indicated by inverting bits 7, 9, 11, 13, 15 (I-bits) of the pointer word to allow 5-bit majority voting at the receiver. Three positive justification bytes appear in the fourth row after the H3 bytes in the AU-4 frame containing the inverted I-bits. Subsequent pointers will contain the new offset. The entire process is shown in Fig. 7.5(f).

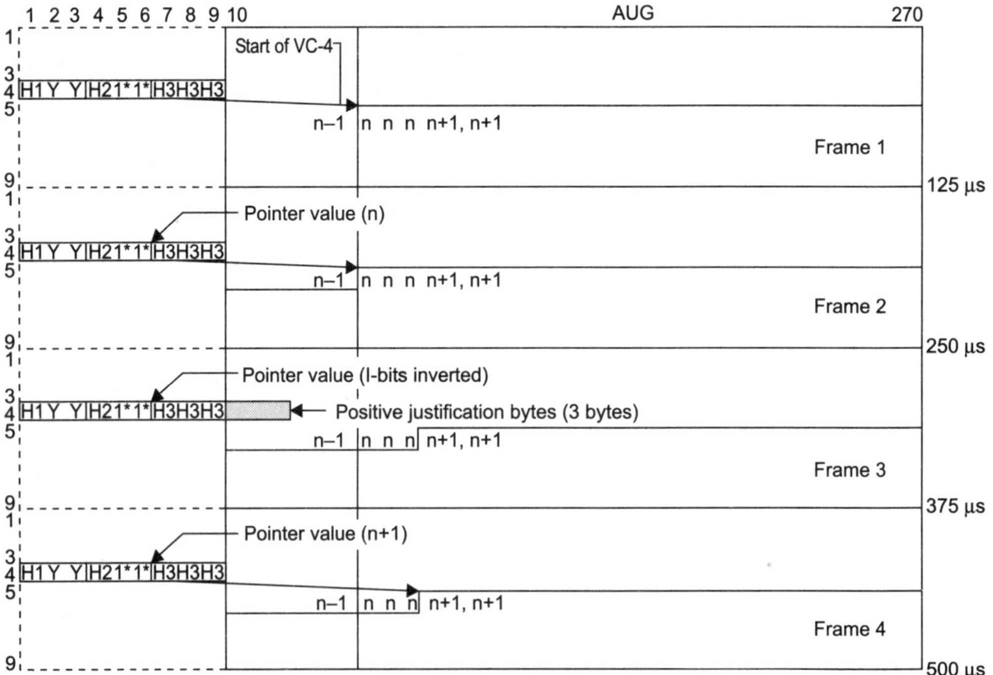

Fig. 7.5(f) *Positive justification and increment of pointer*

The above process is again not a one-time process but happens periodically as the differences in clock times continue. The frequency of the above justification process (with increment of pointer) depends on the difference between the sending NE clock and the receiving NE clock.

The above discussion throws up the following queries.

Q 7.8 If the difference between the NE reference clock and the incoming frame clock is 1 ppm, after how many frames will a pointer adjustment be required?

Let us first calculate the number of frames after which the pointer adjustment will be needed if the difference between the NE clock and the incoming frame clock is 1 ppm.

Let the equipment clock be at 155.52 Mbps, corresponding to a clock interval or bit duration of 6.43004115 ns. Considering a clock variation of 1 ppm, it can be seen that 155.52 Mbps + 1 ppm means 155.52015552 Mbps, a bit rate which corresponds to clock interval or bit duration of 6.43003472 ns.

The difference between the two bit durations is

$$6.43004115 - 6.43003472 = 0.00000643 \text{ ns.}$$

A 1-bit interval of STM-1 frame is 1/155.52 Mbps = 6.43004115 ns.

3 bytes is the cushion for justification, which corresponds to a duration of

$$24 \times 6.43004115 \text{ ns} = 154.3209876 \text{ ns.}$$

This cushion is exhausted in

$$154.3209876/0.00000643 = 24000154 \text{ intervals or}$$
$$24000154/(2430 \times 8) = 1235 \text{ frames.}$$

After exhausting the duration of 3 bytes available as cushion, which corresponds to 1235 frames, the pointer gets incremented for 1 ppm difference of the incoming clock with respect to the standard clock. This implies a *pointer action interval* of 1235 × 125 μs = 154375 μs and a *pointer action frequency* of 1/154375 = 6.48 Hz or 6.48 increments per second.

If the clock difference is more than 1 ppm, the pointer action is more frequent. Obviously, we cannot allow this beyond certain limit as it would cause frequent adjustments of the pointer which would lead to introduction of jitter.

Q 7.9 Are there any limits prescribed for the number of adjustments in a given time? What is the corresponding clock difference in ppm?

According to ITU-T G.707, the pointer event should not be more than 1 in 4 frames. This implies that the cushion of 3 bytes (24 × 6.43004115 ns = 154.3209876 ns) is exhausted in 4 frames or 4 × 2430 × 8 intervals. So, the allowed variation per interval is

154.3209876/(4 × 2430 × 8) = 0.00198458 ns.

Thus, if p is the allowed difference in ppm,

$$\frac{1}{155.52 \cdot 10^6} - \frac{1}{155.52 \cdot 10^6 + 155.52p} = 0.0019845$$

$[1/(155.52 \times 10^6)] - [1/\{(155.52 \times 10^6) + (155.52p)\}] = 0.00198458$ ns.

On solving, we get $p = 300$ ppm.

So, the difference allowed between incoming clock and the equipment clock is 300 ppm, which is quite a big margin.

The role of pointer is not limited to VC-4 of STM-1 (AU-4). In general, pointers play the role of allowing a flexible and dynamic alignment of VC-n within AU-n frame. Thus, the differences in the phases of VC-*n* and SOH as well as differences in frame rates between NEs due to variations in reference clocks are accommodated through the.action of pointers.

Q 7.10 *Why is that the ITU-T Rec. G. 707 prescribes a minimum of 4 frames gap for pointer adjustment?*

To establish frame synchronisation, two frames are required. For justification indication through the inversion of the I or D bits, one frame is required. To set a new value for pointer (by incrementing/decrementing), one more frame is required. Thus, a total of four frames are required and hence a minimum gap of four frames between pointer actions.

Q 7.11 *Pointer value has an allocation of 10 bits, which implies 1024 possible values (0 to 1023) for AU-4 pointer. But, why are only 783 legal values considered?*

In view of the three-byte cushion for justification, pointer increment/decrement takes place only after exhaustion of the 3 bytes and such increment/decrement of pointer by one unit makes the adjustment for 3 bytes in address space of VC-4.

The total address space of VC-4 is 261 × 9 bytes. This implies 261 × 9/3 = 783 values possible for the pointer—0 to 782, both inclusive. Hence the legal values are only 0 to 782 (total 783) even though the pointer has 10 bits.

Q 7.12 *What are the legal values for AU-3 pointer?*

In AU-3 pointer, the cushion byte is only 1 for justification. So, pointer increment/decrement is by one byte. Address space of AU-3 pointer, i.e.,VC-3, is 85 × 9 bytes. So, the allowed values for AU-3 pointer are 85 × 9 = 765, in the range 0-764.

An analogy for pointer operation

Consider a truck carrying containers placed in the rear-carriage space. If the truck speeds up, container tends to fall-back due to inertia and may even slip-off the rear-carriage space. If truck slows down, container tends to bump towards the driver's cabin due to inertia and may even damage the driver's cabin wall. Let us attach a spring to the container and fix one end to the driver-cabin's wall. When truck speeds up, the spring absorbs the fall-back by stretching and the container will not slip off the carriage space. Similarly, when truck slows down, the spring absorbs the forward bump of the container and driver's cabin will not be hit.

It is needless to say that the spring plays the role of pointer here because the edge of the spring, irrespective of its length, is the start of container!

7.6 Mapping PDH Signals

We talked about coexistence of SDH with PDH, HDLC and ATM. SDH supports mapping of the data frames of other transmission networks onto STM envelope, thereby providing a cover for all under one umbrella.

STM-1 frame can support PDH streams of E-hierarchy in the following combinations:

(1 E4) or (3 E3s) or (2 E3s plus 21 E1s) or (1 E3 plus 42 E1s) or (63 E1s)

In this section, mappings of E4, E3 and E1 onto STM-1 frame are discussed.

7.6.1 Mapping E4 onto STM

It can be seen in Fig. 7.6(a) that the portion of STM-1 frame other than RSOH, MSOH, AU-PTR, VC4-POH has 9 rows with 260 bytes in each row. It is container C-4. We have to explore the mapping of E-4 onto C-4.

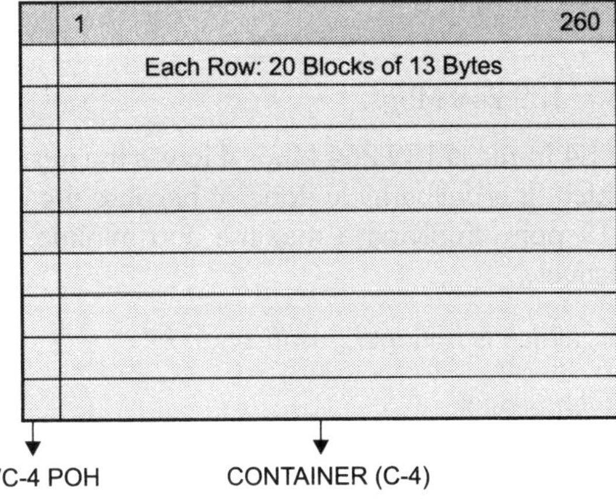

Fig. 7.6(a) *C-4 and VC-4 POH in STM-1 frame*

Refer to Fig. 7.6(b) where one row of the C-4 container (260 byte segment) is depicted as 20 blocks of 13 bytes. Each block of 13 bytes is further divided into 96 D bits (meaning 96 data bits) or 12 bytes of data plus the thirteenth byte, which is either W, X, Y or Z.

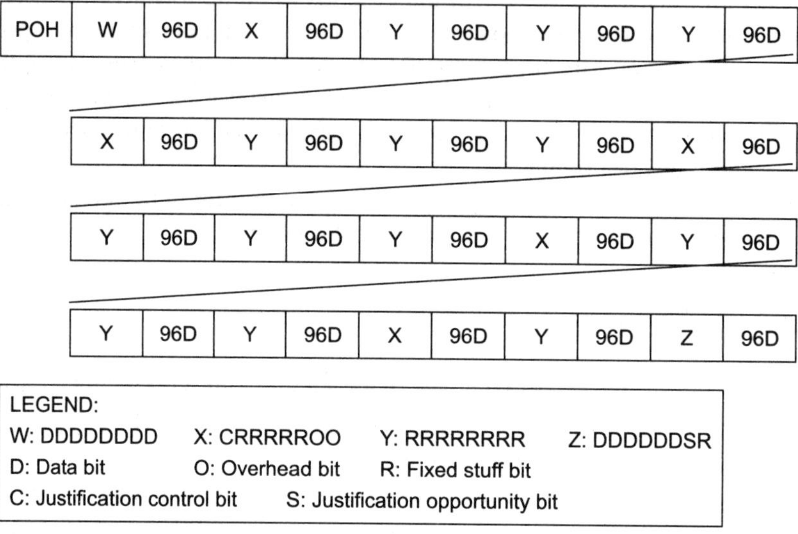

Fig. 7.6(b) *Byte and bit assignment while mapping E4 into C4*

Let us compute the data bit capacity of C-4 while mapping E-4. The total number of data bits in one frame (125 μs) is:

In each row:
(96 × 20) + (8 of W byte) + (6 of Z byte) = 1934 bits

Therefore, in nine rows there are 9 × 1934 = 17046 bits. Since the STM frame duration is 125 microseconds, in one second there are 8000 frames. Thus, the bit rate works out to

8000 × 17046 = 139.248 Mbps.

The actual bit rate of E4 frame is 139.264 Mbps. How is the mismatch overcome? This, in fact is not a mismatch. It is deliberately done so because the clock of E4 has acceptable tolerance of ±15 ppm, implying a maxima and minima of 139.266 Mbps and 139.262 Mbps respectively.

Use of D bit in Z-byte, which is optional (justification bit):

If we use D bit in Z-byte, we get

(96 × 20) + (8 of W byte) + (7 of Z byte) = 1935 bits/row,
In nine rows of a frame, 9 × 1935 = 17415,
In one second (8000 frames), 8000 × 17415 = 139.320 Mbps,

which is greater than the maxima of E4 clock. In other words, normally, the offered bit rate is 139.248 Mbps. To carry leftover bits of E4, which depends on current running rate of E4 clock, additional bit is offered in Z-byte on demand. This additional bit is the justification bit since it performs the role of matching the flow rate of C-4 to current running rate of E-4. The use of justification bit or otherwise is conveyed through justification control bit (JCB), which is C bit in X-byte. Since there are five X bytes, we have 5 JCBs per row.

Q 7.13 *We have discussed the drawbacks of justification in PDH. Why are we again resorting to the same in SDH?*

We resort to justification in PDH for bit-stream integration at each hierarchical level in each tributary. There are plesio or plural streams, clocks, JCBs and JBs to be integrated and jitter manifesting itself as 'multiplexing jitter' and 'waiting time jitter' on each tributary as well as in the integrated stream.

In SDH, justification is carried out to match the C-4 rate to the current running rate of E-4, i.e., we are reproducing the intake E4 bits carried by C-4 container at the receiving end at the intake rate. In other words, SDH network offers a pipe which delivers the output as per the intake. If the current intake is slower (negative variation of E4 clock), the corresponding output (counting the propagation delay) is slower. If current intake is faster (positive variation of E4 clock), the corresponding output (counting the propagation delay) is also faster. The PDH elements which are exchanging E4 through the SDH pipe get a transparent connection. They do not feel the existence of SDH pipe between them. Hence, justification is not an evil in SDH.

Q 7.14 *Why do we require justification in each row and associated JCBs?*

The STM frame is transmitted row-wise, hence we put JBs and JCBs row-wise. Moreover, the capacity offered in one C-4 is 17046 to 170415 bits, which can take about six E-4 frames with different start and end points, implying the need for JBs /JCBs availability at different points in the frame. We have already checked the data rates and verified the throughput we get by using justification bit in each row. Thus, JBs/JCBs are required row-wise.

Furthermore, as the clock rate handled is high, five JCBs are used for proper majority decision at the receiving end.

Q 7.15 *Why have we put so many packing bits or stuff bits in?*

Use of stuff bits and staggering of data bits in C-4 gives additional protection against bursty errors. Also, any waiting time jitter due to the JBs and JCBs being placed in a row is dampened in the clock cycles corresponding to stuff bits.

7.6.2 Mapping E3 to STM

Refer to Fig. 7.6(c). The C-4 container is divided into three TUG-3s (tributary unit group-3). Each TUG has 86 columns and they are shown as column interleaved to signify the order of transmission of bytes. Two columns in the beginning of C-4 contain fixed stuff bytes. So, in total there are (3 × 86) + 2 = 260 columns, which is also the total columns of C-4.

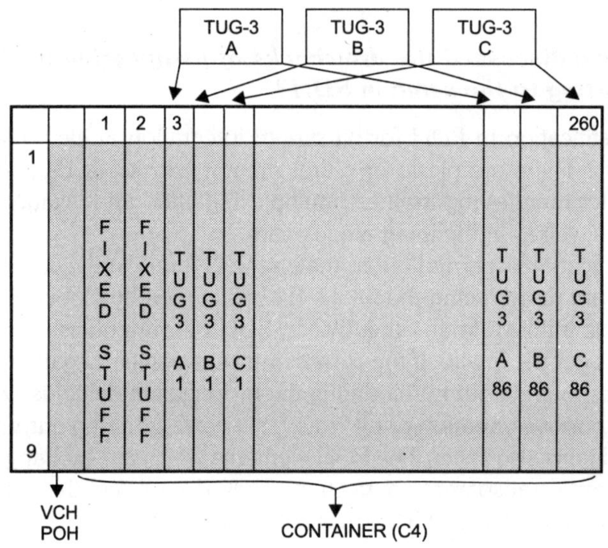

Fig. 7.6(c) *Column interleaved TUG3s in C-4 for mapping E3*

A TUG-3 is elaborated in Fig. 7.6(d). In each of the 86 columns of a TUG-3, one column is for TU-3 pointer, one is for VC-3 POH and the rest are for storing data. The data storage area is C-3 or container-3. Each container has three sub-frames. In each sub-frame there are 84 columns and 3 rows.

Refer to Fig. 7.6(e) where each sub-frame of C-3 is shown. (The (84 × 3)-bytes area of each sub-frame is shown in two parts to accommodate all the 84 columns). Beneath each column, the bit count of each cell in the respective column is given. Let us count total bit capacity of C-3. For this, the byte allocation in each sub-frame of C3, as shown in Fig. 7.6(e) is explained.

- Each sub-frame has 84 columns × 3 rows, which for convenience sake has been shown in 2 parts.
- The last row shown in the figure gives the byte count, which totals to 84.
- X byte stands for stuff byte (RRRRRRRR), R being the stuff bit.

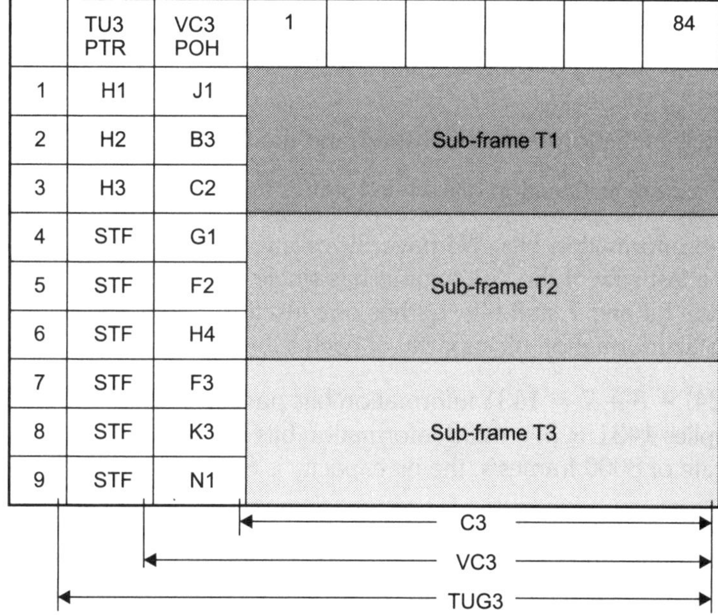

Fig. 7.6(d) *TUG-3, VC-3, C-3 and sub-frames in C3*

X	3 8I	X	3 8I	X	3 8I	X	3 8I	X	X	3 8I	X	3 8I	X	3 8I	X	3 8I	X	3 8I	X	C
X	3 8I	X	3 8I	X	3 8I	X	3 8I	X	X	3 8I	X	3 8I	X	3 8I	X	3 8I	X	3 8I	X	C
X	3 8I	X	3 8I	X	3 8I	X	3 8I	X	X	3 8I	X	3 8I	X	3 8I	X	3 8I	X	3 8I	X	C
1	3	1	3	1	3	1	3	1	1	3	1	3	1	3	1	3	1	3	1	1

3 8I	X	3 8I	X	3 8I	X	3 8I	X	3 8I	X	X	3 8I	X	3 8I	X	3 8I	X	3 8I	X	3 8I	X	C	3 8I
3 8I	X	3 8I	X	3 8I	X	3 8I	X	3 8I	X	X	3 8I	X	3 8I	X	3 8I	X	3 8I	X	3 8I	X	C	3 8I
3 8I	X	3 8I	X	3 8I	X	3 8I	X	3 8I	X	X	3 8I	X	3 8I	X	3 8I	X	3 8I	X	3 8I	X	C	q1q2 8I
3	1	3	1	3	1	3	1	3	1	1	3	1	3	1	3	1	3	1	3	1	1	3

Fig. 7.6(e) *Byte allocation in sub-frame of C-3 for mapping E3*

- 3.8I means 3 information bytes
- C byte is of the format RRRRRRC1C2, R being the stuff bit and C1 and C2 the JCBs.
- Q1 byte is of the format RRRRRRRS1 and the Q2 byte, S2IIIIIII
- S1 and S2 are justification bits while I stands for information bit

8I implies 8 information bits. We have 59 cells with $3 \times 8I = 24$ information bits. The cell in the last row of the last column has single 8I with q1 and q2. While q1 has one justification bit and 7 stuff bits, q2 has one justification bit and 7 information bits. Hence, the total information bit capacity of each sub-frame is

$(59 \times 24) + 8 + 7 = 1431$ information bits per sub-frame.
This implies $1431 \times 3 = 4293$ information bits in C-3.
At the rate of 8000 frames/s, the bit capacity is $8000 \times 4293 = 34.344$ Mbps.

E3 rate is 34.368 Mbps. The mismatch takes care of variations in the incoming E3 rate. We have two justification bits in q1 and q2 in each sub-frame, which enhance the capacity to 4299 bits. This enhanced capacity gives a bit rate of 34.392 Mbps.

$34.392 \text{ Mbps} > 34.368 \text{ Mbps} + 20 \text{ ppm}$

Hence E3 gets accommodated in one C-3.

There are three C3s in one C4; therefore, one C4 or STM-1 frame can take in three E3s.

7.6.3 Mapping of E1

E1 bit rate is 2.048 ±50 ppm. If the POH bytes are kept in each container, it would amount to wastage of payload area. Instead, if we map 4 E1s into a container multiframe, i.e., a C-12 multiframe and add POH to make it a VC-12 multi-frame, this multiframe can be sent in 500 μs. This process reduces the ratio of overhead bits to useful bits while maintaining the required transmission rate.

Each C-12 takes four E1s but spread across four consecutive frames (or VC-4s). Addition of overhead bytes, V5, J2, N2, K4 to C-12 makes it a VC12 multiframe, which has duration of 500 μs. The structure of VC-12 multiframe is shown in Fig. 7.6(f). Various bytes/bits of VC-12 are listed below.

V5: VC12 POH
J2: Path trace
N2: NW operator bit (for order wire or a monitoring channel)
K4: APS (for carrying automatic protection switching messages)
R: Stuff bit

O: Overhead bit
C1,C2: Justification control bits (JCBs)
S1, S2: Justification opportunity bits
D: Data (information) bit

The total number of information bytes is 32 + 32 + 32 + 31 = 127 with 127 × 8 = 1016 bits in them. In addition, we have 7 D bits with S2 in byte number 108, which make the total number of information bits 1023 in 500 μs, i.e., a data rate of 2.046 Mbps.

1	V5	In one VC-4
2	RRRRRRRR	
3–34	32 Info. Bytes	
35	RRRRRRRR	
36	J2	In next VC-4
37	C1 C2 0 0 0 0 R R	
38–69	32 Info. Bytes	
70	RRRRRRRR	
71	N2	In next VC-4
72	C1 C2 0 0 0 0 R R	
73–104	32 Info. Bytes	
105	RRRRRRRR	
106	K4	In next VC-4
107	C1 C2 RRRRR S1	
108	S2DDDDDDD	
109–139	31 Info. Bytes	
140	RRRRRRRR	

Fig. 7.6(f) *VC-12 (multiframe) structure*

If the two justification bits, S1 and S2 also contain information bits, we get 1025 information bits in 500 μs, making the data rate equal to 2.050 Mbps. Thus, 2.048 Mbps ±50 ppm gets suitably accommodated in different kinds of bit rates that are offered.

Whenever a virtual container is made, a pointer is required for dynamic adjustment of the time lead/lag of the incoming frame clock and NE reference clock for enabling the floating of the VC. The pointer bytes are V1, V2, V3 and V4 . Of these, V1 and V2 are pointer value bytes and V3 the cushion for justification; V-4 is spare. The addition of pointer to VC-12 makes it TU-12. The pointer in TU-12 locates VC-12; thus VC-12 is also a floating container. It is the virtual container associated with E1 while VC-11 is the virtual container associated with T1. The pointer bit-field interpretation of TU-2, TU-11, TU-12 is shown in Fig. 7.7.

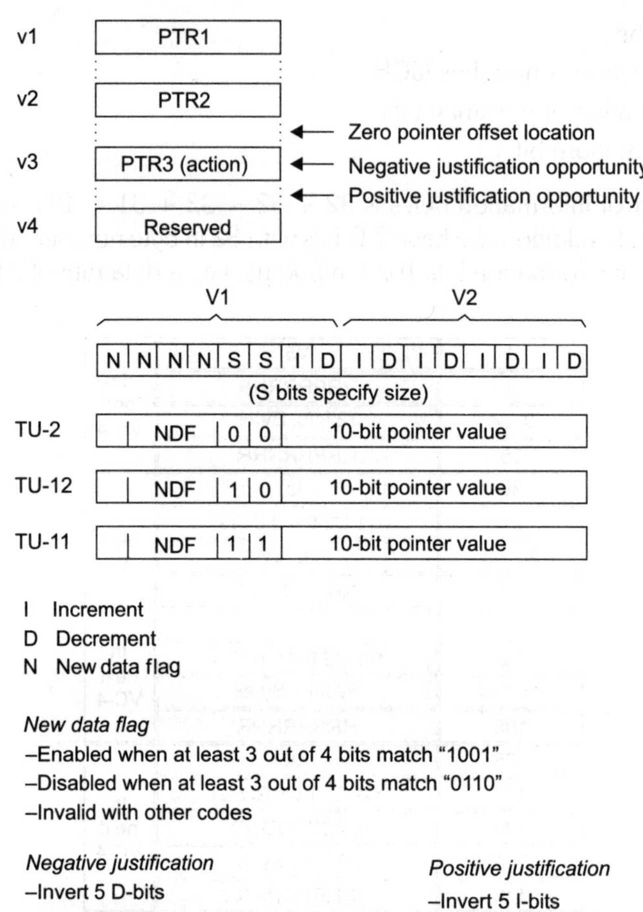

Fig. 7.7 *TU-2/TU-11/TU-12 pointer interpretation*

V1-V2 taken together is a 16-bit sequence. Of this, bits 7-16 give the pointer value. This is a binary number which indicates the offset from V2 to the first byte of VC-2, VC-12 or VC-11. The range of offset is different for each of the tributary unit sizes. The normal range of these pointer values (in decimal) are:

TU-2: 0 to 427 TU-11: 0 to 103 TU-12: 0 to 139

The pointer bytes are not counted in the offset calculation. Figure 7.8 illustrates the pointer/offsets/justification bytes in TU-2, TU-11 and TU-12.

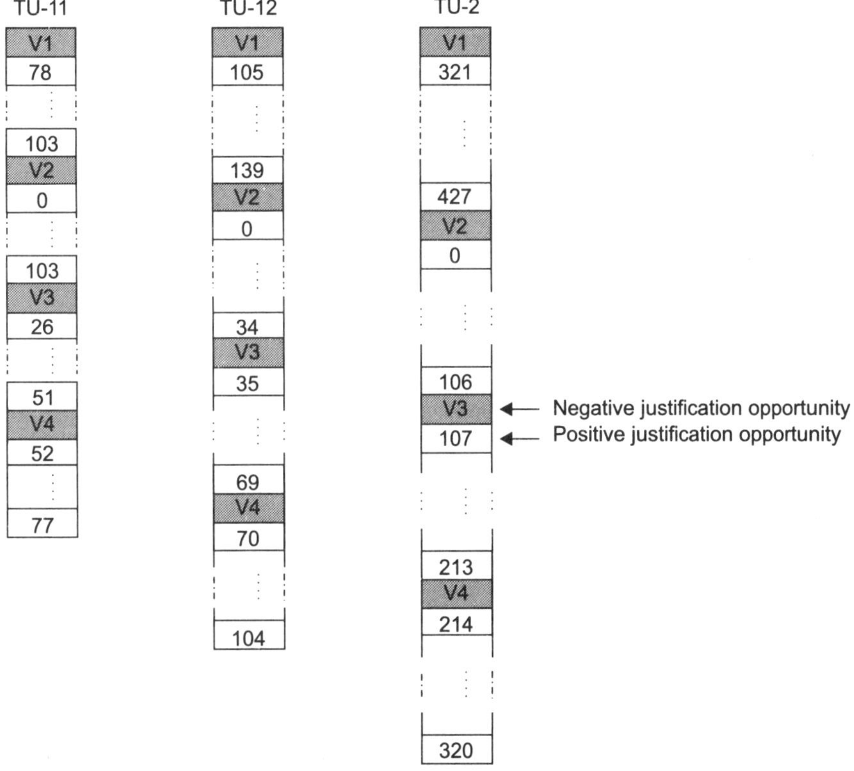

Fig. 7.8 *Pointer bytes and justification bytes in TU-2, TU-11, TU-12*

Q 7.16 *How is justification handled in TU-2, TU-12 and TU-11?*

The TU-2, TU-12 and TU-11 pointers are used to frequently justify VC-2, VC-12 and VC-11 in the same way as in the case of VC-4/VC-3. A positive justification opportunity immediately follows the V3 byte. Also, byte V3 serves as the negative justification opportunity byte such that when the opportunity is taken, V3 is overwritten by data. This is also shown in Fig. 7.8.

The indication of whether or not a justification opportunity has been taken is provided by the I- and D-bits of the pointer in the current TU multiframe. The value contained in V3 when not being used for a negative justification is not defined. The receiver ignores the value contained in V3 whenever it is not used for negative justification.

7.7 Mapping ATM

Asynchronous transfer mode (ATM) is a high speed switching network architecture created in the early 1990s to support B-ISDN. It can be used to carry voice, data and video separately or simultaneously over the same network path. An ATM network carries traffic as number of cells, known as ATM cells. Each cell has a 53 octet size (53 bytes) in which 48 are information bytes and 5 are header bytes. The fixed cell size eases buffer management as well as throughput.

ATM was designed to use SONET as its carrier, hence SDH, which also emanated from SONET, supports ATM. Consider a C-4 container which has 260 × 9 bytes (= 2340 bytes). The number is not an exact multiple of 53. Then, how can ATM cells be packed into an STM frame?

Remember the H4 byte in VC-4 POH. It points to the first byte of fresh ATM cell in the sixth row of the STM frame.

Refer to Fig. 7.9. The number of bytes in C-4 of an STM frame up to the end of the fifth row is 5 × 260 = 1300. ATM cell size being 53, 24 ATM cells get accommodated in this space (since 24 × 53 = 1272 < 1300 but 25 × 53 = 1325 > 1300). The left-over space of 28 bytes (1300–1272) accommodates 28 bytes of the 25th ATM cell. The remaining bytes of the 25th ATM cell (53 – 28 = 25) get accommodated in the sixth row.

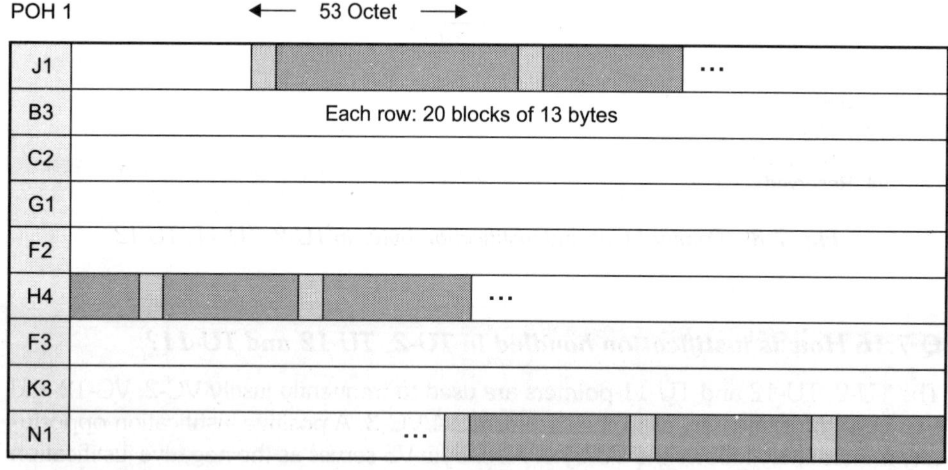

Fig. 7.9 *Mapping ATM cells into VC-4*

A fresh ATM cell, therefore, commences at the 26th byte in the sixth row of C-4. Hence, H4 contains the value 26 in the first STM frame carrying ATM traffic.

In a similar way, H4 value can be worked out for the second STM frame. Up to end of the fifth row of the second STM frame, the capacity is 9 rows of the first frame and

5 rows of the second frame, making it a total of 14 rows and a capacity of $14 \times 260 = 3640$ bytes. In this space, 68 ATM cells get accommodated ($68 \times 53 = 3604$) and in the balance portion ($3640 - 3604 = 36$), 36 bytes of the 69th cell get accommodated. The remaining 17 bytes of the 69th cell get accommodated in the sixth row. The fresh ATM cell, i.e., the 70th ATM cell, commences at the 18th byte in the sixth row. Hence, H4 has value 18 in the sixth row of the second STM frame. Likewise, the H4 value can be computed for any STM frame.

The formula for value of H4 in VC-4 of the *N*th STM frame is given by

$$\text{H4} = 54 - \text{Modulo}_{53}\,[(N-1) \times 260 + 1300]. \tag{7.1}$$

ATM cells can be carried in C-3 also. H4 in VC-3 performs the same function as H4 in VC-4. Since each row in C-3 has 84 bytes, the formula for the value of H4 in VC-3 of the *N*th STM frame is given by

$$\text{H4} = 54 - \text{Modulo}_{53}\,[(N-1) \times 84 + 420]. \tag{7.2}$$

7.8 Structure of STM-N frame

STM-1 has data rate of 155.52 Mbps. Higher rates in SDH hierarchies, viz. STM-4, STM-16 and STM-64 support data rates of $N \times 155.52$ Mbps, where $N = 4$, 16 or 64. With the frame duration of 125 μs, higher data rate is possible by taking *N* times the bytes of STM-1 frame. The structure of STM-N frame is shown in Fig. 7.10.

Regarding payload, we get $N \times$ payload area of STM-1. But, SOH also has the capacity *N* times that of STM-1 of which some of the bytes may not be used. As an example, SOH of STM-4 frame is shown in Fig. 7.10.

<table>
<tr><td colspan="2">12 A1 bytes</td><td colspan="3">12 A2 bytes</td><td>4 J0 bytes</td><td>8 No. × bytes</td></tr>
<tr><td>B1</td><td rowspan="2">Not used</td><td>E1</td><td colspan="2" rowspan="2">Not used</td><td>F1</td><td>11 No. × bytes</td></tr>
<tr><td>D1</td><td>D2</td><td>D3</td><td>Not used</td></tr>
<tr><td colspan="7">AU POINTER</td></tr>
<tr><td colspan="2">12 B2 bytes</td><td>K1</td><td colspan="2"></td><td>K2</td><td rowspan="4">Not used</td></tr>
<tr><td>D4</td><td></td><td>D5</td><td colspan="2"></td><td>D6</td></tr>
<tr><td>D7</td><td></td><td>D8</td><td colspan="2"></td><td>D9</td></tr>
<tr><td>D10</td><td></td><td>D11</td><td colspan="2"></td><td>D12</td></tr>
<tr><td>S1</td><td>11 Z1</td><td colspan="2">11 Z2</td><td>M1</td><td>E2</td><td>11 No. × bytes</td></tr>
</table>

Fig. 7.10 *Structure of STM-N frame*

STM-4 has four numbers of STM-1s. The various bytes in SOH are:

- A1 bytes: $4 \times 3 = 12$
- A2 bytes: $4 \times 3 = 12$
- B2 bytes: $4 \times 3 = 12$

- J0 bytes: 4 × 1 = 4
- Z1 bytes: 4 × 3 – 1 = 11
- Z2 bytes: 4 × 3 – 1 = 11
- AU pointer: 4 × 9 = 36 bytes

Q 7.17 *In STM-N frame, why do some of the bytes have N-times the allocation of STM-1 while others have the same allocation as in STM-1?*

As the frame size increases and transmission rate increases, it is better to have longer FAS. Hence, A1 and A2 become *N* times long. B2 byte, which is for monitoring errors between MUXes, is *N* times long to facilitate error monitoring, container-wise (STM-N can accommodate *N* C-4s).

NMS bytes, viz., D1 to D12, remain the same (i.e., they do not get 'multiplied by *N*') because in a network, some STM-1 paths can be derived from STM-N main route using ADMs and DXCs and, all NEs have uniformity of connection with NMS.

APS bytes, K1, K2, carry APS commands; these are only one-byte long. E1, F1 and E2, the order wire bytes, remain one-byte long as they support only one channel. Parity check-sum result is always one byte, as computed for all the bytes in the previous frame; hence, B1 remains one-byte long. M1 conveys the count of interleaved bit blocks detected in error over the MUX section and remains one-byte long.

S1 has to carry sync status information, i.e., the rank of clock used by the NE sending the frame, and is one byte long. J0 is used to transmit the access identifier over the regenerator section so that the section receiver can verify its continued connection to the intended transmitter. Since more NEs will be present in STM-N network, a longer identifier sequence is required. Hence, J0 is *N* bytes long in STM-N frame.

Z0, Z1 and Z2 are reserved for future use. The remaining portion is therefore allocated for them and both Z1 and Z2 are (3*N* – 1)-bytes long.

Q 7.18 *In STM-N frame, why is that every field in POH has N bytes or N times the allocation in STM-1?*

The payload of STM-N can be treated as *N* different VC-4s. Hence, each has its own POH. In case the payload is treated as combined, the excess bytes in POH are dummies.

Q 7.19 *In STM-N frame, why is the pointer field 9N bytes long?*

As the addressable space has become *N* times the space in STM-1 frame, a longer pointer is required to cover the entire address space. Also, in case each of the *N* C-4s is treated separately, they have their individual 9-bytes-long pointers which get accommodated in the 9*N*-byte long pointer field. The treatment of containers, individually or as a combined set, gives rise to the concept of concatenation. Let us explore the same in the next section.

7.9 Concept of Concatenation

7.9.1 Continuous concatenation

When we have to transmit bit rates in excess of the container capacity (certainly not in excess of rated capacity of NE), we need to combine containers and ensure their arrival at the destination without split-up. For example, the B-ISDN requirement of 622.08 Mbps can be supported by STM-4 network. But the maximum capacity container is only C-4, which is approximately 150 Mbps. So, the C-4 container of each of the 4 STM-1s in STM-4 frame is to be chained by a process known as 'concatenation'. The combined (concatenated) container is termed VC-4c. A single column POH (one byte for each field) would suffice for the concatenated container VC-4c; the remaining bytes in POH area of STM-4 frame are filled with stuff bits or dummy bits.

Since the concatenated container can float in the entire STM-4 frame, the address space required is high and hence there is a requirement for longer pointer. AU-4 pointer of each of the VC-4 has 10 bits containing the address of the payload commencement. As the addressable space has increased four-fold, we need to club together all the four pointers and use the same as address locator. Recall from section 7.5 that H1-H2 of pointers taken together as 16 -bit word has NDF of 4 bits, SS of 2 bits and a pointer value 10 bits long. The SS bits are set to 00 to signify concatenated-container operation.

Figure 7.11(a) illustrates the continuous concatenation of 4 VC-4s, denoted as 4 VC-4c. It may be noted that

1. Concatenation can be at the VC-4, VC-3 or VC-12 level.
2. *N* VC-Xc means continuous concatenation of *N* containers of the type VC-X.

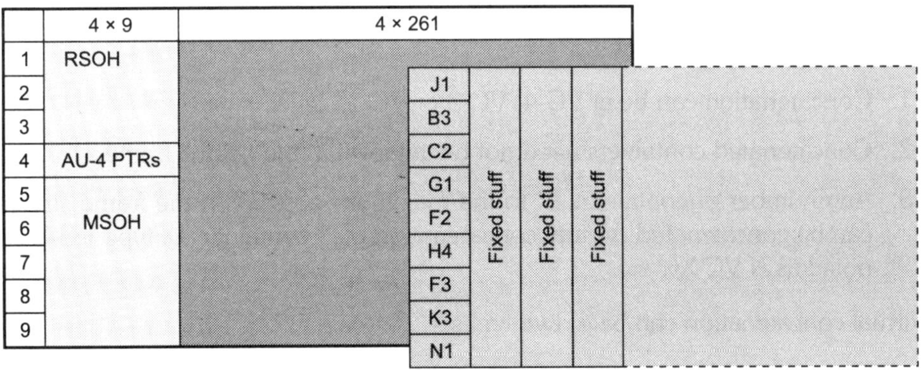

Fig. 7.11(a) *Continuous concatenation*

7.9.2 Virtual concatenation

Let us consider a different situation wherein, we may offer bandwidth or containers between two NEs, not necessarily adjacent containers on the STM frame and not necessarily going along the same path. For example, consider four different VC-4s meant for the same destination in STM-16 frame—the 1st, 4th, 7th and 14th. They may start at a point, go along different paths and arrive at their common destination. This is done so to provide route diversity and ensure some minimum bandwidth availability for the subscriber. In all cases, each VC-4 needs individual POH as well as individual pointer. However, the destination NE identifies the concatenation through the SS field of the H1-byte of the pointer. In this case of concatenation, the SS are set to 11. This process is known as 'virtual concatenation' or 'packet concatenation', as the containers are free to travel as individual packets but are put together at the destination, i.e., they are virtually chained. Virtual concatenation of VC-4 is indicated by VC-4v.

Figure 7.11(b) illustrates virtual concatenation of 7 VC-4s, denoted by 7 VC-4v.

	4 × 9	4 × 261			
1	RSOH				
2		J1	J1	J1	J1
3		B3	B3	B3	B3
4	AU-4 PTRs	C2	C2	C2	C2
5		G1	G1	G1	G1
6	MSOH	F2	F2	F2	F2
7		H4	H4	H4	H4
8		F3	F3	F3	F3
9		K3	K3	K3	K3
		N1	N1	N1	N1

Fig. 7.11(b) *Virtual concatenation*

It may please be noted that–

1. Concatenation can be at VC-4, VC-3 or VC-12 level.
2. Concatenated containers need not be adjacent in the overall frame.
3. Any number of containers (2 to the maximum possible in the frame structure) can be concatenated. Virtual concatenation of *N* containers of type VC-X is denoted as *N* VC-Xv.

Virtual concatenation can be of two types:

- Byte-interleaved virtual concatenation
- Packet virtual concatenation

In byte-interleaved virtual concatenation, the bytes of the incoming traffic can be put in the VCs by rotation and re-aligned at the destination. This will ensure complete utilisation of the bandwidth offered. In packet virtual concatenation, the VCs are filled packet-wise. Even if a VC is not available, the traffic goes on with reduced bandwidth till flow control problems in mapping the traffic arise. These two types of concatenations find application in the offer of bandwidth for LAN to LAN transport service across SDH networks. The concept is treated in detail in chapter 9.

7.10 Synchronisation in SDH Network

7.10.1 Principles of synchronisation in SDH network

In the ideal scenario, every NE in a synchronised network should get their clock reference directly from the central clock reference. It is practically impossible to supply central clock to each NE as we require an exclusive jitter free circuit for each NE— which are in hundreds to few thousands in number—and, such a circuit should have hot-standby arrangement, again jitter free via different paths. Therefore, a hierarchical synchronisation signal distribution is resorted to in SDH networks. Such a plan also allows any NE to change clock reference in case of path faults (disconnections) or equipment failures along the path.

Let us understand the principles of synchronisation in SDH network.

1. A synchronisation plan is made for SDH network as per the number of nodes and their locations. This plan indicates the way each NE is synchronised.
2. There are 4 hierarchical levels requiring a supply of synchronised clock: PRC, SSU (transit node), SSU (local node) and SEC.
3. A high-precision atomic clock unit, known as the primary reference clock (PRC) unit, conforming to ITU-T recommendation G.811 provides synchronisation to the first node hierarchical level.
4. The PRC is doubled or tripled on different sites for availability. In India, in some of the SDH networks, it is doubled (one at Bombay and the other at Delhi), one being the master and the other the slave. These PRC nodes provide synchronisation to all the NEs down the trail.
5. In small scale networks, the clock signal obtained through GPS can be used as reference. Also, when investment on a second PRC is to be avoided, we can take clock reference through GPS and use it as a standby. But, when we offer bandwidth over such network to customers, stability may not be satisfactory. The BER could be high due to repeated pointer adjustments. In such cases, we have to use subordinate synchronisation supply units (SSUs) to filter the jitter accumulated down the trail.
6. Even in large scale networks, though the reference is a high-precision one from PRC, as the signal passes down the trail, degradation takes place due to justification and jitter on the en-route NEs. Hence, a further hierarchical structure is used

in which the signal is passed on by the subordinate synchronisation supply units (SSU), filtering the accumulated jitter.

7. The features of SSU are given below.

   SSUs are located at strategic nodes. They are designed to filter the accumulated jitter and wander. In other words, they refine the clock received from PRC down the trail. SSU clock is in accordance with ITU-T Rec. G.812. There are two qualities of SSU clock:

   - SSU(T) level, implying a reference clock of transit node quality (part of syn chronisation chain).
   - SSU(L) level, implying reference clock of local or last node quality (as the last SSU in the synchronisation chain).

8. We also find SEC (synchronisation equipment clock) in the hierarchy:

   - Normally, NEs between PRC and first SSU or between adjacent SSUs subsequently in a PRC trail are termed SEC, i.e., secondary equipment clock reference points.
   - SEC does not imply that the NE uses its own references. All NEs 'look' to wards PRC (and to the nearest SSU towards the PRC side) for synchronisa tion reference.
   - If any NE termed SEC loses its reference clock, it can enter hold-over mode or free-running mode.

     i. In hold-over mode, NE uses the last average of the phase locked frequency as the reference clock.

     ii. In free-running mode, NE uses its own generator/oscillator timing reference.

9. Synchronisation signal paths are the same as those used for SDH communications.

10. Rules concerning deployment of PRC, SSU and SEC are given below.

    a. There can be up to ten SSUs in a PRC trail.

    b. There can be up to 20 NEs between SSUs.

    c. However, the maximum number of NEs in one PRC trail should not exceed 60.

11. In synchronisation reference distribution, it is to be ensured that there are no sync loops, i.e., no NE shall receive a sync signal it has generated as a reference via a sequence of slave clocks.

12. Stability and precision are defined for each hierarchical level in ITU-T recommendations. These are summarised in Table 7.6.

Table 7.6 *Parameters of PRC, SSU and SEC as per ITU-T recomnendation*

Source	ITU-T Rec.	Accuracy/Offset /Drift	
PRC (primary reference clock, an atomic clock)	G.811	Accuracy:	10^{-13} (Cesium) 10^{-11} (Rubidium)
SSU (transit node)	G.812 T	Accuracy: Offset: Drift:	5×10^{-9} 5×10^{-11} 10^{-9}/Day
SSU (last node)	G.812 L	Accuracy: Offset: Drift:	5×10^{-7} 5×10^{-8} 2×10^{-8}/Day
SEC	G.813	Accuracy: Offset: Drift:	4.6×10^{-6} 5×10^{-8} 5×10^{-7}/Day

Figures 7.12(a) and 7.12(b) illustrate the synchronisation hierarchy.

Let us now study the synchronisation mechanism.

7.10.2 Synchronisation mechanism in NE

Synchronisation mechanism in NE is shown in Fig. 7.12(c). An NE can use any of the following as clock reference:

1. T0, which is NE's internal clock
2. T1, which is 2 MHz signal derived from any STM port of the equipment, i.e., obtained from SDH port
3. T2, which is a 2 MHz signal derived from E1 input, i.e., obtained from PDH port
4. T3, which is the clock reference from PRC/SSU

Sync output of an NE, i.e. signal sent by one NE to another, is termed T4. This is feedback (optional) to SSU. If such feedback is available, an NE can run at the last averaged phase-locked frequency even if the SSU reference is lost. This is known as hold-over mode. Else, the NE will be in free-running mode if SSU reference is lost.

SETG in Fig. 7.12(c) refers to phase-locked loop (PLL). The selector selects timing reference according to the SSM (synchronisation status message) algorithm. Squelch inhibits clocks of quality below a specified level. (If there is alarm indication or poor BER indication, the incoming frame clock is discarded.)

Let us now understand what is a synchronisation status message (SSM). Every NE sends an SSM (synchronisation status message) to its neighbouring NE in which the quality level of the reference being used by it is mentioned. This is carried in S1-byte of the MSOH. In fact, the least 4 bits of S1 (b5 to b8) carry the synchronisation status; this is detailed in Table 7.7. The selector (in NE) selects the timing reference as per SSM (synchronisation status message) algorithm presented in Fig. 7.12(d).

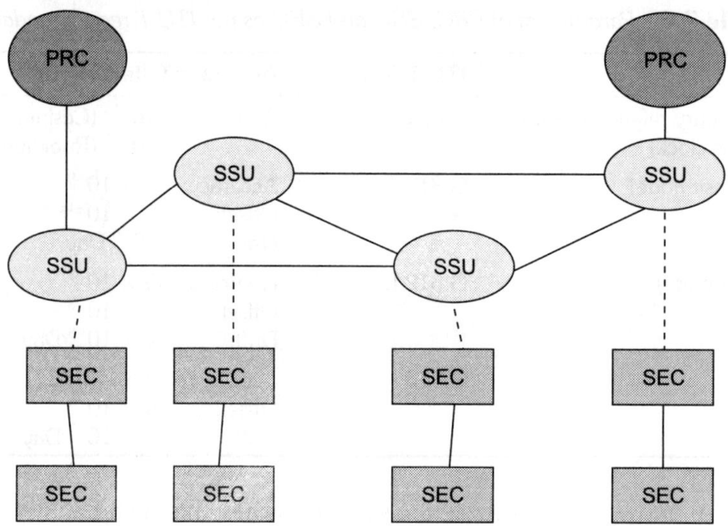

Fig. 7.12(a) *Synchronisation hierarchy structure*

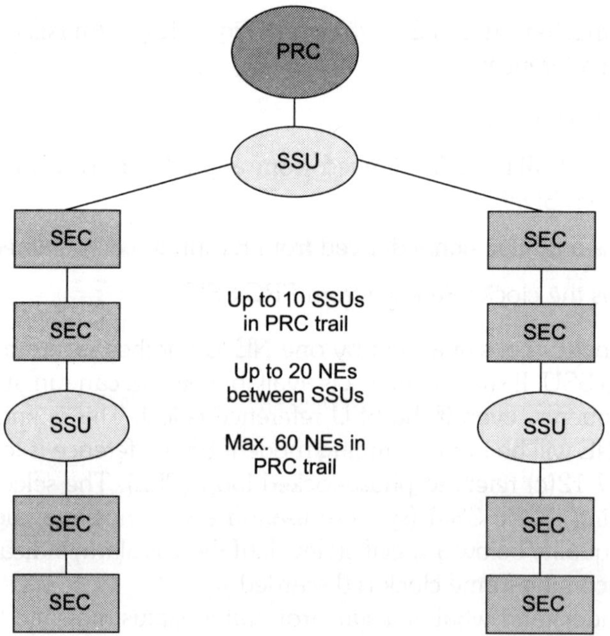

Fig. 7.12(b) *Number of SSUs and SECs in PRC trail*

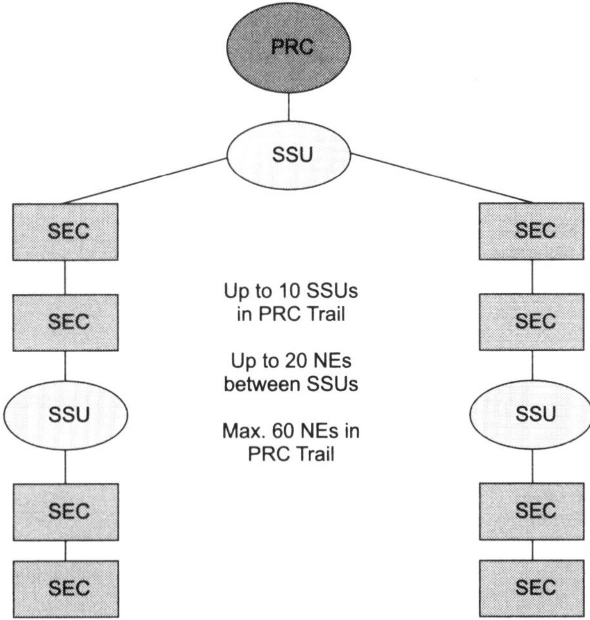

Fig. 7.12(c) *Synchronisation mechanism of NE*

Table 7.7 *Interpretation of S1 byte (synchronisation status byte)*

S1 byte	SDH synchronisation quality level/description
Least 4 bits (b5–b8)	
0000	Quality unknown
0001	Reserved
0010	PRC (ITU-T Rec. G.811)
0011	Reserved
0100	SSU-A (ITU-T Rec. G.812T)
0101	Reserved
0110	Reserved
0111	Reserved
1000	SSU-B (ITU-T Rec. G.812L)[4]
1001	Reserved
1010	Reserved
1011	SEC (SDH equipment clock ITU-T Rec. G.813)[5]
1100	Reserved
1101	Reserved
1110	Reserved
1111	Do not use for synchronisation

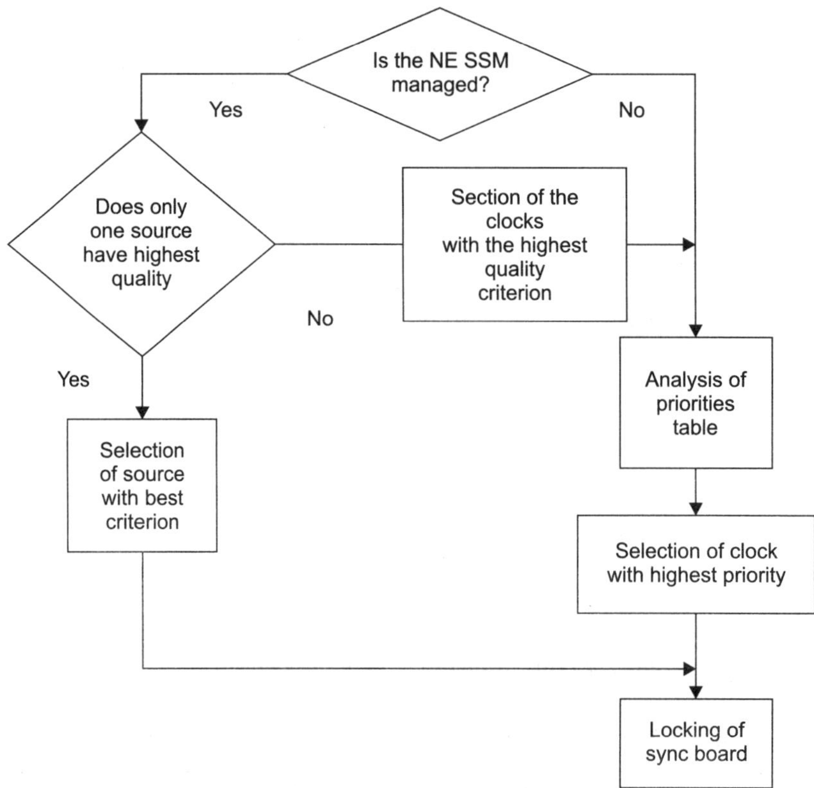

Fig. 7.12(d) *Selection of timing reference by NE*

Q 7.20 *Why do we have limits on the number of NEs associated with PRC/SSU/SEC?*

While extracting the clock from an incoming frame, the jitter due to phase noise on the links is filtered to obtain the clock signal. This process cannot proceed in an ideal manner as we move across number of links or NEs. Hence, clock coming from PRC down the trail cannot retain the best quality after some NEs. We use GPS clock which is also derived from PRC but sent via GPS satellite.

Q 7.21 *In an SDH network, one NE may be connected to several neighbours. How is synchronisation established in such a situation?*

Each NE receives STM frames from its immediate neighbour. The quality of clock used is known through the S1 byte. In any trail, each NE tries to time itself to the PRC(SSU) reference. This way, the sync loops are avoided (principle 11 in section 7.10.1). Once such timing reference to PRC/SSU by every NE in the trail is established, the trail is in stable state.

Afterwards, if there is interruption in the trail due to say, a fibre-cut or equipment failure, the NE may enter free-running mode or hold-over mode. Such an NE cannot supply reference to its neighbouring NE. The neighbouring NEs get back PRC/SSU reference via protection path gradually and synchronisation to PRC/SSU is re-established via protection trail. This process is described in detail with illustration in section 15.2.4 where a comprehensive analysis of jitter and BER degradation due to fibre cut is presented.

For the time being, it may be noted that when any NE takes to a different clock reference, certain justification may be necessary. As already discussed in section 7.5, the allowed difference between the incoming clock and the equipment clock is 300 ppm for the cushion of 3 bytes used for exhausting the justification. Furthermore, payload relocation in real time from NE to NE beyond justification cushion can be accomplished through pointer mechanism. The AU-4 pointer has 783 legal values (0 to 782) covering the entire address space of VC-4 in increments of 3 bytes. Thus, synchronisation is facilitated across all NEs in the network.

7.11 Alarm and Error Processing within an NE

The flow of alarms information is peer-to-peer, i.e., at regenerator level, multiplexer level and HO/LO path level. An NE follows the logic given in Fig. 7.13 for the detection of alarms and generation of alarm information which is sent through SOH bytes. Correlation between SOH bytes and alarm condition is also shown in Fig. 7.13. A summary of alarms with detection criteria is given in Table 7.8.

Table 7.8 *Alarms and detection criteria*

Alarm code	Alarm/Error description	Detection criteria
LOS	Loss of signal	Drop of incoming optical power level
OOF	Out of frame	A1, A2 incorrect for 625 μs
LOF	Loss of frame	If OOF persists for 3 ms (or as set)
RS BIP	RG section BIP error (B1)	Mismatch of BIP8 (covers whole frame)
RS TIM	RG section trace mismatch	Mismatch of trace identifier J0
MS-BIP	MUX section BIP error (B2)	Mismatch of BIP8 (RSOH excluded)
MS-AIS	MUX section AIS	Bits 6,7,8 of K2 = 111 for 3 frames
MS-REI	MUX section remote error indication	Number of detected B2 errors in sink side encoded in byte M1 of the source side
MS-RDI	MUX sec. remote defect indicator	Bits 6,7,8 of K2 = 110 for 3 or 5 frames
AU-AIS	Admin. unit AIS	All '1s' in AU pointer bytes H1, H2

(Continues...)

Table 7.8 *Continued*

Alarm code	Alarm/Error description	Detection criteria
AU-LOP	Admin. loss of PTR	8 to 10 NDF enable, 8 to 10 invalid PTR
HP-BIP	HO path BIP error (B3)	BIP-8 mismatch (covers entire VC-3)
HP-UNEQ	HO path unequipped	C2=0 for 5 frames
HP-TIM	HO path trace mismatch	Mismatch of J1 expected and received
HP-REI	HO path remote error indication	Number of detected B3 errors in sink side encoded in bits 1-4 of G1 of source side
HP-RDI	HO path remote defect ind.	Bit-5 of G1 = 1 for 3, 5 or 10 frames
HP-PLM	HO path payload mismatch	Mismatch of expected and received payload label C2
TU-LOM	Loss of multiframeof TU	Multiframe not recovered for 1 ms
TU-AIS	Tributary unit AIS	All '1s' in TU PTR bytes V1, V2
TU-PLM	TU payload label mismatch	Mismatch of expected and received V5
TU-LOP	Loss of pointer in TU	8-10 NDF enable, 8-10 invalid pointers
LP BIP	LO path BIP error	BIP-8 mismatch (B3) or V5 bits 1 and 2 (covers the entire VC-n)
LP-UNEQ	LO path unequipped	C2=0 in VC-3 for > 5 multi-frames ; for VC2, VC11, VC12, bits 5,6,7 of V5 = 000 for 5 multiframes
LP-TIM	LO path trace identifier mismatch	Mismatch of expected and accepted trace identifier J1 of VC-3 or J2
LP-REI	LO path remote error indication	VC-3: No. of detected B3 errors on sink side encoded as G1 (1-4) of source side VC2, VC11, VC12: If one or more BIP-2 errors are detected on the sink side, bit 8 of V5 of source side is set as 1.
LO RDI	LO path remote defect indication	VC-3: Bit 5 of G1 =1 for 5 frames. VC2,VC11,VC12: Bit 8 of V5 =1 for 3 or 5 or 10 multi-frames
LP-PLM	LO path payload label mismatch	Mismatch of expected and received payload label in byte C2 or V5(bits 5,6,7) accepted

Summary

In this chapter the concepts relating to RG section, MUX section and path is introduced. Features of SDH frame supporting the same are explained. SDH multiplexing process is described with respect to various payloads. Pointer operation and synchronisation are explored.

References

[1] P. V. Sreekanth, *Digital Microwave Communication System*, section 6.7, Bit stream integration in Digital MW Radio, Universities Press, 2003.

[2] NNI for SDH, ITU-T Rec. G707, January 2007.

[3] Timing characteristics of primary reference clocks, ITU-T Rec. G.811, September 1997.

[4] Timing requirements of slave clocks suitable for use as node clocks in synchronization networks, ITU-T Rec. G.812, June 2004.

[5] Timing characteristics of SDH equipment slave clocks (SEC), ITU-T Rec. G.813, March 2003.

[6] Acronyms and abbreviations Section 4, ITU-T Rec. G.707, January 2007.

8

SDH Network Elements and Network

SDH network element (NE) design is guided by the principles of integrated optical interface, scalability and configurability. Let us understand the same before going into the modular details of NE.

8.1 Principles of SDH NE Design

Integrated optical interface

In chapter 6 the inadequacies of PDH were discussed. One of them was the non-availability of bytes/bits for SV and NMS functions. This has been taken care of in the frame structure of SDH, as already described in chapter 7.

When the SV/NMS functions are integrated into an STM frame, the role of transmission equipment is to just provide electrical-to-optical conversion and vice versa. All the interfaces which derive any bandwidth or traffic from the STM frame for correction intended applications, as well as the SV/NMS entities can directly interact with the STM equipment. In fact, the transmission/reception function, viz., the electrical-to-optical/optical-to- electrical function can be simplified as one module and plugged as an interface on to the SDH network element. We can thus do away with the cumbersome OLTE.

Scalability

We get another advantage out of this. Suppose we commission an STM-1 network using G.652D fibre which can operate at 1310 nm as well as 1550 nm but wish to upgrade it to STM-4. We can provide STM-4 interface card replacing STM-1. The already derived STM-1 bandwidth will utilise one STM-1 capacity of the STM-4. We can further equip the NE and reconfigure it to utilise the balance bandwidth available on STM-4.

The same principle applies to upgrading STM-4 to STM-16 or STM-16 to STM-64. Normally, equipment manufacturers provide upgradation facility to the next higher level. In other words, STM-1 equipment initially installed can be upgraded to STM-4. Likewise, STM-4 equipment can be upgraded to STM-16, but this kind of upgradation cannot be sustained beyond one stage because the interface cards required for doing so get multiplied by the upgradation factor, which the equipment may not support for insertion beyond one stage of upgradation. Hence, manufacturers ensure complete scalability by offering products at higher-order STM that support all lower-order STM interfaces through SFP (small four-factor pluggable) transceivers. Details of the same are presented in Appendix A.

The most important factor to be considered is the selection of the optical fibre. The optical fibre used initially should be chosen such that it operates at a higher data rate than the current requirement without the need for reconsidering the link power budget and rise-time budget.

Configurability

Having taken care of scalability, let us turn our attention to configurability which covers the following aspects:

- Flexibility
- Modularity
- Programmability

As discussed in chapter 7, we have four varieties of SDH NEs, viz., TM (terminal MUX), RG (regenerator), ADM (add–drop MUX), DXC (digital cross-connect). However, we need not have different hardware platforms for these varieties; it can be made common by the inclusion/exclusion of some of the modules to get the NE of our choice. Thus, flexibility is taken care.

Also, it will be possible to derive bandwidth as per application (PDH, ATM, LAN connection or concatenated containers) by insertion of suitable access module. Bandwidth access is thus taken care.

Configuration of equipment is possible through LCT (local craft terminal), which is a laptop connected to a serial interface, or through the centralized EMS (element management system) or NMS (network management system) by downloading the configuration information on power-up. Thus, programmability is also taken care of.

8.2 Types of Optical Interfaces on SDH NE

Optical Interface is specified as X.Y.Z

- X can be I, S, L, V or U and denotes haul
  - I for intra station (up to 2 km)
  - S for short haul (15 km)
  - L for long haul (40 km at 1310 nm and 80 km at 1550 nm)
  - V for very long haul (60 km at 1310 nm and 120 km at 1550 nm)
  - U for ultra long haul (160 km at 1550 nm)

- Y can be 1, 4, 16 or 64 and denotes STM level
  - 1 for STM-1
  - 4 for STM-4
  - 16 for STM-16
  - 64 for STM-64
- Z can be 1, 2 or 3 and denotes the fibre type.
  - 1 for 1310 nm over NDSF (G.652 fibre)
  - 2 for 1550 nm over NDSF (G.652 fibre)
  - 3 for 1550 nm over DSF (G.653 fibre)
  - 5 for 1550 nm over NZDSF (G.655 fibre)

The different types of optical interfaces available for STM-1, STM-4 and STM-16 are given in Tables 8.1 to 8.4.

Table 8.1 *Optical interfaces for STM-1*

Code	Wave length (nm)	Fibre	Transmitter	Dispersion (ps/nm)
I.1.1	1310	G.652	LED/MLM	18/25
S.1.1	1310	G.652	MLM	96
S.1.2	1550	G.652	MLM/SLM	296/NA
L.1.1	1310	G.652	MLM/SLM	246/NA
L.1.2	1550	G.652	SLM	NA
L.1.3	1550	G.652	MLM/SLM	296/NA

LED: LED source SLM: Single longitudinal mode laser diode MLM: Multi-longitudinal laser diode

Table 8.2 *Optical interfaces for STM-4*

Code	Wave length (nm)	Fibre	Transmitter	Dispersion (ps/nm)
I.4.1	1310	G.652	LED/MLM	14/13
S.4.1	1310	G.652	MLM	74
S.4.2	1310	G.652	SLM	NA
L.4.1	1310	G.652	MLM/SLM	109/NA
L.4.2	1550	G.652	SLM	Spec. under study
L.4.3	1550	G.653	SLM	NA
V.4.1	1310	G.652	SLM	200
V.4.2	1550	G.652	SLM	2400
V.4.3	1550	G.653	SLM	400
U.4.2	1550	G.652	SLM	3200
U.4.3	1550	G.653	SLM	530

LED: LED source SLM: Single longitudinal mode laser diode MLM: Multi-longitudinal laser diode

Table 8.3 *Optical interfaces for STM-16*

Code	l (nm)	Fibre	Transmitter	Dispersion (ps/nm)
I.16.1	1310	G.652	MLM	12
S.16.1	1310	G.652	SLM	NA
S.16.2	1550	G.652	SLM	Spec. under study
L.16.1	1310	G.652	SLM	NA
L.16.2	1550	G.652	SLM	1600
L.16.3	1550	G.653	SLM	Spec. under study
V.16.2	1550	G.652	SLM	2400
V.16.3	1550	G.653	SLM	400
U.16.2	1550	G.652	SLM	3200
U.16.3	1550	G.653	SLM	530

LED: LED source SLM: Single longitudinal mode laser diode MLM: Multi-longitudinal laser diode

Table 8.4 *Optical interfaces for STM-16*

Code	l (nm)	Fibre	Transmitter	Dispersion (ps/nm)
I.64.1	1310	G.652	MLM	6.6
I.64.2	1550	G.652	SLM	500
I.64.3	1550	G.652	SLM	80
I.64.5	1550	G.655	SLM	Spec. under study
S.64.1	1310	G.653	SLM	70
S.64.2	1550	G.652	SLM	800
S.64.3	1550	G.653	SLM	130
S.64.5	1550	G.655	SLM	130
L.64.2	1550	G.652	SLM	260
L.64.3	1550	G.653	SLM	Spec. under study
V.64.2	1550	G.652	SLM	400
V.64.3	1550	G.653	SLM	Spec. under study

LED: LED source SLM: Single longitudinal mode laser diode MLM: Multi-longitudinal laser diode

8.3 Specifications for Optical Interfaces and Their Explanation

8.3.1 Set of complete specifications

Optical interface of the type X.Y.Z mentions only a few things about the interface. The compete specifications for optical interfaces of NEs in SDH network are summarised below.

- Data rate and interface type
- Tx (transmitter) parameters
  - Wavelength range
  - Optical source
  - Emission spectrum width
  - 20 dB spectral width
  - SMSR
  - Maximum and minimum launched power
  - Extinction ratio
- Optical path (NE to external) parameters
  - Attenuation
  - Chromatic dispersion
  - Optical return loss
  - Dispersion reflectance
- Rx (Receiver) parameters
  - Minimum sensitivity
  - Overload
  - Power penalty
  - Reflectance of receiver

STM rate and type of interface have been discussed in detail in the previous section. Let us understand the remaining parameters.

8.3.2 Transmitter parameters of optical interfaces in SDH NE

Transmitter parameters of optical interfaces in SDH NE are listed in Table 8.5.

Table 8.5 *Transmitter parameters of optical interface*

Tx parameter	Meaning
Wavelength range	Though operation is at 1310 nm or 1550 nm, a range is specified to signify WDM compatibility. Usually, the range is 50 to 100 nm
Optical source	Laser diode – SLM/MLM Single longitudinal mode/Multi longitudinal mode
Emission spectrum width	Spectral width in nm of sustainable longitudinal modes of MLM
20 dB spectral width	Separation in nm between –20 dB points on either side of peak emission wavelength of the SLM
SMSR (side mode suppression ratio)	Ratio of intensities of main (single) mode and largest side mode for SLM laser diode
Launched power	Power output of laser diode
EX or ER	Extinction ratio is the ratio of max. to min. light power representing logic 1 and logic 0, respectively.

A brief on SLM and MLM laser diodes

An MLM or Fabry-Perot (FP) laser diode has mirrors encasing its active region to provide positive feedback and facilitate stimulated emission. The emitted signal has a number of modes of which $\mathbf{l}_{n+1}$, $\mathbf{l}_n$ and $\mathbf{l}_{n-1}$, each with 2 nm separation are sustained. The emission spectrum of MLM has about 4 nm spectral width. 20 dB spectral width has no meaning in this case.

An SLM or distributed feedback (DFB) laser diode has corrugated mirrors encasing its active region and hence a single mode is sustained in emission, i.e., a single spectral line is possible. Its 20 dB spectral width is about 1 mm. Emission spectrum width has no meaning in this case.

Modulators

Modulators on transmitter can be either internal or external. The block diagram of an internal modulator is shown in Fig. 8.1(a). In this arrangement the incoming data stream after line-code conversion gets applied directly to the laser diode (LD) to change its bias.

Internal modulators suffer from the following drawbacks:

- The bandwidth is restricted by the laser diode's relaxation frequency.
- The fast variation in the radiating frequency emitted by the LD (in accordance with the pulse or modulating current) results in the broadening of pulse, a phenomenon known as chirp. This is a serious limitation for high speed transmission.
- The driving current required to launch high power into fibres over long-haul links is high.

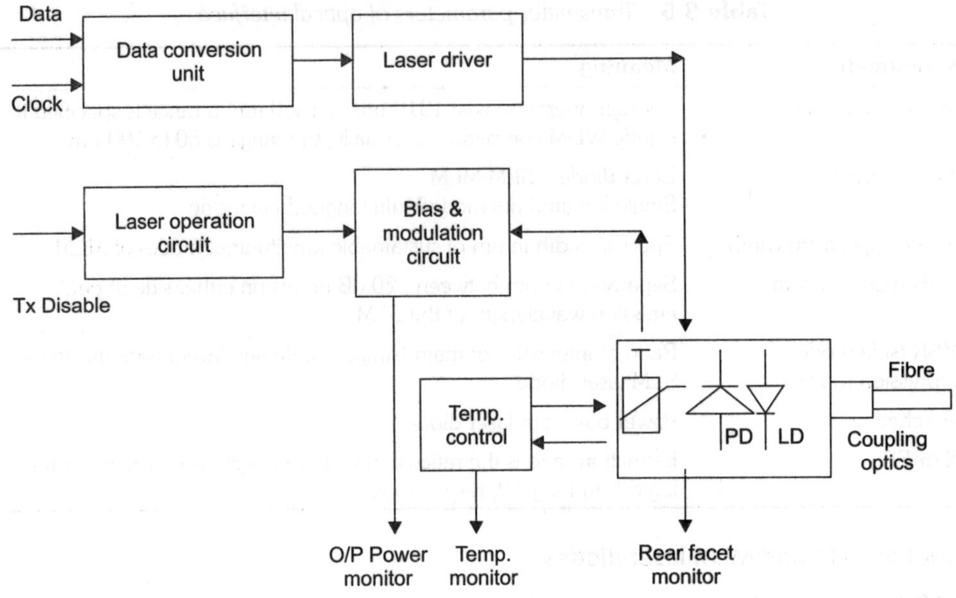

***Fig. 8.1(a)** Transmitter with internal modulator*

Nowadays, for all high speed, long-haul transmitters, external modulators are being used.

The block diagram of a transmitter with external modulator is presented in Fig. 8.1(b). In this arrangement the modulation is carried out as per the incoming 1s and 0s after the laser beam has been emitted by the LD.

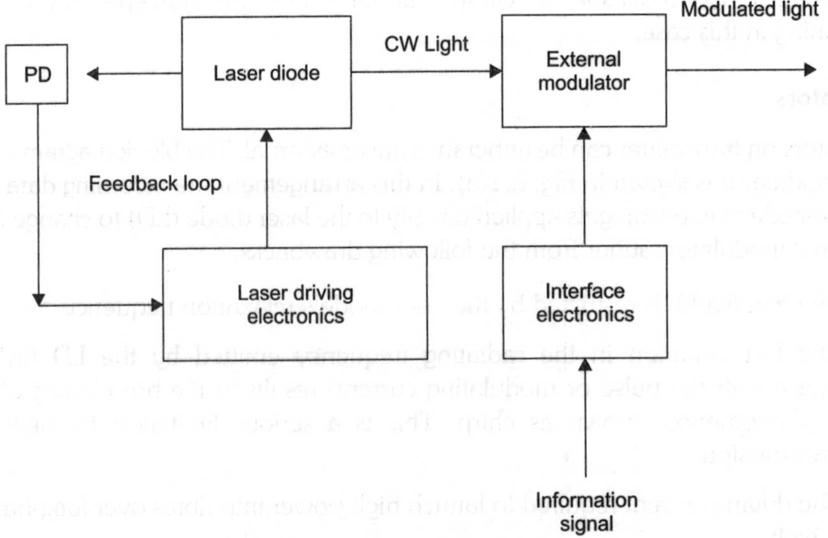

***Fig. 8.1(b)** Transmitter with external modulator*

External modulators have the following advantages over internal modulators.

- The laser diode circuit is not loaded with the extra task of modulation.
- The feedback loop using photodiode maintains a very stable level of power radiated by the LD; hence, there is no problem of chirp.
- The bandwidth of the transmitter is determined by the external modulator.

The study of types of external modulators and their parameters is beyond the scope of this book.

8.3.3 Path parameters

The path parameters are of importance to an SDH NE connected in the optical path (OFC). These are enumerated in Table 8.6.

Table 8.6 *Path parameters of SDH NEs*

Parameter	Meaning
Attenuation	Attenuation in connectors and patch cords
Chromatic dispersion (ps/nm)	Spreading of output pulse due to varying speed (travel time) of different wavelengths in the emitted signal
Optical return loss	Ratio (in dB) of the reflected power to the transmitted power. Its value is negative; the more negative it is, the better.
Dispersion reflectance	Ratio (in dB) of the reflected transmitted pulse to the reflected dispersion pulse

8.3.4 Receiver parameters of optical interfaces of SDH NE

The receiver parameters of optical interfaces of SDH NE are enumerated in the Table 8.7. Typical sensitivities and overload parameters of commonly used receivers are given in Table 8.8.

Table 8.7 *Receiver parameters of optical interface*

Rx parameter	Meaning
Sensitivity	Minimum power required at the receiver to detect the signal at a particular data rate and ensuring a BER of 1×10^{-12}
Overload	Maximum input power that the receiver can accept
Power penalty	Nonlinear effects contribute to signal impairment and hence additional amount of power will be needed at the receiver to maintain the same BER as in their absence. This additional power required is known as power penalty. Nonlinear effects include variations of RI, polarisation mode dispersion, nonuniform gain of amplifiers, group velocity dispersion and inelastic scattering (scattering due to interaction between optical signals and molecular or acoustic vibrations in the fibre).
Reflectance	Input-power-to-reflected-power ratio

Table 8.8 *Receiver types commonly used and their sensitivities and overload parameters*

Data rate	Receiver type	Sensitivity	Overload parameter
STM-1	Pin FET	–36 dBm	–7 dBm
STM-4	Pin FET	–32 dBm	–7 dBm
STM-16	Pin FET	–23 dBm	–3 dBm
STM-16	APD	–34 dBm	–8 dBm
STM-64	Pin FET	–18 dBm	–1 dBm
STM-64	APD	–24 dBm	–6 dBm

The above discussion covers the trans and receive functions of the optical interface in SDH equipment. To facilitate scalability and configurability as discussed in section 8.1, SDH equipment or NEs are made available as higher-order DXCs so that they can be configured as ADMs, TMs, DXCs and lower-order interfaces can be supported through SFPs. The capacity of DXCs is expressed in 'N' gig into VC-n granularity, which implies that the NE can support N/(BW of VC-n) cross-connects in total, in all directions put together. The different makes of SDH equipment, the interfaces available for them and details of how aggregates, add–drops and cross-connects are achieved are presented in Annexure A. This will also serves as a NW design guide.

8.4 SDH Network Topology and Protection Mechanisms

8.4.1 General Topology of SDH Network

SDH NEs, viz., TM, RG, ADM and DXC are connected to provide path between TMs and diversions at intermediate stations through ADMs. Different segments are cross-connected through DXCs. The topology is usually ring to provide for path diversity. Rings are interconnected through DXCs. The various topologies of SDH network is illustrated in Fig. 8.2.

Such large networks with several segments and rings call for protection mechanism against cuts in fibres and equipment failures. These are required on span (point-to-point) basis as well as ring basis.

8.4.2 Classification of protection mechanisms

Between two connection points (NEs), there are working trails and protection trails. In normal condition the traffic is carried on the working trail. If the working trail fails or if its performance falls below a required level, the traffic is carried over/selected from the protection trail.

SDH trail protection can be on the multiplex section layer or the path layer (line or path in the case of SONET). Accordingly, ITU-T Rec. G.841 covers two basic types of protection mechanisms:

1. *Multiplex section layer protection:* Multiplex section layer (line in the case of SONET) operates on the entire set of connections (all VC-4s or STM-1s) in the aggregate signal.
2. *Path layer protection:* Path layer protection scheme operates on individual paths or connections in the network. Consider an STM-16 signal which has 16 VC-4s (and STM-1s). A path layer protection scheme treats each STM-1 connection independently and switches them independently.

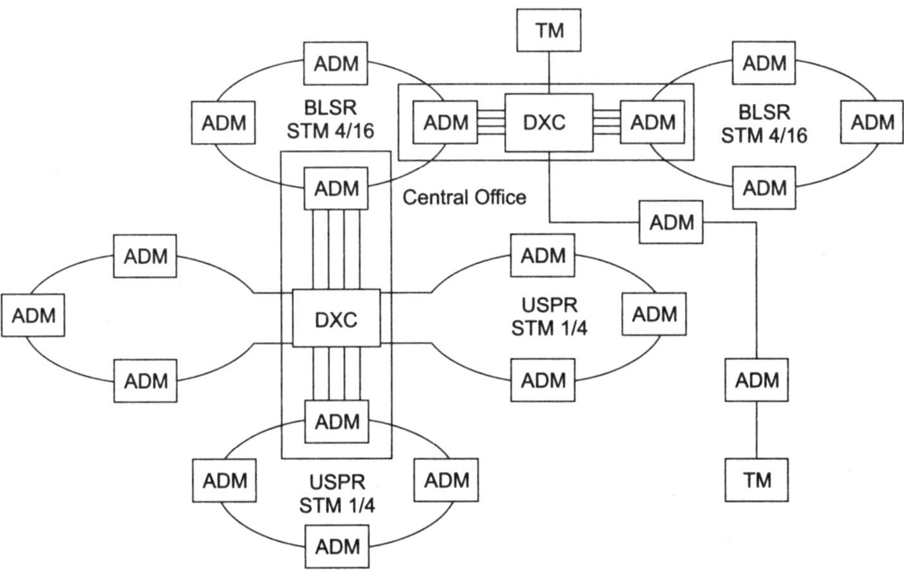

Fig. 8.2 *Topology of SDH network*

The multiplex section protection (line protection in the case of SONET) can be on point-to-point basis (also called linear-basis—do not confuse 'linear' with 'line') or ring-basis. In case of point-to-point (linear) protection, we have two varieties: 1+1 and 1:1 (1:*N* is a generalised form of 1:1). Again, under each of theses we have the revertive and non-revertive varieties.

In the case of SDH rings, we have two varieties:

- Multiplex section-shared protection ring (2 fibre) or MS-SPRing (2),
- Multiplex section-shared protection ring (4 fibre) or MS-SPRing (4).

In the case of SONET rings, we have two varieties:

- Bidirectional line switched ring, 4 fibre (BLSR/4),
- Bidirectional line switched ring, 2 fibre (BLSR/2).

Path layer protection scheme is only on ring basis. Path layer ring protection is known as UPSR (unidirectional path switched ring) in the case of SONET and sub-network connection protection (SNCP) in the case of SDH. In case of SNCP, we have two varieties: inherent monitoring and non-intrusive monitoring.

Figure 8.3(a) gives an overall picture of protection mechanisms in SDH networks, while Fig. 8.3(b) provides the same for protection mechanisms in SONET networks. The equivalent terms associated with SONET are given in Fig. 8.3 for information purpose only. These are not part of ITU-T Rec.G.841.

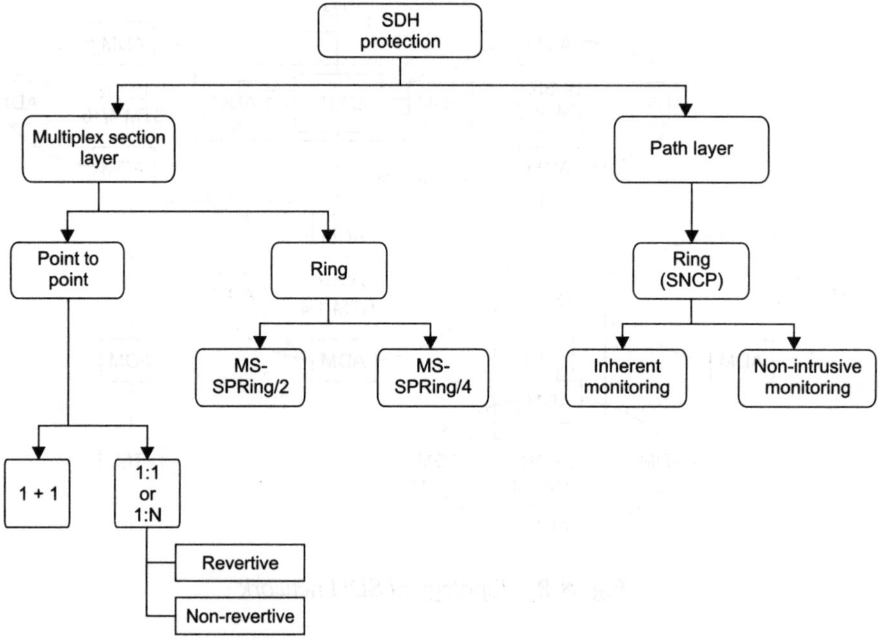

Fig. 8.3(a) *Protection mechanisms for SDH network*

8.4.3 Point-to-point protection in MUX section layer of SDH (line layer of SONET)

There are two types of protection schemes for point-to-point links:

i. 1 + 1 Protection: The signal is transmitted over two fibres (usually different routes) simultaneously.

ii. 1 : 1 Protection: The signal is transmitted over a single fibre (working) under normal conditions but switched over to the protection fibre in case of failure. It offers the following advantages over 1+1 protection:

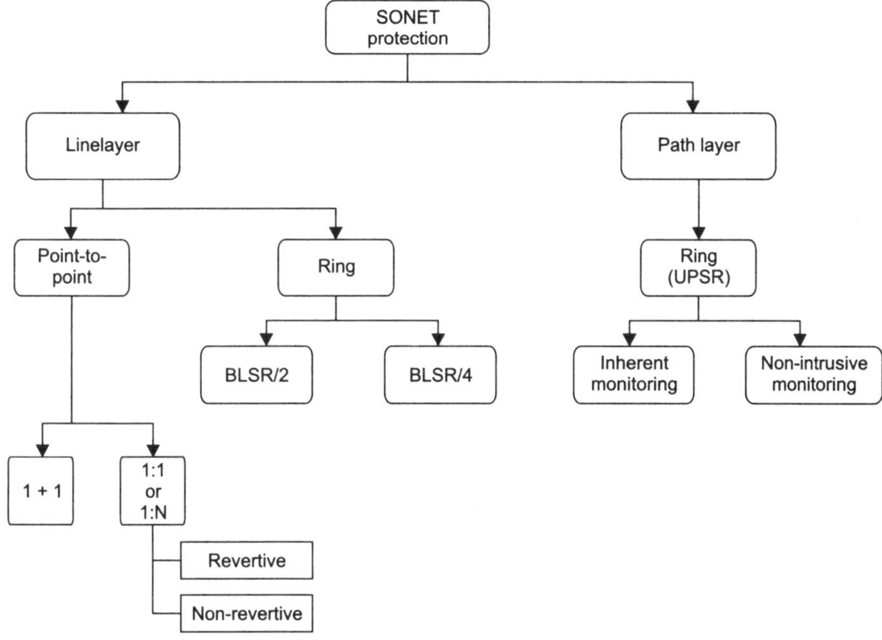

Fig. 8.3(b) *Protection mechanisms for SONET*

- Low priority traffic can be carried on protection fibre and can be discarded when the main traffic is switched over to the protection fibre.
- It can be upgraded to 1: *N* (*N* working fibres (paths) sharing a common protection fibre (path)).

The three schemes are illustrated in Fig. 8.4.

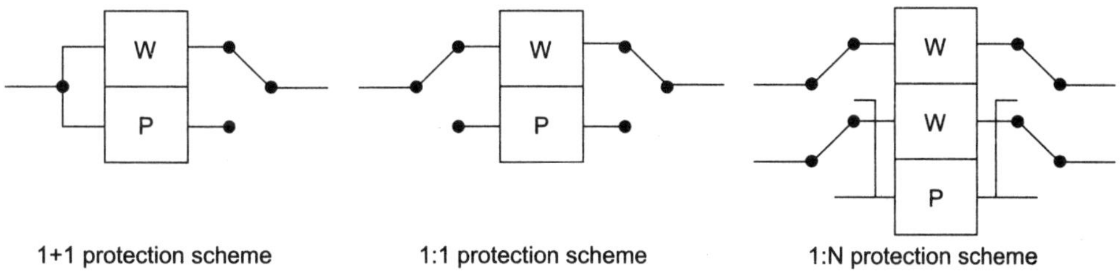

Fig. 8.4 *Point-to-point protection schemes in MUX/line layer of SDH/SONET*

8.4.4 Basics of self-healing rings in SDH and SONET

Ring topology provides two separate paths between any pair of nodes that do not have any nodes or links in common except the source and destination nodes. Hence, it is '2-connected'. Because of OAM&P features supported in SDH (and SONET) hierarchy, rings in SDH (and SONET) network are self-healing, i.e., in case of a cut in the fibre, it will be possible to switch to alternate path automatically.

Rings can be classified in two ways:

1. *Based on direction of working traffic*
   - Unidirectional ring carries traffic in only one direction in the ring (either clockwise or anti-clockwise).
   - Bidirectional ring carries traffic in both directions (clockwise as well as anti-clockwise).
2. *Based on protection mechanism:* Protection mechanism can be line-switched or path-switched.

The following ring architectures are mostly deployed in SONET.

- 2-fibre unidirectional path switched rings (UPSR)
- 4-fibre bidirectional line switched rings (BLSR/4)
- 2-fibre bidirectional line switched rings (BLSR/2)

The ring architectures generally deployed in SDH are given below.

- Sub-network connection protection (SNCP)
- Multiplex section shared protection ring on 4 fibres (MS-SPRing/4)
- Multiplex section shared protection ring on 2-fibre (MS-SPRing/2)

Q 8.1 *When more combinations are possible, why are only three are deployed in SDH?*

According to ITU-T Rec.G.841, the service in a network must be restored within 60 ms of a failure. Out of this period, 10 ms is set aside as detection time. The remaining 50 ms is the budget available for sending APS messages to other nodes, propagation delays, switching times and frame synchronisation after the switch-over to protection path takes place. The propagation delay is about 5–10 μs per NE and the physical line delay about 5 μs per km.

With combinations other than those mentioned above, it is difficult to remain within the limit of 50 ms when a number of NEs are in the optical path.

Q 8.2 A path is a point-to-point connection. Hence, can we call UPSR 1+1 protection?

In a way, they appear to be same because in UPSR one fibre carries the working traffic (clockwise) while the other carries protection traffic (anti-clockwise). There is subtle difference though. While both are point-to-point protection mechanisms, UPSR is in path layer whereas 1+1 is at line level (MUX layer in SDH).

In fact, UPSR of SONET operates on ring topologies while its equivalent in SDH, the SNCP (sub-network connection protection) operates on ring/mesh topology.

8.4.5 Features of UPSR of SONET

Refer to Fig. 8.5 showing UPSR. Node P sends traffic to node Q simultaneously on two fibres. On one fibre it is in the clockwise direction and in the other it is in the anti-clockwise direction. For duplex communication on OFC, we need one fibre for Tx and another for Rx. Then how is node-A sending traffic on both?

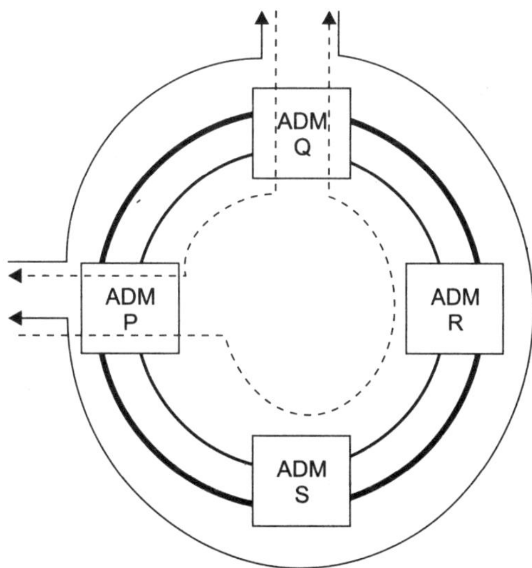

Fig. 8.5 *UPSR on four nodes*

On examining Fig. 8.5, it can be noted that:

1. The outer fibre is the working fibre and the inner the protection fibre.
2. The working connection from P to Q is shown as a continuous arc in clockwise direction (shorter path).
3. The working connection from Q to P is shown as continuous arc in clockwise direction (longer path).

4. Protection connection from P to Q is shown as broken arc in anti-clockwise direction (longer path).
5. Protection connection from Q to P is shown as broken arc in anti-clockwise direction (shorter path).

Let us now examine the sharing of bandwidth. Consider STM-4 link, i.e., four VC4s on each fibre. These are shared by the nodes to send their data. If P has Tx data for Q on the clockwise fibre, say in two VC4s, the corresponding duplex link of Q to P, i.e., Q's Tx data to P or Rx data of P is in two VC4s on the anti-clockwise fibre. We have two VC-4s leftover on each fibre.

On the protection path, we need to send Tx data of P. This occupies the leftover two VC-4s on the anti-clockwise fibre; the corresponding duplex protection path of Q to P is through the remaining two VC-4s of the clockwise fibre. Thus, duplex communication is achieved on the main and the protection paths.

However, there is no spatial reuse of fibre capacity as protection bandwidth cannot be shared between connections. Because of this constraint, UPSRs are popular in exchanges for accessing network connection (up to STM-4) when the access is hubbed to central office.

8.4.6 Features of SNCP of SDH

Sub-network connection protection is performed at the path level (VC-n). Refer to Fig. 8.6 showing SNCP arrangement.

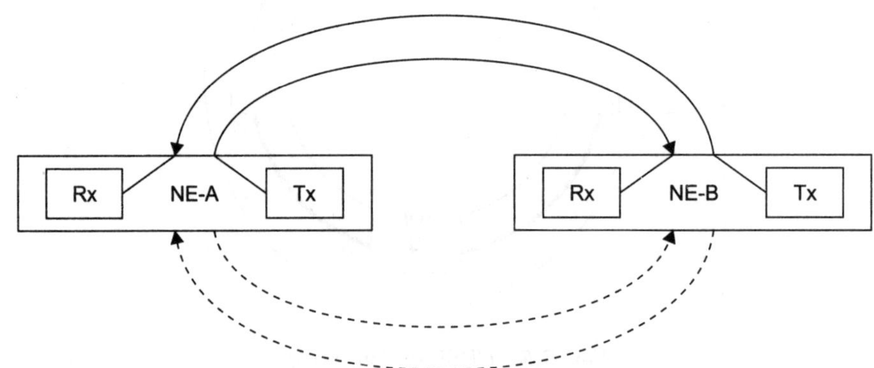

Fig. 8.6 *SNCP arrangement*

Features of SNCP:

1. It is a dedicated protection (1+1), i.e., there is protection path on a different fibre for the working path (VC-n).
2. SNCP is available in duplex. In Fig. 8.6, the solid line from A to B and B to A (follow the arrows) represent the working paths and the dashed lines the

protection paths. Only the end nodes of VC-n are shown as they are terminations for that particular VC-n. The same arrangement can be made for any VC-n at its terminating nodes.

3. At nodes A and B there is provision to select signal from either the working or protection path, i.e., switch-over between the two is allowed.
4. There are two operating modes:
   - Revertive: Traffic is switched back when the fault is rectified.
   - Non-revertive: No switching back on rectification of fault.
5. SNCP can be with:
   - Intrusive monitoring (SNCP/I): Alarms, TU/AU-AIS, loss of pointer, loss of frame monitoring
   - Non-intrusive monitoring (SNCP/N): BER and BIP are monitored in addition to the monitoring of alarms, TU/AU-AIS, LOP and LOF.
6. Switching can be automatic or through external commands.
   - The criteria for automatic switching are:
     - LOS, LOF, LOP, AU/TU-AIS in case of SNCP/I
     - LOS, LOF, LOP, AU/TU-AIS, AP/LP-UNEQ, HP/LP-E-BER, HP/LP-DEG, HP/LP-TIM, B3/BIP-2-SD in case of SNCP/N
   - External commands associated with switching are:
     - Protection mechanism lock
     - Forced manual switch
     - Manual protection switching
     - Forced switching to the normal channel
     - Manual switching to the normal channel

8.4.7 Features of BLSR/4 of SONET

Refer to Fig. 8.7(a) showing BLSR/4 (bi-directional line switched ring on 4 fibres). It has two pairs of fibres in ring topology.

1. The outer pair carries working traffic in duplex mode. (One fibre in the pair carries traffic from A to B and the other from B to A.)
2. The inner pair carries protection traffic in duplex mode. (One fibre in the pair carries traffic from A to B and the other from B to A.)
3. Protection is available at line level of SONET.

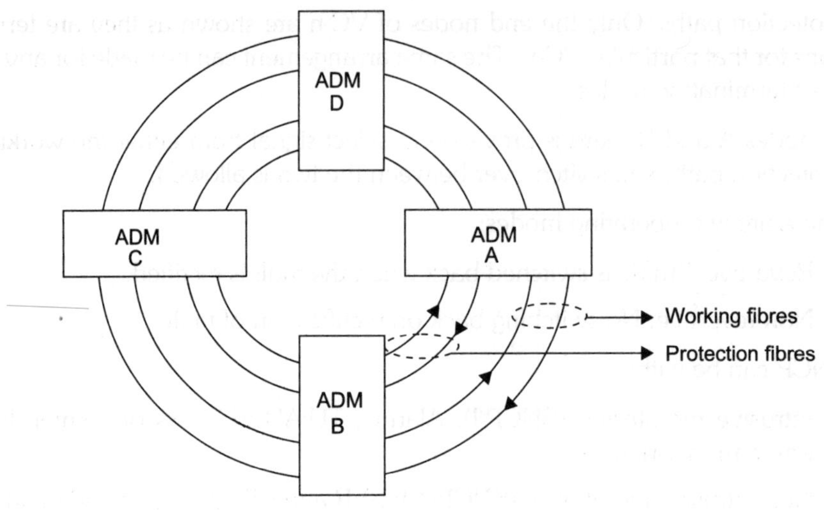

Fig. 8.7(a) *BLSR/4*

In BLSR/4, protection is by span switching or ring switching. In span switching if a node fails on the working fibre (Tx or Rx), the traffic is routed onto the protection fibres between the two nodes on the same span as shown in Fig. 8.7(b). In this figure the continuous line arcs with arrows at its ends depict the working traffic from C to D and from D to C while the dashed line arcs with arrows at its ends depict protection traffic from C to D and from D to C.

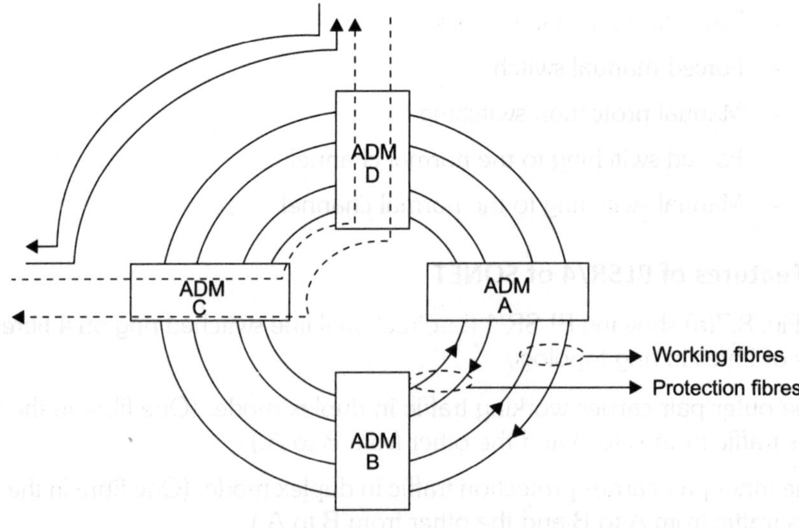

Fig. 8.7(b) *Span switching in BLSR/4*

Span switching can be used to restore traffic in the event of a cut in the fibre if the protection fibre on that span is separately routed, which is usually not the case.

In fact, in case of a fibre-cut, protection is by ring-switching, as illustrated in Fig. 8.7(c). Since all fibres are in the same cable, all get cut in the case of a cable-cut . Continuous line arcs with arrows at the ends show working traffic from C to D and D to C. Dashed-line arcs with arrows at the ends show protection traffic from C to D and from D to C.

8.4.8 Features of MS-SPRing/4 of SDH

Similar to BLSR/4 of SONET, we have MS-SPRing/4 or multiplexed section shared protection ring (4 fibre) of SDH working in multiplex-section layer. Out of four fibres in the ring shown in Fig. 8.8,

1. The outer two fibres shown as solid-filled arrows are for working traffic (observe the duplex connection).
2. The inner two fibres shown as hollow arrows are for protection traffic (observe the duplex connection).
3. In case of a fault condition on the working link alone (Tx/Rx failure of a node or a cut in the working fibre alone, which is not usually the case unless protection fibres are taken separately on the same span), span switching is resorted to, i.e., the traffic is switched on to protection fibres on the same span. For example, B to C connection will be switched over to protection fibres in the event of Tx/Rx failure or working fibre failure on B-C span. The switch-over is shown by the dashed line shown across ADM 'B'.

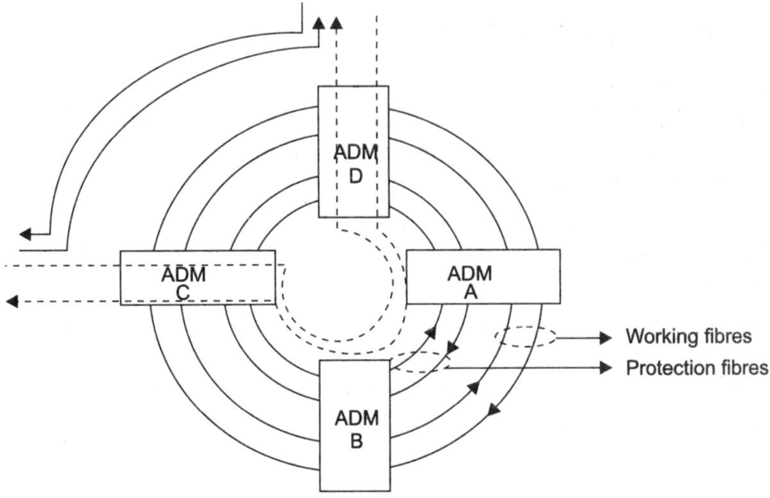

Fig. 8.7(c) *Ring switching in BLSR/4*

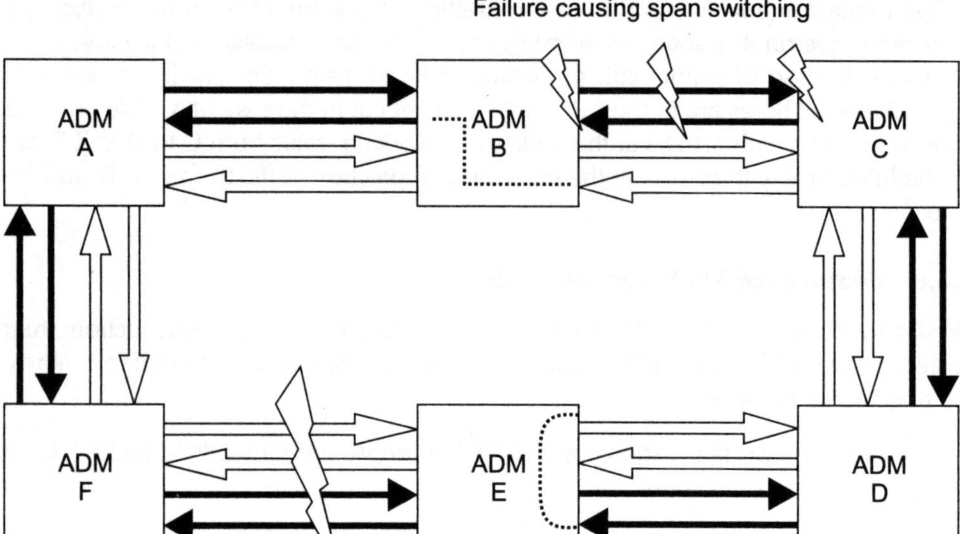

Fig. 8.8 *MS-SPRing/4 of SDH with span-switching and ring-switching*

4. Fault condition on both working and protection fibres causes ring switching. For example, a cut in all fibres between E and F causes bridging at E as shown by the dashed-line.

5. Switching criteria are AIS/LOS/LOF on the multiplex section.

8.4.9 Features of BLSR/2 of SONET

BLSR/2 or bi-directional line-switched ring on two fibres is a self-healing ring operating in the line layer of SONET. Refer to Fig. 8.9 showing BLSR/2 arrangement. Both the fibres carry working traffic but half the capacity is reserved for protection purpose. Please note that 'reserved' is the key-word here.

From Fig. 8.9 it can be seen that the working traffic A-B-C-D occupies half the bandwidth on the outer fibre while the other half is reserved for protection traffic, D-C-B-A. Similarly, the working traffic D-C-B-A occupies half the bandwidth on the inner fibre while the other half bandwidth is reserved for protection traffic A-B-C-D.

In normal condition, this bandwidth can be used to carry low-priority traffic which will be discarded in case of a link or node failure when protection traffic will be carried on it. Span switching is not possible here; only ring switching is feasible.

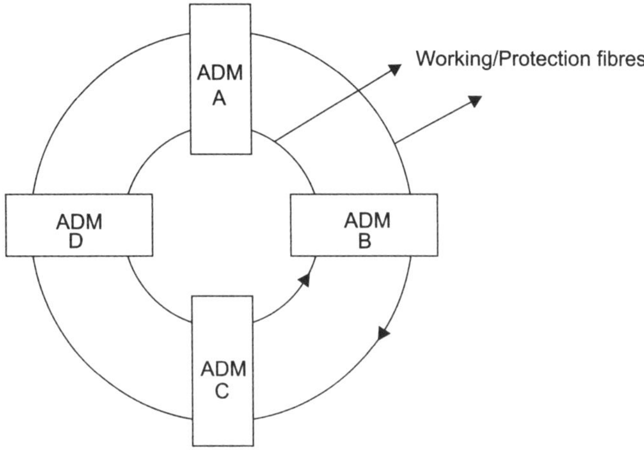

Fig. 8.9 BLSR/2 arrangement

Q 8.3 What is the difference between UPSR and BLSR/2?

UPSR offers 1+1 protection, because the working traffic is connected to both fibres. In BLSR/2, it is 1:1 protection, where, though half the capacity on each fibre is reserved for protection traffic, it is connected only in case of a failure on the main link. When there is no failure, low-priority traffic is carried on the spare capacity.

8.4.10 Features of MS-SPRing/2 of SDH

Refer to Fig. 8.10 showing MS-SPRing arrangement on 2 fibres.

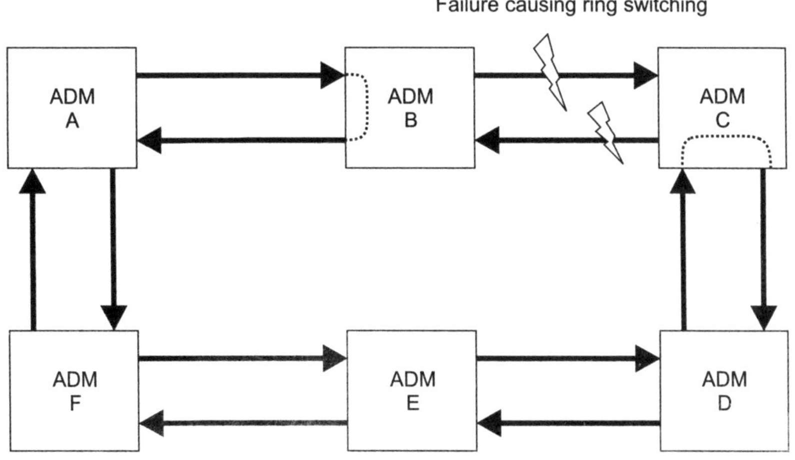

Fig. 8.10 MS-SPRing/2 of SDH with ring switching

Features of MS-SPRing/2:

1. Each fibre simultaneously carries normal and protection channels. For example, if it is STM-16 ring, eight AU4s are for the working traffic while eight AU4s are for the protection traffic.
2. In case of a failure, say between B and C in Fig. 8.10, bridging is done at nodes B and C as shown.
3. Ring switching is feasible but not span switching.

8.4.11 Limitations of self-healing rings

Limitation on number of nodes

Recall from the discussion on APS message bytes (K1 and K2) that the number of bits used for identifying nodes are 4, which means the maximum number of nodes in BLSR or MS-SPRing is 16.

Problems in handling node failures

Handling node failures seems simple, but it is not so. Node failure is viewed by neighbours on either side of the failed node as link failure and they resort to restoration mechanism which can lead to deadlocks. The neighbours of failed node should exchange messages to find out whether both have found link failures and, if so, invoke correct procedure of restoration. So, BLSR is complex to implement although high on the promise. UPSR, on the other hand, though not quite so promising, is simpler to implement because of its 1+1 strategy.

Limitation on length of ring

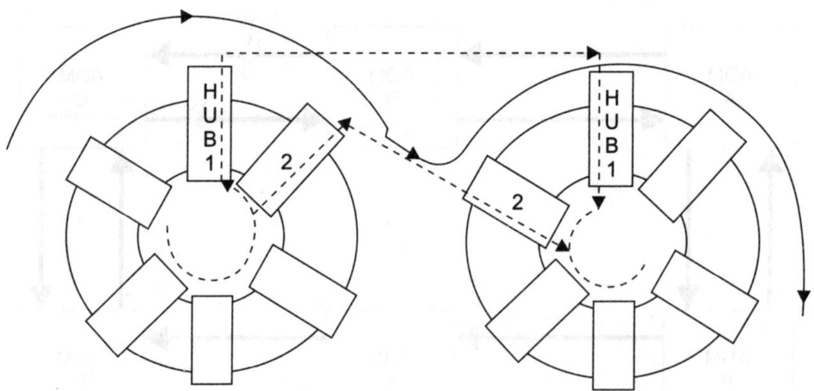

Fig. 8.11 *Interconnection of rings through dual-homing*

Given the 60 ms limit on the restoration time, the maximum length of a ring is limited to 1200 km. However there is some relaxation in the case of undersea OFC rings which run into few thousands of kilometres.

VoIP (voice over internet protocol) subscribers experienced more varying delays in voice communication over internet in January/February 2008 due to snapping of Mediterranean sea OFC cables. These cables supported VoIP traffic on SDH frames, and though ring protection was available, restoration time through APS beyond the normal limit of 60 ms, which were deliberately permissible, had a bearing on VoIP traffic. If the restriction of 60 ms. delay was stringently followed, there could not have been ring protection network, and traffic would not have been supported at all. Better to receive later than never!

Interconnection of rings

Refer to Fig. 8.2 once again. Some of the ADMs or DXCs are members of two rings which are interconnected. It is not that simple to interconnect rings. Even if we keep two separate NEs (ADM or DXC) and connect them through cables, in case of failure of the interconnecting NE or the fibre connecting the two NEs, both rings fail and protection switching in the two rings can create contrary traffic flows with the result that the interconnection remains broken. A way out of this situation is via dual-homing.

In dual homing the interconnection is at two nodes as shown in Fig. 8.11. Nodes marked Hub-1 and Hub-2 in the two rings are interconnected. If one of them fails, the other can take over as indicated by the dotted lines showing the protection path. Also, restoration is possible in case of cable cuts between two nodes. The NEs which are used for dual homing use 'drop-and-continue' of traffic rather than add–drop feature, i.e., the traffic dropped also continues towards the next ADM which is part of the dual homing scheme. To be clearer, the ADM at Hub-1 in Fig. 8.11 drops traffic as well as continues towards the ADM at Hub-2 in the clockwise direction. The same is done by ADM at Hub-2 in the anti-clockwise direction towards the ADM at Hub-1. Additional bandwidth is used up between the hub nodes to support this scheme.

8.4.12 Comparision of self-healing rings

- 4-fibre protection rings can simultaneously handle one transmitter failure on each span in ring topology whereas 2-fibre rings can handle only one such failure.
- 4-fibre rings can handle a cut in all fibres on a span without losing bandwidth. 2-fibre rings handle cuts in fibre by sharing bandwidth.
- Ring management is more complicated in 4-fibre rings than in 2-fibre rings because of the multiple protection mechanisms to be coordinated.

- Also, BLSRs (MS-SPRings) are more complicated than UPSRs (SNCPs) as they require extensive signalling between the nodes.

A comparison of the features of protection mechanisms is presented in Table 8.9.

Table 8.9 *Comparison of protection mechanisms*

Parameter	SONET: UPSR	BLSR/4	BLSR/2
	SDH: SNCP	**MS-SPRing/4**	**MS-SPRing/2**
Fibre pairs	1	2	1
Tx–Rx pairs per node	2	4	2
Protection type	Dedicated	Shared	Shared
Link failure handled on	Path	Span/Ring	Ring
Node failure handled on	Path	Ring	Ring
Working capacity	Half the STM bandwidth provided on fibre	Complete STM bandwidth provided on fibre	Half the STM bandwidth provided on fibre
Restoration speed	Fast	Slow	Slow
Complexity	Moderate	High	High

8.5 Network Protection Objectives

So far, we discussed several protection options for SDH networks. Let us go through the network protection objectives prescribed in ITU-T Rec. G.841. There are ten objectives, as elaborated below.

1. **Switch time**

   In a ring with no extra traffic, all nodes in idle state (no detected failures, no active automatic or external commands and receiving only idle K-bytes) and with less than 1200 km of fibre, the switch (ring and span) completion time for a failure on a single span should be less than 50 ms. On rings under all other conditions, the switch completion time can exceed 50 ms (the specific interval is under study) to allow time to remove extra traffic or negotiate and accommodate coexisting APS requests.

2. **Transmission delay**

   There is no network objective on transmission delay.

3. **Hold-off times**

   There is no network objective on hold-off times.

4. **Extent of protection**

   a. For a single point failure, the ring will restore all the traffic that would be passing through the failed location had no failure occurred.

   b. The ring should restore all the traffic possible even under conditions of multiple bridge requests of the same priority (including the combination of forced-switch-ring and signal-fail-ring).

5. **Switching types**

   Bidirectional protection switching should be provided.

6. **APS protocol and algorithm**

   a. The switching protocol should be able to accommodate up to 16 nodes on a ring.

   b. In order to provide an additional degree of protection for the four-fibre ring, a mechanism to perform span switching should be provided in the APS protocol.

   c. The APS protocol should be optimal for the AU-3/4 level of operation.

   d. The APS protocol and associated OAM&P functions should accommodate the ability to modify and upgrade the ring. In particular, adding and deleting nodes from the ring should be accommodated.

   e. A deterministic process should be used to avoid misconnecting traffic.

   f. All spans on a ring should have equal priority. Therefore, no higher priority spans will exist which would allow ring bridge requests for that span to override (automatically) other span switches of the same type (e.g., signal-fail, signal-degrade, or forced-switch).

   g. The state of the ring, i.e., normal or protected state, should be known at each node.

   h. A span bridge request should have higher priority than a ring bridge request of the same type.

   i. If a ring switch exists and a failure of equal priority occurs on another span requiring a ring switch (including the combination of forced-switch-ring and signal-fail-ring), then, if the priority of the bridge request is signal-fail (ring) or higher, both ring switches should be established resulting in the ring segmenting into two separate segments.

   j. AUG squelching should be done at the switching nodes.

7. **Operation modes**

   a. Revertive switching should be provided. A switch should revert only to the working channels and not to a different set of protection channels.

   b. The ring APS signalling should provide protection switching for both 2-fibre and 4-fibre bidirectional MS shared protection rings.

8. **Manual control**

   The following externally initiated commands should be supported:

   Lockout of protection - Span
   Forced switch - Span
   Forced switch - Ring
   Manual switch - Span
   Manual switch - Ring
   Exerciser - Span
   Exerciser - Ring.

9. **Switch initiation criteria**

   The following automatically initiated commands should be supported.

   Signal failure: Protection   Signal failure: Span   Signal failure: Ring
   Signal degrade: Span   Signal degrade: Ring   Signal degrade: Protection
   Reverse request: Span   Reverse request: Ring
   Wait-to-restore and no request.

10. **Ring utilisation criteria**

    Time-slot interchange will allow better utilisation of bandwidth of the ring. If time-slot interchange (TSI) is allowed, the traffic having a time-slot interchange through the failed location may or may not be restored. It is for further study whether TSI should be allowed, and if allowed, whether traffic having time-slot interchange through the failed location will be restored.

8.6 Protection for IP Networks

Internet service is always a best-effort service rather than a best-quality service. This is due to routing methods used by IP. IP packet routing is distributed, next-hop-based dynamic routing. If we are keen on faster restoration when some paths fail, we will have to somehow do 'path fixing', i.e., send a pilot packet on a route followed by other packets between the same source and destination in a particular session. This scheme is supported by SDH on traffics of all protocols that can be mapped onto SDH containers. Hence, it is termed multi protocol label switching (MPLS). The paths are called LSPs (label switched paths) and protection schemes discussed in previous sections can be used. MPLS implementation on SDH is discussed in detail in chapter 9.

There is another angle to protection of IP traffic. Routers exchange 'hello' packets every 10 seconds. A fault is declared if three successive 'hello' packets are missed, i.e., it takes 30 seconds to detect failure on one link. Imagine the time it will take to find failure at IP level over several links! Even if we increase the frequency of hello packets and detect link failure early (10 seconds as in the case of core routers), re-establishing another LSP and consequent actions would still cause delay. The best way out of this situation is to detect link failure at 'optical level' and inform IP layer directly. So, we need an optical layer to encapsulate the SDH traffic, of course retaining all the frames and methods. This led to the concept of 'managed light path service'.

The concept of managed light path services and feasibility of supporting multiple wavelengths on OFC (each carrying an STM signal) and also achieving network functions at optical level led to the evolution of OTN (optical transport network) and optical transport hierarchy (OTH). This evolution is discussed further in chapter 11 and protection mechanism for optical networks is discussed in chapter 15.

8.7 Element Management and Network Management

8.7.1 Use of craft terminal for management

Management of NE means preparing NE for OMA&P (operations, maintenance, administration and provisioning) functions. In a small scale network, it can be done using 'craft terminal' or CT. CT has the OMA&P software downloaded into the management card of an NE through a serial interface.

If the CT is used locally, it is called 'local craft terminal' or LCT. If CT is used remotely through another NE on the network, it is called 'remote craft terminal' or RCT. In such a case, the DCC bytes of STM frame carry the management information. Refer to Fig. 8.12 where CT is LCT for NE-A and RCT for other NEs. In this arrangement, the maximum number of NEs that can be managed is 32.

EQ 8.4 Why cannot more than 32 NEs be managed using CT?

As discussed in the answer to the preceding question, the failure of NE to be detected in 10 ms DCCr provides for data rate of 192 kbps.

8.7.2 Use of EMS or NMS server and LAN for management

When the STM network is medium scale (number of NEs more than 32), EMS (element management system) or NMS (network management system) is used for management of the network. EMS or NMS is on server supported by LAN and gets connected to STM network through the gateway NE. Refer to Fig. 8.13 where the EMS or NMS is connected to NE-A, the gateway NE, and the client view is provided through GUI to carry out OMA&P functions. It may be noted that the flow of management information is through DCC only.

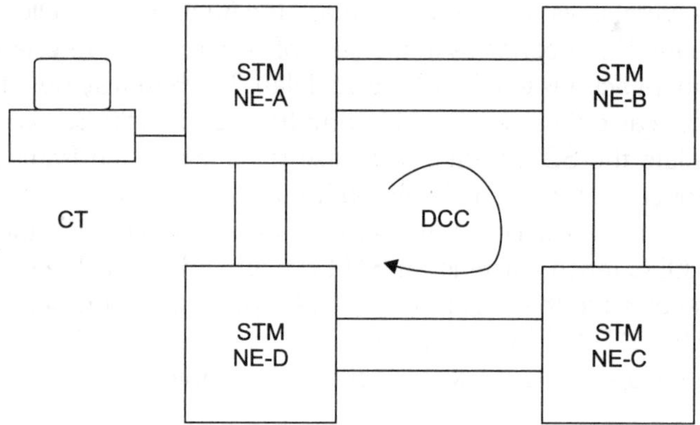

Fig. 8.12 *Use of CT for NE Management*

Q 8.5 *What are the differences between EMS and NMS?*

OAM&P functions for large number of nodes interconnected on an OFC network are done through network management system or NMS. It handles essentially the following processes:

Fault management
Configuration management
Accessibility and accounting management
Performance management
Security management

Some of the manufacturers provide an element management system, where configuration management is mandatory; others listed above are optional.

8.7.3 EMS or NMS with DCN support for network management

Management of large scale STM networks over vast geographical spread with the arrangement discussed in the previous section is not possible in case there are fibre cuts/ poor reception quality on any hop or interruption of management information flow through DCC due to which some NEs become unreachable. To manage such networks, a supporting data communication network or DCN is required. Also, on such STM networks, implementation of NMS or EMS requires servers, preferably at two different locations, so that one site can be the master and the other a standby for disaster recovery. DCN connectivity locations and server locations are to be planned strategically to ensure redundancy. The same is discussed in this section.

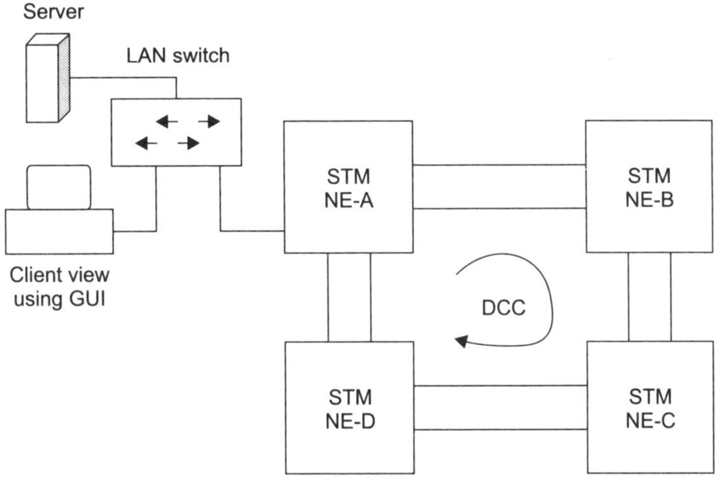

Fig. 8.13 *Use of EMS or NMS server and LAN for management*

OFC cable comes with many fibres: 6 or 12 or 24 or 48. OFC network operators commission OFC rings/networks with equipment of different makes on different fibre-pairs. Connections in DCN plan can be E1s (or VC-12s) or Ethernet; DCN plan for network with NEs of one make can be provided on networks with NEs of another make. DCN connectivity locations are so chosen that nodes of different makes are available at that point and the DCN easily mappable from one node to the other.

At the server location the NMS data finds its way onto the OFC network through the gateway node and gets carried to all other nodes through the DCC bytes of the STM frame. If there are fibre cuts, DCC flow is uninterrupted as the DCN channels come to the rescue. Resilience of the network depends on how well the network is organised into regions/areas and the redundancy in DCN channels connecting such areas.

DCN implementation has varied options:

1. Use of routers with E1(VC-12) connectivity
2. Use of V-LAN switch and router with E1/ETH DCNs
3. Use of L3 switch with ETH DCNs

Use of routers with E1 (VC-12) connectivity

Refer to Fig. 8.14(a) showing the network divided into four regions. Routers of each region are interconnected through E1s making the DCN. Servers at the master site and diaster recovery site are connected through a LAN switch which also support client view. In fact, for each region there will be more than one DCN connection at different locations; these are not shown in Fig. 8.14 for the sake of compactness of illustration.

A decision on the number of DCNs and their location has to take into consideration the NE population and future expansion so that congestion on DCN is avoided.

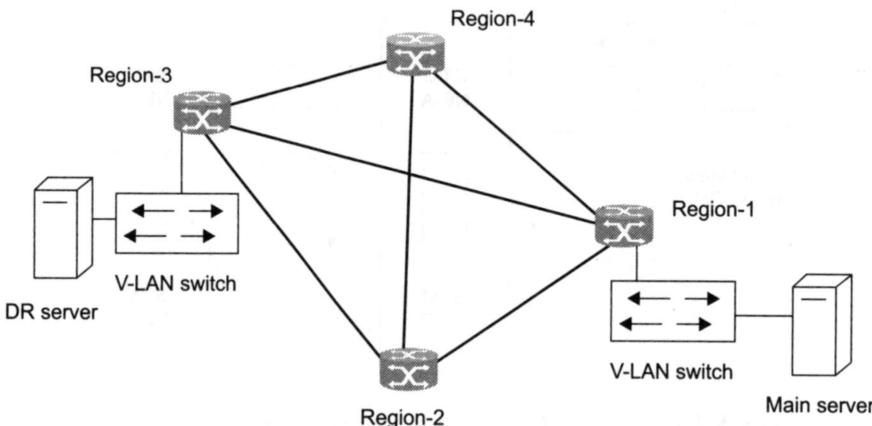

Fig. 8.14 *DCN implementation with routers and E1(VC-12) connectivity*

Use of V-LAN switches and routers with ETH connectivity

Refer to Fig. 8.15. The STM NE or node has Ethernet interface card and DCN channels are provided as ETH connections (shown as dotted lines) to three different areas in Region-1. The management information is further carried to NEs in an area on DCC. The NMS port of gateway NE of each area is connected to the ETH port where the DCN is connected. In case the NE does not support ETH interface, E1 to ETH converters are to be used at either end of the DCN channel. The figure also shows DCN connections between routers of different regions as dashed lines. Each region has similar arrangement of V-LAN switch and DCN on ETH to different areas in that region. The main server is in one region, the DR server in another.

Use of L-3 switch and ETH DCN connectivity

A variant of the arrangement described in previous subsection using L3 switch only and no routers is shown in Fig. 8.16. This arrangement is possible with IP-based (private IP, of course) EMS or NMS implementation. Here the gateway node's NMS port is connected to one of the ports of L3 switch. ETH connections from other ports are carried through ETH interface card ports to different areas as DCN channels. Server and client view are connected to L3 switch. However, in the server, separate IP subnets are to be provided for each area as indicated so that in case of interruption in the OFC ring, the gateway node of each area can ping to its subnet process in the server. To facilitate the same, the metric for subnet-process selection in the server needs to be defined.

One can define the metric for subnet-process selection in the server.

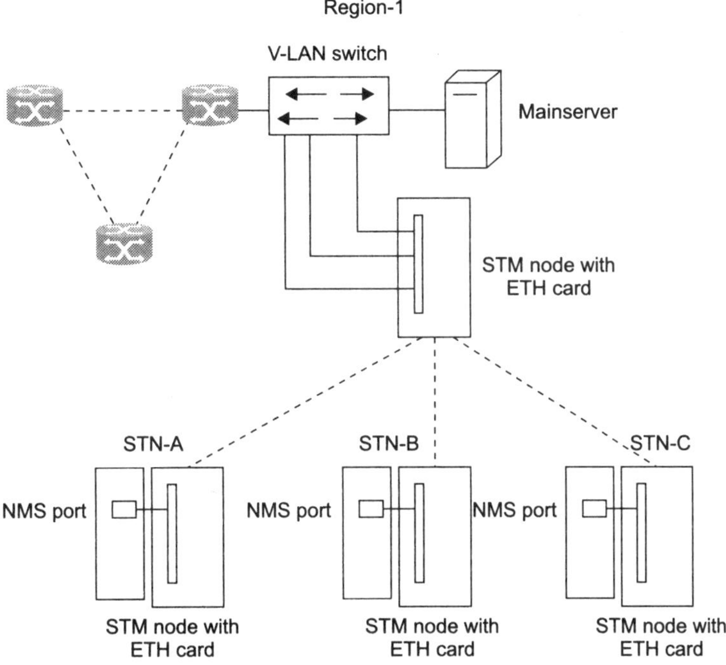

***Fig. 8.15** DCN implementation with routers, V-LAN switch and ETH connectivity*

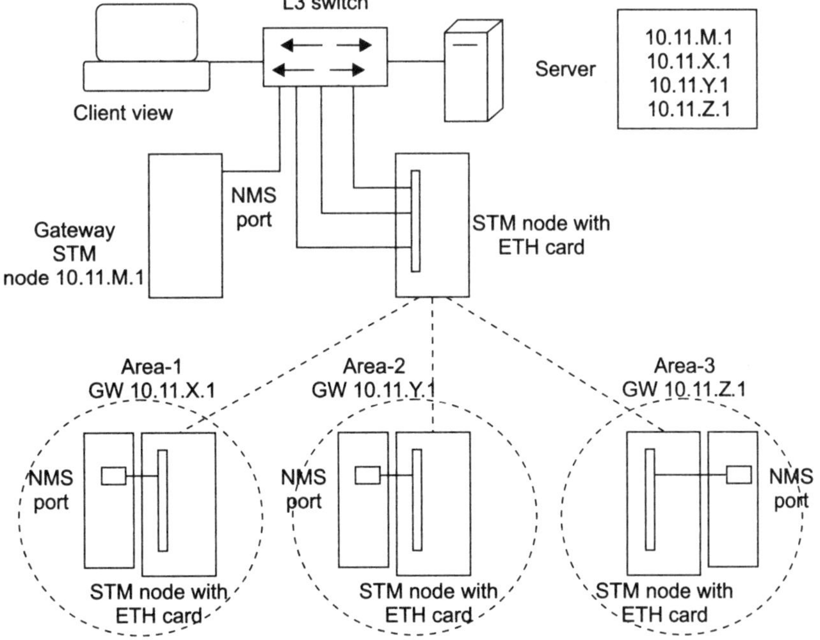

***Fig. 8.16** DCN implementation with L3 switch and ETH connectivity*

8.8 ASON (Automatically Switched Optical Network)

ASON or automatically switched optical network, is a new concept in the management of transport networks which facilitates dynamic policy-driven control of SDH or OTH network based on signalling between the user and components of the network. Its aim is to automate the resource and connection management within the network. It is an alternative/supplement to NMS based management. Details of ASON are presented in chapter 15 as inputs on OTH are needed before discussing ASON.

Summary

In this chapter the principles of SDH NE design are presented. It covered optical interface varieties in SDH NE, parameters of transmitter, receiver and media interfaces. Parameters of commonly used sources and detectors, types of NEs, classification of protection mechanisms of SONET and SDH (UPSR and BLSR of SONET; MS-SPRing, and SNCP of SDH) are explained along with network protection objectives.The chapter also cover protection of IP networks over SDH and implementation of DCN for network management systems.

References

[1] Types and characteristics of SDH network protection architecture, ITU-T Rec. G.841, October 1998.

9

Applications of SDH Network

9.1 Introduction

The rigours of multiplexing process in SDH and OAM&P in SDH network have been discussed in detail in chapters 7 and 8 respectively. In this chapter, services and applications supported on SDH network are discussed. From the user's perspective, SDH network is a network with one-point multiplexing and transporting support for PDH, ATM and any serial-bit or byte-oriented traffic over 'digital pipes' of chosen capacity across NEs with various protection options that can offer diverse services such as:

1. LAN-to-LAN data transportation
2. MPLS virtual private networks (VPN)
3. Internet backbone connection for distributing bandwidth to exchanges for broadband internet connectivity
4. Transportation for VoIP service
5. Transportation for next generation networking (NGN)

The key to getting any of these services is to tap the bandwidth or get connected to the desired pipe through a suitable interface plugged on the SDH NE. In the following sections of this chapter, we look at how the above services are supported and the interfaces involved in the system.

9.2 LAN to LAN Transport or Ethernet over SDH

9.2.1 Why LAN to LAN transport over SDH?

Organisations need interconnectivity for their LANs for units located in different cities or in far-flung locations within the same city. One option to meet this need is to use leased lines. If the leased line bandwidth is equal to rated speed of LAN, we can ensure smooth flow of data. However, the use of leased lines for LAN-to-LAN connectivity has the following disadvantages:

1. Wastage of bandwidth/resources as traffic flow for most of the time may not be to equal the full capacity of leased line
2. Tariffs for leased lines are high.

One can think of a 'pipe' with lower capacity on a network and incorporate flow control on the same to achieve the desired objective. SDH networks, with its containers of various capacities and concatenation options, make the 'pipe' selection possible. Refer to Fig. 9.1 illustrating LAN-to-LAN transportation service across SDH. Interconnectivity for LANs at locations A and H can be decided in terms of the required bandwidth. Based on the same, containers and concatenation options on SDH network can be worked out. In this scheme:

- The LANs can be 10 Mbps/100 Mbps/1 Gbps Ethernet.
- The bandwidth (BW) subscribed need not be equal to LAN data rate because the traffic between LANs at different locations may not be high at all times.
- SDH networks offer BW in terms of containers and their multiples (concatenations, e.g., N×VC-4, N×VC-3, N×VC-12).
- Each container can move over a common path or on different paths on the SDH network and have standby (protection) paths.
- If, for some reason, some containers are unavailable, the service can continue with reduced BW.

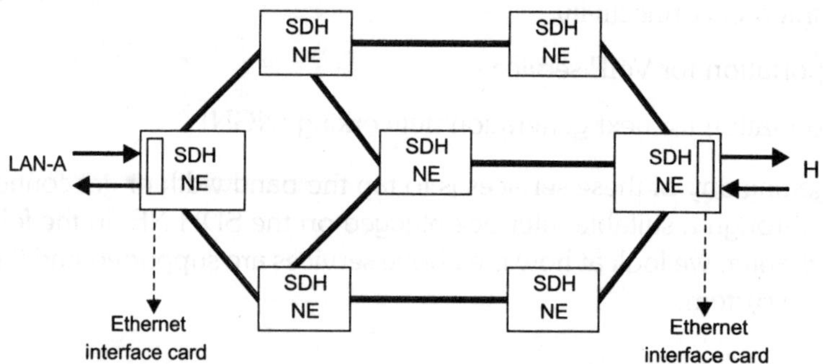

Fig. 9.1 *LAN-to-LAN Transportation over SDH*

9.2.2 What needs to be done for Ethernet over SDH

SDH network elements equipped with SDH to LAN mapping units known as 'Ethernet interface cards' can provide connectivity, LAN-to-LAN or Ethernet to Ethernet. However, this is an oversimplification of the design issues.

Carrying ETH over SDH has more to do than plugging-in 'ETH to SDH' boards at either end. The design issues involved are:

- Support for certain basic features
- Identification of SDH containers and end-to-end SDH path
- Mapping ETH to containers as per standard framing procedure
- Flow control
- Resolving latency issues
- Selection of appropriate boards/configuration

These are explained one by one.

Support for basic features

The basic features to be supported are:

Transparency: ETH frames are to be forwarded to output port with no protocol termination (whatever mapping we do is to be unmapped).

Low latency: Low end-to-end delay to be ensured. It is possible as SDH nodes in between facilitate continuous data flow.

High availability: Availability of LAN-to-LAN transport to be high (possible due to ring protection).

High quality: QoS in terms of BER, synchronisation, etc., to be high. (These are natural qualities of SDH networks.)

Identification of SDH containers and end-to-end path

Ethernet speed can be 10 Mbps/100 Mbps/1 Gbps. The SDH containers are identified depending on the desired bandwidth, and if the bandwidth requirement is not met with a single container, concatenation of containers to be resorted to. The available containers are VC-12, VC-3, VC-4 and the path to be set up as normal VC path or concatenated VC paths are N×VC-12; N×VC-3; N×VC-4.

Mapping of LAN's MAC traffic onto SDH frame

Ethernet's MAC frame (the frame handled at the media access control layer) is of variable length (min. length 64 bytes and max. length 1518 bytes) whereas SDH frame is of fixed length. Also, MAC data is discontinuous in nature whereas SDH traffic is continuous. Hence, Ethernet MAC frames can be mapped onto SDH virtual containers only through encapsulation algorithms. The encapsulation algorithm as per ITU-T Rec. G.7041 for mapping Ethernet MAC frames to SDH virtual containers is called the generic framing procedure (GFP). GFP is discussed in detail in section 9.2.3.

Flow control

To economise on resources and reduce tariffs, the connectivity provided can be less than Ethernet rate. This may not pose a problem most of the time as the traffic from LAN at one place to LAN at another place may not be high. However, a flow control mechanism has to be in place to tackle situations when the traffic shoots up. Flow control can be on the input side or the output side or both, or initiated externally. Flow control is discussed in detail in section 9.2.4.

Latency

While transporting LAN traffic across SDH network, certain delays are involved. These are: physical line delay on OFC cable, processing delay in SDH network element and ETH board latency. ETH board latency is the mapping (time) delay of Ethernet MAC frame onto SDH container/concatenated containers and de-mapping of the same back to Ethernet MAC frame. Buffer sizes chosen at input/output should be commensurate with the total delay accounted for in the network and should not result in any flow control violation. Latency is discussed in detail in section 9.2.5.

Selection of appropriate boards/configuration

The Ethernet interface card of the STM equipment is chosen according to the speed of the LAN (10 Mbps/100 Mbps/Gigabit) and the ports configured through EMS or NMS as per the containers and concatenation chosen. Also, the path has to be defined with the the desired protection option.

9.2.3 Generic framing procedure (GFP)

Ethernet's discrete and of variable length MAC frames are mapped onto the continuous and fixed length SDH containers as per GFP (generic framing procedure) defined in ITU-T's recommendation G. 7041. Before we examine this procedure in detail, let us recall IEEE 802.3 LAN's MAC frame format (refer to Fig. 9.2). The length of Ethernet MAC frame is variable from 64 bytes (minimum) to 1518 bytes (maximum). Preamble is the training sequence of alternate 1s and 0s that allows the physical signalling circuitry to reach its steady state synchronisation with the received frame's timing. The start frame delimiter or SFD of one byte (10101011) indicates the actual start of the frame. The minimum data field in Ethernet MAC frame is 46 bytes. If the information to be conveyed is less than 46 bytes, the 'pad' field is filled up to make the 'data + pad' length to 46 bytes. If the data field is 46 bytes or more, there is no need to fill up the pad field. Hence, the pad field is considered to be of size 0 to *n* bytes, where *n* is number of bytes needed to make 'data + pad' equal to 46 bytes. The maximum length of data field is 1500 bytes.

FCS or the frame check sequence is obtained through CRC-32 procedure (applied from destination address to pad). Preamble and SFD are not considered in the length of MAC frame. Hence, the minimum length of Ethernet MAC frame is the minimum length of data field plus other field lengths except preamble and SFD, i.e., 46 + 18 = 64 bytes. Similarly, the maximum length of Ethernet MAC frame is 1500 + 18 = 1518 bytes.

Preamble (7 bytes)	SFD (1 byte)	Destin. address (6 bytes)	Source address (6 bytes)	Length/ Type (2 bytes)	Data (Max. 1500 bytes)	Pad (0 to n bytes)	FCS (4 bytes)

Fig. 9.2 *IEEE 802.3 MAC frame format*

Let us understand now the requirements of GFP algorithm.

1. The GFP should support all three types of Ethernet, i.e., 10 Mbps/100 Mbps/1 Gbps Ethernet.
2. When there is no data to be transported at all, there is no MAC frame on the Ethernet. This is okay for a LAN. But, when such LANs are interconnected over SDH network, silence of LAN sends AIS on VCs of SDH which is undesirable. So, GFP should be able to send 'silence indicator' or a 'filler packet' in the case of absence of MAC frame.
3. The GFP should distinctly indicate normal packet (or its length) and filler packet through a field in the frame.
4. Whatever header is being added by GFP, it is preferable to have its own error check sequence since it is a fresh attachment.
5. If there is continuous flow of MAC frames of full length from the LAN, the GFP can attach its header for the first MAC frame and then give an extension indication to economise on the attachment of overheads.
6. There should be a provision to carry failures data and indicate the same to the far end.
7. Attachment of overall FCS can be optional as per media characteristics.

In accordance with the above requirements, GFP is defined for all three types of Ethernets. Also, for each type, three different GFP frame formats are defined:

- GFP without FCS
- GFP with FCS
- Extended GFP

GFP without FCS

Encapsulation of MAC frame using GFP without FCS is illustrated in Fig. 9.3(a), where,

PLI: Payload length indicator if the value is 0000H, no payload; it is a filler packet.

CHEC: Core header error control field (CRC-16)

Type: Type field of payload header

PTI – 3 bits (payload type identifier)
0: Frame has client data
1: Frame has far-end failures data

PFI – 1 bit (payload FCS indicator)
0: For GFP without FCS
1: For GFP with FCS

EXI – 4 bits – Extension header identifier

UPI – 8 bits – Used payload identifier

THEC: Type header error control field

PLI (2 bytes)	CHEC (2 bytes)	Type (1 byte)	UPI (1 byte)	THEC (2 bytes)	Payload field for ETH data

Fig. 9.3(a) *Frame format for GFP without FCS*

As can be seen, the GFP frame can carry MAC frame of variable length and indicate the same to the far end through PLI. If there is no MAC traffic, the GFP sends filler frame and indicates the same by setting PLI to 0000H. It can extend the framing to save on the overheads by setting the EXI field. It has exclusive error check field in the form of CHEC for the GFP header. GFP frame can also send failures data when there is no user traffic (it is better to do so rather than send filler bytes every time) and indicate the same through the PTI field.

GFP with FCS

The frame format for GFP with FCS is presented in Fig. 9.3(b).

PLI (2 bytes)	CHEC (2 bytes)	Type (1 byte)	UPI (1 byte)	THEC (2 bytes)	Payload field for ETH data	FCS (4 bytes)

Fig. 9.3(b) *Frame format for GFP with FCS*

The purpose of each field is explained below.

PLI: Payload length indicator (if value is 0000H, no payload; it is a filler packet).

CHEC: Core header error control field (CRC-16)

Type: Type field of payload header:

PTI – 3 bits (payload type identifier)
0: Frame has client data
1: Frame has far-end failures data

PFI – 1 bit (payload FCS indicator)
0: For GFP without FCS
1: For GFP with FCS

EXI – 4 bits – Extension header identifier

UPI – 8 bits – User payload identifier

THEC: Type header error control field

Payload field: 64 to 1518 bytes to carry Ethernet frame bytes

FCS: Frame check sequence for entire frame

It may be noted that GFP with FCS is functionally the same as GFP without FCS. The presence of four FCS bytes for overall frame is the only difference.

Extended GFP

The frame format of extended GFP is shown in Fig. 9.3(c).

PLI (2 bytes)	CHEC (2 bytes)	Type (1 byte)	Status (5 bytes)	THEC (2 bytes)	Payload (ETH data)

Fig. 9.3(c) *Frame format for extended GFP*

The purpose of each field is explained below.

PLI: Payload length indicator (if value is 0000H, no payload; it is a filler packet).

CHEC: Core header error control field (CRC-16)

Type: Type field of payload header

PTI – 3 bits (payload type identifier)
0: Frame has client data
1: Frame has far-end failures data

PFI – 1 bit (payload FCS indicator)
0: For GFP without FCS
1: For GFP with FCS

EXI – 4 bits – Extension header identifier

Status bytes: 1 for path, 2 for link and 2 for circuit

THEC: Type header error control field

Payload field: 64 to 1518 bytes to carry Ethernet frame bytes.

9.2.4 Flow control

As already mentioned in section 9.2.2, we are not providing for LAN-to-LAN connectivity at the rated speed of LAN. Though LAN-to-LAN traffic is less, at times it can be more, which leads to buffer saturation. Flow control is necessary to avoid such a situation. When excessive traffic is received, the transmitter has to be slowed down through a suitable control mechanism.

Flow control is required in the situations indicated in Table 9.1.

Table 9.1 *Flow control mechanism*

Situation	Flow control method
Input traffic is faster than uploading rate on SDH network	Input flow control
Downloading at destination is slower than traffic flow	Output flow control
Downloading at destination is slower and input traffic is still continuing	Output flow control and input flow control
LAN at destination has problem and requests pause in traffic	External flow control

Before we study the flow-control methods, let us understand the transportation of LAN traffic across SDH network. Refer to Fig. 9.4 illustrating buffering in LAN-to-LAN transportation. The sequence of operations is given below.

1. Client sink termination function adapts client interface.
2. All payload frames received are stored in the input buffer.
3. Source SDH adaptation function draws the frames from memory, emptying the cluster and making it available again.
4. Frames are then mapped to SDH VC applying GFP encapsulation.
5. SDH paths act as the traffic pipes carrying continuous data flow.
6. At the output port, output buffer is filled with payloads received from SDH pipes.
7. Client source at the output port draws from memory and forwards them to the client.

Buffer sizes are so chosen that they compensate for the delays (physical line delay + NE processing delays + ETH board latency) involved in receiving flow-control request. Buffer sizes used depend on the equipment design.

Since it is not possible to explain all models from different manufacturers, Alcatel's STM-16 equipment is taken as an example to explain the intricacies involved in flow control and latency issues and the buffer sizes.

Input and output buffer size : 80 frames without concatenation
: 528 frames with N×VC-3
: 2064 frames with N×VC-12

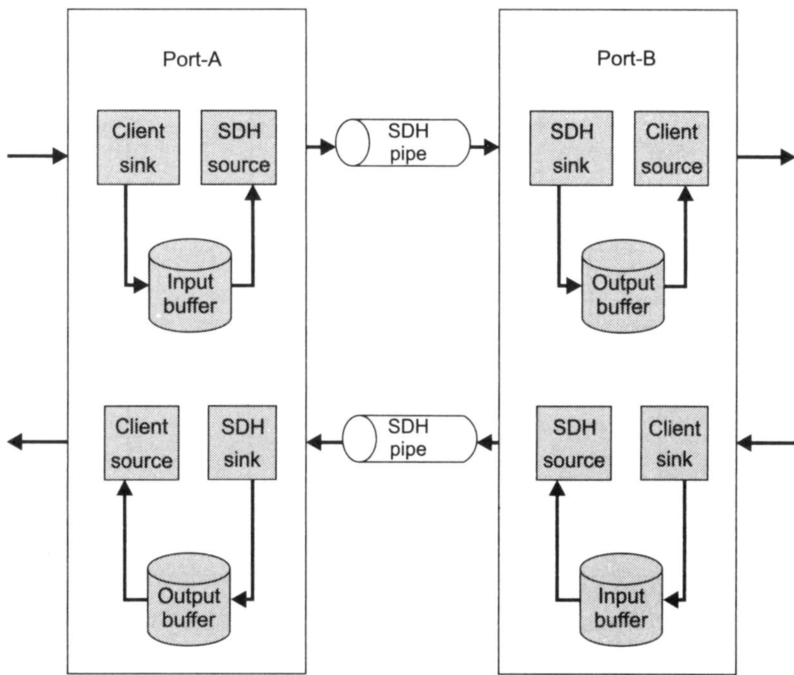

Fig. 9.4 *Buffering in LAN-to-LAN transport over SDH*

Input flow control

Refer Fig. 9.5 showing the input flow control mechanism. Input flow control is enabled in every GFP. It has the following features.

1. Size of the input buffer is 80 frames. When input buffer of port-A crosses a fixed threshold of 48 frames, a pause request is sent back to the transmitter.
2. The transmitter pauses till resume command is given.
3. There is time lag between identification of the threshold condition and the transmitter getting the command to pause. The 32 frames margin in the buffer (between full size 80 and threshold level 48) will hold any frames coming in the meanwhile.

4. No input frame is thrown away till 80 frames are filled, which is not possible as the transmitter would have paused by then.
5. Transmitter is requalified when the input buffer of port-A falls below the 44-frames threshold.

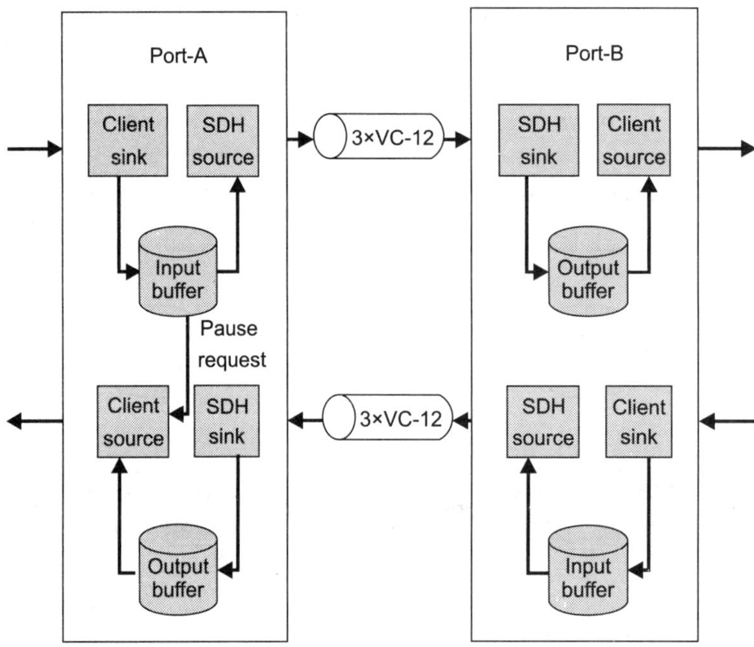

Fig. 9.5 *Input flow control in LAN-to-LAN transport over SDH*

Output flow control

Refer to Fig. 9.6 showing the output flow control mechanism. It is enabled in extended GFP only.

Features of the output flow control are:

1. The output buffer size is 80 frames without concatenation; 528 frames for VC-3 or N×VC-3 concatenation; 2064 frames for VC-12 or N×VC-12 concatenation.
2. When the output buffer of port-B crosses a fixed threshold (48 frames without concatenation or 272 frames for N×VC-3 or 1040 frames for N×VC-12), a pause request is sent back to the SDH source of port B and then through SDH pipe to the SDH source of port A.
3. This can cause queue filling of input buffer of port-A. So, input flow control is also used.

4. No output frame is lost till the filling up of output buffer, which is not possible as the transmitter would pause by then.
5. The Transmitter is re-qualified when the output buffer of port-B falls below resumption level (44 frames without concatenation or 240 frames for N×VC-3 or 1008 frames for N×VC-12).

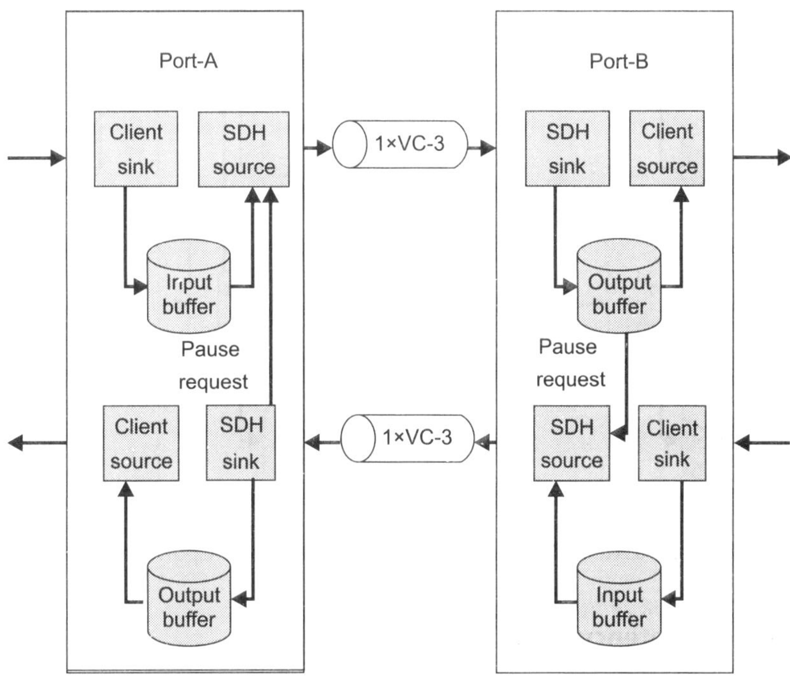

Fig. 9.6 *Output flow control in LAN to LAN transport over SDH*

External flow control

Refer to Fig. 9.7 showing external flow control mechanism performed at the output port. When a pause activity is requested from an external client, the output transmitter is stopped for a period. The output buffer levels prescribed for node-B ETH card in the output flow control hold good for client-B output buffer in external flow control. All types of equipment (NEs or their ETH cards) may not support external flow control. User needs to check the suitability. One question any serious reader would think of is what is the logic in choosing pause level and resumption level for various buffers?

Assume the worst-case scenario, i.e., the LAN is sending maximum-sized frames in quick succession. By the time the pause request reaches LAN (node in the LAN originating the frames), the LAN should not send more of them so as not to cause overflow of the buffer. The question boils down to finding the time taken for the pause request to reach the LAN node. For this, we require the latency or processing delay of each card for each type of container. This data is useful in answering not only the above question but also the question of overall differential delay between two LANs.

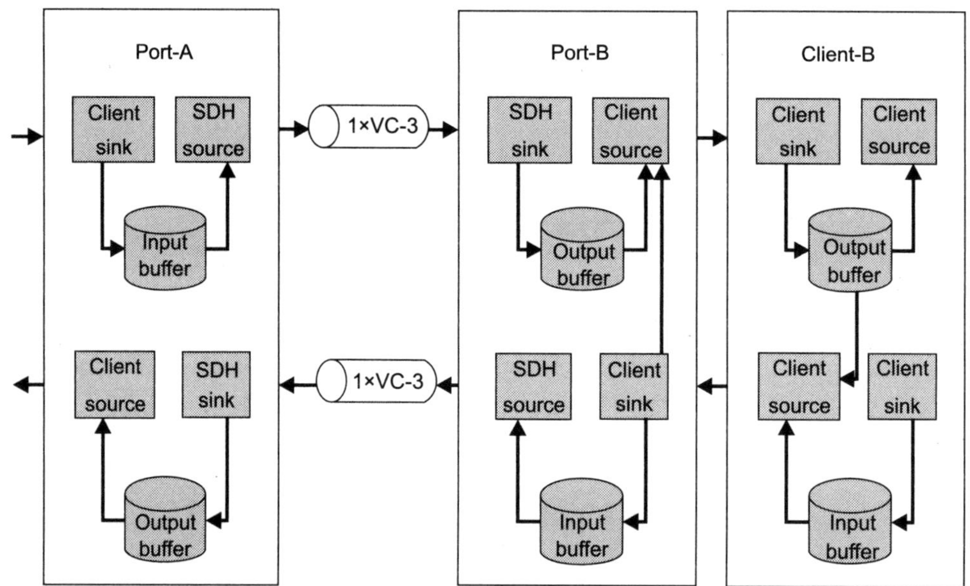

Fig. 9.7 *External flow control in LAN-to-LAN transport over SDH*

9.2.5 Resolving latency issues

As already explained in section 9.2.2, ETH board latency is the mapping (time) delay of Ethernet MAC frame onto SDH container/concatenated containers and de-mapping back to Ethernet MAC frame. Let us examine the latency data for each type of Ethernet board in Alcatel STM-16 equipment for various container options.

The latency data for gigabit Ethernet for VC-4 mapping is given in Table 9.2.

Table 9.2 *Latency data for VC-4 mapping of gigabit Ethernet frames*

Size of MAC frame being mapped	Latency
64 byte (minimum MAC frame)	193 μs
512 bytes (average MAC frame)	201 μs
1518 bytes (maximum MAC frame)	205 μs

The latency data for fast Ethernet for VC-4, VC-3 and VC-12 mapping is given in Table 9.3.

Table 9.3 *Latency data for VC-4, VC-3 and VC-12 mapping of fast Ethernet frames*

Size of MAC frame being mapped	Latency for VC-4	Latency for VC-3	Latency for VC-12
64 byte (min.)	101 μs	122 μs	517 μs
512 bytes (avg.)	125 μs	147 μs	2108 μs
1518 bytes (max.)	178 μs	317 μs	5841 μs

The latency data for mapping fast Ethernet frames on to concatenated containers is given in Table 9.4.

Table 9.4 *Latency data for 2×VC-3 and 8×VC-12 mapping of fast Ethernet frames*

Size of MAC frame being mapped	Latency for 2×VC-3	Latency for 8×VC-12
64 byte (min.)	277 μs	657 μs
512 bytes (avg.)	1462 μs	3452 μs
1518 bytes (max.)	3997 μs	5824 μs

The following series of queries on end-to-end delay, latency issues and buffer sizes make these concepts clear. The buffer sizes with respect to end-to-end delays also get verified.

Q 9.1 Compute the end-to-end delay for LAN-to-LAN transport (fast Ethernet) over 400 km of network with 20 NEs using different ETH board and concatenation options.

Physical line delay = 400 km × 5 μs/km = 2000 μs

NE delay = 20 NE × 10 μs/NE = 200 μs

For end-to-end delay, we consider an average size MAC frame as LAN-to-LAN traffic is intermittent and frame size is variable. For fast Ethernet, latency data (for average MAC frame) are:

VC-4 : 125 μs
VC-3 : 147 μs
VC-12 : 2108 μs
2×VC-3 : 1462 μs
8×VC-12 : 3452 μs

Total end-to-end delays for various cases are obtained by adding 2000 + 200 μs to respective latencies. The results are:

VC-4 : 2325 ms
VC-3 : 2347 ms
VC-12 : 4308 ms
2×VC-3 : 3662 ms
8×VC-12 : 5652 ms

Q 9.2 *What is the best choice for fast Ethernet connection?*

Use of VC-4 for fast Ethernet is a waste of bandwidth.The allowed differential delay given in Table 9.1 is exceeded in case of 2×VC-3, hence it is not suitable. Single VC-12 implies flow control is too often, and hence it is not considered. Among 8×VC-12 or 2×VC-3 or VC-3 or VC-12, the one taking the least bandwidth is VC-12. In customer-specific situations, where high data flow is expected, VC-3 or 2×VC-3 can be considered.

Q 9.3 *What are the advantages with concatenation?*

Concatenation not only increases bandwidth but also gives the options of packet concatenation and virtual concatenation.

In packet concatenation, MAC frames are put frame by frame onto VCs in the concatenated group in rotation and the VCs are sent on different paths. In case of non-availability of any path, fewer VCs (of available paths) will be used and there will be no disruption to traffic.

In virtual concatenation, MAC frames are put byte by byte onto VCs in the concatenated group in rotation and these VCs are sent on the same path. Ring protection of path ensures that there is no disruption to traffic (unless there are interruptions on both work and protection paths).

Q 9.4 *In packet concatenation, when frames travel on different paths, are there any problems associated with different delays on different paths?*

The allowed differential delays are as per choice of the container. According to ITU-T Rec. G.783, these are as given in Table 9.5.

Table 9.5 *Allowed differential delay for various containers*

Container used	Allowed differential delay
VC-4	3 ms
VC-3	35 ms
VC-12	256 ms

Q 9.5 Coming back to the very first point of the deliberations, what is the logic in choosing buffer sizes, pause level and resume level for the buffers used in the input and output flow controls?

Transmission time for each maximum size MAC frame =

Frame size/Data speed = 1518 × 8 /100 × 10^6 = 121.44 μs

Input flow control without concatenation

From pause level to full size, we have 32 frames. So, a pause request has to be noticed and frame transmission stopped before the cushion of 32 frames is filled-up. The time taken to notice a pause request at the input port is the de-mapping delay, which is half the latency time.

When there is no concatenation, we notice from Table 9.2 that the maximum latency is for VC-12 at 5841 μs. So, the maximum time to notice a pause request is 5841/2 = 2920.5 or say 2921 μs.

Taking the time taken to fill-up one frame as 121.44 μs, the time for filling up the 32-frames cushion is 121.44 × 32 = 3886.08, or say 3886 μs, which is more than the pause request notice time of 2921 μs. Hence the choice of buffer size is okay.

It may please be noted that the pause level is 48 frames and resume level 44 frames to avoid hysteresis in the pausing–resuming–pausing action. Also, if the transmission is not resumed quickly after falling below the set threshold, too many frames will queue up at the client sink. To avoid this, the resume level is chosen close to the pause level.

Output flow control without concatenation

From the previous discussion it is clear that notification time for pause request is 2921 μs without concatenation. In case of output flow-control, such notification should come from the output buffer back to the client source. So, we have to add end-to-end delay also. This means that in the cushion time of 3886 μs available, we are left with only 3886–2921 = 965 μs only for end-to-end delay. Such low value is possible in only small scale networks.

As already mentioned, output flow control is enabled for extended GFP only, where traffic is heavy. Hence, it is clear that concatenation option along with output flow control is a must in extended GFP.

Q 9.6 Can we verify suitability of buffer sizes for various concatenation options?

Let us consider N×VC-3. In this case, with the full buffer size as 528, pause level as 272, the cushion available is 256 frames. In terms of time, this is 256 × 121.44 μs = 31088.64 μs.

De-mapping time for pause request = ½ (Board latency for N×VC-3)

= ½ × 3997 (from Table 9.4) = 1998.5 μs.

The available margin for end-to-end delay on network is 31088.64 – 1998.5 = 29090.14 μs, which is a very good margin!

In the case of N×VC-12, with full buffer size 2064, pause level as 1040, the cushion available is 1024 frames. In terms of time, this is 1024 × 121.44 μs = 124354 μs.

De-mapping time for pause request = ½ (Board latency for N×VC-12)
= ½ × 5824 (from Table 9.3) = 2912 μs.

The available margin for end-to-end delay = 124354 – 2912 =121442 μs, which is a fantastic margin!

Q 9.7 Why can't we resort to packet concatenation for gigabit Ethernet while using VC-4?

Messages require realignment in packet concatenation, which in turn require large buffers especially while handling gigabit traffic on large containers like VC-4. Hence, we use only virtual concatenation in gigabit Ethernet.

9.2.6 Selection of ETH boards and their configurations

STM equipment manufacturers support three types of Ethernet interfaces:

10 Mbps Ethernet interface
100 Mbps Ethernet interface
1 Gbps Ethernet interface

Each card has a number of ports. Therefore, more than one Ethernet interface can be provided end-to-end, i.e., more LANs (of different customers) can be interconnected across the same pair of ports, as also different LANs of one customer at more than two places.

The total BW that can be derived at an interface is within the available SDH NW bandwidth. After checking for total latency (physical line delay + NEs delay + latency of ETH board for concatenation options under consideration), we can finalise the choice of concatenation. The configuration to be adopted card-wise and port-wise for concatenation, the GFP and the flow control options also are to be chosen by observing the performance of the interface.

9.3 MPLS over SDH Network

MPLS stands for multi protocol label switching. Though it was designed initially to enhance route look-up speed by a router, today, it is essentially used for traffic engineering. In this section we discuss the evolution of MPLS, operations in MPLS and the kind of support it receives from SDH.

9.3.1 Evolution of MPLS

Traditional routers performed route look-up via a software which was not competent to handle traffic in large-scale, large-bandwidth networks. For efficient routing at layer-3 in large meshed IP networks, routing by IP switching was felt necessary. The Internet Engineering Task Force (IETF) studied the available routing solutions and adopted ATM concepts for IP routing. In this approach, the ATM switches perform traffic forwarding in hardware (not software), i.e., an overlay network solution. The forwarding paths are pre-established through the ATM switch network and the traffic flows are directed based on the ATM cell header. Four different flavours were developed in this approach:

- LANE (ATM LAN emulation)
- IPOA (IP over ATM)
- NHRP (next hop resolution protocol)
- MPOA (multi-protocol over ATM)

The above approach had the following drawbacks. It required specialists in ATM and IP to run the setup—switches and routers could not share the responsibility for engineering the traffic flows. Besides, there was considerable wastage of bandwidth. The ATM cell had fixed size of 53 octets (5-byte header and 48-byte payload); the transmission of ATM header implied significant wastage of bandwidth. In one cell it was 10%, and for a 64-byte IP packet, two ATM cells which were needed implied that of the 106 bytes, 42 bytes or about 40% bandwidth was wasted. On an average, about 20% bandwidth was wasted.

Despite the drawbacks, ISPs were willing to go for the above approach as it offered higher interface speeds.

With the advent of SDH in the 1990s, bandwidths of STM-1 and higher order became available. Though ATM could be taken over SDH (as described in section 7.7, the hardware responsible for the segmentation and re-assembly (SAR) was costly for higher-order STM data rates. Also, the use of ATM at higher order STM rates implied large scale wastage of bandwidth–at the STM-16 data rate, 20% wastage meant about 500 Mbps!) Moreover, by end of 1990s, routers with STM channeling were being manufactured.

So, IETF, borrowing the ATM cell header idea, proposed the strategy of using tag on packet for aggregate route-based IP switching or MPLS. This is essentially the use of tag or fixed label on every IP packet to facilitate look-up faster than the longest match look-up used by IP. Each label is used as look-up index on a 'label forwarding table' to obtain forwarding information, and the packet is assigned an LSP (label switched path)

Before MPLS could find widespread production and deployment, there was yet another development—manufacturers of routers started using ASIC (application specific ICs) to quicken routing, i.e., routing in hardware or routers with speeds of switches. This could have made MPLS extinct, but it survived on the following strengths:

1. MPLS provides a clear separation between routing, i.e., control and forwarding or moving of data. This segregation allows the use of single forwarding algorithm for multiple services and traffic types.

2. MPLS is futuristic, i.e., should ISPs generate new revenue generating applications, MPLS forwarding infrastructure can remain the same and new services can be built by simply changing the way the packets are assigned to LSPs based on a combination of destination address and type, QoS requirements, IP multicast group and VPN identifier, to mention a few.
3. MPLS facilitates 'traffic engineering'. Traffic engineering is the process of selecting the paths chosen by the data traffic in order to balance the traffic load on various links, routers and switches in the network. In MPLS, the router analyses the packet header just once–before the packet enters the engineered path. Based on this analysis, the router assigns a fixed length label (32-bit MPLS header, as we will see later in this section) before the IP header. Routers in the path use labels as look-up indexes on the label-forwarding table and decide on the exit interface and label operation.

With this idea of MPLS, let us look into MPLS terminology and operation.

9.3.2 MPLS terminology and operation

Figure 9.8 shows a network with MPLS implementation.

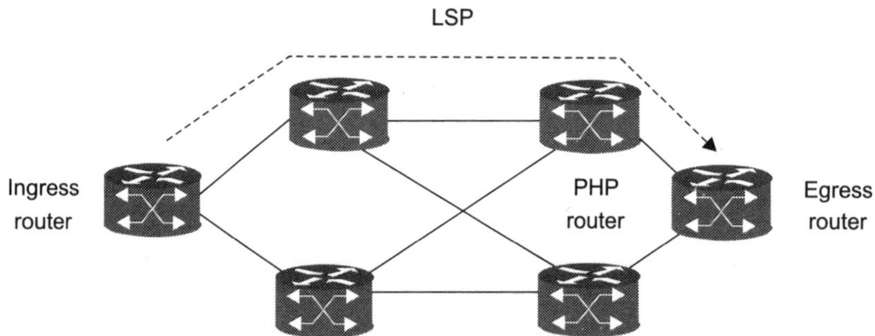

Fig. 9.8 MPLS network and terminology

Label switched path (LSP)

Each network path created by MPLS protocol is a label-switched path (LSP). It is unidirectional unlike the ATM VC, which is bi-directional. Because of the unidirectional nature of LSP, we have to establish LSPs in both directions, which may seem like an additional burden. But, this gives us the total control of the traffic engineering in the network.

Label switched router (LSR)

Each IP router that supports MPLS protocol is called a label switched router (LSR). An LSR understands MPLS header and the encoding carried in it. LSR forwards the packets through established LSP.

There are four types of LSRs.

Ingress router: It is the entry point router for LSP. It is obvious that for any LSP there is only one ingress router.

Transit router: All routers located in the middle of an LSP are called transit routers. Since the TTL or time to live field is 8-bit in size, the maximum number of hops is 255. Leaving out ingress and egress routers, we have 253 transit routers in the LSP.

Egress router: It is the end point of LSP. Since it is end point of LSP, IP route look up is to be performed by it. For any LSP there is only one egress router.

Penultimate router: As the name suggests, it is last but one router on the LSP. It has to perform label pop operation to remove MPLS header and after consulting MPLS switching table, forward the remaining packet, which is the normal IP packet, to the egress router.

Q 9.8 Why have a PHP router? Can't we do these jobs at the egress router?

A router can be an egress router for many LSPs terminating on it. The egress router has to work on the IP address. If the strategy of PHP router is not in place, the egress router will have to perform two operations, viz. MPLS label pop and IP look-up to forward packets on all the LSPs terminating on it. This can be severe burden because even in a small-to-medium-scale network, there can be 50 to few hundreds of LSPs terminating on the egress router. It defeats the very purpose of MPLS.

MPLS header

Figure 9.9 shows the 32 bit long MPLS header.

L2 Header	MPLS header (32 bit)	IP packet

Label (20 bits)	CoS (3 bits)	S (1 bit)	TTL (8 bits)

Fig. 9.9 *MPLS Header structure*

MPLS header consists of the following four fields:

Label (20 bits): Identifies the packet to a particular LSP. As the packet flows from one LSR to the next on the LSP, this label changes.

CoS (3 bits): These three bits were intended for class of service bits, but so far, they have not been determined. They are used to indicate queuing priority.

Bottom of stack bit (1 bit): This field facilitates the use of stack of labels. It is set to 1 for the bottommost label in the stack. When LSR finds this bit as 1, it means that after popping the label, no more MPLS labels and unlabelled packets remain. In other words, if S=1, it implies that the IP header is following the MPLS header; S=0 implies stack label is present.

Time to live or TTL (8 bits): This field contains a limit on the number of router hops the MPLS packet can travel. It gets decremented on each hop and if it falls below 1, the packet gets dropped.

Q 9.9 How does TTL facilitate MPLS operation?

TTL prevents looping of MPLS packets in the network. If this field is not there, it is not known when the egress (exit) point is reached. This field is same as TTL field of IP header. An IP header's TTL value is copied at entry point of the LSP, and at the exit point, where it has already become 1, it is copied back to IP header. So, the exit point router knows that it is the last hop.

Before we go into the details of the usage of labels, let us discuss packet processing in MPLS.

Packet processing in MPLS

Figure 9.10 shows packet forwarding in an MPLS network. As can be seen, the LSP is from Mumbai to Kolkata. All down stream routers have assigned label values to their upstream neighbours. Kolkata assigned 3 to Visakhapatnam, a PHP router. Visakhapatnam has assigned 900 to Vijayawada, Vijayawada has assigned 600 to Secunderabad, Secunderabad has assigned 500 to Pune and Pune has assigned 300 to Mumbai. IP packet forwarding takes place as described below.

1. IP packet arrives at ingress router of Mumbai.
2. Mumbai router performs IP route look up on the destination IP address. Due to MPLS protocol enabling, it finds that the next hop for the route is the LSP to Pune. An MPLS header is added to the packet with label 300 and the packet is forwarded to Pune.
3. Pune router receives MPLS packet with label 300. It performs look-up on the MPLS forwarding table and finds a swap operation. In a swap operation, label 300 is removed and replaced with label 500. The packet gets forwarded to Secunderabad.
4. Secunderabad router receives MPLS packet with label 500. It performs look-up on the MPLS forwarding table and finds a swap operation. In the swap operation, label 500 is removed and replaced with label 600. The packet gets forwarded to Vijayawada.

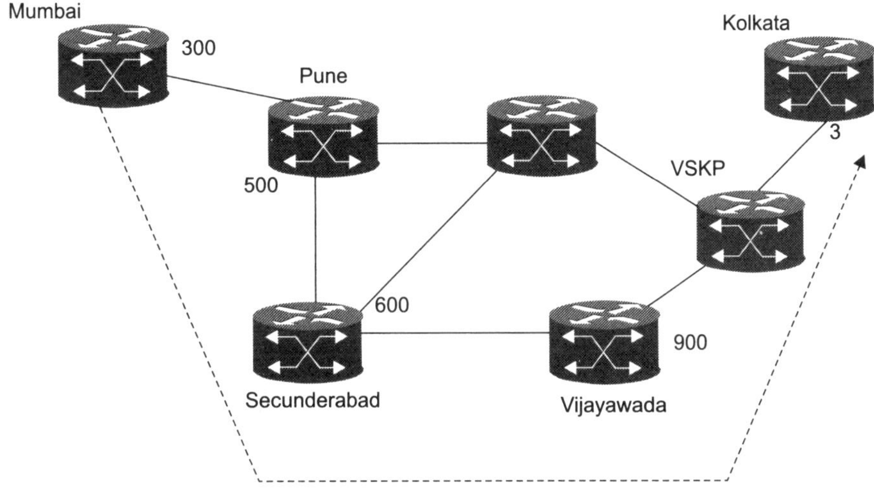

Fig. 9.10 *Packet forwarding in MPLS*

5. Vijayawada router receives MPLS packet with label 600. It performs look-up on the MPLS forwarding table and finds a swap operation. In the swap operation, label 600 is removed and replaced with label 900. The packet gets forwarded to Visakhapatnam (VSKP).
6. Visakhapatnam router receives the MPLS packet with label 900. It performs look-up on the MPLS forwarding table and finds a pop operation (since Visakhapatnam router got label 3 assignment by its downstream router and hence it is a PHP router). So, Visakhapatnam router removes the MPLS header from the packet and forwards the IP packet to Kolkata.
7. Kolkata router receives IP packet. It performs a routing table look-up and forwards the packet to the next hop, say the customer edge router.

Label values as per IETF recommendations

Labels 0 to 15 are reserved as per IETF's RFC3032 (see Table 9.6). Labels 16 to 1023 and 10,000 to 99,999 are reserved for statically assigned labels.

Stacking labels

MPLS supports 'tunnels within tunnels' through stacking labels. Label stacking allows an LSR to switch a bundle of individual LSPs without having to carry state information about the individual LSPs which are in the bundle.

Static and dynamic LSP

MPLS label switched path is established by static or dynamic method. A static LSP is established by configuring each router along the LSP explicitly. At the time of configuration, it is decided which way the LSP should lead and what labels are to be used, and the resources are assigned accordingly. No signalling protocol is operational in static LSP and no state information is required to be maintained. Static LSP is simple to operate and requires few resources.

Dynamic LSP is created and maintained by signalling protocol without user intervention. Only the ingress router is configured with the information pertaining to LSP; all other LSRs receive signalling messages during the establishment process.

The user's control on the setup parameters of LSP depends on the signalling protocol used. There are two such protocols: RSVP (resource reservation protocol) and LDP (label distribution protocol).

Table 9.6 *Label values as per IETF's RFC3032*

Label value	Nomenclature	Purpose
0	Explicit null for IPv4	Indicates that the label stack must be popped and the forwarding of packet is to be based on IPv4.
1	Router alert label	It provides a means of communication between MPLS modules in different routers.
2	Explicit null for IPv6	Indicates that the label stack must be popped and the forwarding of packet is to be based on IPv6.
3	Implicit null	This is a label that LSR can assign and distribute but does not appear in encapsulation.
4 to 15	Reserved	Reserved for future use.

9.3.3 Dynamic LSP using RSVP

RSVP is a generic signalling protocol designed by IETF to be used by applications to request and reserve resources hop by hop across the internetwork, specific to quality of service (QoS) requirements. RSVP was designed in such a way that end hosts reserve capacity in network. Each router receives the resource reservation request, establishes and maintains the necessary state for the data flow (if the requested resources are available) and forwards the resource reservation request to the next router along the path.

In RSVP, the ingress router initiates a '*path*' message and sends it downstream to the egress router. It contains information about the required resources for the connection. In accordance with this information, each router en-route maintains a database of the intended current reservations, known as '*soft state*'.

When the 'path' message reaches the egress router, the actual reservation of the resources begins. The egress router sends '*Resv*' message upstream to the ingress router. As each router en-route receives the Resv message, its soft state information is updated to indicate firm reservation of resources for the LSP. Once the ingress router receives the Resv message, an LSP is established. An established LSP remains operational as long as soft state stays active.

Q 9.10 *How are the soft states of routers retained in RSVP?*

Each router sends *Path* and *Resv* messages to its neighbours for all current states every 30 seconds. This informs the neighbours of the active paths and ensures maintenance of their own soft states.

In addition, RSVP messages shown in Table 9.7 are used for the indicated purposes.

Table 9.7 *RSVP messages and their purposes*

Message	Direction	Purpose
Path tear	Downstream to egress router	To remove established *Path* soft state for all routers receiving this message
Resv tear	Upstream to ingress router	To remove established *Resv* soft state for all routers receiving this message
Path err	Upstream to ingress router	To denote an error along established path (but no soft state is removed by routers receiving the message)
Resv err	Downstream to egress router	To denote an error along established path (but no soft state is removed by routers receiving the message)
Resv conf	Sent to egress router from each router along the LSP when asked by egress router in its *resv* message.	To confirm that *Resv* message was received

MPLS extensions to RSVP

For MPLS implementation, RSVP not only provides router-to-router signalling but also message contents called *objects*. These objects are included in the exchange of path and reservation messages as they make their way back and forth from the ingress router to the egress router in the MPLS domain. There are more than 50 objects for MPLS implementation. Prominent among them are given in Table 9.8. (A discussion of all the 50 is beyond the scope of the book.)

9.3.4 LDP (label distribution protocol)

As discussed in the previous section, RSVP is a generic protocol with extensions for MPLS implementation. LDP was designed specifically for MPLS.

LDP detects neighbours using *Hello* protocol. Once a neighbour is found, an LDP router establishes a TCP session with that peer to exchange label information. Each set of established LDP peers generates a set of labels for inclusion in the LDP database. The LDP database is advertised throughout the network. Every LDP router consults the LDP database to find out the label required to reach every other router in the network.

Table 9.8 *MPLS objects in RSVP*

Object	Purpose	Facility we get
Explicit route object (ERO)	Used to tell downstream routers how to setup the LSP to the destination. ERO can be strict or loose.	Facilitates to force the RSVP-signalled LSP to go through certain nodes along the path to the egress router, i.e., imparts traffic engineering capability
Label request object (LRO)	Used to request label on a segment from the downstream router in the downstream path message.	When a path message containing label request arrives at an LSR, the LSR allocates a label and stores it as part of the path state. When the corresponding reservation message arrives, the label is placed in its label object. Path messages are ingress to egress, and turn around and come as reservation messages from egress to ingress routers. Thus, these two objects support distribution of labels from downstream nodes to upstream nodes.
Label object (LO)	Used to report the label value chosen in the reservation message sent to the upstream router.	
Record route object (RRO)	Used to record the actual sequence of routers and IP addresses that the LSP uses to reach the destination.	When RRO is present in a path message, each node along the downstream path to the egress LSR records its outgoing IP address in the RRO. IF RRO finds itself recording same IP address over and again, a loop in LSP is detected and error message sent to LSR and path message is dropped.
Session attribute object (SAO)	Used to carry the control information for the LSP as it is set up as well as the name for LSP.	It is useful to signal the LSP's priority, preemption and fast re-route protection option.
T spec object (TSO)	Adapted to carry the details of the LSP for link management.	Used to indicate requested bandwidth as well as min. and max. packet size to be used over LSP.

Q 9.11 *In which layer is MPLS implemented?*

The MPLS implementation using MPLS header with the label structure as explained earlier is all right within a single autonomous system. It is a layer-2 implementation. If MPLS VPN is to be implemented across different autonomous systems (different service providers), it has to be L3 VPN. These two methods of MPLS implementation viz. layer-2 implementation and layer-3 implementation using LDP are discussed in the next section.

9.3.5 Layer-2 implementation of MPLS

Layer-2 implementation of MPLS is depicted in Fig. 9.11. Terms/entities associated with layer-2 implementation of MPLS and their meanings/functionalities are explained below.

Attachment circuit (AC)

It is the physical or virtual circuit attaching a CE device (layer-2 switch or router) to a PE router. AC can be frame relay, VC on SDH, Ethernet port, Ethernet VLAN, ATM VC, ATM virtual path, HDLC circuit or even simply a PPP connection.

Pseudo-wire (PW)

PW is a connection between two PEs (provider edge devices) that connects two ACs of the same type or of different types. PW carries the actual layer-2 traffic from the AC by encapsulating it in the MPLS.

Provider equipment (PE)

Pseudo-wire edge equipment, also called provider equipment (PE), provides end customer attachment and service delivery. Edge routers are PEs. The PE at the entry end is known as ingress router and that at the exit end the egress router.

Tunnel LSP

Pseudo-wires are supported on tunnel label-switched path (LSP) in each direction between two pseudo-wire-capable edge devices. The situation is like having a tunnel between two cities to avoid traffic congestion with multiple roads in the tunnel. Different traffics originating in the city take different roads in the tunnel.

Each tunnel LSP carries multiple PWs identifiable by their PwiDs (pseudo-wise identifier). Each PW is a virtual circuit. Let us say, there are *N* different customers requiring connectivity between sites A and D. As and when the traffic of each customer originates a session, one of the pseudo wires is allotted as virtual circuit or LSP. When the allotted PW usage is over for that session, it is available for allotment to a different customer/session. In the case of SDH, the tunnel is an HO container and PW is an LO container. Thus, the containers are no longer leased but shared.

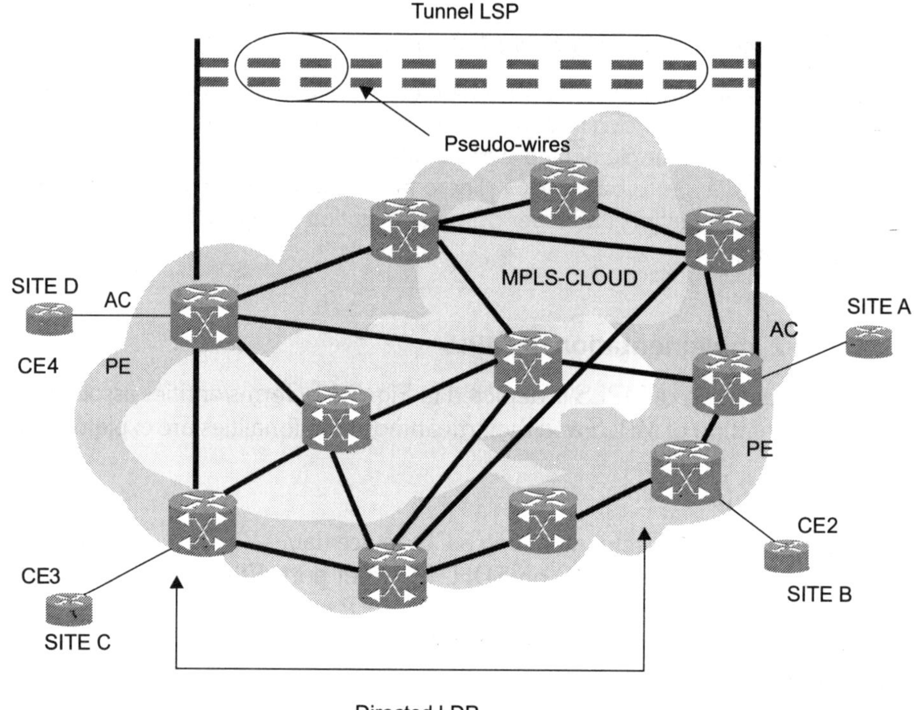

Fig. 9.11 *Layer-2 implementation of MPLS*

An MPLS label identifies the 'tunnel LSP and PWiD and signals them between a given pair of ingress and egress PE devices.

Tunnel LSPs are created by:

- Distributing labels automatically using label distribution protocol (LDP) or border gateway protocol (BGP).
- Deploying traffic engineered tunnels.
- Setting static labels along the path between PEs.

Directed LDP session

A directed LDP session runs between PEs that provide pseudo-wire services. This service provides the signalling for pseudo-wire service setup and status monitoring.

Virtual circuit label

It is a unique, locally assigned label for each AC. Virtual circuit labels are signalled across the directed LDP sessions to the appropriate remote PE device.

Pseudo-wire identifier (PWid)

It identifies each pseudo-wire uniquely through a unique 32 bit ID.

Pseudo-wire forwarding equivalent class (FEC)

A pseudo wire setup is signalled between PE devices using LDP downstream session. An LDP label mapping message consisting of 320 bytes of information (comprising tunnel label, PW type, PWiD, FEC TLV (type length value), label TLV, etc.) is sent from the ingress PE to the egress PE. Further details of LDP label mapping message and FEC TLV are given in Appendix B.

How is PW created?

First of all, ACs (attachment circuits) are to be established at each PE router through command line interface of router or through EMS. Having established ACs, pseudo-wire connections need to be configured, including end-points of directed LDP session (i.e., ingress and egress points of PW). There are five steps in this process as illustrated in Fig. 9.12(a). These are:

1. PW configuration entered on ingress PE.
2. Ingress PE starts LDP session with egress PE (PE2) if one does not already exist.
3. PE1 allocates VC label for new interface and binds to configured PWid.
4. PE1 sends label mapping message containing PW FEC TLV and label TLV as described under pseudo-wire forwarding equivalent class.
5. PE2 receives PW FEC TLV and label TLV that matches local PWid.

It may be noted that PE2 also carries out the above steps so that bidirectional label/PWid mappings are established. The format of LDP message sent by PE to convey information for a pseudo-wire is shown in Fig. 9.12(b).

Pseudo-wire packet (frame) flow

The steps in creation of pseudo-wire packet and its flow are:

1. MPLS header is prefixed to a packet with a *push* operation at the ingress node. The label is added immediately after layer 2 encapsulation header.
2. The PDU comes from CE1 to PE1 on an AC.
3. PDU + VC label is sent to the core PE.
4. Tunnel label (outer label) is attached to the core PE to identify the tunnel LSP.
5. PDU + VC label + tunnel label goes on to the tunnel LSP's PW.
6. At the egress node, labels are noted and removed. Based on the noted labels, the PDU is put on the concerned AC.

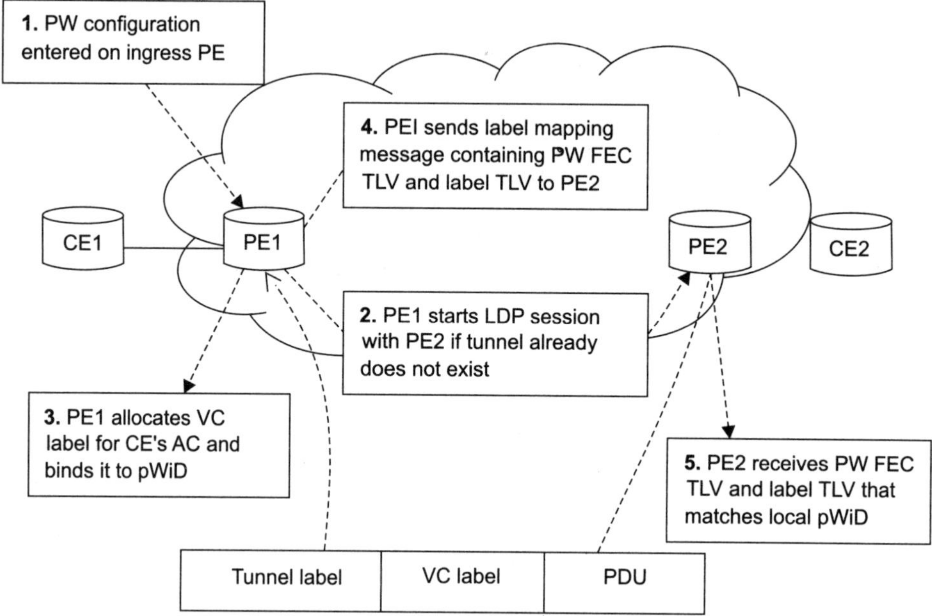

Fig. 9.12(a) *Pseudo-wire (PW) signalling and setup*

Fields				Purpose
0 (2 bit)	Label mapping (24 bit: 0×0400)	Message length (16 bits)		LDP label mapping message
Message ID (32 bit)				
0 (2 bit)	FEC (24 bit: 0×0100)	Length (16 bit)		FEC TLV header
PW TLV (8 bits)	C (1 bit)	PW-Type (15 bits)	PW Info length (8 bits)	PWid FEC element
Group ID (32 bit)				
PW ID (32 bit)				
Interface parameters				
0 (2 bit)	Generic Label (24 bit: 0×0200)	Length (16 bits)		Label TLV header
Label				
Optional parameters				Optional

Fig. 9.12(b) *Format of LDP message sent by PE to ccnvey PW information*

9.3.6 Layer-3 implementation of MPLS

Layer-3 implementation is shown in Fig. 9.13. Terms/entities in layer-3 implementation of MPLS and their meanings/functions are as follows.

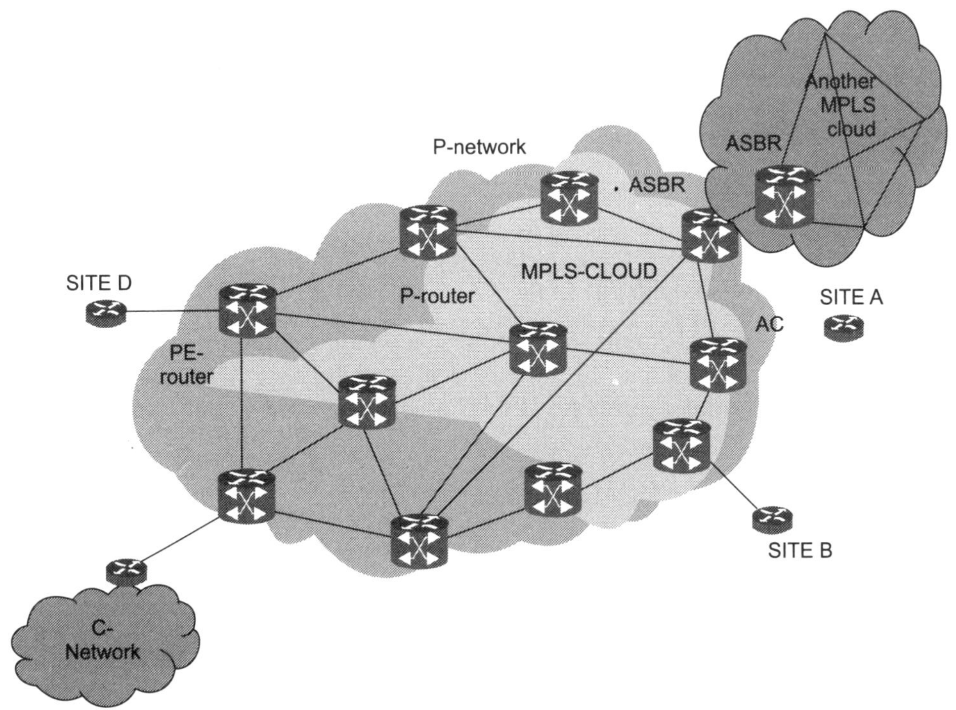

Fig. 9.13 *Layer-3 implementation of MPLS*

Provider network (P-network) is the core MPLS-IP network. Autonomous networks can also be interconnected.

Provider router (P-router) is an MPLS/IP router deployed within the MPLS network with no edge attachments.

Provider edge router (PE-router) is the service provider's edge router that provides VPN end customer attachment and service delivery.

Autonomous system boundary router (ASBR router) provides attachment between autonomous routers.

Customer network (C-network) is administered by end-user attached to layer-3 MPLS VPN service.

Customer edge router (CE-router) provides a gateway between C-network and P-network. It can be administered by user/service provider.

Stages in layer-3 implementation of MPLS

Layer-3 implementation of MPLS involves the following:

1. Creation of customer-specific VRF in PE
2. Customer-to-service-provider routing exchange
3. Label allocation at PE router
4. Advertisement of route across IP/MPLS backbone
5. Import of remote routing information into VRFs
6. Forwarding layer-3 MPLS VPN packets

Creation of customer-specific VRF in PE

Since a PE router supports multiple customers, separation of their routing states or instances is a must. In PE routers, customer-specific routing information is stored in VRF (virtual routing/forwarding instance). VRF is a separate virtual router inside a router with the following information:

- IP packet forwarding and routing table
- A set of interfaces that use the forwarding table
- A set of rules that control the import and export of routes to and from the VPN-specific routing table
- A set of routing protocol peers that inject routes into the VPN-specific routing table

The above segregation of VRF customerwise in the PE gives the following benefits:

1. It will not be possible for one VPN customer to gain access to another.
2. As the customer uses his own registered IP address space or allotted addresses by service provider, his routing information does not clash with that of another customer supported by the service provider. Among the routes listed for him in the VRF, the best ranked one can be picked up whenever the packet flow starts.

Customer-to-service-provider routing exchange

Once the CE router has physical connectivity with the PE router, it has to exchange routing information with the service provider. This is possible in two ways, static and dynamic configuration, as already discussed in section 9.3.2.

It may be noted that a PE can run multiple dynamic routing protocols of different customers concurrently. It is not necessary to have the same type of dynamic routing protocols at a PE for all customers.

Label allocation at PE router

To forward layer-3 MPLS VPN packets, the PE router has to refer to the context of incoming packets from the attached customer sites or labels of packets coming from other routers in the backbone network.

For the packets from the customer site, label allocation can be in two ways:

i. If the customer has single connection to the PE router, a single label is attached.

ii. If the customer has several attachments, a label that represents the entire VRF is allocated.

For packets from other routers, labels are examined and then the packets are forwarded.

Advertisement of route across IP/MPLS backbone

Having received routing information from the attached customer sites through dynamic routing protocol or static configuration, a PE router has to send the routing information to other PE routers in the network for subsequent distribution to remote customer sites in the same VPN. Since multiple customers use the same address space, PE creates VPN prefix to the IP addresses (it is 64 bit route distinguisher) as shown in Fig. 9.14.

Route distinguisher (64 bits)	IPv4 address (32 bits)

Fig. 9.14 *Route distinguisher attachment*

Route distinguisher has different formats as shown in Fig. 9.15(a–c).

i. 2 bytes for the type, 2 bytes for asynchronous system number (ASN), 4 bytes for local administration assigned number (LAAN).

ii. 2 bytes for type, 4 for IPv4 address, 2 bytes for assigned number.

iii. 2 bytes for the type, 4 bytes for asynchronous system number, 2 bytes for assigned number.

Type (0 × 00)	Autonomous system number (2 bytes)
Local administrator assigned number (4 bytes)	

Fig. 9.15(a) *2-octet ASN with 4-octet assigned number*

Type (0 × 01)	IP address
IP address (contd.)	Assigned number (2 bytes)

Fig. 9.15(b) *Ipv4 address with 2-octet assigned number*

Type (0 × 02)	Autonomous system number (2 bytes)
Autonomous system number (contd.)	Local administrator assigned number (2 bytes)

Fig. 9.15(c) *4-octet ASN with 2-octet assigned number*

Import of remote routing information into VRFs

Whenever routing information updates are received from a CE, the same are to be forwarded to other PEs for storing in the concerned VRFs. This is done by sending 'route target' information, which informs the receiving PE router into which VRFs the routes are to be imported. The formats are shown in Fig. 9.16(a–c).

Type (0 × 00 or 0 × 40)	Sub-type	Global administrator
Local administrator		

9.16(a) *2-octet AS specific extended community*

Type (0 × 01 or 0 × 41)	Sub-type	Global administrator
Global administrator (contd.)		Local administrator

9.16(b) *IPv4 address specific extended community*

Type (0 × 02 or 0 × 42)	Sub-type	Global administrator
Global administrator (contd.)		Local administrator

9.16(c) *4-octet AS specific extended community*

Fig. 9.16 *Route target message formats*

Forwarding layer-3 MPLS VPN packets

Forwarding of layer-3 MPLS VPN packets is illustrated in Fig. 9.17. Layer-3 MPLS VPN packet forwarding is implemented via label switching with 2-level label stack imposed on original IP packet by the ingress router. The outer label (OL) is assigned by LDP (the other option being RSVP) router. The inner label or the VPN label (VL) represents the label assigned to the customer route itself by the originating (egress) PE router and corresponds to the label assigned to the IP address of the remote PE. The outer label is removed by PHP router whereas VPN label is removed by egress router.

When layer-3 MPLS is implemented across autonomous system boundaries, VRF to VRF connections are made through ASBRs. PE routers are attached via multiple physical or logical interfaces, each of which is associated with a given VPN (via a VRF). So, each PE router treats the adjacent PE router as a CE router and layer-3 MPLS mechanisms are used for route distribution within each autonomous system boundary.

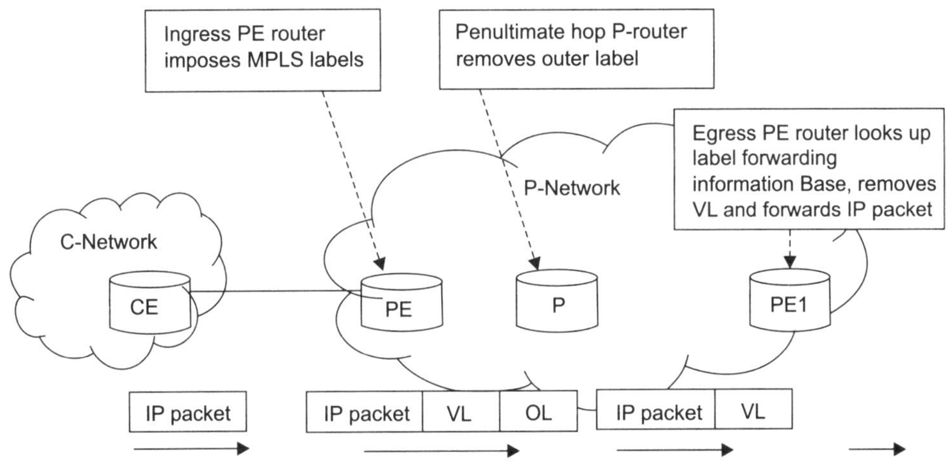

Fig. 9.17 *Forwarding layer-3 MPLS VPN packets*

9.3.7 SDH network support for MPLS

Routers supporting MPLS (e.g., CISCO or Juniper) have either E1 or VC3 or STM-1/higher order channelling (to be connected to SDH network side) and provide ETH interface cards—fast Ethernet/gigabit Ethernet/10 GB Ethernet—to interface on the LAN side.

Such routers connected with E1 or VC3 or STM-1/higher order channeling have the LSPs of E1 or VC3 or STM bandwidth carried on SDH network. They obviously have work and protection paths in the SDH ring and are managed through NMS.

In the third generation transportation network namely OTN, where labels can be correlated to wavelengths, the use a PXC (photonic or optical cross connect) switch, i.e., optical to optical payload processing (O/O/O), reduces further the delay as well as variations in delay.

9.4 Internet Backbone Connectivity

One of the major applications of SDH is the internet backbone connectivity. But for this we would not be enjoying broadband internet facility, despite the availability of ADSL (asymmetric digital subscriber l line).

As already explained in chapter 1, ADSL deployment on subscriber line from exchange offers 1.5 Mbps upstream (minimum of 640 kbps) and 8 Mbps downstream (minimum of 2 Mbps). But aggregates of such bandwidths to various subscribers require much larger bandwidth 'pipes' from exchanges to the access server. This requires a tiered way of aggregating bandwidth from the exchanges, which is possible only through SDH network.

The arrangement to provide broadband internet facility at subscriber's end using ADSL modems is shown in Fig. 9.18(a). An ADSL modem at the subscriber's premises works on the same pair that brings in the voice connection. A data splitter segregates data traffic from voice traffic. A landline phone, a PC with Ethernet card, DSL modem, data splitter and a broadband account (with user name and password) are the required elements to establish a connection at the customer promises. It is also necessary to have the connection software loaded on the customer's PC.

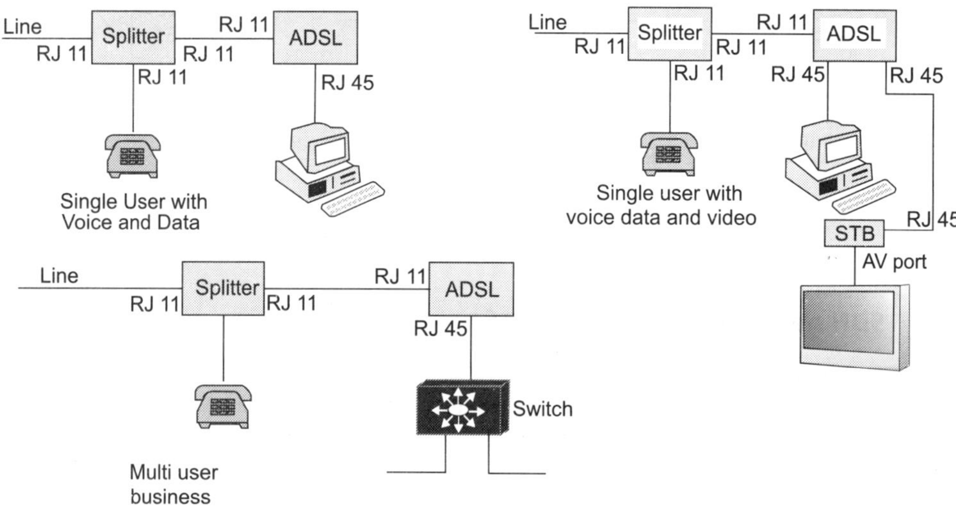

Fig. 9.18(a) *ADSL modem and data-splitter installation at customer's premises*

Let us turn our attention to aggregating the bandwidth to broadband internet access server.

While the customer side equipment are only ADSL modem and data splitter, the provider's equipment are many, DSLAM being just one of them. These equipment and their role are enumerated below.

1. **Digital subscriber line access multiplexer (DSLAM)**
   DSLAM has an in-built data splitter which splits voice and data. While voice follows the normal conventional path through exchange, data is aggregated and uplinked through the Ethernet port (gigabit Ethernet for 480 port DSLAM and fast Ethernet for lower DSLAM).

   DSLAMs will in general be collocated with existing PSTN exchanges which provide the last-mile access to customers over copper wire up to an average span lengths of 5.5 kms. DSLAM sizes vary considerably: 480 ports, 240 ports,

120 ports, 64 ports, 48 ports and 24 ports. Uplink of DSLAM is on an Ethernet interface which converts ATM traffic to Ethernet.

2. **LAN switch**
   LAN switch is for aggregating multiple DSLAM and providing a common uplink to the core.

3. **Broadband remote access server (BRAS)**
   Broadband remote access server is the first intelligent device in the whole chain. It terminates the customer session, authenticates and allots IP addresses and keeps track of user sessions for billing along with RADIUS.

4. **Subscriber service selection system (SSSS)**
   SSSS is the customer's interface screen. When a customer logs in, he or she will be welcomed with this customised screen from where it is possible to select various range of service. This provides on-demand service without manual intervention.

5. **RADIUS**
   This, in conjunction with BRAS authenticates the customer, uploads customer profile in the SSSS and keeps track of billing. (Remote authentication dial-in user service (RADIUS) is a networking protocol that provides centralised authentication, authorisation and accounting (AAA) management for computers to connect and use a network service.)

6. **LDAP (SSSC)**
   It stores the customer database which include the username, password and the default services a customer can subscribe to. LDAP (lightweight directory access protocol) is an application protocol for querying and modifying directory services running on TCP/IP. It is defined in IETF standard RFC 4510.

7. **Provisioning system**
   This is the most critical component for ensuring quick delivery of service. It ensures end-to-end provisioning of service right from DSL CPEs to DSLAM to switch to BRAS to LDAP.

Let us trace the connection from the subscriber premises to the broadband internet server.

The ADSL gets connected to DSLAM (digital subscriber line access multiplexer) port at an exchange on tier-II LAN switch. From there, the gigabit Ethernet connection takes the connection to tier-1 gigabit Ethernet aggregate switch. From tier-1 switch, the connection is to the router attached to the broadband remote access server. The complete connection is shown in Fig. 9.18(b). We can note the support provided by SDH from DSLAM to tier-2 switch, tier-2 switch to tier-1 switch and also ring protection for (add–drop MUXes) providing connectivity to DSLAMs.

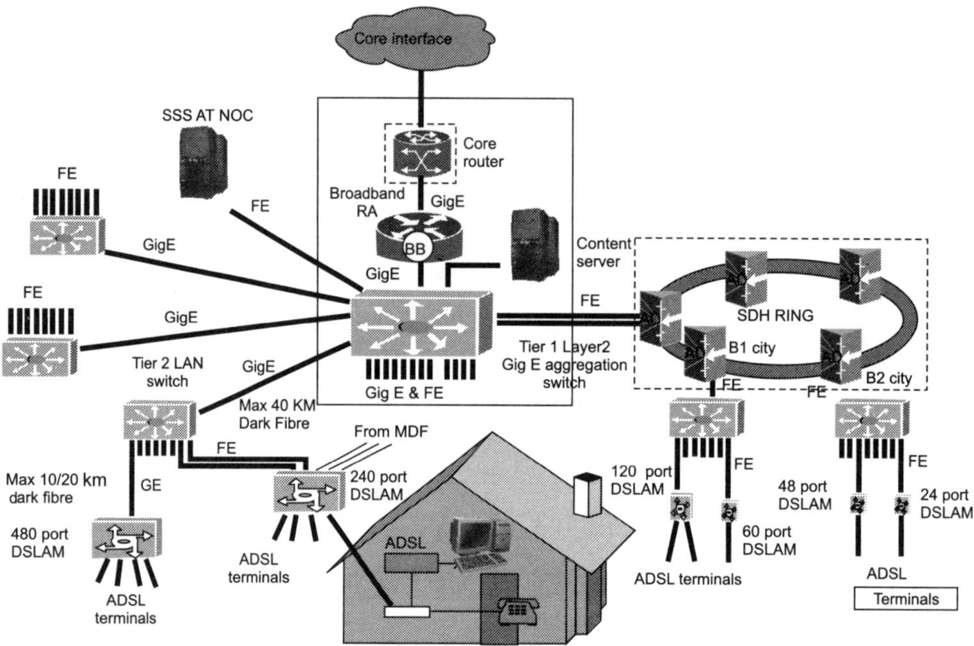

Fig. 9.18(b) *SDH support for broadband service*

Refer to Fig. 9.19(a) & (b) showing a single user connection in bridge mode and router mode and Fig. 9.19(c) & (d) showing multi-user connection in bridge mode and router mode.

Whatever be the mode, single or multi-user connection, tier-2 and tier-1 switches ultimately provide connectivity with internet backbone through B-RAS. These are through Ethernet interface via SDH network (gigabit Ethernet for 480 port DSLAM and fast Ethernet for lower DSLAM).

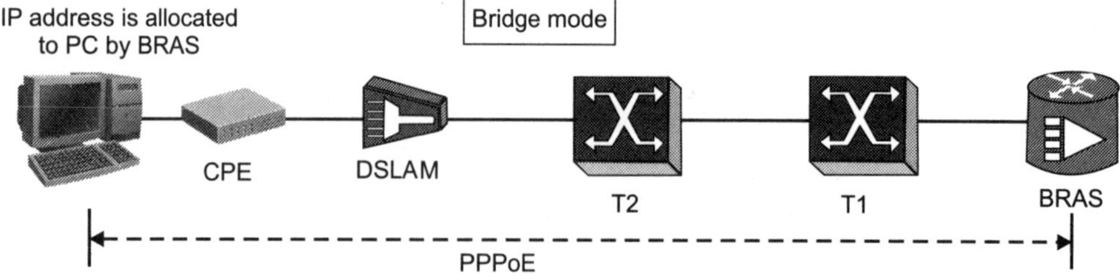

Fig. 9.19(a) *DSL connection in bridge mode – single user*

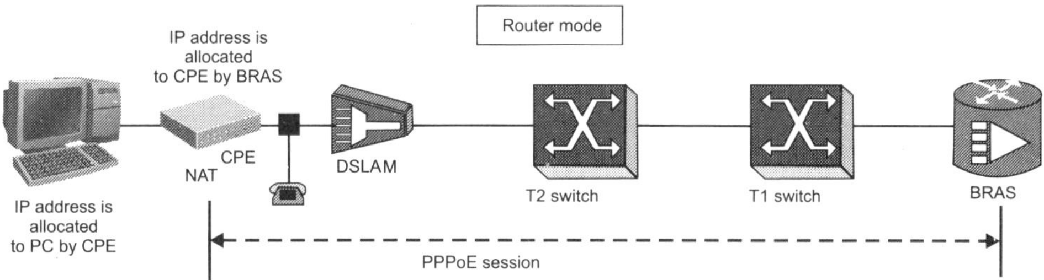

Fig. 9.19(b) DSL connection in router mode – single user

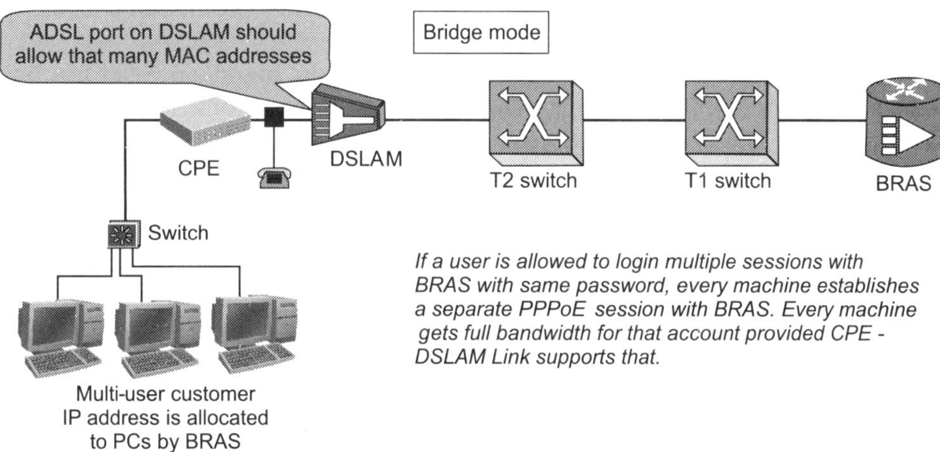

Fig. 9.19(c) DSL connection in bridge mode – multi-user

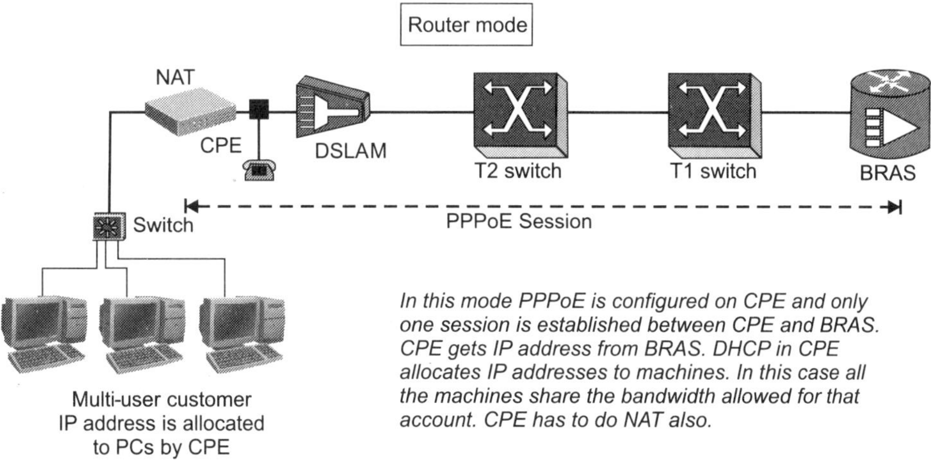

Fig. 9.19(d) DSL connection in router mode – multi-user

9.5 VoIP on SDH Network

9.5.1 What is VoIP?

VoIP refers to the set of technologies that allow voice traffic to be carried over Internet protocol (IP) networks as IP packets. VoIP enables users to avoid long-distance telephone charges by placing local calls to the ISP (Internet service provider) and getting the encapsulated voice traffic to move over the Internet. Though this approach is simple, latency levels of internet make it unattractive. Another way is to implement it over a firm's intranet, which is a costly solution and adds up new equipment in the network. Whatever be the way we implement VoIP, the packet loss should not exceed 15 % and the latency should not be more than 700 μs. SDH holds great promise in this role with its protection mechanisms and low latency.

Organisations having backbone transmission networks based on SDH are better placed to implement VoIP, especially when telephone connections are to be given at several places. It is not economical to provide the large number of exchanges to do that. Digital PBX with few lines can be connected to VoIP gateway, which in turn is connected to a router. The router gets E1 connectivity supported on SDH backbone. As SDH maps E1 to VC-12, protection paths can be provided to ensure availability. Since IP packets carrying voice traffic are basically carried on SDH network rather than internet, latency is not a problem.

9.6.2 VoIP implementation

The key to implementing VoIP is the VoIP gateway, a device that converts voice signals into IP packets for transmission over IP backbone networks and converts the received IP packets to voice signals. The VoIP gateway is connected to digital PBX on one side and router on the other. It can also be connected to the LAN switch port to co-exist with nodes of a LAN. Refer to Fig. 9.20(a) showing VoIP implementation using routers.

VoIP gateway can support PBX with lines from 8 to 960. Some vendors offer them on one chassis. In this implementation, the PBXs close to the VoIP gateway can be connected on the IP backbone and those far away can be connected to the on SDH backbone. Connectivity with SDH NE is through E1.

Please refer to Fig. 9.20(b) showing VoIP implementation using Ethernet connectivity with LAN switches. If the SDH NE has ETH card, we can connect it directly to LAN switch and take MAC frames. If the SDH has no ETH card, we can use E1–ETH converter and provide E1 connectivity with SDH NE.

9.6.3 VoIP gateway protocols and their evolution

Signalling is a vital ingredient of telecommunications infrastructure because it enables various network components to communicate with each other to set up and tear down calls. VoIP is no exception to this. Since the very beginning of the VoIP, the focus has been on signalling protocols from the point of view of implementation as well as evolution. So far, VoIP has gone through three stages in terms of signalling protocol implementation: pre-commercial, PC-centric, carrier-grade and finally the evolution converged in NGN switch (next generation networking). Rather, the NGN switch emerged out of these evolutions of VoIP and its gateway and paved the way for convergence of PSTN (public switched telephone network), PLMN (public land mobile network) and IP networks. In the following sections, we trace these events.

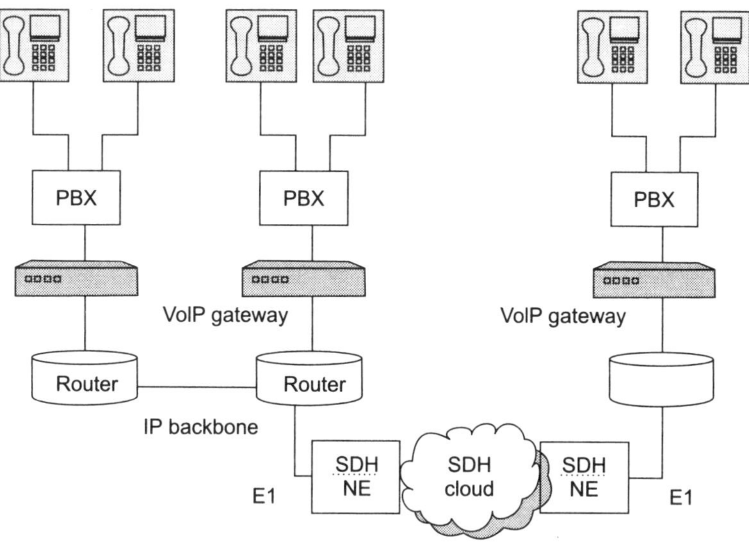

Fig. 9.20(a) *VoIP implementation using routers*

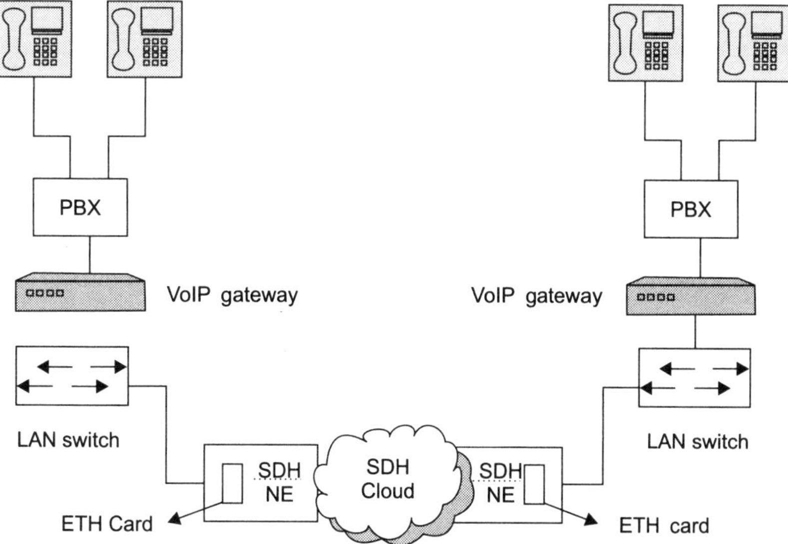

Fig. 9.20(b) *VoIP implementation with ETH connectivity on LAN switches*

9.6.4 VoIP in pre-commercial stage

In the pre-commercial stage, the objectives of VoIP were to run audio and video conferencing over the Internet. Interworking with the PSTN was of minor interest. Work on this was pioneered by two groups under IETF (Internet Engineering Task Force) – audio/video transport (AVT) working group which produced the real-time transport protocol (RTP) and multiparty multimedia session control (MMUSIC) working group which designed a family of protocols for multimedia conferencing over the Internet. The most prominent of these is the session initiation protocol (SIP).

SIP is a signalling protocol that controls the intiation, modification and termination of interactive multimedia sessions, viz. audio or video calls between two or more participants including chat sessions, game sessions. SIP is a text based protocol similar to HTTP (hyper text transfer protocol), an SMTP (simple mail transfer protocol). In contrast to centralised switches of PSTN, SIP is a peak-to-peak protocol. Initial flavour of SIP is defined in RFC 2543 (1999) by MMUSIC working group of IETF. Now SIP with extensions for instant messaging, presence, event modification was defined in June 2002 (IETF's RFC 3261). Details of SIP are available in Annexure B.

9.6.5 VoIP in PC-centric stage

Proprietary solutions

The release of commercial VoIP software in 1995 marked the beginning of PC-centric stage. This software allowed calls between multimedia PCs via internet. All the signalling and control functions resided in the PCs. However, interoperability was very poor due to proliferation of products. To ensure proper take-off for the fledgling VoIP, ITU started work on standardising VoIP signalling protocols in May 1995, which culminated in an interoperable solution.

Interoperable solution

In June 1996, the ITU-T Study Group 16 issued Rec. H.323 (Ver.1, latest version as on date being Ver.6 issued in the year 2006)[1]. ITU-T Rec. H.323 is a standard for real-time videoconferencing over non-guaranteed quality of service (QoS) LANs. The market acceptance for this standard was so great that by the end of 1996, most PC client software vendors were moving toward building H.323 compliant products.

9.6.6 ITU-T Rec. H.323

Features of H.323 network A network implementing H.323 protocol consists of number of *zones* interconnected via a WAN. Each zone has a single H.323 gatekeeper (GK), a number of H.323 terminal endpoints (TEs), a number of H.323 gateways (GWs), and a number of multipoint control units (MCUs) interconnected via a LAN. A zone can span a number of LANs in different locations or just a single LAN. The only requirement is that each zone contain exactly one GK, which acts as the administrator of the zone. The functionality of each component of the architecture is defined as follows.

1. **Terminal:** An H.323 TE is an endpoint in the network which provides for real-time two-way communications with another H.323 terminal, GW, or MCU. This communication consists of control, indications, audio, video and/or data between two terminals. A TE may set up a call to another TE directly or through GK.
2. **Gatekeeper:** The GK is an H.323 entity in the network that provides address translation and controls access to the network for H.323 terminals, GWs, and MCUs. GK may also provide other services to the TEs, GWs, and MCUs such as bandwidth management and location of GWs. GK is optional in H.323 systems.
3. **Gateway:** An H.323 GW is an endpoint in the network that provides real-time two-way communications between H.323 TEs on the packet-based network (and GK, if available).
4. **Multipoint control unit:** MCU is an endpoint in the network that provides the capability for three or more TEs and GWs to participate in a multipoint conference.

Signalling and control protocols of H.323

H.323 is an umbrella of the following signalling and control protocols for multimedia call processing (Table 9.9).

H.323 Interworking with the PSTN

Even though H.323 was designed for multipoint multimedia conferencing over packet networks, its usage has been primarily driven by VoIP applications and interworking with the PSTN has been a focus from the very beginning. Unlike SIP, the GW to the PSTN has been an integral part of the H.323 architecture.

Interworking with PSTN usually concerns three call set up scenarios: H.323 TE to phone; phone to H.323 TE; and phone to phone via intermediate H.323 networks. In all these cases, an H.323 GW is involved in connecting the PSTN with the Internet. Generally speaking, a GW needs to provide the following functionalities.

- **PSTN interfaces:** This function includes the PSTN signalling interface that terminates signalling protocols such as ISDN Q.931 and the PSTN media interface that terminates media streams such as pulse code modulation (PCM) voice streams.
- **VoIP interfaces:** This function includes the VoIP signalling interface that terminates H.323 (including RAS, Q.931 and H.245) and the packet media interface that handles RTP.
- **Signalling conversion:** This function typically enables connectivity between ISDN Q.931 signalling and H.323 signalling for call control.
- **Media transformation:** This function typically enables translation between 64 kb/s PCM streams and RTP streams of various speeds.

- **Connection management:** GW must internally coordinate between signalling flows and media transformations. This involves creating, modifying, and deleting the association between the PSTN and Internet flows during the lifetime of a call.

Table 9.9 *Signalling and protocols under H.323 umbrella*

Entities involved	Protocol	Remarks
Between TE or GW and GK	RAS (registration admission & status) (Note: RAS is used only when GK is present.)	TE or GW can use RAS to discover a GK, register/unregister with a GK, requesting call admission and bandwidth allocation and clearing a call. A GK can use RAS for inquiring the status of TE or GW. There is also a mechanism for GKs to communicate with each other for address resolution across multiple zones. GKs can communicate with each other for address resolution across multiple zones.
Between two TEs	Q.931	For call set up and teardown between two H.323 TEs, Q.931 is used. Variation of Q.931 is used for PSTN. H.323 only uses a subset of the Q.931 messages in ISDN and a subset of info. elements (IEs). All H.323-related parameters are encapsulated in the user-user IE (UUIE) of a Q.931 message.
Between TEs (for negotiation or connection control)		H.245 is used for connection control (i.e., open and close logical channels) between end-points, negotiation of media processing capabilities such as audio/video codecs for each media channel between end points.
Transport of VoIP packets	RTP (Real Time Transmission Prtocol)	RTP is used as the transport protocol for packetised VoIP in H.323. It is taken from IETF and is associated with real-time control protocol (RTCP).

As can be seen, the redeeming feature of H.323 since the beginning has been interworking with PSTN, since bypassing of PSTN telephone charges was the main economic driver for VoIP. Consequently, there was a proliferation of H.323 gateway products that enabled phone calls to be made across PSTN and the Internet.

Q 9.12 With the availability of H.323 gateways, entire telephony network could have been made IP based. What held it back?

H.323 Gateways had limitations in the following aspects:

1. **Scalability:** H.323 gateway handles signalling conversion, call control and media transcoding in one box, which poses scalability problems for large-scale deployment. The maximum number of lines an H.323 GW can support is a few thousands. This is small when compared to a regular telephone switches with tens of thousands of lines.
2. **SS7 connectivity:** The earlier versions of H.323 GWs did not have SS7 connection capability and connected to PSTN switches via ISDN trunks, i.e., the connection to PSTN switches was not seamless.
3. **SS7 services:** Without SS7 connectivity, VoIP cannot provide the same rich set of services as enabled by SS7.
4. **Availability:** When a GW is down, all active calls through the GW disappeared. There was no mechanism in H.323 for failure control.
5. **User friendliness:** Most VoIP services require that a subscriber dial the phone number to connect to the GW and then dial the number of the destination of the call. This procedure is called two-stage dialing. This is largely a result of lack of SS7 connectivity.

With the above limitations of H.323, VoIP industry realised the need to provide carrier-grade services.

9.6.7 VoIP in carrier-grade stage

In May 1998, the concept of a decomposed gateway was introduced to overcome the scalability limitation of GW. In a decomposed gateway, call control is taken care of by 'media gateway controller (MGC)' and media transformation is taken careof by 'media gateway (MG)'. Along with the concept of decomposed gateway, media gateway control protocol (MGCP) came into being in 1998.

MGCP is a device control protocol developed by IETF to control devices such as media gateways (MG) and integrated access devices (IAD). The difference between MGCP and other multimedia control protocol systems is that MGCP allows the endpoints in the network MGCs (media gateway controllers) or call agents or soft switches to control the media gateway. In other words, MGCP is modelled as a master-slave protocol, MGC being the master and MG being the slave.

These devices use text messages in the MIME format (multipurpose Internet mail extensions) to set up, manage and terminate multimedia communication sessions (layer 5 in a communications system). The text is formatted according to SDP (session description protocol) and placed in a SAP (session announcement protocol) packet. That is, telephone signalling messages are sent via user datagram protocol (UDP) packets with a SAP header and a text payload (telephone number, email, protocol, connection information, bandwidth, etc.). Once signalling is completed, real time protocol (RTP) packets are sent via UDP between the callers.

Q 9.13 ***In the light of the limitations of H.323 and IETF's initiative on MGCP, what were the revisions taken up by ITU-T?***

In the first version of H.323, a gatekeeper kept track of the IP addresses to phone number mapping for routing calls between VoIP GWs. In subsequent versions, an alternative gatekeeper was introduced to provide alternative routing capabilities. The concept of an assigned gatekeeper was introduced in H.323 Ver.6 (released in the year 2006), which is an extension of the alternative gatekeeper mechanism. An assigned gatekeeper is the one with which an endpoint should register out of a list of available alternative gatekeepers. In the event of an assigned gatekeeper failing, the endpoint switches to an alternative gatekeeper. However, either the endpoint or the current gatekeeper would monitor the status of the assigned gatekeeper and, once the original assigned gatekeeper is available, the endpoint will transit back to the same. This allows the network operator to maintain a preferred provisioned state. While alternative gatekeepers may be geographically dispersed, endpoints in general register with an alternative gatekeeper that is physically closest. Whatever be the registration preference of the network operator, the assigned gatekeeper concept gives more control on resource utilisation in a network.

ITU-T has been working on improvised versions of H.323 on one hand to give more facilities and flexibilities to the already existing H.323 base. On the other, it joined with IETF to develop a protocol to provide SS7 based call-routing and services so as to converge on the next generation networking (NGN). ITU-T SG-16 and IETF jointly issued H.248 or 'Megaco' in June 2000 itself.

In the next section, the evolution of the NGN switch is presented, followed by a description of the features of H.248/Megaco which serves the cause of NGN switch.

Appendix B gives more details on SIP, H.323, enhancements in Ver.6 of H.323, differences between H.323 and SIP, which a reader may browse before moving on to the following discussion.

9.7 Evolution of NGN Switch

9.7.1 Need to go beyond H.323 and MGCP

The need to go beyond H.323 and MGCP for supporting the delivery of telephony, multimedia and innovative value added services in a cost-effective manner led to the evolution of the next generation networking switches. The challenges that led to this evolution were quite daunting. An NGN switch has to be implemented in such a way that:

1. There is no need to expand the existing PSTN switches which have exhausted their lifetime.
2. The total cost of ownership is reduced.
3. The time to market services and increase business opportunities is less.
4. An operator's capability to scale investment in telecom infrastructure is improved.
5. Market demand is followed with minimum risk.

The NGN switch succeeds in meeting all the above challenges.

There are several other aspects of this evolution. In addition to the switch, NGN encompasses seamless multiple access, next generation IPv6 based network, next generation of cellular networks (3G and beyond), next generation automatic switching optical network. These are covered in detail in chapter 14. For the present, we confine our discussions to NGN switch features.

9.7.2 Features of NGN switch

NGN switches are three-layered switches interconnected over PDH or SDH or IP or MPLS cloud and intended to deliver full range of telephony, multimedia and innovative value-added services with improved scalability. The three-layered switch is not hierarchical like the conventional PSTN switch.

Refer to Fig. 9.21 showing three-layered switch architecture. It has a media adaptation layer comprising of media gateways and access devices, a call control and signalling layer comprising soft switches and signalling gateways and a service layer comprising feature servers. There is standardisation between layers. Readers may note that the three entities, viz. media gateways, soft switches, feature servers and their functions put together is a three-layered switch. The soft switch alone is not a three-layered switch!

<table>
<tr><td>Layer-3
(Service layer)</td><td colspan="3">Service logic</td></tr>
<tr><td rowspan="3">Layer-2
(Control layer)</td><td rowspan="3">TDM
Sigg.</td><td>Call control</td><td rowspan="3">Signalling
over IP</td></tr>
<tr><td>Signalling inter-working</td></tr>
<tr><td>Resource management</td></tr>
<tr><td>Layer-1
(Media layer)</td><td>Tele-phony</td><td>Switching fabric</td><td>Media
adaptation</td></tr>
</table>

Fig. 9.21 *Three-layered switch*

9.7.3 Functions of media gateway

As can be seen from the three-layered architecture, the media gateway, which is at layer-1, has to perform media adaptation and telephony functions. It has ports to interface with TDM networks and IP network to provide switching fabric. Its functions are:

1. Convert the information from time slots that are used with time division multiplexing systems into packets that are used with IP networks.
2. Perform media processing functions such as cancelling line echos, managing jitter buffer and compensating for packet loss to assure overall quality of service of the connection.
3. Insert appropriate signals such as dialling and call progress tones, comfort (background) noise, and so on into the media.
4. Include the ability to detect specific call events such as on/off hook states and voice/video/multimedia activity.

Stated simply, the functions of media gateways are: to provide a physical interface to the first network another physical interface to the second network, and carry on signal processing functions that are required to make all the protocol conversions between these two networks under the control of the media gateway controller. It is not the functions that are carried out which make the MGW so special but the requirements it meets in doing so. These are listed below.

1. Toll quality voice encoding/decoding with low rate compression/expansion algorithms matching mean opinion score of PSTN's services
2. High performance silence suppression giving dual benefit of minimisation of bandwidth (to work on access and backbone network segments) and reduction in processing throughput (to work on WAN routers)
3. Superior echo cancellation to provide pleasing conversational speech
4. Highest compression ratio achieved through packet aggregation such that consumption of IP bandwidth and processing resources is the lowest
5. Bandwidth efficient and high picture quality FAX transmission
6. Highly efficient and transparent signal handling for multimedia call processing
7. High performance jitter buffering (minimum up to 300 ms)
8. Smart substitution of lost packets in voice transmission
9. Low latency
10. Smart packet priority selection mechanism to reduce outgoing packet payload without perceivable voice degradation in case of temporary bandwidth constriction

Media gateways come with TDM network interface (4 or 8 ports per card–port can be E1 or T1) and total ports can be up to 72. On the network side, interfaces available are FE or STM-1 so that MGW can be connected to IP/MPLS network or SDH network. On network side, the interface has redundancy.

9.7.4 Media gateway controller

While MGW operates in layer-1 or media layer or transport plane, MGC operates in layer-2 or control layer or call control and signalling plane. In brief, MGC is responsible for establishment and tear-down of media connections. Its functions can be classified into two groups:

1. Call agent functions (to provide call control and call state maintenance operations)
2. Management functions (to manage network resources)
3. Interworking functions (to manage connections needed to/from a network that deploys another type of networking protocol).

These functions are elaborated below.

1. Call agent functions
    a. Originating and terminating signalling messages to and from end user stations, other MGCs or networks (such as PSTN).
    b. Maintaining call state information for every call on the media gateway.
2. Management functions
    a. Negotiation of media parameters viz. selection of a particular audio or video codec, bandwidth requirement, etc.
    b. Managing network resources, such as ports on the media gateway, the available network bandwidth, and so on.
3. Interworking functions
    a. Providing policy functions such as access capabilities and permissions for endpoints.
    b. Interfacing with feature servers for functions related to application, accounting or routing in support of the call.

9.7.5 Feature server

The feature server operates in layer-3 or service and application plane and provides the functions to control applications that are offered on the network. It takes care of provisioning, billing, network operations, authentication, lawful intercepts, etc.

9.8 Signalling and Control Protocol for NGN Switch: ITU-T Rec.H.248/Megaco

9.8.1 Features of H.248/Megaco

While the later versions of H.323 confined themselves to providing additional features and flexibility, H.248/Megaco used the NGN switch approach for the functional decomposition of GWs to overcome the deficiency of monolithic H.323 GWs in carrier-grade VoIP deployments. The benefits reaped are as elaborated below.

- **Scalability:** The bottleneck of scalability for H.323 GWs is media transformation. Since the MGC and MG are in separate boxes and use one MGC to control multiple MGs, there are, in effect, virtual H.323 GWs that can handle more lines.
- **SS7 connectivity:** This has been done by connecting the SG function to the SS7 network.
- **Availability:** Decoupling the MGC from the MG increased the availability in the sense that multiple MGCs can be used to control a single MG. If one MGC fails but the call states are kept in stable storage, one can apply traditional failover procedures to switch to another MGC. Graceful failover ensures that active calls in the MGs are not lost.
- **One-stage dialling:** This is achieved through support for SS7 connectivity.

H.248 is a master/slave communications protocol with two basic components viz. terminations and contexts. Simply stated, terminations are media connections called physical (time slots) and ephemeral (IP flow). Contexts are associations between terminations which can be added or deleted.

A termination sources and/or sinks one or more data streams. Data streams can be telephony, RTP streams, or MP3 streams. Terminations have properties such as the maximum size of a jitter buffer, which can be inspected and modified by the MGC.

A context is an association among a collection of terminations. There is a special type of context, called the *null context*, which contains all the terminations that are not associated with any other termination. For example, in a decomposed access gateway, all the idle lines are represented by terminations in the null context.

Contexts are created and released by the MG under the command of MGC. A context is created by adding the first termination; it is released by removing (subtracting) the last termination. In other words, terminations are placed into contexts which are defined as when two or more termination streams are mixed and connected together.

H.248/Megaco uses a series of commands to manipulate terminations, contexts, events, and signals. The SS7 network provides many unique features such as *69 (automatic callback), 800 (toll free service), look ahead call routing to route calls to any call centre or agent in the world and emerging services such as CRBT (caller ring back tones), music and other information services.

9.8.2 Call analysis and processing by MGC

The processes by which MGC analyses and routes calls are discussed in this section. These processes depend on the information that is provisioned in the dial plan and the routing plan in the MGC.

Call analysis[2]

Refer to Fig. 9.22 showing the sequence of actions in call analysis. The calling number (A-number) and the called number (B-number) are contained in the ISDN user part (ISUP) initial address message (IAM), which is included in the signalling information field (SIF) of an SS7 message signal unit (MSU), or in the ISDN PRI setup message as shown on the left in Fig. 9.22.

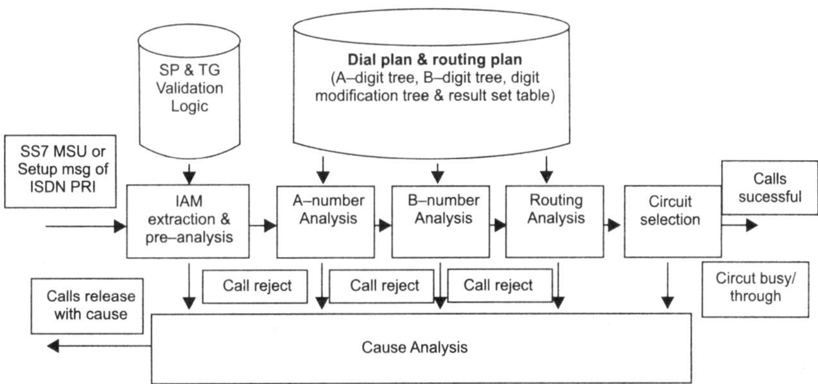

Fig. 9.22 *Call analysis and routing by MGC*

The IAM is just one part of the ISUP information that the signaling link terminal (SLT) transmits to an associated MGC. In B-number pre-analysis, the called number (B-number) undergoes varying degrees of analysis, depending on other data in the IAM or ISDN PRI setup message.
The sequence is explained below:

1. MG has 'point code' known as SP code allotted by regulatory authority so that mere media connectivity does not allow call progress and connectivity is authenticated. Trunk Group (TG) allotment is specific to call-type, national/international and incoming/both-ways. SP code and TG information is checked by MG as part of authentication.

2. Each branch under the A-number and B-number analysis trees represents a path used by the call-processing engine to process the calling or called number. Number modification can be used to add, delete, or modify specific digits of the calling number (A-number) or the called number (B-number) to support regional or private dialing plans.

3. The dial plan also provides for backward information requests, in case backward request messaging is needed.
4. After completion of number analysis, results set table entries are obtained. (Tables are pre-configured with dial plan).
5. Call release at any stage is notified with cause code. If number analysis is successful and circuit is available (time slot to take call available and called number is not busy), call is successful.
6. The dial plan can be changed dynamically at any time and the change will be effective with the next call.

Call routing

The goal is to establish a connection or circuit between the calling number (A-number) and the called number (B-number). Terms related to call routing which are defined in this context are:

1. **Trunk:** A trunk (or circuit) is a single TDM voice channel (DS0). It is a physical connection between two points through which a call can be established.
2. **Trunk group:** A trunk group is a collection of identical trunks (or circuits) that connect the same two points and are controlled by the same signalling channel. MGC trunk groups are often arranged exactly the same as the trunk groups on the switches on the opposing ends of the packet network for the sake of simplicity.
3. **Route:** The route defines the path that a call uses. It might be a collection of trunk groups with the same destination or a logical path over a packet network fabric.
4. **Route list:** Route list is a collection of routing alternatives that can be used to transport a call between the origination and the destination points. Individual routes within a route list can connect the same two origination and destination points, but over different physical paths.
5. **Dial plan:** Dial plan is the primary determinant of how a call is routed from its origination to its termination through an MGC-controlled packet-switched network.

Q 9.14 *What is the significance of dial plan design? How is it done?*

Dial plan design provides the basic functionality to meet customer requirements for calling and for called number analysis. It also allows the creation of permutations of combined open and closed numbering plans, as well as the ability to easily modify them. The MGC dial plan design provides interface to real time database. It is the final step in provisioning the MGC node.

To create a dial plan, one must first determine what routes, trunk groups and trunks have been provisioned in the MGC. The provisioning information on the MGC gives a logical representation of the actual routes, trunk groups and trunks on the media gateways that are controlled by the MGC.

The routing plan developed during provisioning of the MGC determines which trunks in a media gateway are assigned to a specific trunk group and which trunk groups are assigned to which routes. The routes determine which originating and terminating points are serviced by the MGC.

With this information in hand, one can use the procedures detailed in MGC manuals to create a dial plan specifying what types of service requests can be accommodated by the MGC based on the call setup information, the calling number (A-number), the called number (B-number), and the routes, trunk groups and trunks (individual circuits) available on the attached media gateways.

9.8.4 Call routing sequence

Figure 9.23 illustrates the sequence of events that occur in routing a call from its origination (TDM network A) to its termination (TDM network B).

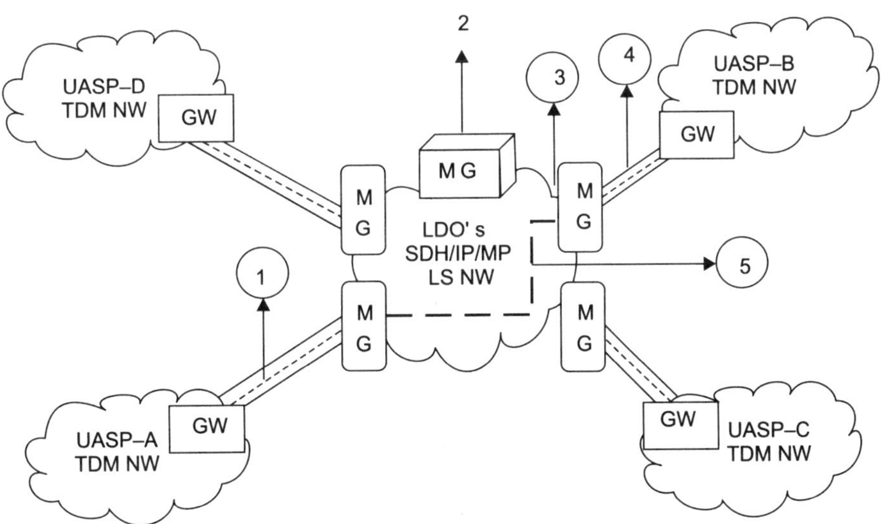

Fig. 9.23 *Call routing sequence*

UASP (universal access service provider) means PSTN/PLMN provider. LDO (long distance operator) is the carrier provider, i.e., transmission NW provider. UASP's NW is connected through gateway (GW) switches to MG of the LDO's NW.

If LDO's NW is an SDH NW, each MG (also MGC) is connected through TDM ports to the nearest SDH node (not shown in figure) in such a way that each MG has ports identified for connectivity with all other MGs and MGC, i.e., links are ready to other MGs and MGC opened as per call requirements and call sequence. For a call from UASP-A to UASP-C:

1. The originating TDM NW's GW selects TG and circuit, and opens connection to MG of LDO.
2. MG sends received signalling message to MGC. MGC reads IAM and performs number analysis as illustrated in Fig. 9.22.
3. MGC identifies route (to which MG call is to be connected) and exchanges signalling messages with the terminating MG.
4. MGC selects TG and circuit on terminating MG's egress so that the circuit is connected to UASP-C's gateway switch.
5. The link (TDM circuit) between originating MG's TDM port to the terminating MG's TDM port, which is already available over SDH cloud is activated by MGC so that end-to-end circuit is available and the call takes place.

If the LDO's network is SDH, each MG's connectivity to all other MGs and MGC (with redundancy arrangement) implies cumbersome connections as the network scales up. If LDO has IP/MPLS over SDH, then each MG needs to be connected to MPLS router through FE link and the routing of signalling message packets and call packets is taken care of over the MPLS cloud. The scaling-up is also very smooth.

9.9 SDH Support for NGN Switch Interconnection

9.9.1 SDH support for 3-layered switch

As already mentioned in section 9.7.3, the MG (media gateway) has TDM ports of E1 to interface to PSTN switch and STM/FE ports to connect to SDH/IP/MPLS networks. NGN switch interconnection can be set up across SDH/IP/MPLS network as shown in Fig. 9.24. The interface point for MG in SDH/IP cloud is the STM node and in MPLS cloud, it is the MPLS router (which is in turn connected to STM node). Signalling links between soft-switch (MGC) and MGs at different locations, shown as dotted lines, are taken on SDH/MPLS cloud.

9.9.2 Control switch – a new approach

MGC and feature server put together is the 'control switch', which is the latest way of implementing NGN switch. The term control switch should not be confused with three-layered switch which is for the entirety of NGN switch, whereas the control switch is for layer-2 and layer-3 of NGN switch put together.

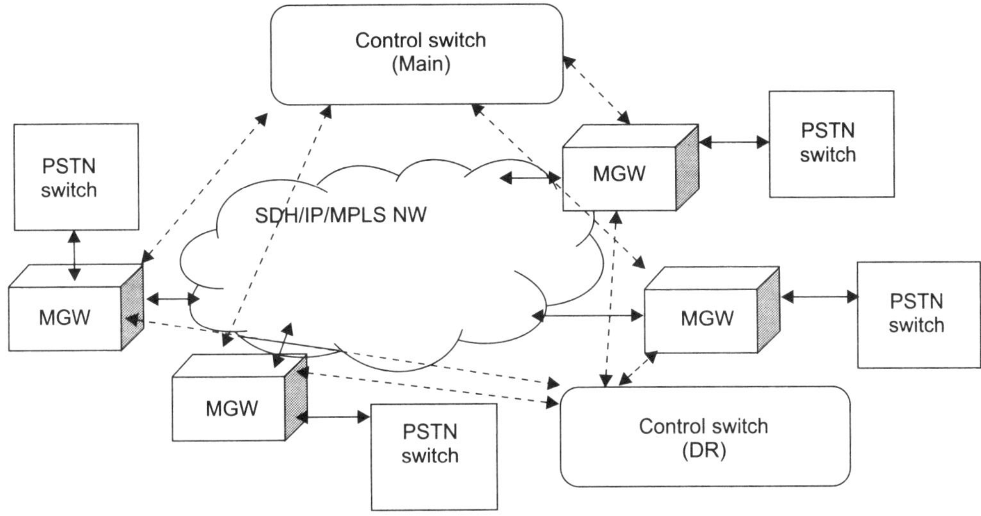

Fig. 9.24 *NGN across SDH/IP/MPLS cloud*

The control switch (in the strict sense it is a not switch but a switch plus server) is a distributed IP-network based system providing traditional switch functions of call control, call routing, signalling gateway and media device control in addition to back-office functions for supporting provisioning, billing and network operations. It has several software elements which interwork as shown in Fig. 9.25.

These can be considered in three groups:

1. Control logic
2. Policies and service logic
3. Management modules

On the network side, the control switch operates on three resources: media gateway, signalling gateway and border gateway.

Elements in control logic

1. Signalling gateway control: Takes care of signalling functions associated with call management.
2. Call control element (CCE): Provides protocol processing and resource management.
3. Sigtran signalling element (SSE): It arbitrates provisioning requests from EMS, manages application service provider processes, manages redundancy logic, provides ISUP encode-decode functionality, reports events to EMS.

4. IP call element (ICE): It is support element to CCE for optimising VoIP based protocols, whereas CCE is concerned with resource management protocols and PSTN protocols.

5. Interconnect border control function: It provides for topology hiding, media component and border gateway control.

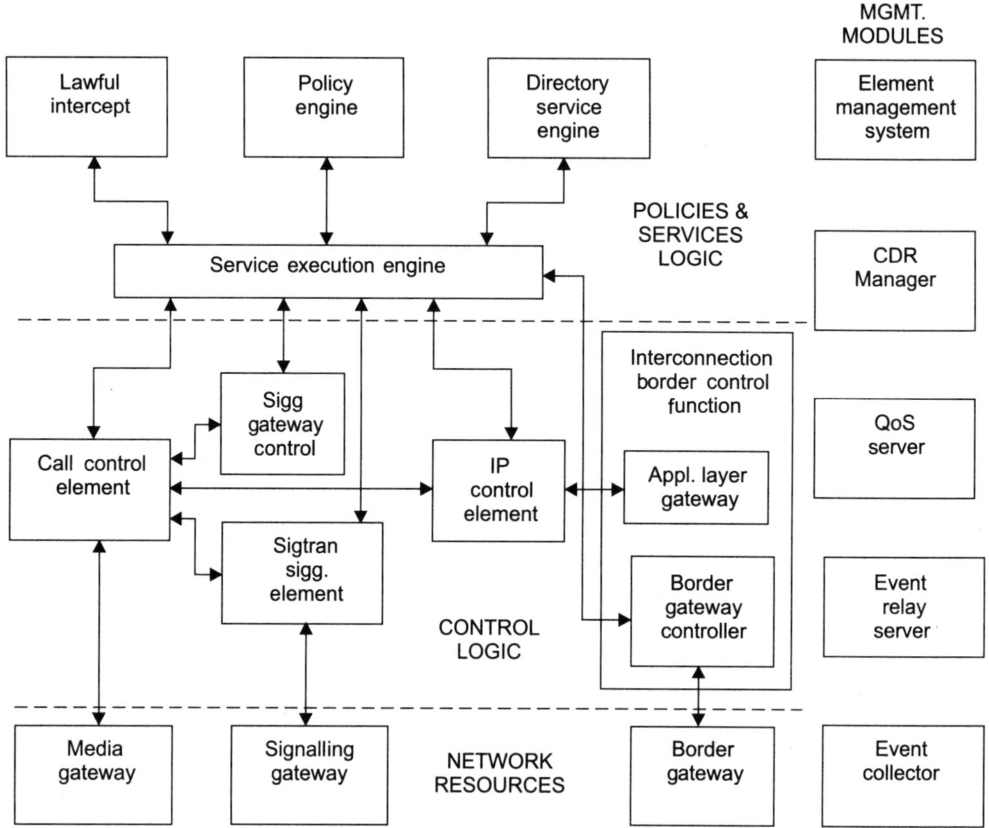

Fig. 9.25 *Control switch elements and their interworking*[3]

Elements in policies and services logic

1. Policy engine: Takes care of network policy administration and answers queries by SEE on call treatment, by checking with the database.

2. Directory service engine: It holds large databases and performs screening and translation services.

3. Service execution element: It is based on object oriented engine and with adapters for legacy signalling (SS7, PRI, CAS) as well as MGC related protocols (H.323, SIP, MGCP, H.248).

4. Lawful intercept: Enables connectivity to law enforcing agencies for transfer of captured data and administration of surveillance.

Management modules

1. Event collector: Collects call related information and passes it on to the event relay server.
2. Event relay server: Prepares billing as per call-related information and stores both.
3. QoS server: Performs statistics collection and analysis on real-time. For example, answer seizure ratio, latency, packet loss, jitter, etc.
4. CDR (call data record) manager: Generates CDRs for control switch billing and data analysis functions.
5. EMS (element management system): Responsible for provisioning of all components of the control switch. It initialises CCE, SG, PE, EC and CDR manager. It supports multiple aspects of system monitoring.

9.9.3 SDH support for control switch

Figure 9.26 shows SDH support for control-switch based NGN switch. The media gateway (MG) has TDM ports of E1 formed into trunk group (TG) to interface to the gateway (GW) switch of UASP's TDM network over SDH access network. The interface point for MG in MPLS cloud is the MPLS router (which is in turn connected to STM node). The control switch is duplicated to facilitate disaster recovery. The main and disaster recovery (DR) control switches are also on MPLS cloud, through MPLS routers. The control switches, (main as well as DR) and MGs exchange all call related IP packets the over MPLS cloud.

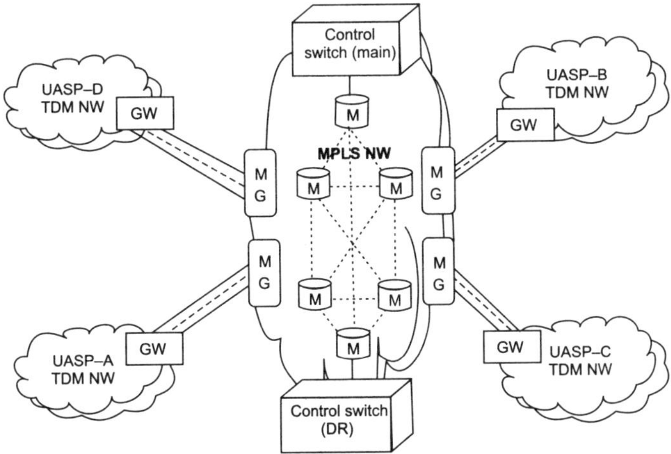

Fig. 9.26 *SDH Support for control-switch based NGN*

Summary

In this chapter, higher-end applications of SDH viz. LAN-to-LAN transport, MPLS VPN, Internet backbone, VoIP and NGN are discussed. Practical aspects like selection of VC, concatenation, latency issues in LAN-to-LAN transport; layer-2 and layer-3 implementation in MPLS; and equipment deployment in internet bandwidth distribution for DSL are presented. SDH support for implementation of VoIP is discussed. The evolution of VoIP gateway and emergence of NGN switch are explained. Implementation of NGN across SDH/MPLS cloud are presented.

References

[1] Real time video conferencing over non guarantied QOS LAN, ITU-T Rec. H.323 ver. 6, (2006).

[2] Mac call analysis and routing function, ITU-T Rec. Y. 2001, Overview of NGN (2004).

[3] General architecture of control switch and interworking of elements, ITU-T Rec. Y. 2001, Overview of NGN (2004).

10

Reliability Issues in SDH Networks

10.1 Introduction

In chapter 5 we studied how noise and ISI are tackled in digital transmission networks based on PDH hierarchy and operating on digital MW as well as OFC. We also studied ITU-T recommendations on control of jitter and wander in PDH networks. Long-haul SDH networks are not operational on digital radio for the following reasons.

1. The higher bandwidth associated with the higher data rate of STM envelope poses spectral allocation problems on multiple carriers with diversities.
2. Equalisation over such a wide band to keep phase errors and ISI under check is a daunting task
3. Design of wideband radios with optimal power plant on several hops is a constraint.

Few STM-1 and STM-4 radios are deployed on short hops but not on long-haul networks. This leaves us with reliability issues of long-haul SDH networks on OFC. Methods of tackling noise in OFC and ISI through adequate system margin and through proper rise time analysis were already presented in sections 5.2.2 and 5.3.2. These concepts apply to PDH as well as SDH networks, the only difference being in the transmission rate. In this chapter, we will study the control of jitter and wander on SDH networks.

In SDH network, jitter and wander accumulates in the transmission path due to the jitter and wander generated in each equipment in the network, be it multiplexers/ demultiplexers, cross-connects, clocks or line systems. Excessive jitter and wander can cause signal impairment because of the generation of bit errors, slips and buffer related problems. It is therefore necessary to set limits on the maximum magnitude of jitter and wander, and the corresponding minimum jitter and wander tolerance at network interfaces in order to guarantee acceptable quality in transmitted signals.

The ITU-T recommendations regarding the control of jitter and wander on SDH networks, cover the following aspects:

1. Network limits for output jitter and wander at any hierarchical interface
2. Network limits for jitter and wander at synchronisation interfaces
3. Jitter and wander tolerance for STM-N input ports
4. Limits for intrinsic jitter and wander
5. Jitter and wander transfer function

Since the applications of SDH are many and varied (LAN-to-LAN transport, VoIP, MPLS and ATM in addition to support of E1/E3/E4 streams and voice and data services carried on them), the consequences of such impairment, in general, will depend on the particulars of the service and on the characteristics of the terminating equipment. The quality of service (QoS) is another factor for concern, with different premiums being paid by customers for different classes of services. Therefore, there is a need to quantify the manifestation of jitter in terms of BER.

In sections 7.5 and 7.10, while studying pointers and synchronisation in SDH network, we discussed three modes of operation of the receiver viz. ideal mode, stressed mode and hold-over mode. In ideal mode, random phase variations of reference clock cause pointer adjustments in isolated cases. In stressed mode, short term impairments of clock cause the NEs to switch clock reference, resulting in pointer adjustments in a persistent way till the impairment is overcome. In hold-over mode, fibre-cut or node failure causes NEs to enter hold-over mode, and the periodic adjustments of pointers over several NEs cause accumulation of phase errors, jitter and BER. It may be noted that the protection schemes (APS) discussed in chapter 8 ensure the availability of MUX section/VC-path despite problems on a trail, but the manifestation of jitter is due to pointer adjustments that continue in an NE after APS, which contributes to BER.
Hence, there is need to study

- jitter–BER relationship
- how jitter manifests in all the three modes of operation of SDH NE (ideal/stressed/hold-over) and the consequent BER projections and effects like call-dropping, flow control problems and concatenation problems
- solutions for the same.

These issues, covered in this chapter, are not generally found in standard books. Readers are invited to ponder on the intricacies of these problems and the solutions put forth here, as these have been presented with a practical insight obtained from having conducted simulations/tests. It is possible that they offer clues to tackling similar problems being faced by users/service providers in networking.

Let us start the analysis with ITU-T recommendations on the control of jitter and wander in SDH networks.

10.2 Network Limits for Output Jitter and Wander at Any Hierarchical Interface

10.2.1 Sources of jitter in SDH networks

The types of jitter in SDH networks and the causes for the same are summarised in Table 10.1.

Table 10.1 *Type and causes of jitter in SDH network*

Type	Causes
Mapping jitter	Mapping of PDH streams into STM payloads requires stuffing. This results in jitter when signal is de-mapped.
Pointer jitter	The timing of transported payload containers must match that of the outgoing frame. This is done by incrementing/decrementing the pointer. Locating the container at changed address and clocking out data from there causes jitter.
Intrinsic jitter	Intrinsic to the equipment.
Wait-time jitter	Clock made to wait for de-justification.
Pattern jitter	Due to ISI in time-domain, which is pronounced for certain patterns.
Phase-noise jitter	Thermal noise and drift in the oscillator cause variation in clock period that manifests itself as jitter.

The various types of jitter and their magnitudes are given below. This only gives an idea of relative peak-to-peak values of the various types of jitter.

Mapping jitter	: 0.4 UI
Single pointer adjustment	: 0.6 UI
Triple pointer bursts	: 1.3 UI
Phase transient	: 1.2 UI
Periodic patterns in traffic being mapped onto containers	: 1.0 UI
Periodic pattern and single pointer adjustment	: 1.3 UI

10.2.2 Sources of wander in SDH networks

When data is transported via an SDH network, the wander accumulated depends on the total dynamic filling of all intermediate signal processing buffers in the SDH network. The buffer fill of a single NE depends on the relative wander between the incoming data and the read clock. The reference clock used by the NE can be the recovered clock through PLL if no pointer processing is involved, or an external source if pointer processing is involved.

Hence the buffer fill and, consequently the wander, depends on the reference clock phase noise and transient effects due to pointer processing or temperature effects in PLL.

10.2.3 Network limits for jitter

ITU-T Rec. G.825 prescribes the maximum permissible levels of jitter at hierarchical interfaces in an SDH network. According to the recommendations, jitter, as measured over a 60 seconds interval using the specified measurement filters, should not exceed the limits specified in Fig. 10.1.

Interface	UI (ns)	Jitter measurement bandwidth ($f_1 - f_4$)	Peak-to-peak amplitude A_2(UI pp)	Jitter measurement bandwidth ($f_3 - f_4$)	Peak-to-peak amplitude A_1(UI pp)
STM-1	6.43	500 Hz to 1.3 MHz	1.5	65 kHz to 1.3 MHz	0.15
STM-4	1.61	500 Hz to 1.3 MHz	1.5	65 kHz to 1.3 MHz	0.15
STM-16	0.402	1 kHz to 5 MHz	1.5	250 kHz to 5 MHz	0.15
STM-64	0.100	20 kHz to 80 MHz	1.5	4 MHz to 80 MHz	0.15

Fig. 10.1 *Network limits for output jitter at STM interfaces*

It may be noted that

a. The limits given in the above table should be met for all operating conditions, regardless of the number of equipment preceding the interface. In other words, these limits are compatible with the input tolerance limits of the equipment in the hierarchy where the output under consideration is applied.

b. The highpass measurement filters given in Fig. 10.1 has first-order characteristic and a roll-off of 20 dB/decade.

c. The lowpass measurement filters in Fig. 10.1 have a maximally-flat Butterworth characteristic and a roll-off of –60 dB/decade.

As can be seen, two network limits are specified in Fig. 10.1 for all STM rates—one for wideband measurement filter ($f_1 - f_4$) and the other for a highband measurement filter ($f_3 - f_4$). The limits in Table 10.1 means that

- the timing jitter as measured over a 60-second interval with a bandpass filter of lower cut-off frequency f_1 and a minimum upper cut-off frequency f_4 should not exceed A_2 unit intervals, peak-to-peak (A_2 = 1.5).
- the timing jitter as measured over a 60-second interval with a bandpass filter of lower cut-off frequency f_3 and a minimum upper cut-off frequency f_4 should not exceed A_1 unit intervals (A_1 = 0.15).

Q 10.1 *What is the significance of f_1, f_3, f_4, A_1 and A_2?*

The value of f_1 reflects the narrowest timing circuit cut-off frequency expected in an NE. The timing circuit may time a regenerator's output signal and could be implemented as a PLL. Jitter at frequencies higher than the bandwidth of this PLL will be partially absorbed by the PLL's buffer. The portion not absorbed could cause transmission errors due to buffer spill. Jitter at frequencies lower than this bandwidth will simply pass through without affecting the transmission performance. The value of f_1, therefore, represents the narrowest bandwidth that might be used in this output timing circuit.

The value of f_3 is related to the bandwidth of input timing acquisition circuitry. Jitter at frequencies higher than this bandwidth will constitute alignment jitter and will cause an optical power penalty. This high frequency jitter must therefore be limited to the same degree as that of the equipment specifications limit optical power penalty through jitter tolerance.

The value of f_4 reflects reasonable measurement limitations and is listed to establish the minimum measurement bandwidth requirements. f_4 is chosen to include all expected, significant alignment jitter. A value one decade beyond the widest expected cut-off frequency is chosen.

The values of A_1 and A_2 are directly related to the input sinusoidal jitter tolerance. These parameters have built-in margin and are on the higher side because

(i) sinusoidal jitter represents the worst case jitter with respect to input jitter tolerance,

(ii) accumulated SDH line jitter will not be sinusoidal, but noisy.

10.2.4 Network limits for wander

Network limits for wander at STM-N interfaces are specified in ITU-T recommendation G.823. As already mentioned in section 10.2.4 while explaining sources of wander, the buffer fill and consequently the wander in an NE depends on the reference clock phase noise and transient effects (due to pointer processing) or temperature effects in PLL. It may appear that we have a tough battle on hand against wander. However, it is not so because:

- Under normal conditions, buffer fills remain almost constant due to stable network synchronisation performance.
- The fluctuating part of buffer fills contribute randomly to the accumulation of wander because of the lack of correlation between buffers of different NEs.

Still, a limit is required to obeyed, which is specified by ITU-T in Rec. G.823 as 18 μs in a single connection over a duration of 24 hours.

Q 10.2 *How is the figure of 18 μs arrived at?*

The answer to this question is quite involved. A model was used for simulations to generate noise in the clock inputs of all SDH equipment along the data path and in the transmitting and receiving slip-buffers of the TM. Refer to Fig. 10.2.

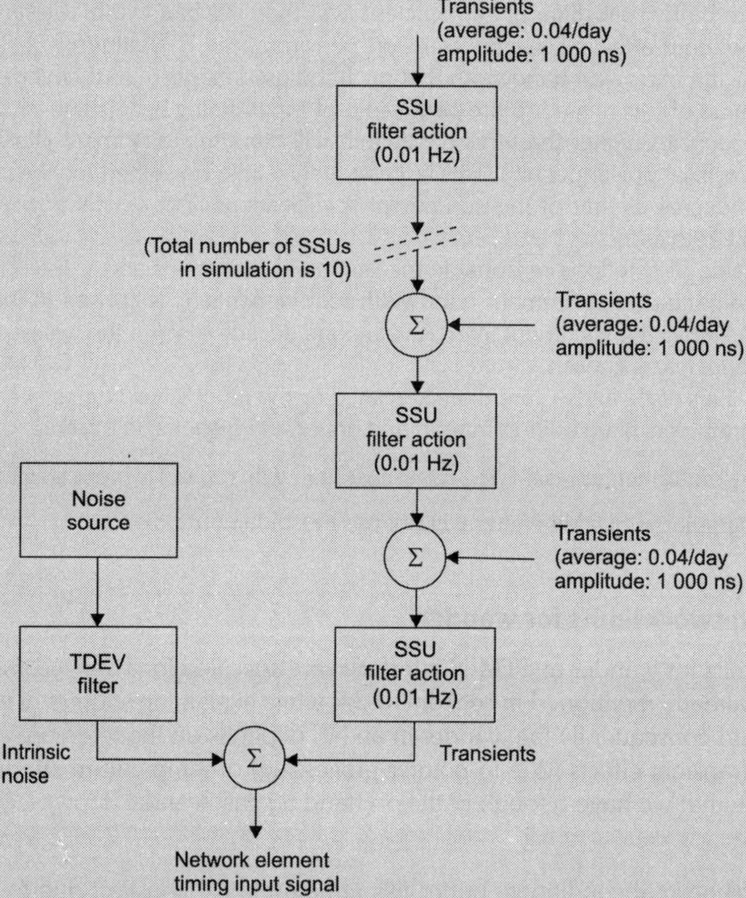

Fig. 10.2 *Simulation model for measuring wander due to clock noise and transients on SDH network*

Intrinsic noise and transients are generated separately. The intrinsic noise for 1 PRC (primary reference clock), 10 SSU (synchronisation supply unit) followed by 20 SEC (SDH equipment clock) is based on the data from ITU-T Rec. G.811, 812 and 813.1 PRC.

Wander due to clock noise and transients was found to be 15 ms in the simulations. To this, the following were added:

- Diurnal wander caused by environmental influence on OFC, which is considered 1 ms over one SDH island (taking OFC link of 6000 km length, subject to a temperature change of 2°C and with a temperature coefficient of 85 ps/km/°C).
- Mapping wander of 2.048 Mbps signals into VC-12 was accounted as 2 ms. This is based on the consideration that the VC-12 mapping wander is 4 UI (corresponding to 2 ms at 2.048 Mbps).

Thus, we get the limit of 18 ms for wander.

The discussion so far was concerned with wander caused in SDH elements and its accumulation. Wander of reference clock caused by pointer processing in NE gets transferred to the PDH signal at the mapping and de-mapping nodes. We have to consider the worst case scenario, i.e., mapping/de-mapping when the NE is operating on SEC. The related phase transient is 240 ns. Using this data, let us estimate the wander for an SDH network segment at the mapping and de-mapping nodes.

At the de-mapping node

In addition to the transfer of wander from the SEC to the PDH signal being de-mapped, there is the stuffing transient, besides which there may be a single pointer justification just before the de-mapping. The resulting wander (maximum) can be due to

Bit stuffing	:	60 ns
SEC phase transient	:	240 ns
TU-3 pointer justification	:	160 ns
Total	:	460 ns

At the mapping node

We have to consider SEC phase transient and the effect of double pointer justification (intermediate node pointer justification plus pointer justification at the de-mapping node). Hence, the wander (maximum) can be:

SEC phase transient	:	240 ns
Double pointer justification	:	320 ns
Total	:	560 ns

We need not add the two above, as the reference clock switching is a rare event and happens at one end at a time. Hence, an output wander of 500 ns is considered as typical for a single SDH network.

If N networks are cascaded, the output wander would work out to $500N^{1/2}$.

To summarise:

- On SDH networks the wander limit is 18 ms in 24 hours for SDH NEs taken together;

- The maximum output short-term wander on PDH interfaces working through SDH network is 1000 ns.

10.3 Network Limits for Jitter and Wander at Synchronisation Interfaces

One might wonder why we are discussing jitter and wander limits at synchronisation interfaces when we have already covered STM interfaces. It may be noted that STM interfaces are synchronous interfaces. Synchronisation interfaces are the clock signals (PRC, SSU, SEC and PDH clocks) used to synchronise NEs. These signals leave a trail as they pass from the starting node down the network, but become more and more out of sync as they pass through each NE. So, every NE can check out if any reference that it is getting has exceeded the network limit prescribed, or how close it is to the limit and decide on which reference to use.

Let us see the limits prescribed in ITU-T recommendations.

10.3.1 Network limits for jitter on synchronisation signals (clock signals)

Network limits for jitter on timing signal specify the maximum allowable high frequency noise components in the signal. These are specified in ITU-T Rec. G.823 and tabulated in Table 10.2. Jitter, as measured over a 60 second interval, should not exceed the specified limits, when using the specified measurement filters.

Table 10.2 *Network limits for jitter on clock signals*

Output interface (clock)	Jitter measurement bandwidth ($f_1 - f_4$)	Peak-to-peak amplitude A_2 (UI pp)	Jitter measurement bandwidth ($f_3 - f_4$)	Peak-to-peak amplitude A_1 (UI pp)
PRC	20 Hz to 100 kHz	0.05	–	–
SSU	20 Hz to 100 kHz	0.05	–	–
SEC	20 Hz to 100 kHz	0.5	49 Hz to 100 k Hz	0.2
PDH	20 Hz to 100 kHz	1.5	18 kHz to 100 kHz	0.2

10.3.2 Network limits for output wander at synchronisation interfaces (clock signals)

An NE can get clock references from two neighbours which might be derived from the same synchronisation source (e.g., PRC or SSU) but have arrived on different paths. At very low frequencies, synchronisation networks are transparent to wander. So, the two clock references received by an NE from two neighbours over different paths may even have opposite phase (worst-case variation). Hence, the tolerance for such differential phase variation in the frequency range where the NE is affected should be higher than the network limit for absolute wander. Accordingly, the following wander tolerance limits are specified for synchronisation signals.

PRC interface output wander limit

Network limit for wander at the output interface of PRC, expressed in MTIE (maximum time interval error), is given in Table 10.3(a) and shown graphically in Fig. 10.3(a).

Table 10.3(a) *PRC interface output wander limit in MTIE*

Observation interval (t seconds)	MTIE requirement (ns)
$0.1 < t \leq 1000$	$25 + 0.275\,t$
$t > 1000$	$290 + 0.01\,t$

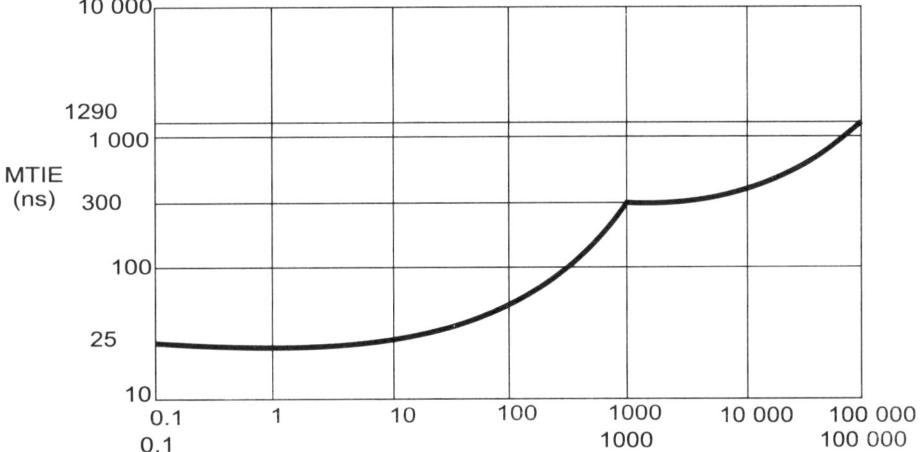

Fig. 10.3(a) *Graphical representation of PRC interface output wander limit*

SSU interface output wander limit

The network limit for wander at the output interface of SSU, expressed in MTIE, is given in Table 10.3(b) and shown graphically in Fig. 10.3(b).

Table 10.3(b) *SSU interface output wander limit in MTIE*

Observation interval (t seconds)	MTIE requirement (ns)
$0.1 < t \leq 2.5$	25
$2.5 < t \leq 200$	$10\,t$
$200 < t \leq 2000$	2000
$t > 2000$	$433\,t^{0.2} + 0.01\,t$

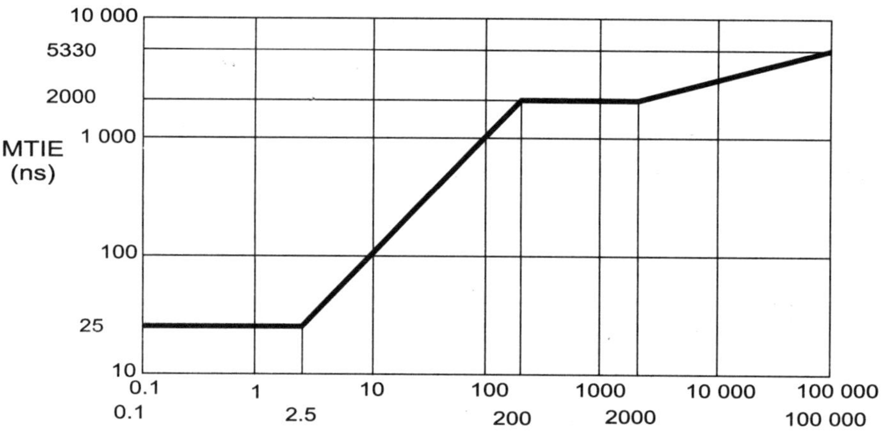

Fig. 10.3(b) *Graphical representation of SSU interface output wander limit*

SSU interface output wander limit

Network limit for wander at the output interface of SSU, expressed in MTIE, is given in Table 10.3(c) and shown graphically in Fig. 10.3(c).

Table 10.3(c) *SSU interface output wander limit in MTIE*

Observation interval (t seconds)	MTIE requirement (ns)
$0.1 < t \leq 2.5$	250
$2.5 < t \leq 20$	$100\, t$
$20 < t \leq 2000$	2000
$t > 2000$	$433\, t^{0.2} + 0.01\, t$

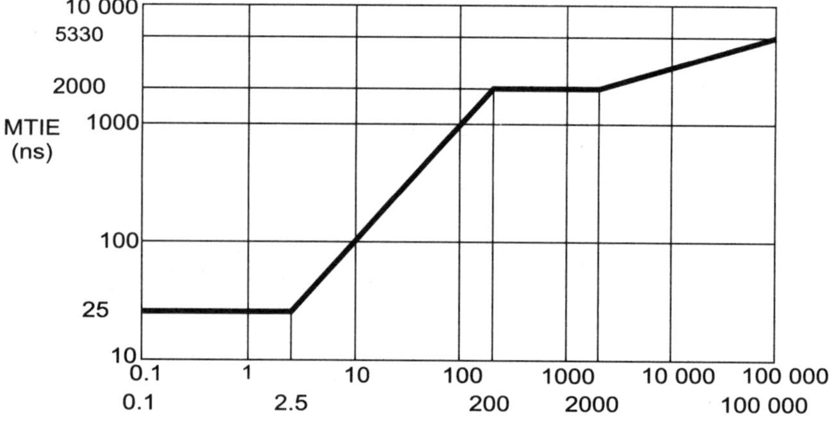

Fig. 10.3(c) *Graphical representation of SSU interface output wander limit*

PDH synchronisation interface output (clock) wander limit

The network limit for wander at PDH synchronisation interface output (clock), expressed in MTIE, is given in Table 10.3(d) and shown graphically in Fig. 10.3(d).

Table 10.3(d) *PDH synchronisation interface output (clock) wander limit in MTIE*

Observation interval (t seconds)	MTIE requirement (ns)
$0.1 < t \leq 7.3$	732
$7.3 < t \leq 20$	$100\,t$
$20 < t \leq 2000$	2000
$t > 2000$	$433\,t^{0.2} + 0.01\,t$

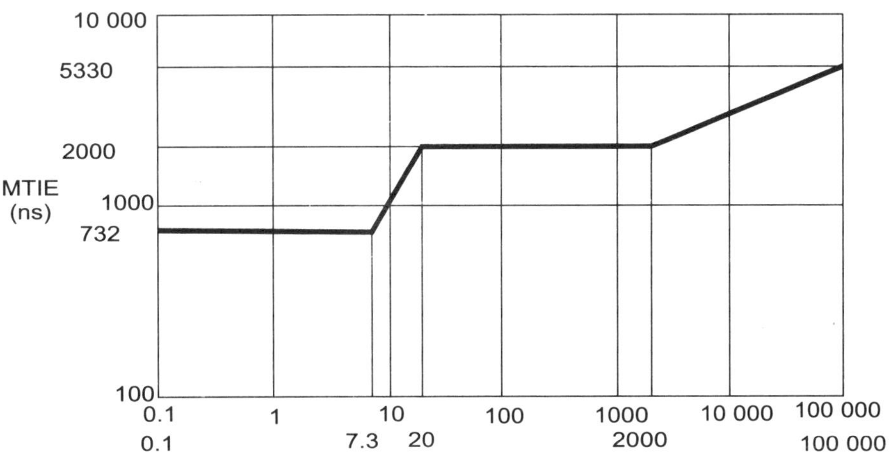

Fig. 10.3(d) *Graphical representation of PDH synchronisation interface output (clock) wander limit in MTIE*

It may be noted that these limits can be expressed in TDEV (time deviation) instead of MTIE (maximum time interval error). Interested readers can refer to ITU-T Rec. G.823 on this.

Q 10.3 *We are specifying wander by MTIE and TDEV parameters. Which is a better measure?*

MTIE or maximum time interval error is the maximum peak-to-peak delay variation of a given timing signal with respect to an ideal timing signal within an observation time ($t = n\,t_0$) for all observation times of that length within the measurement period (T). TDEV or time deviation on the other hand, is a measure of the expected time variation of a signal as a function of integration time. TDEV can also provide information about the spectral content of the phase (or time) noise of a signal. TDEV is expressed in units of time.

MTIE is considered useful for capturing the phase transients in a timing signal since it describes the maximum phase variation of a timing signal over a period of time. However, MTIE is inadequate to show the underlying noise on the timing signal because of its sensitivity to phase transients. Random noise is better characterised by TDEV, which is an RMS power estimator instead of a peak estimator. TDEV tends to remove transients in a timing signal, and is therefore a better estimator of the underlying noise processes. However, transients and periodic components should be removed from data prior to calculating TDEV. This is not appropriate for measurements on network interfaces since there is no previous knowledge of the types of disturbances experienced in the timing signals. So, the TDEV results from processing of raw phase data may not exactly reflect the random noise processes in a timing signal on a network interface, but they can provide a good estimate.

Hence the choice of MTIE limits in the previous section concerned with network limits for wander. (Mathematical analysis of MTIE and TDEV is presented in ITU-T Rec. G.810).

10.3.3 Measurement of output wander at STM interfaces

From the above discussion, it is clear that the wander budget is of great importance in SDH network and we need to measure the quantum of wander on the outputs carrying synchronisation. When the STM output is PRC-traceable and is used to carry synchronisation, its wander is measured by comparing its phase with that of another PRC. The test setup for measuring MTIE of wander of such STM output is shown in Fig. 10.3(e).

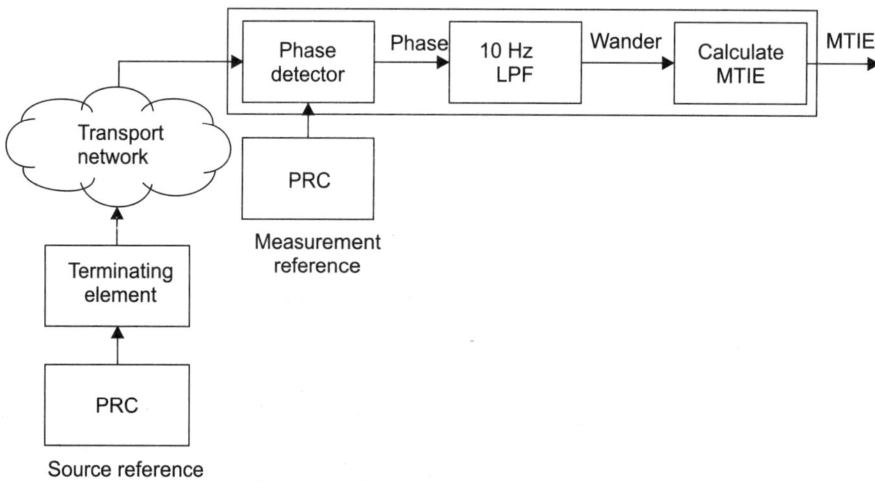

Fig. 10.3(e) *Measurement of wander (MTIE) at STM output*

10.4 Jitter and Wander Tolerance for Input Ports

Since SDH equipment map PDH signals also, the jitter and wander tolerances of PDH input ports should be as per ITU-T Rec. G.823, which was discussed in detail in chapter 5. We will discuss jitter and wander tolerance limits for STM interfaces (input ports).

STM interfaces should accommodate input signal that has

1. Electrical characteristics in ITU-T Rec. G.703.
2. Frequency offset of 4.6 ppm if the SDH network is optimised for E-hierarchy and deployed with G.813 clock.
3. Jitter and wander tolerance as per ITU-T G.825.

The jitter and wander tolerances of an SDH interface as specified in ITU-T Rec. G.825 indicate the minimum level of phase noise that the input port should accommodate in such a way that:

- No alarms are caused
- No slips are caused
- No bit errors are caused except for optical STM-N interfaces at jitter frequencies above f_p (f_p is 6.5 kHz for STM-1, 25 kHz for STM-4, 100 kHz for STM-16 and 400 kHz for STM-64), where an equivalent 1 dB optical power penalty should not be exceeded.

Let us examine these tolerances.

10.4.1 Input jitter tolerance of STM ports

The levels of jitter that must be accommodated by STM-1, STM-4, STM-16 and STM-64 interfaces are given in Tables 10.4(a–d) and also shown graphically in Fig. 10.4(a–d) respectively. It may be noted that 1 UI for STM-1 is 6.43 ns, for STM-4 it 1.61 ns, for STM-16 it is 0.402 ns and for STM-64 it is 0.100 ns.

Table 10.4(a) *Input jitter tolerance at STM-1 port*

Frequency (Hz)	Tolerance level (peak-to-peak)
$10 < f \leq 19.3$	38.9 UI
$19.3 < f \leq 68.7$	$750/f$ UI
$68.7 < f \leq 500$	$750/f$ UI
$500 < f \leq 6.5$ k	1.5 UI
6.5 k $< f \leq 65$ k	$9800/f$ UI
65 k $< f \leq 1.3$ M	0.15 UI

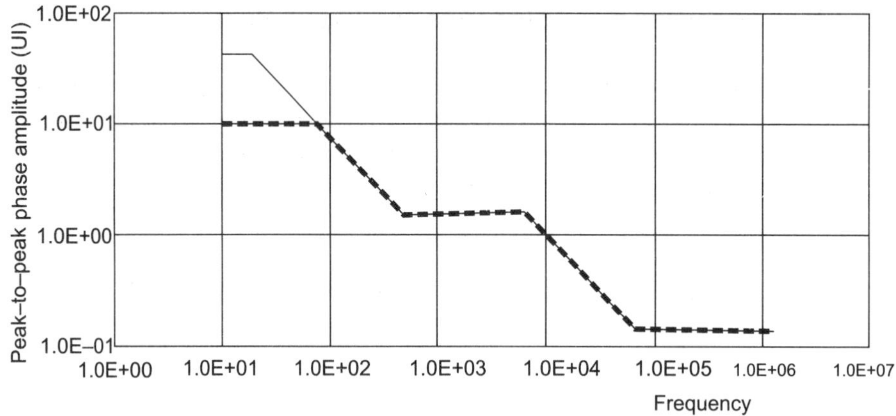

(**Note:** Dashed line shows tolerance limit for T1 based NW for f < 68.7 Hz)

Fig. 10.4(a) *Input jitter tolerance at STM-1 port*

Table 10.4(b) *Input jitter tolerance at STM-4 port*

Frequency (Hz)	Tolerance level (peak-to-peak)
$9.65 < f \leq 100$	1500/f UI
$100 < f \leq 1000$	1500/f UI
$1k < f \leq 25k$	1.5 UI
$25k < f \leq 250$ k	38000/f UI
$250k < f \leq 5$ M	0.15 UI

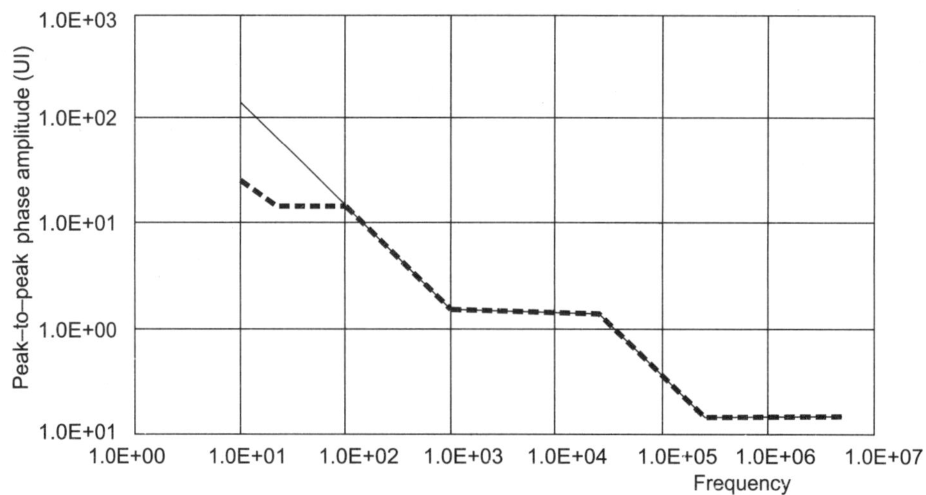

(**Note:** Dashed line shows tolerance limit for T1 based NW for f < 100 Hz)

Fig. 10.4(b) *Input jitter tolerance at STM-4 port*

Table 10.4(c) *Input jitter tolerance at STM-16 port*

Frequency (Hz)	Tolerance level (peak-to-peak)
$10 < f \leq 12.1$	622 UI
$12.1 < f \leq 500$	7500/f UI
$500 < f \leq 5k$	7500/f UI
$5k < f \leq 100k$	1.5 UI
$100k < f \leq 1M$	15000/f UI
$1M < f \leq 20$ M	0.15 UI

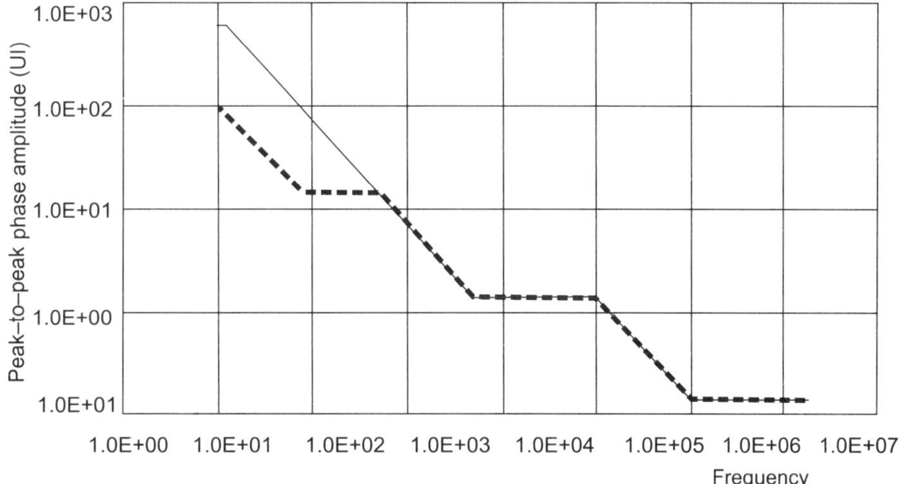

(**Note:** Dashed line shows tolerance limit for T1 based networks for frequencies less than 500 Hz)

Fig. 10.4(c) *Input jitter tolerance at STM-16 port*

Table 10.4(d) *Input jitter tolerance at STM-64 port*

Frequency (Hz)	Tolerance level (peak-to-peak)
$10 < f \leq 12.1$	2490 UI
$12.1 < f \leq 2k$	30000/f UI
$2k < f \leq 20k$	30000/f UI
$20k < f \leq 400k$	1.5 UI
$400k < f \leq 4M$	600000/f UI
$4M < f \leq 80$ M	0.15 UI

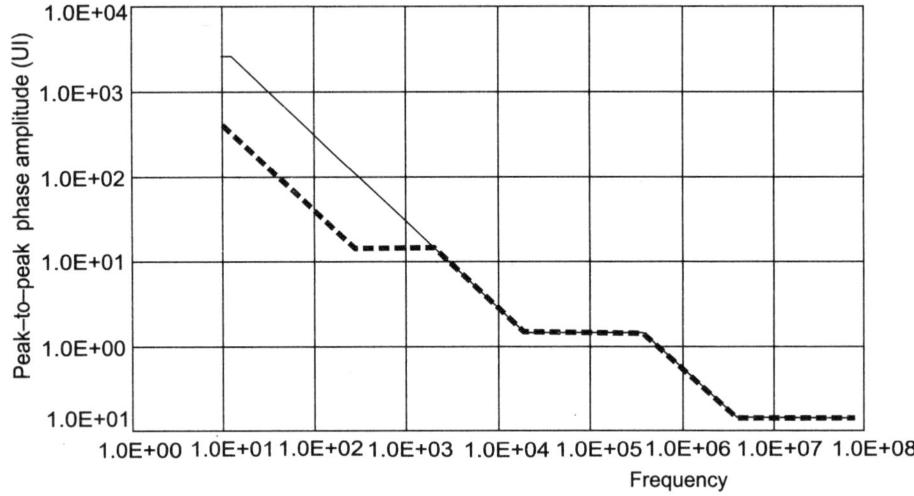

(**Note:** Dashed line shows tolerance limit for T1 based NW for *f* < 2 kHz)

Fig. 10.4(d) *Input jitter tolerance at STM-16 port*

10.4.2 Input wander tolerance

SEC input wander tolerance expressed in MTIE is shown in Table 10.5(a) and Fig. 10.5(a).

Table 10.5(a) *Input wander tolerance in MTIE*

Observation interval	MTIE limit
$0.1 < \tau \le 2.5$ s	0.25 μs
$2.5 < \tau \le 20$ s	$0.1\ \tau$ μs
$20 < \tau \le 400$ s	2 μs
$400 < \tau \le 1000$ s	$0.005\ \tau$ μs

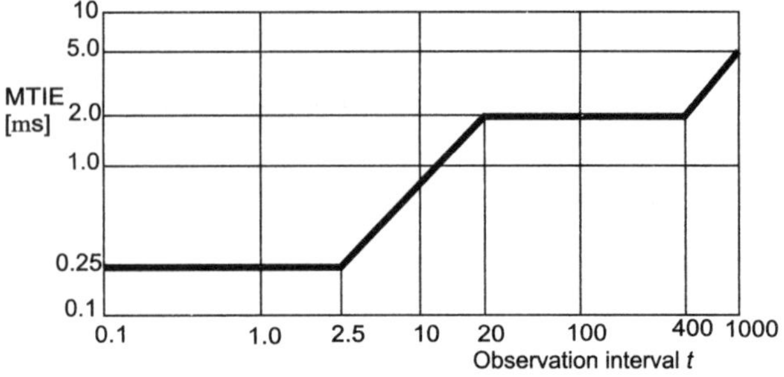

Fig. 10.5(a) *Input wander tolerance in MTIE*

SEC input wander tolerance expressed in TDEV is shown in Table 10.5(b) and Fig. 10.5(b).

Table 10.5(b) *Input wander tolerance in TDEV*

Observation interval	TDEV limit
$0.1 < \tau \leq 7$ s	12 ns
$7 < \tau \leq 100$ s	$1.7\ \tau$ ns
$100 < \tau \leq 1000$ s	170 ns
$400 < \tau \leq 1000$ s	$0.005\ \tau$ μs

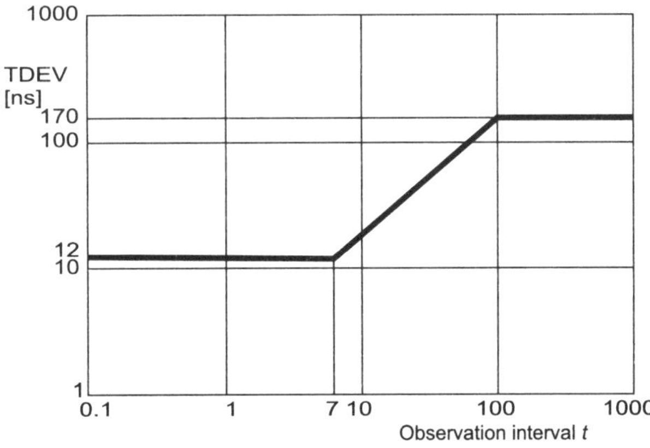

Fig. 10.5(b) *Input wander tolerance in TDEV*

10.5 Intrinsic Jitter

As already mentioned in Chapter 5, intrinsic jitter is the output jitter in the absence of input jitter. A type-A NE on SDH network, i.e., an NE optimised for E hierarchy, should not have intrinsic jitter at the STM-N output in excess of the values shown in Table 10.6 for the frequency range mentioned.

Table 10.6 *Intrinsic jitter limits for STM-N equipment*

STM-N output	UI for the STM-N	Measurement bandwidth ($f_1 - f_4$)	Jitter in UI (peak-to-peak)	Measurement bandwidth ($f_3 - f_4$)	Jitter in UI (peak-to-peak)
STM-1	6.43 ns	0.5 k Hz to 1.3 M Hz	0.3	65 k Hz to 1.3 M Hz	0.1
STM-4	1.61 ns	1 k Hz to 5 M Hz	0.3	250 Hz to 5 M Hz	0.1
STM-16	0.40 ns	5 k Hz to 20 M Hz	0.3	1 M Hz to 20 M Hz	0.1
STM-64	0.10 ns	20 k Hz to 80 M Hz	0.3	4 M Hz to 80 M Hz	0.1

10.6 Jitter and Wander Transfer Function

The jitter and wander transfer function is defined as the ratio of jitter (wander) on the output STM-N signal to the jitter (wander) applied on the input STM-N signal versus frequency. Let us use the term 'jitter transfer function' for convenience to represent both jitter and wander. The jitter transfer function of an NE on SDH network optimised for E hierarchy is shown graphically in Fig. 10.6 and the parameters for the same are specified in Table 10.7.

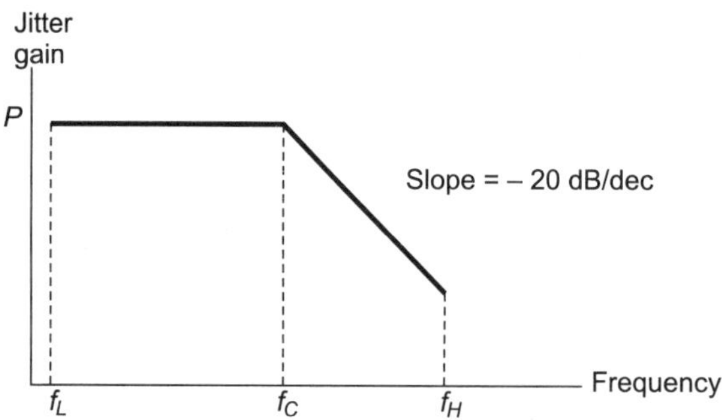

Fig. 10.6 *Jitter transfer function for STM-N output*

Table 10.7(a) *Parameter values on jitter transfer function*

STM-N level	f_L (k Hz)	f_C (k Hz)	f_H (k Hz)	P (dB)
STM-1	1.3	130	1300	0.1
STM-4	5	500	5000	0.1
STM-16	20	2000	20000	0.1
STM-64	10	1000	80000	0.1

Q 10.4 What is the significance of f_L, f_C and f_H? Can we apply any input jitter and expect the equipment to obey the above transfer function?

The jitter transfer measurement is made over the frequency range f_L to f_H. The lower frequency f_L is set to $f_C/100$ (f_C is the corner frequency shown in Fig. 10.6) and f_H is defined as the lower of either $100 \times f_C$ or the maximum frequency specified for the low pass filter function for measurement of jitter at each of the defined rates in Table 10.7(a).

We cannot apply any jitter at an input and expect the equipment to obey the transfer function. Remember that an equipment has to obey input tolerance requirements as well as intrinsic tolerance requirements, i.e., the resultant of the transferred jitter at the output as the input jitter and the intrinsic jitter should together obey the output jitter limits.

The input jitter mask or the input jitter which is applied to measure transfer function is shown in Fig. 10.7, with its parameters given as per Table 10.7(b).

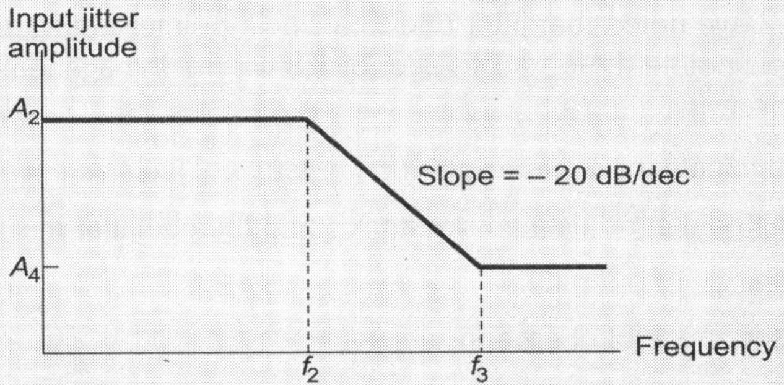

Fig. 10.7 *Input jitter mask to be applied for transfer function measurement*

Table 10.7(b) *Parameter values for input jitter mask while measuring transfer function*

STM level	A_3 (UI)	A_4(UI)	f_2 (k Hz)	f_3 (k Hz)
STM-1	1.5	0.15	6.5	65
STM-4	1.5	0.15	25	250
STM-16	1.5	0.15	100	1000
STM-64	1.5	0.15	400	4000

10.7 Impairments and Interruptions Related to Jitter and BER

In OFC networks based on SONET/SDH, regular operations like mapping and de-mapping which are part of the multiplexing process cause jitter. Periodic patterns, if any, also cause jitter. Scrambling techniques in frame formation and the high-Q PLL of clock extraction circuit ensure neutralisation of such jitter and compliance to ITU-T recommendations discussed in the previous sections. But, the jitter problem is not yet completely attended to. There are other problems faced by networks such as random phase variations in reference clock, short-term impairments, long term interruptions (e.g., fibre cuts, node failures) which occur at multiple locations. Random phase variation in reference clock causes sporadic movement of the pointer and can cause single pointer/triple pointer adjustment and associated jitter, but it has no cumulative or chaining effect.

The situation is more complex in case of short-term impairments/long term interruptions. In such situations, survivability features of the network through ring protections (MS-SPRing and SNCP, as covered in chapter 8) ensure continuity of traffic through protection bandwidth, either 1+1 arrangement or 1:1 arrangement with APS. But, NEs on the trails/rings involved in such restoration process change their clock reference. This in turn causes pointer adjustments in a cascade and the resultant jitter and BER have a cumulative effect on the payloads till the NEs settle to the new scenario.

To summarise, there are random/persistent/periodic adjustment of pointers in different situations.

In section 10.2, we noted that jitter due to a single pointer adjustment is about 0.6 UI and that a triple pointer burst causes jitter of 1.3 UI. So, the agenda before us in this chapter is very clear.

1. Derive an expression to represent BER in terms of jitter
2. Predict the pointer adjustments in an NE and thereby jitter in:
    - Ideal mode of operation
    - Stressed mode of operation
    - Hold-over mode of operation
3. Compute the increase in BER as per the predicted jitter till the NEs settle down
4. Compare with actual measurements taken.
5. Predict the ill-effects of such pointer adjustments on voice/data/video circuits carried over SDH
6. Propose solutions

These are covered now, one by one.

10.7.1 Derivation of expression for BER in terms of jitter

The manifestation of jitter as error depends on two sequential decisions made by receiver:

First decision: When to sample?

Second decision: What is received, 1 or 0?

The second decision is dealt first, assuming perfect symbol synchronisation, i.e., no jitter. Later, the first decision will be dealt with; rather, it is analysed how the error in the timing of the sampling instant (which is jitter) affects the difference between the zero and one levels and thereby increases the probability of bit errors. Then, applying probability theory and using the results of the two decisions, we find expression for BER in terms of jitter.

The one or zero decision

Assuming perfect synchronisation between the bit stream and the bit clock, the cause of erroneous decision is the receiver noise, which, in optical receivers is thermal noise and shot noise. Thermal noise is characterised by Gaussian distribution and shot noise by Poisson distribution. However, the inaccuracy due to assumption of Gaussian distribution is negligible[1] and hence, we can assume the noise to be additive white Gaussian noise (AWGN) over the band of operation.

Let the mean and variance of the Gaussian output for pulse 1 be V_{S1} and σ^2_{ON} and that for pulse 0 be V_{S0} and σ^2_{OFF}. Let the decision threshold be V_{th}. Let us assume equal number of 1s and 0s over a sufficiently large period of observation, and also set the threshold for equal error probability for the 0 and 1 pulses.

$$P(e) = P(\text{Recd. level} > V_{th} \text{ when level 0 is sent}) + P(\text{Recd. level} < V_{th} \text{ when level 1 is sent})$$

$$P(e) = P[V(t) > V_{th} | V_{S0}]\, P[V_{S0}] + P[V(t) < V_{th} | V_{S1}]\, P[V_{S1}], \tag{10.1}$$

where $P(e)$ is the probability of error and $P(x \mid y)$ represents the probability of x given y.

$$P[0] = P[1] = 0.5.$$

Then, equation (10.1) can be reduced to[2]

$$P(e) = \frac{1}{\sqrt{2\pi\sigma^2_{ON}}} \int_{V_{th}}^{\infty} \exp\left[\frac{-(V - V_{S1})^2}{\sigma^2_{ON}}\right] dv = \frac{1}{\sqrt{2\pi\sigma^2_{OFF}}} \int_{V_{th}}^{\infty} \exp\left[\frac{-(V - V_{S0})^2}{\sigma^2_{OFF}}\right] dv \tag{10.2}$$

$$\text{Define } Q = \frac{(V_{th} - V_{S0})}{\sigma_{OFF}} = \frac{V_{S1} - V_{th}}{\sigma_{ON}} \tag{10.3}$$

$$BER = P(e) = P_e(Q) = \frac{1}{\sqrt{\pi}} \int_{Q/\sqrt{2}}^{\infty} \exp\left[-x^2\right] dx = \frac{1}{2}\left[1 - \text{erf}\left(\frac{Q}{\sqrt{2}}\right)\right], \tag{10.4}$$

where erf(x) is the error function[2]. The above simplified expression for BER in terms of Q means the following:

- relative to the noise at V_{S0}, V_{th} to be at least Q standard deviations above V_{S0}, or
- relative to the noise at V_{S1}, V_{th} to be no more than Q standard deviations below V_{S1}

to have the desired BER. Let us find out the Q and SNR requirement for BER standards of 1×10^{-9} and 1×10^{-12}.

For BER of 1×10^{-9}, we get $Q = 6$ (approximately), which implies an SNR of 12, i.e., 10.8 dB.

For BER $= 1 \times 10^{-12}$, we get $Q = 7$, which implies an SNR of 14, i.e., 11.46 dB.

Decision regarding sampling instant

Essentially, jitter causes deviation of the sampling instant from the ideal, which can in turn influence the quality of decision made on zero/one. The decision of the sampling circuit depends on how the input value at the sampling instant compares with the threshold value. In other words, the location or the timing of the sampling instant affects the difference between the zero and one levels and thereby increases the probability of bit errors. It is important to note that timing of the sampling instant has no effect on the noise.

We can represent the timing of the sampling instant by t_s, as shown in Fig. 10.8.

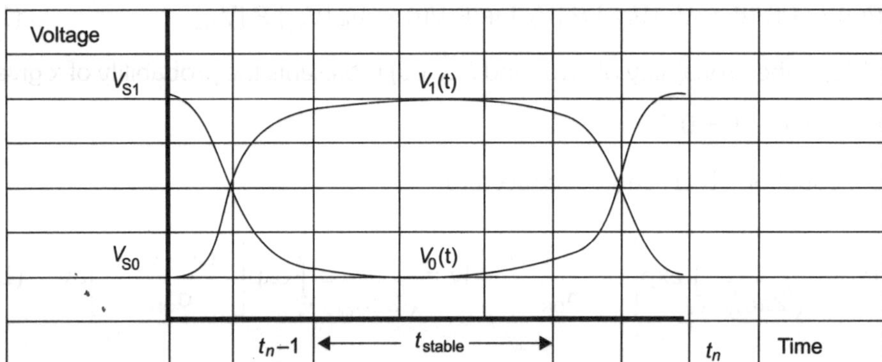

Fig. 10.8 *Timing of the sampling instant*

We can see from Fig. 10.8 that the difference between V_{S1} and V_{S0} is maximised (and thus the bit error ratio is minimised) when the bit is sampled in the stable region, t_{stable}. If, due to jitter, the bit is sampled between t_{n-1} and t_{stable} or t_{stable} and t_n, then the difference between V_{S1} and V_{S0} will be less than the maximum, resulting in an increased probability of bit error. When the bit is sampled exactly at the transition time, the probability of bit error is 50% (assuming 50% mark-density random data).

Predicting the BER caused by jitter

Bit errors occur in the absence of jitter effects (even with optimum sampling) as a consequence of amplitude noise, as already discussed in the previous section. Jitter adds the dimension of time to the problem of predicting bit error ratio (BER). When jitter is considered, the calculated probability of bit error, P_e must take into account the probability of bit error at each potential sampling instant, $P_e | t_s$, as well as the probability that each sampling instance actually occurs, $P(t_s)$.

Hence, BER due to jitter can be computed (plotted) with the following steps.

1. Plot $P_e | t_s$ according to equation (10.4).
2. Plot $P(t_s)$.

3. Multiply $P_e|t_s$ and $P(t_s)$ instant-by-instant and find the area under the desired portion of UI (jitter magnitude).

Probability of error at each sampling instant ($P_e|t_s$)

By equation (10.4), P_e is determined by the noise (σ^2_{ON}, σ^2_{OFF}) associated with the input waveforms and the difference between the zero and one levels, ($V_{S1} - V_{S0}$). Noise can be measured directly and its magnitude generally stays constant for long periods of time. The sampled signal amplitude, however, may vary depending on the sampling instant. This means that P_e will be different at different times, depending on the sampling instant.

An example of the relationship between the sampling instant and P_e is illustrated in Fig. 10.9. Figure 10.9(a) represents the eye diagram of the data signal at the input to the sampling circuit in a receiver. From a data-referenced perspective, which is easier to appreciate, the horizontal position of data eye diagram is fixed and the relative horizontal movement of the clock signal (data referenced jitter) is observed. The time of interest to us is the time between the zero crossings, i.e., one unit interval (UI).

Figure 10.9(b) is a plot of $V_1(t) - V_0(t)$ versus time.

In the example presented here, the level of logic 1 is chosen as 1.85 V. For BER of 1×10^{-12}, the SNR required is 14. Since ${V_{rms}}^2/\sigma^2$ = SNR =14, for BER of 1×10^{-12}, σ is found to be 0.25. It may be noted that the values chosen for V are immaterial as we are concerned with the SNR, the Q required for BER of 1×10^{-12} and how BER degrades with change of value of V from the chosen logic 1 level in the interval of observation.

As the voltage level of the received signal falls to a value say, V_j, the signal power would be ($V_j^2/1.85^2$) times the signal power at logic 1 level and the SNR would be ($V_j^2/1.85^2$) times SNR at logic 1 level (= ($V_j^2/1.85^2$) × 14). We can compute the SNR for different V_js and also *for* Q = ½ SNR using equation (10.4). Using this value of Q, we can find the error function value and thus $P(e)$ or BER for the V_j under consideration. Table 10.8 presents BER for some of the V_js used in the plot of Fig. 10.9(b).

Figure 10.9(c) is a plot of $P_e|t_s$ computed as above. This plot shows the probability that the bit will be erroneously detected as a function of the sampling instant.

Table 10.8 *BER computation from V_js*

V_j	SNR	Q	$P(e)$
1.85	14	7	1E-12
1.48	8.96	4.48	1E-8
1.11	5.04	2.52	1E-5
0.74	2.24	1.12	1E-3
0.37	0.56	0.28	1E-2

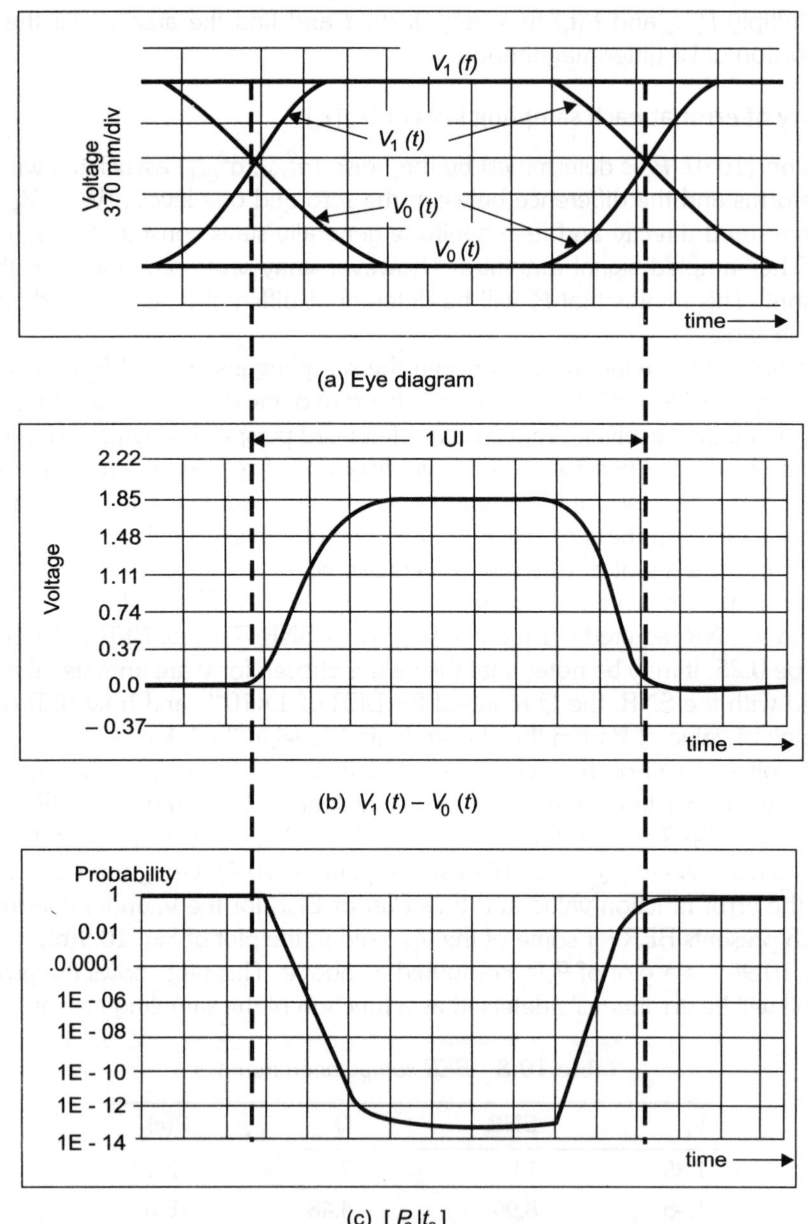

Fig. 10.9 *Computing Pε at each sampling instant*

Sampling time probability [$P(t_s)$]

Figure 10.10 shows sampling instant PDF (jitter histogram) on both linear and logarithmic vertical scales. This PDF represents the timing probability of the recovered bit clock relative to the data at the input to the receiver sampling circuit. It may be noted that, even though it is highly probable that the bit will be sampled around the centre of the UI, there is still a finite probability, due to jitter, that the bit will be sampled in the vicinity of the bit transition or beyond.

BER due to jitter

We now have the required data to compute BER due to jitter. In order to do this, we can apply the statistical definition of conditional probability to compute the probability of bit error over the full range of sampling times.

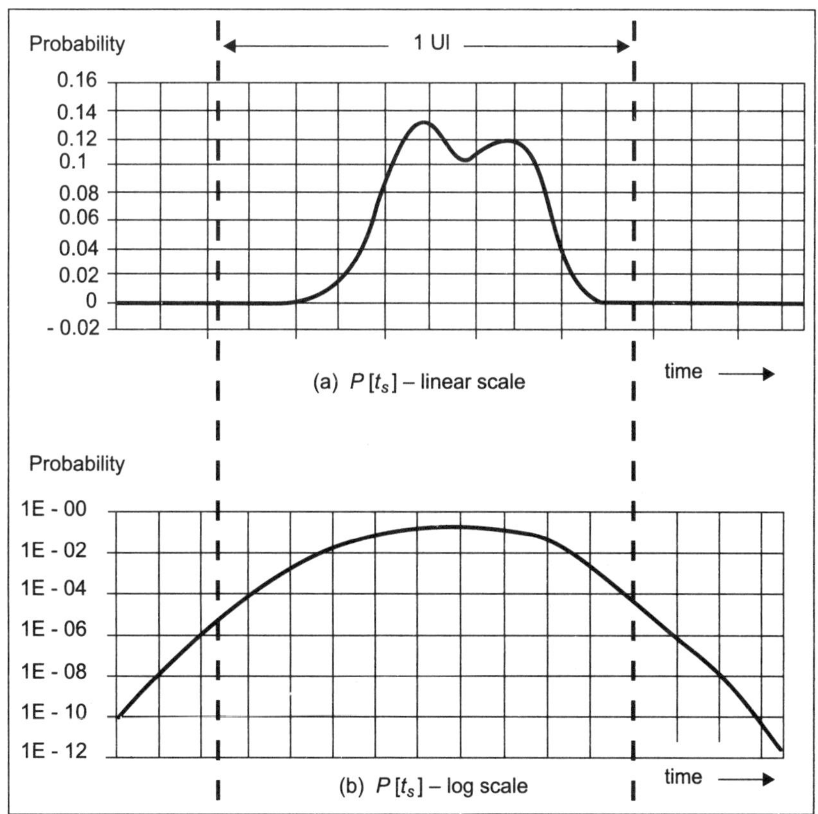

Fig. 10.10 *Sampling instant probability density function*

$$P(e, t_s) = [P_e | t_s].[P(t_s)] \tag{10.5}$$

By equation (10.5), we can do a point-by-point multiplication of the jitter-free plot of Fig. 10.9(c) (i.e., $P_e | t_s$) and sample the instant PDF of Fig. 10.10 (i.e., $P_e | t_s$). The result is the total probability of bit error distributed over the full range of possible sampling time instants. This result is plotted in Fig. 10.11. It is interesting to note that most of the bit errors occur near the bit transition times, and that these errors are caused by the relatively improbable extremes in sampling instant deviations, i.e., jitter.

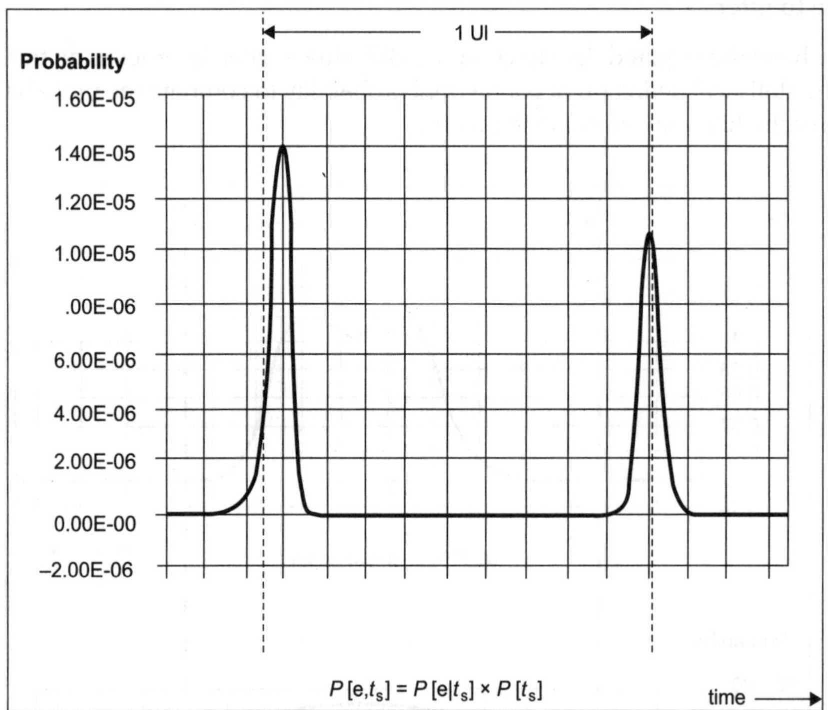

Fig. 10.11 *Probability of bit error over the range of possible sampling instants*

The cumulative P_e (or BER) can be computed by integrating the result of equation (10.5) or by finding the cumulative area under the curve in Fig. 10.11.
To summarise:

1. In order to make accurate predictions of jitter BER, it is important to consider the effects of finite rise and fall times and also amplitude noise, rather than brush them under the carpet of convenient assumptions.
2. If jitter is small enough, the resulting timing deviations of the sample clock will be confined to the stable region of the bit period and will have no effect on the BER.

3. The stable region of the bit period can be increased (and thus susceptibility to jitter decreased) by decreasing the rise and fall times and/or decreasing the noise.

If we find out the jitter in NEs due to random phase variations/short term impairments/ long term interruptions, we can estimate the BER using results of section 10.7.1. To estimate jitter in NEs, we have to take a closer look at pointer adjustments in situations of clock impairment. Once we know the pointer adjustments in each NE, we can find the jitter contribution per NE as well as the overall contribution of jitter to predict the BER.

10.7.2 Estimating pointer adjustments in different modes

In section 7.5, three modes of operation of NE were explained – ideal, stressed and hold-over. Let us understand the pointer adjustments in these modes.

Pointer adjustment in ideal mode

In ideal operation the receiver clock experiences no interruptions of the input timing reference. However, random phase variations of considerable amount cause pointer adjustments in SDH payloads and thereby jitter, in an isolated way. The number of such variations cannot be formulated mathematically, but observations on operational networks have indicated that there can be one such variation every 100–120 second in a trail so such variations per day can be 800 on an average, causing as many pointer adjustments. Though jitter per pointer adjustment is 0.6 UI, there is no cumulative effect in this case, as each of such events is isolated.

Q 10.5 In section 10.2, it was mentioned that mapping jitter is 0.4 UI and single pointer adjustment jitter is 0.6 UI. What is the basis for this?

These values of jitter were found using test sequences developed by ANSI which were later included in ITU-T Rec.G.783. They are known by G.783 test sequence for finding the mapping jitter and 87/3 sequence for finding the single pointer adjustment jitter.

The first test sequence (G.783 sequence) emulates the situation when there is no pointer activity. This test requires the frequency of the tributary signal to be varied in order to find the maximum jitter due to the pulse stuffing ratio used for maping the tributary signal into the VC, i.e., the 'mapping jitter'. The result was 0.4 UI.

The second test sequence has a single pointer movement every 30 seconds. This allows time for the effects of each pointer movement to completely subside before another pointer movement is applied—we thus measure 'Single pointer movement jitter'. The sequence is illustrated in Fig. 10.12.

When an NE cross-connecting VC-4 payload loses its lock to the master timing source in the SDH network, it enters hold-over mode and uses the internal reference working at the last average rate. The frequency difference between this reference of NE and the master timing source will have the effect of generating evenly spaced pointer movements, *T* seconds apart. However, rather than continuous, evenly-spaced pointer movements, the equipment produces a sequence in which a pointer movement occurs during the first 87 periods but no pointer is produced during the last 3 periods, i.e., an "87/3" sequence. This effect is the fall-out of positioning SOH bytes in the SDH frame.

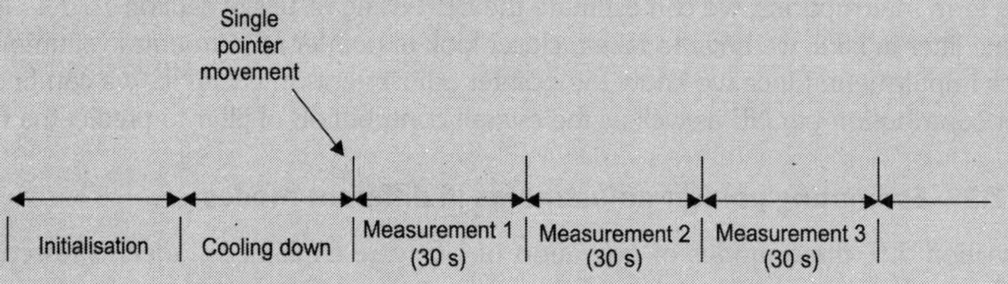

Fig. 10.12 *Single pointer movement test sequence*

To get pointer movement similar to this practical situation, the clock reference is varied by 4.6 ppm (maximum allowed variation according to ITU-T G.813) every 30 seconds, and for every 87 pointer activities, 3 gaps (no pointer activity) are created to facilitate positioning of SOH bytes.

Q 10.6 *When we are interested in finding jitter due to single pointer activity, what is the need to use 87/3 sequence for testing purposes?*

We have used the sequence only to get the average value over some observations rather than stick to the result of a single observation. The result is found to be 0.6 UI jitter for each pointer movement (average taken over several observations).

Pointer adjustments in stressed mode and hold-over mode of operation

Stressed mode of operation reflects the performance of a receiver clock under actual network conditions where short interruptions of the timing reference or significant degradations can be expected. The number of such incidents can be 1 to 100 per day! Each such incident can perpetuate a series of pointer adjustments till the temporary impairment disappears or till the protection mechanism offers master reference via a different trail. The number of NEs involved in pointer adjustment depends on the network topology, protection mechanism being used and the location of disturbance point in the trail.

In hold-over mode of operation the clock received by an NE is interrupted due to fibre-cut or node interface failures. An NE uses its own reference clock in the hold-over mode, which has 4.6 ppm deviation with respect to the master reference clock. This causes periodic pointer adjustments till the protection mechanism of the network re-establishes reference in the entire trail. Here also the number of NEs involved in pointer adjustment depends on the network topology, protection mechanism being used and location of disturbance point in the trail.

Hence, we can analyse stressed mode and hold-over mode in one go by considering impairment or interruption as an 'incident' at a certain location in the trail. But we have to do the analysis separately for different protection mechanisms in the network, viz., MS-SPRing and SNCP.

In section 8.4.8 and 8.4.9, while discussing MS-SP Ring, it was mentioned that the maximum number of NEs in the ring is 16. In SNCP, there is no such restriction. However, in case of SNCP we have to remember that in each sub-net the number of NEs between two SSUs can be 20 or less. Let us examine re-synchronisation in MS-SPRing and SNCP after an 'incident'.

Re-synchronisation in MS-SPRing after incident (impairment/fibre cut/nodal interface failure)

Refer to Fig. 10.13 (a–c) showing the clock references in NEs in different stages. In Fig. 10.13(a), NEs are in a stable state. NE1 has got the PRC reference, so in the STM frame sent by NE_1 to NE_2, S1 byte carries PRC rank. NE_2 derives this clock and uses it as reference. Likewise, PRC reference is given down the trail and all the *N* NEs around the ring ($N \leq 16$) are using the PRC.

NE_2 sends frame to NE_1 in duplex direction. This frame is prepared as per PRC reference received by NE_2 from NE_1 in the forward direction. Now, NE_2 cannot give PRC rank to the frame it is sending to NE_1. Such ranking would confuse NE_1 and create a sync loop. So, the S1 byte of the STM frame sent by NE_2 to NE1 is given rank 1111, implying that it is not to be used for synchronisation. Likewise, the frames of all NEs give rank 1111 for the duplex direction frames.

As the trail is considered from NE_1 to NE_N, all NEs in a trail look towards the same direction for reference: NE_N uses the reference of NE_{N-1}, although it gets PRC from NE_1 also. NE_1 gets PRC from NE_N, but uses the original PRC reference.

Consider the situation after the 'incident'. The NE after the cut enters hold-over mode, which means that it generates frame as per the last average clock. Now, this NE can no longer claim that it is PRC. It gives the rank 'SEC' (synchronous equipment clock) in S1 byte of the frame it is sending to the next NE down the trail. The next NE also enters hold-over mode and sends SEC rank. This goes on till NE_N at the closure of the ring, back to NE_1. This situation is shown in Fig. 10.13(b). Note that the clock rank message from NE_N to NE_1 contains 1111 in Fig. 10.13(b), whereas it was PRC in Fig. 10.13(a).

As NE_N has access to PRC from NE_1 and SEC from NE_{N-1}, it chooses PRC and gives this reference to NE_{N-1}. This continues around the ring till the NE adjacent to the cut is reached. As the NEs progressively get PRC, they withdraw SEC and give 1111 rank in their messages. This situation is shown in Fig. 10.13(c).

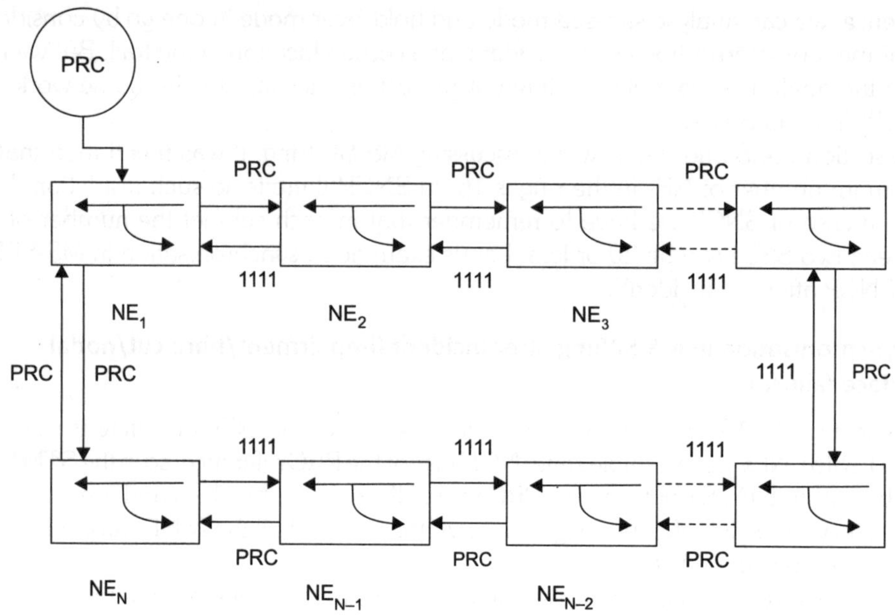

Fig. 10.13(a) *Clock references and clock rank messages in MS-SPRing in a stable state*

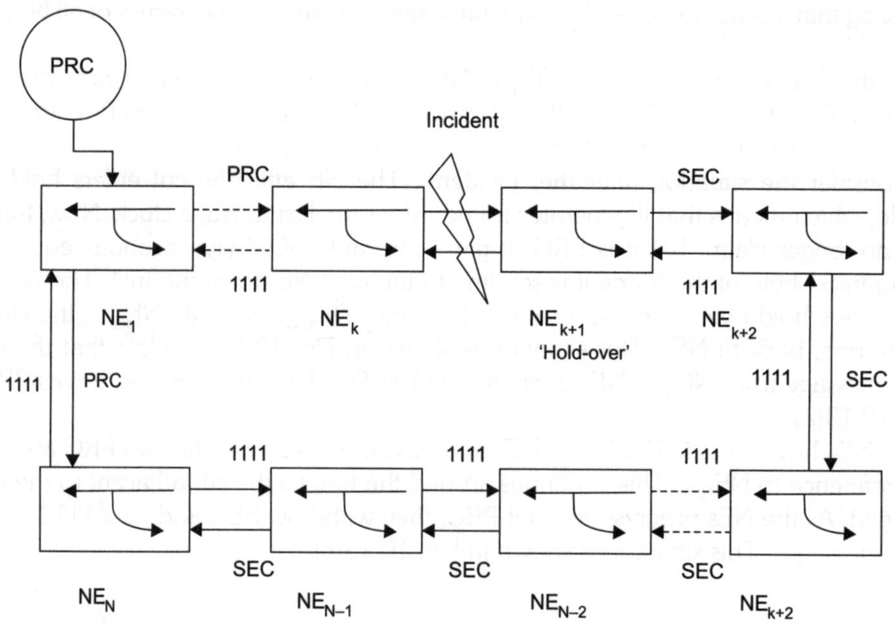

Fig. 10.13(b) *Clock references and SEC rank messages in MS-SPRing after fibre-cut*

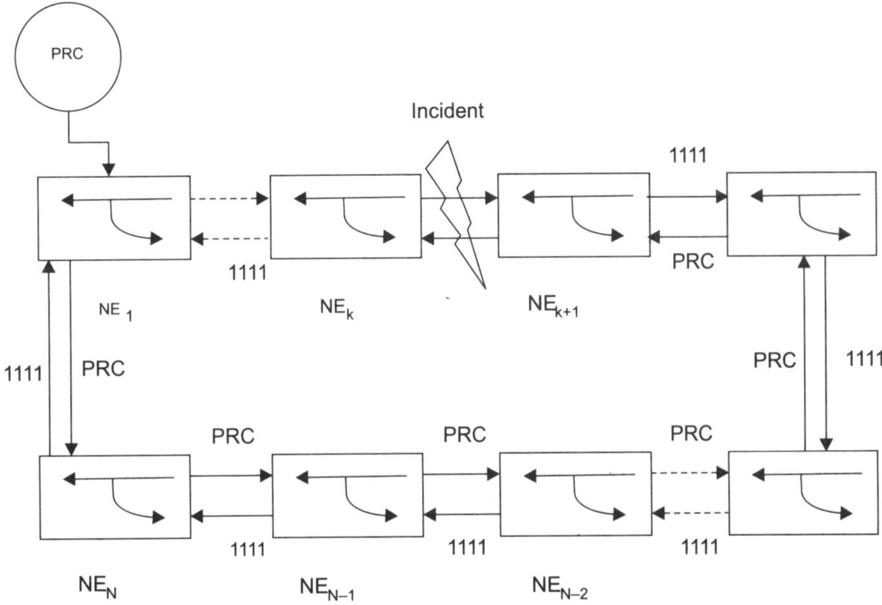

Fig. 10.13(c) *Resynchronisation to PRC in MS-SPRing after fibre-cut*

It may be noted that

1. After the incident, we have two trails, NE_1 to NE_k and NE_N to NE_{k+1}. The trail NE_1 to NE_k has PRC synchronisation for all NEs whereas trail NE_N to NE_{k+1} has got PRC reference for all NEs through the re-synchronisation process described above.
2. The change of clock references from the situation in Fig. 10.13(a) to that in Fig. 10.13(b) and later to the situation in Fig. 10.13(c) is a progressive one.
3. If we have N NEs in MS-SPRing ($N \leq 16$) and the fibre-cut is after NE_k, we have $(N - k)$ clock reference adjustments by the time the '1111' clock message goes from NE_N back to NE_1 in the ring. Later, we have another $(N - k)$ adjustments by the time PRC reference goes back to the NE adjacent to the cut, i.e., NE_{k+1}. The total adjustments are therefore $2(N - k)$ in MS-SPRing, where N is the number of NEs ($N \leq 16$) and the fibre-cut is after k NEs.

Re-synchronisation in SNCP after incident

In SNCP, the working and protection paths are on two different trails. Refer to Fig. 10.14(a) showing the NEs in stable state, i.e., all synchronised to PRC/SSU prior to a cut. NE_1 has got the PRC reference, so in the STM frame sent by NE_1 to NE_2, the S1 byte carries the PRC rank. NE_2 derives this clock and uses it as reference. In a similar manner, the PRC reference is given down the trail and all NEs use PRC. As the clock reference travels down the trail, jitter and phase noise accumulate, which are filtered out by SSU at NE_n ($n \leq 20$). Note that the SSU is not giving any different reference to NE*n*. It is only cleaning up the PRC reference received by it from the previous NE.

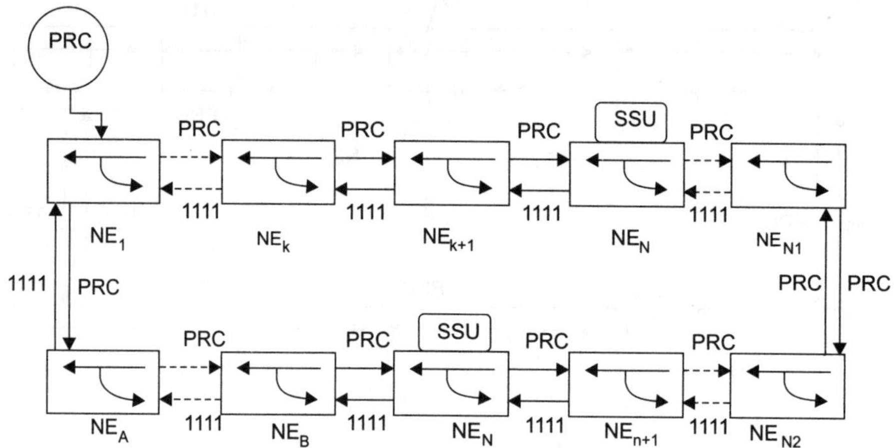

Fig. 10.14(a) *Clock references and clock rank messages in stable state*

The PRC, and later the SSU references are taken down the trail on the protection path in the same manner. It may be noted that number of NEs on both paths can be different. NE_{N1} and NE_{N2} give PRC reference to each other but use the reference obtained from the previous NEs in their respective trails.

Let us examine the change of references after the incident. The situation after the incident is shown in Fig. 10.14(b).

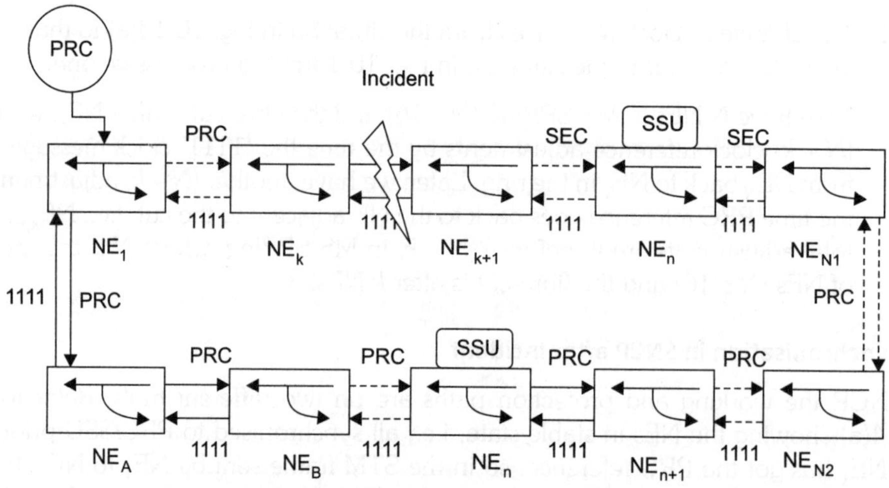

Fig. 10.14(b) *Clock references and clock rank messages after fibre-cut*

Let us follow the sequence of events after the cut next to NE_k.

a. NE_{k+1} enters hold-over mode and sends SEC rank down the trail, up to NE_{N1}.

b. NE_{N1}, which has been receiving VC-*n* via the working trail, receives alarms in the VC-*n* and hence chooses VC-*n* on the protection trail.

c. NE_{N1} synchronises to protection trail (to PRC reference given by NE_{N2}) and the traffic is restored.

d. APS message is sent back to the NE_1 originating the VC-*n*.

e. The two trails prior to the cut are (i) NE_1 to NE_{N1} and (ii) NE_A to NE_{N2}. The two trails after the cut are (i) NE_1 to NE_k and (ii) NE_A to NE_{N2} to NE_{N1} to NE_{k+1}.

The number of clock reference adjustments are $N_1 - k$ in Fig. 10.14 (b) prior to synchronisation of NE_{N1} to protection trail. When the situation reverts to the original condition after the disappearance of the 'incident', the NEs get back to the status in Fig. 10.14(a). So, another $(N_1 - k)$ pointer adjustment occur during such reversal. Therefore, the total number of pointer adjustments is $2(N - k)$ for SNCP (considering revertive) and $(N - k)$ for SNCP (considering non-revertive), where N is the number of NEs in the trail of VC-*n* and k is the incident-occurring NE.

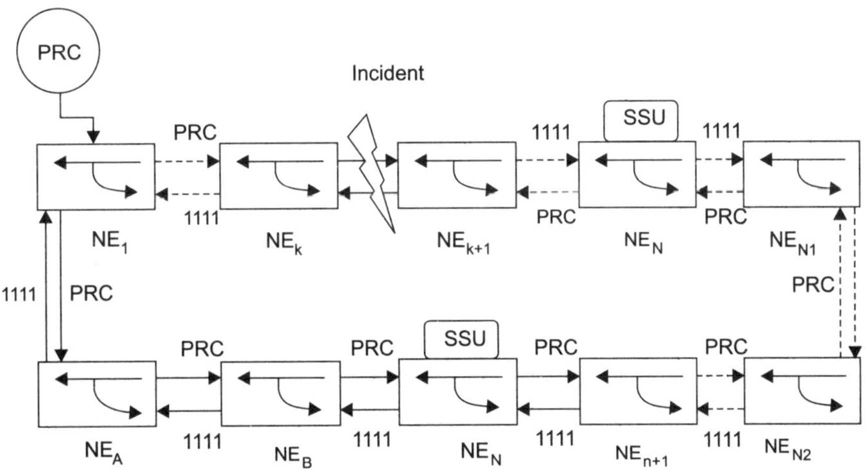

Fig. 10.14(c) *Clock references and clock rank messages after fibre-cut*

Reverting to the original working trail generally happens after stabilisation, i.e., 3 to 4 minutes after the availability of the working path. Hence, the second set of $(N - k)$ pointer adjustments are far apart from the first set and can be considered to have independent effect on jitter and BER.

To sum up:

1. In ideal mode, random phase variations causing pointer adjustment happen 800 times a day, each causing a jitter of 0.6 UI in an isolated way without any cumulative effect on payload.
2. Stressed mode of operation, which is due to short-term impairments in clock (causing change of reference by NE) can happen 1 to 100 times a day, and each such incident causes $2(N-k)$ pointer adjustments in cascade in the case of MS-SPRing and $(N-k)$ in the case of SNCP (non-revertive), and another $(N-k)$ if SNCP reverts to original working trail. Such adjustments cause cascading of jitter, which is 0.6 UI per NE. This shoots up BER and also causes associated problems on payloads.
3. Hold-over mode of operation happens for every fibre-cut/node failure and each such incident causes $2(N-k)$ pointer adjustments in cascade in case of MS-SPRing and $(N-k)$ in case of SNCP (non-revertive) and another $(N-k)$ if SNCP reverts to original working trail. Such adjustments results in cascading of jitter, which is 0.6 UI per NE. This shoots up BER and brings about associated problems on payloads.

10.7.3 Estimating increase in BER due to pointer adjustments in different modes

Ideal mode

Let us now estimate BER due to 0.6 UI jitter caused by single pointer adjustment.

We have to find the area under 0.6 UI from Fig. 10.11. Reckoning of this 0.6 UI is important. Usually, the sampling instant is taken at the centre so that the area under second peak gives BER, which is approximately 0.5×10^{-6}. If the sampling instant shifts towards right in Fig. 10.11, both peaks may get included for BER consideration, which means a BER of 1.2×10^{-6}. Additional errors can also be estimated for these worsened BER (otherwise 1×10^{-12}).

In single pointer movement, 24 bits of justification are offered in one step to compensate for the accumulated phase lead/lag. Such a pointer movement results in 0.6 UI jitter and BER of 0.5×10^{-6} to 1.2×10^{-6}. Applying this error rate over one frame, we get the results given in Table 10.9(a).

Sometimes, triple pointer burst can also happen, which causes jitter of 1.3 UI, i.e., the minimum BER equal to the area under both peaks of Fig. 10.11 and the maximum BER equal to the area under both peaks plus the first peak. This works out to a minimum BER of 1.2×10^{-6} and a maximum BER of 1.9×10^{-6}. Additional errors in case of triple pointer burst are tabulated STM- rate-wise in Table 10.9(b).

Table 10.9(a) *Additional errors due to jitter caused by isolated single pointer movement*

STM rate	**No. of bits per frame**	**BER due to single pointer movement jitter**		**Additional errors**	
		Min. BER	**Max. BER**	**Min.**	**Max.**
STM-1	19440	0.5×10^{-6}	1.2×10^{-6}	0.00972	0.023328
STM-4	77760	0.5×10^{-6}	1.2×10^{-6}	0.03888	0.093312
STM-16	311040	0.5×10^{-6}	1.2×10^{-6}	0.15552	0.373248
STM-64	1244160	0.5×10^{-6}	1.2×10^{-6}	0.62208	1.492992

Table 10.9(b) *Additional errors due to jitter caused by isolated triple pointer movement*

Additional errors	**No. of bits per frame**	**BER due to triple ptr. jitter**		**Additional errors**	
		Min. BER	**Max. BER**	**Min.**	**Max.**
STM-1	19440	1.2 E-6	1.9 E-6	0.023328	0.036936
STM-4	77760	1.2 E-6	1.9 E-6	0.093312	0.147744
STM-16	311040	1.2 E-6	1.9 E-6	0.373248	0.590976
STM-64	1244160	1.2 E-6	1.9 E-6	1.492992	2.363904

As discussed in section 10.7.3, in the ideal mode of operation of an NE, random phase variation of clock causing single pointer adjustment in isolated way happens 800 times a day. Out of this, about 10% cause triple pointer adjustments. So, by taking 720 single pointer adjustments and 80 triple pointer adjustments (maximum errors in each case) and rounding off to nearest integer, we get additional errors per day as given in Table 10.9(c).

Table 10.9(c) *Additional errors per day due to random phase variations*

STM rate	**Additional errors due to single pointer adjustment (720 per day)**	**Additional errors due to triple pointer adjustment (80 per day)**	**Total additional errors per day**
STM-1	18	30	48
STM-4	72	120	192
STM-16	288	480	768
STM-64	1152	1920	3072

Stressed mode

In section 10.7.3, it was mentioned that each incident causing stressed mode operation causes $2(N - k)$ pointer adjustments in cascade in MS-SPRing and SNCP (revertive). Additional errors per incident in a day considering the maximum number of such incidents (100 based on observations) are given in Table 10.10(a).

Table 10.10(a) *Additional errors in stressed mode per incident and per day*

STM rate	Additional errors due to single incident	Additional errors in a day (100 incidents)
STM-1	$2(N - k)$ 0.023328	$2(N - k)$ 2.3328
STM-4	$2(N - k)$ 0.093312	$2(N - k)$ 9.3312
STM-16	$2(N - k)$ 0.373248	$2(N - k)$ 37.3248
STM-64	$2(N - k)$ 1.492992	$2(N - k)$ 149.2992

Suppose the NEs fail to re-synchronise to alternate clock reference due to another incident or bad phase transients. There would then be periodic adjustments of pointers till the NEs settle to stable state. According to ITU-T Rec.G.813, the allowed variation of SEC is 4.6 ppm with respect to PRC. Hence, for every 33.5 ms, there would be adjustment of pointers in cascade (for a 1 ppm variation, adjustment of pointer takes place every 1235 frames as studied in section 7.5.3. Hence for 4.6 ppm it would be every 268 frames, i.e., every 33.5 ms). This implies a high BER till NEs settle. For example, errors over 60 s interval are shown in Table 10.10(b).

Table 10.10(b) *Additional errors with NEs in free-running mode*

STM rate	Additional errors due to one set of cascaded adjustments of pointers after single incident	Additional errors in 60 s
STM-1	$2(N - k)$ 0.023328	$2(N - k)$ 42
STM-4	$2(N - k)$ 0.093312	$2(N - k)$ 168
STM-16	$2(N - k)$ 0.373248	$2(N - k)$ 672
STM-64	$2(N - k)$ 1.492992	$2(N - k)$ 2688

Hold-over mode

In hold-over mode, additional errors per incident like fibre-cut/node failure after restoration via protection trail would be those caused by one set of pointer adjustments in cascade. These are indicated in Table 10.11. But, if re-synchronisation fails in the trail due to bad SEC somewhere, NEs continue in free-running mode. In such a situation there would be periodic adjustments of pointers resulting in accumulation of errors. For example, such accumulation over 60 s interval is given in the right-most column of Table 10.11.

Table 10.11 *Additional errors per incident in hold-over mode*

STM rate	Additional errors due to one set of cascaded adjustments of pointers in NEs in hold-over mode	Additional errors in 60 s if NEs fail to re-synchronise
STM-1	$2(N-k)$ 0.023328	$2(N-k)$ 42
STM-4	$2(N-k)$ 0.093312	$2(N-k)$ 168
STM-16	$2(N-k)$ 0.373248	$2(N-k)$ 672
STM-64	$2(N-k)$ 1.492992	$2(N-k)$ 2688

10.7.4 Comparison of predictions with measurements

In ideal mode

In the ideal mode, errors due to single pointer adjustment cannot be measured in isolated cases but can be ascertained over a period. During that period other modes of operation are excluded. So, the readings were taken on a MUX section over 10 different 1-hr intervals (to exclude other incidents). The total random phase variations were about 300 (considering 800 per day). The predicted and measured results are tabulated in Table 10.12(a). This shows that the measurements are pretty close to predictions.

Table 10.12(a) *Predicted and measured results in ideal mode*

STM rate	Total additional errors per day	Predicted for 10 hours	Actual measurement
STM-1	48	20	18
STM-4	192	80	70
STM-16	768	320	303

In stressed mode (SNCP)

We have to take measurements exclusively for the stressed mode for ruling out random phase variations. So stressed-mode is created by forcing SEC for an NE and measuring the errors due to resultant pointer adjustments in cascade. Measurements are taken for 20 different observations in succession (on SNCP with N=40, incident at k=20) and the results tabulated in Table 10.12(b). This shows that the measurements are fairly close to predictions.

Table 10.12(b) *Predictions and actual measurements in stressed mode*

STM rate	Total additional errors per incident with N=40, k=20	Predicted for 20 incidents	Actual (measured)
STM-1	1	20	15
STM-4	4	80	69
STM-16	16	320	306

In hold-over mode (SNCP), continuing after incident

Measurements made in this case check the predictions when NEs fail to re-synchronise and when the pointer adjustments are taking place periodically. In case of SNCP, considering a trail of 40 NEs with the incident occurring at k = 20, the measurements are taken over 60 s interval keeping the NE's k to N in the SEC. The test is repeated 20 times and the average readings are taken.

Table 10.12(c) *Periodic adjustment of pointers in cascade – errors predicted and measured*

STM rate	Additional errors in 60 s if NEs fail to re-synchronise	Predictions for N=40, k=20	Actual (measured)*
STM-1	$2(N - k)$ 42	1680	1590
STM-4	$2(N - k)$ 168	6720	6846
STM-16	$2(N - k)$ 672	26880	24729

***Note:** Since high errors can disconnect traffic, the trail was continued to function despite high the BER by suitably setting the BER threshold.

10.7.5 Effect of pointer adjustments on payloads

From the discussions of Section 10.7.3 and the measurements presented in section 10.7.4, the following inferences can be drawn:

1. Random phase variation of clock in ideal mode causing pointer adjustment in isolated way does not contribute to bit errors significantly.
2. In case of incident like short term impairment of clock in the trail, though the traffic is restored on protection bandwidth, the NEs adjacent to the incident at path termination enter the hold-over mode. They seek re-synchronisation with PRC via alternative trail and in this process, there is a cascade of pointer adjustments. This cascade goes on periodically till successful re-synchronisation, i.e., till the NEs settle down. (Another such incident or phase transition on SEC of NE in hold-over mode prolongs the re-synchronisation process.) This causes piling up of errors.
3. In case of an interruption like fibre-cut when traffic is restored through protection bandwidth, the NEs adjacent to the interruption enter hold-over mode.

They seek PRC re-synchronisation via alternate trail and in this process, there is a cascade of pointer adjustments. This cascade goes on periodically till successful re-synchronisation. Another such incident or incidents like short term impairment or phase transition on SEC of NE in hold-over mode prolongs the re-synchronisation process which causes piling up of errors.

Cases 2 and 3 above cause concern not only because of the high error count, but also on account of the effects they have on payloads. Clearly, the duration of such degradation is more than 60 ms, the time limit prescribed for complete restoration and hence can lead to interruption of services. Also, they can cause any of the alarms listed in Table 7.11 and interrupt services.

The effect is disturbing on voice traffic as voice circuits get disconnected when the interruption is more than 60 ms which causes call-dropping. Coming to data payloads, flow control mechanisms like GFP described in chapter 9 ensures that loss of data frames is minimal (except in loss detection duration after the incident). But, piling up of errors for long due to periodic adjustment of pointers in cascade from the NE adjacent to the incident to the path-terminating NE can cause severe flow-control problems especially on SDH pipes with fewer VC-ns. In case of video traffic, we may find still pictures, glitches, audio–video non-synchronisation, including absence of one of them.

Q 10.7 Can we predict when there would be VC-12 disturbance due to effect of cascaded pointer adjustments?

TU-12 pointer, i.e., pointer associated with VC-12 or E1 payload, requires adjustment if 8 justification bits of TU-12 is offered (recall discussion on VC-12 and TU-12 from section 7.6.3). Since TU-12 duration is 500 ms and of 144 bytes, i.e., 1152 bits, TU-12 pointer adjustment (VC-12 disturbance) through 8 justification bits implies a timing difference of 3.5 ms ($8 \times 500/1152$).

10.7.6 Solutions to the problem

From Tables 10.9, 10.10, we can draw the following inferences.

1. When we have longer trails and fewer protection connections at intermediate nodes, i.e., when $(N - k)$ is large, we have more bit errors.
2. Re-synchronisation time over entire trail even after re-establishing VC-n at the path-terminating NE via protection trail and resuming transmission increases as (N – k) increases.
3. Cascade effect of pointer adjustment in $(N - k)$ NEs may cause interruptions in payload.

4. If NEs continue in hold-over mode after the incident, errors pile up due periodic adjustment of pointers.
5. Such piled-up errors also increase as $(N - k)$ increases.

There are two ways to tackle the problem:

(i) *Reduce $(N - k)$:* This is possible with shortened synchronisation chains and multiple PRCs, which call for judicious planning of the network. On the one hand, we cannot have large trails with few protection paths, while on the other, too many protection connections make the network meshed with large number of cross-connects and larger inter-ring traffic, which means additional cost and complexity. Clearly, it is a challenge to design survivable, resilient multi-ring *SDH* networks.

(ii) *Minimise NE jitter in stressed mode:* There are three approaches to doing this.

  a. Reduce BER contribution of pointer adjustment jitter by shifting sampling point in Fig. 10.4 so that 0.6 UI goes free of the BER peaks.
  b. Use appropriate de-jitter unit so that stressed mode jitter is reduced.
  c. Re-time incoming data clock using elastic buffer to reduce stressed mode jitter.

We will discuss these solutions in the following section.

10.8 Design of Survivable and Resilient Multi-Ring SDH Networks

10.8.1 Design of multi-ring architecture overlaid on mesh network

The challenge before us is SDH ring design problem with an interconnected multi-ring architecture overlaid on a mesh network. Let us divide the problem and conquer.

The first problem is the *SDH ring selection*. Since inter-ring traffic increases the network cost and complexity, we aim to minimise the inter-ring traffic in selecting the rings, considering total nodes on the network. We follow heuristic approach for networks with fewer nodes and an algorithmic approach based on graph theory for network with several nodes.

The second problem is the *mapping of these rings onto the physical mesh topology*. Once the logical topology is decided, mapping of SDH rings onto the mesh (physical) topology is to be made. This is done in an algorithmic way rather than a heuristic way. The objective of this mapping is to minimise the total fibre cost in the network.

SDH ring selection

The design objectives of SDH ring selection are:

1. Each ring should satisfy 2-connectivity, i.e., there should exist two physically diverse paths between any two nodes on the ring. This ensures 100% restoration on a single failure.

2. The number of nodes on each ring should not exceed a maximum of 16 in the case of MS-SPRing, in accordance with conformity to SSU/SEC number restrictions as discussed in section 7.10 in case of SNCP.

3. Minimisation of inter-ring traffic.

Inter-ring traffic is the amount of the traffic between two demand nodes belonging to different rings, i.e., dual-homed nodes. Digital cross-connects used in such situations have to incorporate ADM functions to provide connectivity between rings, as DXCs are connected to ADMs. More the inter-ring traffic, higher is the cost of cross-connect equipment and greater the complexity of the network.

For ring selection, a draft plan showing rings can be made. Then, some rings can be merged by suitably selecting DXCs. But while doing so, too many merges should not be there because it would violate the first design objective of 2-connectivity. Figs. 10.15(a), (b) and (c) illustrate the philosophy. While merging of rings as done in Fig. 10.15(b) is all right, the merger of Fig. 10.15(c) is not correct as it leaves one node unconnected.

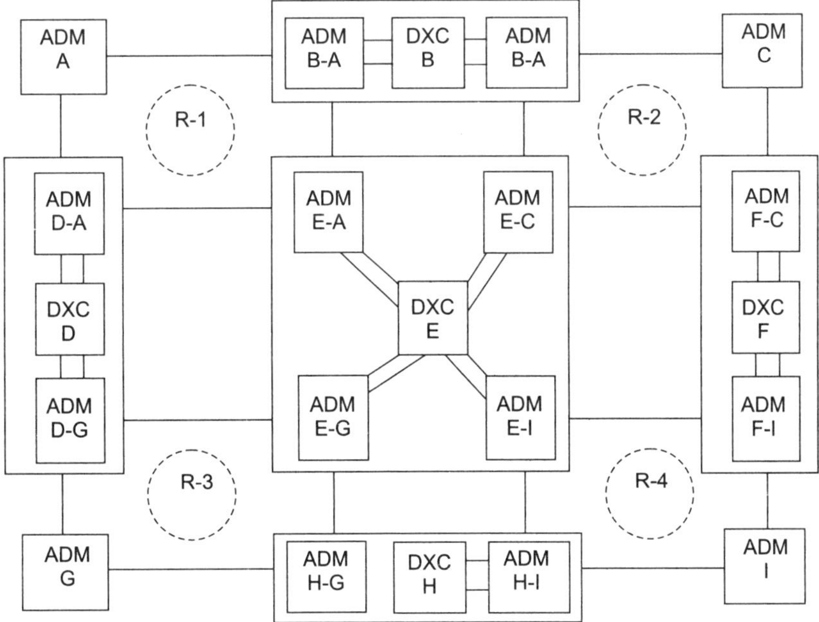

Fig. 10.15(a) *Interconnected rings with more DXCs*

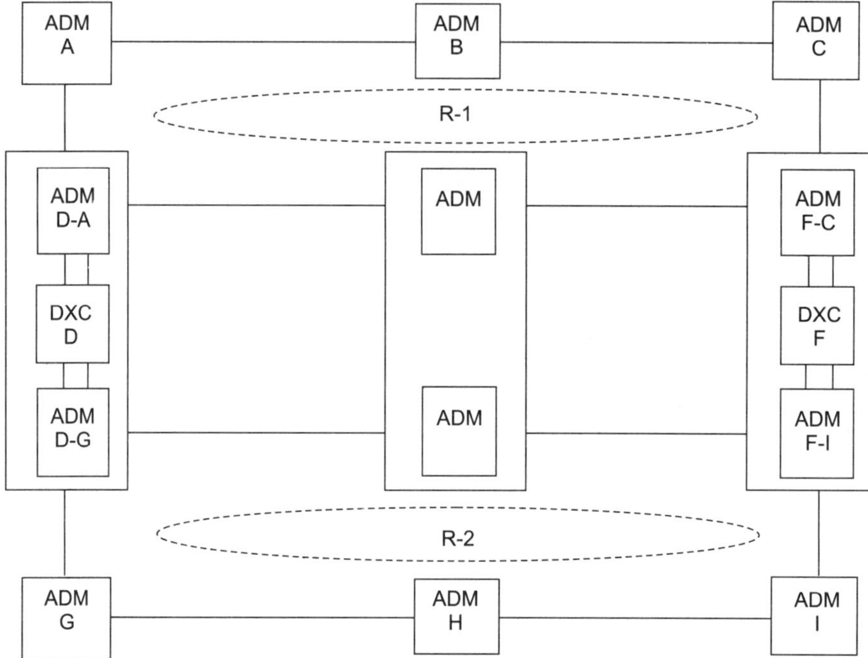

Fig. 10.15(b) *Interconnected rings with reduced DXC*

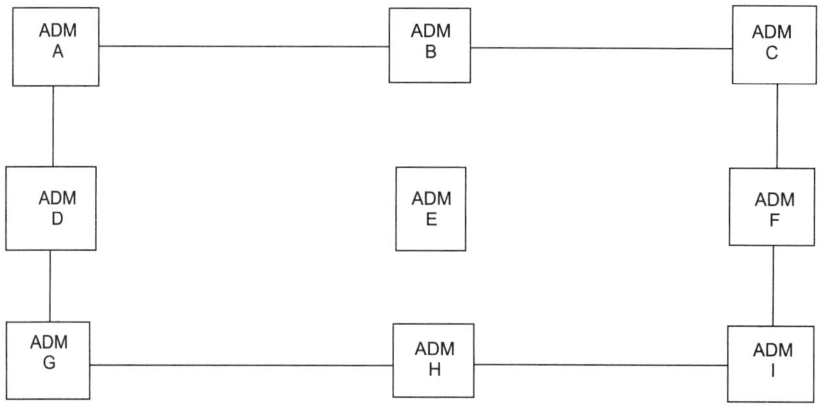

Fig. 10.15(c) *Unwarranted reduction of DXC resulted in disconnection of node-E*

For network with several nodes, an algorithmic approach based on graph theory is described briefly below.

1. *Obtaining the logical graph:* Connect geographically-close SDH nodes to each other. The logical graph so obtained should be planar.

2. *Finding the faces:* Use faces of the planar logical graph as a starting solution. To determine a face, starting from one node the edges have to be followed in clockwise or counter-clockwise direction. The procedure is followed until the first node is reached. The number of faces F of a planar graph G with V vertices and E edges is given by $F = E - V + 2$ (Euler formula).

3. *Merging faces:* Once the faces are known and demands between SDH nodes known, fix the routing such that each route passes through minimum number of rings. For example, between source node S and destination node D,

   a. Calculate *k*-shortest paths between S and D.

   b. Construct a trellis graph for finding the path with the minimum number of rings.

   c. Since some links may belong to more than one face, we determine the routing which achieves the minimum number of ring crossings.

   d. After fixing the route for each demand of rings-situation on hand, start merging the rings in order to minimise the inter-ring traffic. To do so, a face adjacency graph is to be constructed. This is done by assigning a node corresponding to each face and connecting the faces that share a link in the logical topology with a link in the face adjacency graph. For example, for the planar graph in Fig. 10.16(a), face adjacency graph is shown in Fig. 10.16(b).

Fig. 10.16(a) *Planar graph*

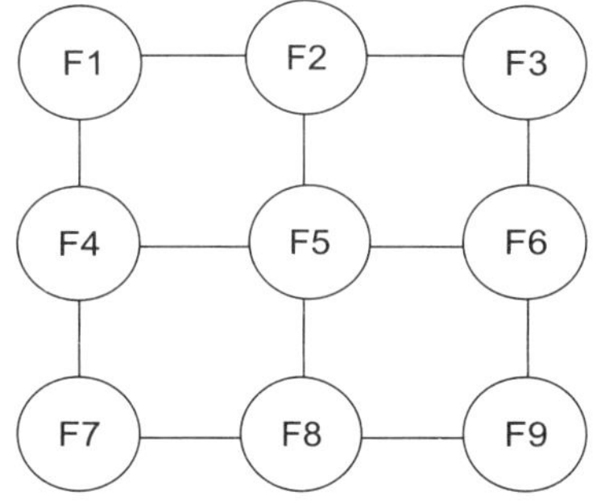

Fig. 10.16(b) *Adjacency graph*

   Planar graph in Fig. 10.16(a) has 9 faces. Corners of each cell (face) are nodes; thus there are 16 nodes. In adjacency graph, show F1 to F9 and connect adjacent Fs' by common link, as shown in Fig. 10.16(b).

   e. Carry out merging of faces by combining two adjacent faces by eliminating the shared link. While doing so, look for all feasible face mergers but obey the rules of 2-connectivity and maximum nodes in the ring.

Mapping of ring to physical mesh network

Mapping of SDH rings onto the mesh physical topology is to be formulated as a mixed integer linear programming (MILP) problem in order to guarantee proper operation of SDH ring protection. In fact, as we will see chapter 14, with the introduction of optical protection layer, each link of the ring gets mapped onto an exclusive light path. A light path is a wavelength in WDM plan, which should be a link and node disjoint from all other light paths so that any wavelength among the available wavelengths on the WDM plan can be repeated. The objective of this mapping is to minimise the total fibre cost in the network.

MILP is quite an involved process. A full length discussion of the same is beyond the scope of this book. Interested readers may go through references [1] and [2]. To formulate the problem in mathematical terms, we introduce a number of definitions.

Suppose the logical network topology is represented by a directed graph $G = (V, E)$, where V is the set of SDH nodes in the logical topology and E the set of edges. The decision variables are defined as follows.

y_{jk} = 1 if there exists a logical link from SDH node j to k, i.e., $(j, k) \in E$
= 0 otherwise.

$Z^{rs}{}_{jk}$ = 1 if there exists flow on logical link $(j, k) \in E$ resulting from traffic carried between SDH nodes r and s,
= 0 otherwise.

Let the physical network topology be represented by a directed graph $G' = (V', E')$, where V' is the set of OXC nodes in the physical topology and E' the set of edges. d_{nm} is the length of the physical link, with $(n, m) \in E'$. The decision variable $X^{jk}{}_{nm}$ is defined as follows.

$X^{jk}{}_{nm}$ = 1 if there exists flow on physical link $(n, m) \in E'$ resulting from the flow on logical link $(p, q) \in E$,
= 0 otherwise.

Objective:

$$\text{Minimise } \Sigma_{(n,m)\in E} \Sigma_{(j,k)} X^{jk}{}_{nm} d_{nm}$$

subject to:

$$\text{cycle constraint: } Y_{jk} + Y_{kj} \leq 1 \ \forall \ (j, k) \in E.$$

The total flow constraints on a logical link:

$$\Sigma_{j\neq k} Y_{jk} = 1 \ \forall \ j \in V$$

$$\Sigma_{j\neq k} Y_{kj} = 1 \ \forall \ k \in V$$

$$Z^{rs}{}_{jk\,pq} \leq Y_{jk} \ \forall \ (j, k), \ \forall \ (r, s)$$

Flow conservation at each SDH node:

$$\Sigma_{j\neq k} Z^{rs}_{jk} - \Sigma_{j\neq k} Z^{rs}_{kj} \quad = 1 \text{ if } j = r;$$
$$= -1 \text{ if } j = s; \qquad \forall\, j \in V, \forall\, (r, s)$$
$$= 0 \text{ otherwise}$$

Physical link disjoint paths constraint:

$$\Sigma_{(j,k)} X^{jk}_{nm} + X^{jk}_{mn} \leq 1 \qquad \forall\, (n, m) \in E'$$

Node disjoint paths constraint:

$$\Sigma_m \Sigma_{(j,k)} X^{jk}_{nm} = 1 \qquad \forall\, n \in V'$$

Flow conservation at each physical node:

$$\Sigma_m X^{jk}_{nm} - \Sigma_m X^{rs}_{mn} \quad = Y_{jk} \text{ if } n = j;$$
$$= -\,Y_{jk} \text{ if } n = k; \qquad \forall\, n \in V', \forall\, (p, q)$$
$$= 0 \text{ otherwise}$$

Integer constraints:

$$Y_{jk}, Z^{rs}_{jk} \in \{0, 1\} \qquad \forall\, (j, k) \in E$$
$$X^{jk}_{nm} \in \{0, 1\} \qquad \forall\, (n, m) \in E'$$

With the above formulation, it is possible to minimise the total cost of fibre in the network.

Q 10.8 *What is the meaning of constraints expressed above?*

1. Cycle constraint: $Y_{jk} + Y_{kj} \leq 1\ \forall\, (j, k) \in E$ guarantees that two logical links between the same logical nodes do not form a cycle.
2. Total flow constraints (1) and (2) on a logical link implies that for any node on a ring, both the in-degree and out-degrees are 1. Total flow constraint (3) indicates that if there is no SDH logical link between *j* and *k*, then $Y_{jk} = 0$ (i.e., it is not possible to meet traffic demand between those nodes).
3. The flow conservation constraint verifies the flow conservation at each SDH node for logical topology such that each demand is routed from its source to its destination.
4. Physical link disjoint paths constraint implies that each logical link on a ring maps onto physical link disjoint paths.

5. Node disjoint paths constraint implies that mapping is done onto node disjoint paths.
6. Flow-conservation constraint implies flow conservation at each node for the physical topology
7. Integer constraint implies that variables thereof, i.e., Y_{jk}, Z^{rs}_{jk}, X^{jk}_{nm} are logical variables with values 1 and 0.

Note that the mapping algorithm presented above is to be implemented for all rings in the network, separately. The solution obtained as per the discussions in this section provide optimal length of synchronisation chains, adherence to 2-connectivity and maximum node number rules.

Now, we move on to explore solutions related to avoidance/reduction of jitter.

10.8.2 Phase-adjust feature of receiver

As quantified in our discussion and also measured on operational networks, the BER contribution incident becomes significant when it happens in cascade. While we do not have control on pointer adjustment in NEs and the consequent jitter, the ill-effect of jitter can be tackled by ensuring that the sampling instant does not fall in the rise/fall time of the data pulse. Because of asymmetry in any data eye diagram and also factors like setup and hold times of the receiver decision circuit, it is advantageous to shift the sampling time to a position other than the centre of the eye diagram to achieve reduced BER. This strategic positioning of sampling instant can be done using the phase-adjust feature of receivers.

10.8.3 Use of de-jitter

In this circuit, the incoming data is fed into a 32-bit asynchronous FIFO. The write clock of the FIFO is the gapped input clock, which is fed to the de-jitter attenuator consisting of a second all-digital phase locked loop (ADPLL) having a 2.0 Hz loop bandwidth with external reference clock at 65.536 MHz +/– 100 ppm. The filtered clock output of the ADPLL is the read clock of the FIFO. This function is bypassed when the re-timing

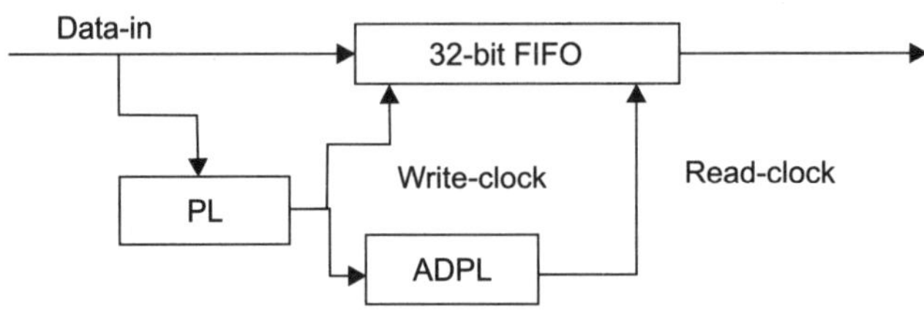

Fig. 10.17 *Use of de-jitter*

function is enabled or while configuring the data path in a pass-through mode. The working of de-jitter is fully transparent—no data is lost or added in the transmission. By this arrangement the transmitter filters the jitter and eliminates gaps in the incoming E1 clock and data.

10.8.4 Use of re-timing

In this arrangement the incoming E1 data is converted to a byte-parallel format and fed into a two-frame-wide elastic buffer. The E1 frame acquisition process activates the write and read control logic of the elastic store. The data is read out of the elastic buffer using an external clock reference input. The reference clock may or may not be at the same frequency as the incoming clock. If the read and write frequencies are different, the elastic store will periodically overflow or underflow. In either case, the read control logic will process a controlled slip of one complete frame in order to re-centre the elastic buffer. This will result in the loss of or repetition of one complete frame.

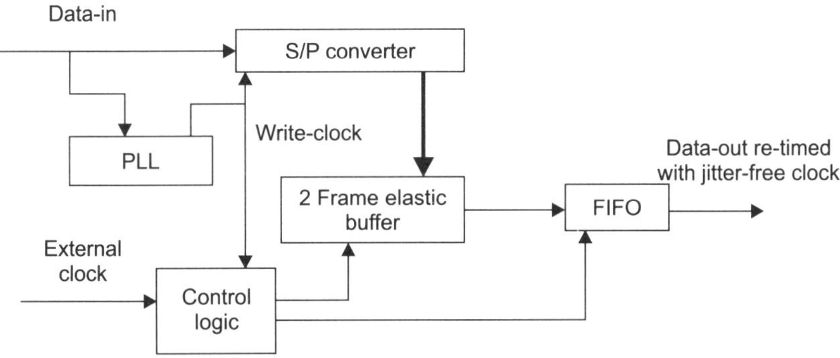

Fig. 10.18 *Use of re-timing to remove jitter*

With this arrangement, all jitter and wander that have set in the incoming clock due to multiplexing/de-multiplexing process in the transmission are eliminated.

Q 10.9 *What is the effect of the above arrangement in test-mode operation?*

The re-timing FIFO may also operate on an unframed 2Mbit/s signal by setting the transmitter to 'Retiming Test Mode'. In this mode the read and write control logic of the elastic store is in full free-running mode and independent of the framing algorithm. As the data is considered to be unframed in test mode, CRC-4 error monitoring is not valid. This mode is not transparent. It can handle a maximum of 26 time slots (208 UI) of wander or low frequency jitter before a frame slip occurs. This controlled frame slip assures that the time-slot assignment is not lost at the output of the device.

> **Q 10.10 *A challenge to designers:***
> *Considering the size of network, let us say we finalised PRCs, SSUs and implemented the network. What are the recovery mechanisms in case of SSU failure? Will not such failures increase the number of NEs between two working SSUs?*
>
> To tide over such crisis, the network is provided with NMS as well as SYNC server. the SYNC server monitors all SSUs (interconnected through routers – links similar to DCN links discussed in section 8.7). A client view of synchronisation mechanism of all NEs and SSUs is provided. In case of failure of any SSU (usually more than one failure creates error accumulation), synchronisation choice of NEs can be explicitly changed to avoid repeated pointer adjustments. Also, any unwanted timing loops in clock synchronisation can be detected.

Summary

This chapter provides a thorough insight into the practical aspects of SDH network design, the ITU-T recommendations on control of jitter and wander in SDH network and the mathematical expression for jitter manifested as BER based on probability theory. As a single pointer adjustment causes jitter of 0.6 UI. The effect of such adjustments in ideal mode (due to phase variations), stressed mode (due to impairments of NE clock) and hold-over mode (due to interruptions like fibre-cut, node failure) are quantified. In stressed mode and hold-over mode, the adjustment of pointers in cascade over NEs from the incident location to the path-termination location, that too periodically till re-synchronisation is achieved, would pile up errors which have adverse effect on payloads of all types. Design techniques such as optimisation of synchronisation chain through ring selection based on graph theory and physical link mapping based on MILP have been presented to counter the above effect. The use of phase-adjust feature of the receiver, de-jitter and re-timing techniques to control jitter in NE are also discussed.

References

[1] Ravi Chandran, A., *Operations Research and Management Science Handbook* (chapter 3, Integer Programming) CRC Press, 2007.

[2] Allan B Tucker, *Computer Science Handbook* (section 15.4.2 MILP), CRC Press, 2004.

PART III
OTH Systems

11

WDM Essentials

11.1 Introduction to WDM

Modal dispersion phenomena restricts data transmission rate or bandwidth in multi-mode fibres because of the optical pulse energy being dispersed in different modes arriving at different phases at the receiving end. In the 1970s, this limitation of multi-mode fibres was overcome to some extent by using graded index fibre. In the 1980s, when the manufacturing of longer drums of single mode fibre cable (2 or 3 km drum) with core diameter as low as 7 microns and cladding of 125 microns became a commercially viable, single mode fibre links got commissioned in a big way. Then, with the advent of SDH and supporting electronics, data rate transmission touched 10 Gbps (STM-64). Even then, this was only 1/100th of the theoretical bandwidth limit of 1 Tbps/carrier or wavelength on a single mode fibre (see section 1.3.4).

When already the highest in TDM, which the electronics could support had been reached, was it possible to go beyond this limit without discarding the already working systems? This was the challenges before telecom engineers then. Was there a need for higher data rates in reality?

Though single mode fibre has no modal dispersion, it suffers from chromatic dispersion and certain nonlinear effects. Chromatic dispersion is of the order of 2 picoseconds per nm of spectral width over 1 km. The electronics cannot operate at more than 10 Gbps. So, on a single mode fibre, using TDM (STM) systems, it is not possible to go beyond 10 Gbps if we are operating on a single wavelength, and the possibility of reaching theoretical limit of 1 Tbps is ruled out. Our aim is to push up the transmission rate, not necessarily on single wavelength. We can achieve transmission rates in multiples of 10 Gbps if we use more than one wavelength.

Let us now come to the second point, i.e., the need for transmission capacities in excess of 10 Gbps. Well, considering the ever increasing subscribers for broadband internet services, B-ISDN, WANs interconnections, etc., single mode fibre-optic systems operating on one wavelength with transmission capacity of 10 Gbps cannot meet the required traffic volume needs. So, OFC networks have to gear up to send more data on same fibres, as several of them are already deployed in STM systems. Since we have exhausted the available TDM spectrum, we have to send optical signals (STM frames) on more than one single wavelength. The process of doing this is via wavelength division multiplexing (WDM). By employing this technique wherever feasible, we can enhance the transmission capacity of fibres already working without resorting to laying of new OFC networks.

The following example highlights the utility of WDM. VSNL (Videsh Sanchar Nigam Ltd.) acquired Tyco Global Network's trans-Pacific submarine cable in the year 2005. Known as TGN-P cable, it has eight fibre pairs, with each pair having capacity of 96 wavelengths and carrying STM-64 (10 Gbps). It is currently equipped with electronics for 64 wavelengths, i.e., a capacity of 640 Gbps. VSNL has planned fullest utilisation i.e. 960 Gbps capacity on each of the eight fibre pair or 7.68 Tbps. In addition, VSNL launched TIC (Tata Indicom Cable), the first fully Indian owned under-sea OFC cable from Chennai to Singapore with 5.12 Tbps capacity (64 wavelengths on each of eight fibre pairs, each wavelength carrying 10 Gbps). This will have links to other submarine OFC cables and enhance the off-shore bandwidth availability.

11. 2 Principle of WDM

Wavelength-division multiplexing (WDM) is based on the following principle: more photons can occupy the same space, which means several light beams at different wavelengths can simultaneously propagate over the same optical path without interference. The principle of operation of a WDM system is shown in Fig. 11.1.

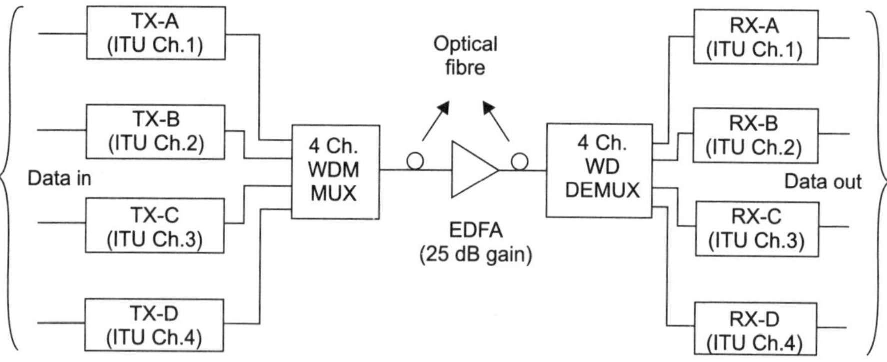

Fig. 11.1 *Basic principle of operation of WDM*

Transmitters A, B, C, and D use different wavelengths, λ_1, λ_2, λ_3 and λ_4. All these signals are combined by a multiplexer and transmitted simultaneously over the same optical fibre. A demultiplexer at the receiver end separates the signals of different wavelengths and directs them to the appropriate receivers. From the transmission standpoint, WDM divides the entire fibre-optic bandwidth into many segments and each signal (wavelength) uses its individual bandwidth segment. In Fig. 11.1, four wavelengths are used. Hence, the transmission capacity is $4 \times 10 = 40$ Gbps. If *n* wavelengths are used, the transmission capacity would be 10*n* Gbps. For example, Pirelli Telecom System's 1.28 Tbps link uses a WDM link with 128 wavelengths.

It may be noted that TDM and WDM are not competitive technologies but supplement each other, as the WDM link has STM (TDM) signal on each wavelength. The following analogy may be useful to differentiate the processes of TDM and WDM. TDM is analogous to increasing the speed of cars moving on a highway (STM-64 is at 10 Gbps whereas STM-4 is at 2.5 Gbps, i.e., packing of more bytes in the frame duration), while WDM is analogous to adding more lanes on the same road (parallel transmission over several wavelengths).

11.3 WDM Varieties

11.3.1 Broadband WDM

The proliferation of broadband internet or ISDN access may tempt one to think that it is the broadband WDM that is offering more transmission capacity. Broadband WDM being employed for this is the variety that uses 1310 nm and 1550 nm wavelengths for a full-duplex transmission. The signal is sent in one direction on 1310 nm wavelength and in the duplex direction on 1550 nm wavelength. This type of WDM has been in vogue for a number of years, and is also known as wideband WDM or 2-channel WDM.

Nowadays, WDM systems are operating in single window, around 1550 nm.

11.3.2 Narrowband WDM or DWDM

Narrowband WDM, also called dense wavelength division multiplexing (DWDM), is the multiplexing of 4, 8, 16, 32, 64, 128 or more wavelengths in the range of 1530 to 1610 nm with a very narrow separation between wavelengths. Today, WDM applies primarily to DWDM. In this book the term WDM refers to DWDM systems and sometimes both terms, WDM and DWDM, are used interchangeably.

DWDM channel spacing may be 12.5 GHz, 25 GHz, 50 GHz, 100 GHz and wider (integer multiples of 100 GHz), with channel count of 4, 8, 16, 32, 64, 128 or more. The network can be several thousands of kilometres with optical amplifiers and regenerators en-route. ITU-T Rec.G.694.1 specifies the channel spacing for DWDM system.

11.3.3 CWDM

Coarse WDM or CWDM uses channel spacing of 3000 GHz (separation of 20 nm) and up to 18 channels in the range of 1270 nm to 1610 nm and target distances of about

50 km on SM fibres (G.652, G.653 and G.655 fibres). Frequencies for CWDM are specified by ITU-T Rec. G.694.2. It may be noted that the transmitters employed in DWDM applications require a control mechanism to enable them to meet the frequency stability requirements of the application in contrast to CWDM transmitters which provide cost-effective transmission through a combination of un-cooled (uncontrolled) lasers, relaxed laser wavelength selection tolerances and wide passband filters. CWDM is usually employed in metropolitan area networks (50 km range), where a variety of clients, services and protocols are required to be supported.

11.4 ITU-T Grid for WDM

11.4.1 ITU-T grid for DWDM

The range of useable wavelengths in DWDM is from 1528.77 nm to 1563.86 nm, determined by the physical parameters of the erbium-doped fibre amplifier in the network. However, within this range if different manufacturers offer transmitters/ receivers at different wavelengths, there would be compatibility problems and even short range DWDM networks will not be possible. The ITU-T prescribed international standards (Rec. G.694.1) for wavelength in DWDM systems stipulates a frequency grid with respect to krypton line at 193.1 THz (1552.52 nm wavelength in vacuum), with channel spacing of 12.5 GHz, 25 GHz, 50 GHz, 100 GHz and wider (integer multiples of 100 GHz). The currently available grid points with spacing of 12.5 GHz, 25 GHz, 50 GHz have evolved by sub-dividing the initial 100 GHz grid spacing by successive factors of two. The wavelengths (frequencies) to be used for WDM should be chosen from this grid as per carrier plan, i.e., 4 wavelengths, 8 wavelengths, 16 wavelengths and so on. Channels can be spaced uniformly or non-uniformly. The current grid has more bandwidth than the bandwidth of EDFAs in order to leave scope for higher bandwidth operations in future.

Since 193.1 THz is the centre frequency of the grid, channel frequencies for various spacing are as given below:

$193.1 + 0.0125n$ for 12.5 GHz spacing
$193.1 + 0.025n$ for 25 GHz spacing
$193.1 + 0.05n$ for 50 GHz spacing
$193.1 + 0.1n$ for 100 GHz spacing

n is a positive or negative integer including 0. For example, ITU-T grid frequencies as specified in G.694.1 with channel spacing of 12.5 GHz, 25 GHz, 50 GHz, 100 GHz are given in Tables 11.1 and 11.2. These are examples over certain region in C-band and L-band and not an exhaustive list. To know the grid frequencies for all bands over the entirety of the bands, readers can refer to ITU-T Rec. G.694.1.

Table 11.1 *C-band grid frequencies and corresponding wavelengths for channel spacing of 12.5 GHz, 25 GHz, 50 GHz and 100 GHz*

Nominal central frequencies (THz) for spacing of				Central wavelength (nm)
12.5 GHz	**25 MHz**	**50 GHz**	**100 GHz**	
-	-	-	-	-
-	-	-	-	-
-	-	-	-	-
195.9375	-	-	-	1530.04
195.9250	195.9250	-	-	1530.14
195.9125	-	-	-	1530.24
195.9000	195.9000	195.9000	195.9000	1530.33
195.8875	-	-	-	1530.43
195.8750	195.8750	-	-	1530.53
195.8625	-	-	-	1530.63
195.8500	195.8500	195.8500	-	1530.72
195.8375	-	-	-	1530.82
195.8250	195.8250	-	-	1530.92
195.8125	-	-	-	1531.02
195.8000	195.8000	195.8000	195.8000	1531.12
195.7875	-	-	-	1531.21
195.7750	195.7750	-	-	1531.31
195.7625	-	-	-	1531.41
195.7500	195.7500	195.7500	-	1531.51
195.7375	-	-	-	1531.60
195.7250	195.7250	-	-	1531.70
195.7125	-	-	-	1531.80
195.7000	195.7000	195.7000	195.7000	1531.90
195.6875	-	-	-	1532.00
195.6750	195.6750	-	-	1532.09
195.6625	-	-	-	1532.19
-	-	-	-	-
-	-	-	-	-
-	-	-	-	-
193.2375	-	-	-	1551.42
193.2250	193.2250	-	-	1551.52
193.2125	-	-	-	1551.62
193.2000	193.2000	193.2000	-	1551.72

(Continues...)

Table 11.1 *Continued*

Nominal central frequencies (THz) for spacing of				Central wavelength (nm)
12.5 GHz	**25 MHz**	**50 GHz**	**100 GHz**	
193.1875	-	-	-	1551.82
193.1750	193.1750	-	-	1551.92
193.1625	-	-	-	1552.02
193.1500	193.1500	193.1500	-	1552.12
193.1375	-	-	-	1552.22
193.1250	193.1250	-	-	1552.32
193.1125	-	-	-	1552.42
193.1000	193.1000	193.1000	193.1000	1552.52
193.0875	-	-	-	1552.62
193.0750	193.0750	-	-	1552.73
193.0625	-	-	-	1552.83
193.0500	193.0500	193.0500	-	1552.93
193.0375	-	-	-	1553.03
193.0250	193.0250	-	-	1553.13
193.0125	-	-	-	1553.23
193.0000	193.0000	193.0000	193.0000	1553.33
192.9875	-	-	-	1553.43
192.9750	192.9750	-	-	1553.53
192.9625	-	-	-	1553.63
-	-	-	-	-
-	-	-	-	-
-	-	-	-	-

Table 11.2 *L-band grid frequencies and corresponding wavelengths for channel spacing of 12.5 GHz, 25 GHz, 50 GHz and 100 GHz*

Nominal central frequencies (THz) for spacing of				Central wavelength (nm)
12.5 GHz	**25 MHz**	**50 GHz**	**100 GHz**	
-	-	-	-	-
-	-	-	-	-
-	-	-	-	-
184.7750	184.7750	-	-	1622.47
184.7625	-	-	-	1622.58
184.7500	184.7500	184.7500	-	1622.69

(Continues...)

Table 11.2 *Continued*

Nominal central frequencies (THz) for spacing of				Central wavelength (nm)
12.5 GHz	**25 MHz**	**50 GHz**	**100 GHz**	
184.7375	-	-	-	1622.80
184.7250	184.7250	-	-	1622.91
184.7125	-	-	-	1623.02
184.7000	184.7000	184.7000	184.7000	1623.13
184.6875	-	-	-	1623.24
184.6750	184.6750	-	-	1623.35
184.6625	-	-	-	1623.46
184.6500	184.6500	184.6500	-	1623.57
184.6375	-	-	-	1623.68
184.6250	184.6250	-	-	1623.79
184.6125	-	-	-	1623.90
184.6000	184.6000	184.6000	184.6000	1624.01
184.5875	-	-	-	1624.12
184.5750	184.5750	-	-	1624.23
184.5625	-	-	-	1624.34
184.5500	184.5500	184.5500	-	1624.45
184.5375	-	-	-	1624.56
184.5250	184.5250	-	-	1624.67
184.5125	-	-	-	1624.78
184.5000	184.5000	184.5000	184.5000	1624.89
-	-	-	-	-
-	-	-	-	-
-	-	-	-	-

Q 11.1 *What is the significance of channel spacing in WDM? Why can it not be any arbitrary value?*

Channel spacing is the minimum frequency separation between two multiplexed signals. The channel spacing, in GHz, relates to the optical wavelength separation in the band of consideration, as described below.

- Spacing of 200 GHz corresponds to about 1.6 nm wavelength separation
- Spacing of 100 GHz corresponds to about 0.8 nm wavelength separation
- Spacing of 50 GHz corresponds to about 0.4 nm wavelength separation
- Spacing of 25 GHz corresponds to about 0.2 nm wavelength separation
- Spacing of 12.5 GHz corresponds to about 0.1 nm wavelength separation

This may suggest that by reducing the channel spacing, one can pack more channels in the useable bandwidth. However, it is not so simple. Such densely packed systems require very high frequency accuracy and stability. For a 100 GHz separation, the allowed deviation from nominal central frequency is ±20 GHz, while for a 200 GHz spacing it is ±40 GHz. *Densely packed* WDM networks require monitoring of each wavelength. This became possible with absolute frequency reference (AFR) and absolute frequency controlled multi-carrier generator, technologies that were developed gradually. Hence the initial WDM systems deployed 200 GHz spacing; the 100 GHz spacing (with sub-multiple of 100 GHz) came later with the advent of AFR and absolute frequency controlled multi-carrier generator.

Q 11.2 *What are the various bands in WDM?*

Erbium doped fibre amplifiers (EDFA) are used to amplify WDM optical signal instead of regenerators for individual wavelengths. (Amplifiers have gain flatness over the bandwidth of the amplifier.) EDFAs operate in two bands: C-band (1530–1560 nm) and L-band (1560–1610 nm), also known as the red band and blue band respectively. Tellurium doped fibre amplifiers are used in the 1480–1530 nm band or the S-band. The 1420–1480 nm band is known as S+ band. In this band, Raman amplifiers are used.

11.4.2 ITU-T grid for CWDM

The frequency grid for CWDM is 18 frequencies (wavelengths) with channel spacing of 20 nm as specified in ITU-T Rec. G.694.2. The nominal central wavelengths (nm) for a spacing of 20 nm are given below.

1271	1331	1391	1451	1511	1571
1291	1351	1411	1471	1531	1591
1311	1371	1431	1491	1551	1611

Q 11.3 *What is the reason for choosing 20 nm as the channel spacing in CWDM?*

Effective CWDM realisations with un-cooled lasers and wide passband filters require a nominal central wavelength spacing of not less than 20 nm. The total source wavelength variation is expected to be of the order of (±6 to 7) nm, compatible with current filter technologies. For the guardband, one-third of the channel spacing is required. Therefore, in order to maximise the number of channels (which is 18), 20 nm has been chosen.

11.5 WDM Transmitters

11.5.1 Technical requirements of WDM transmitters

WDM transmitters radiate as many wavelengths as a WDM system plan needs. This is achieved by combining in one transmitter many individual laser diodes radiating at fixed wavelengths. Another way is to use tuneable transmitters capable of radiating any desired wavelength. Tuneable transmitters and receivers are more expensive than their fixed counterparts. The transmitters for WDM communications have to meet a set of specific requirements in terms of quality, stability and reliability of the radiated light. For example, a 50 GHz spacing between individual channels requires a line-width of at least 0.5 GHz to avoid inter-channel crosstalk. Transmitter requirements in WDM networks are summarised below.

1. *Quality of generated light*

   a. The line width has to be as narrow as possible. Indeed, to attain 40-channel multiplexing, we need a channel spacing of 100 GHz and therefore, the line-width cannot be more than 1 GHz in order to avoid channel crosstalk. Modern DFB (distributed feedback) lasers have line widths of the order of tens of megahertz (typical line width is 10 MHz and the maximum is 40 MHz).

   b. The side-mode suppression ratio (SMSR), the ratio of the amplitude of the main lobe to largest side lobe, has to be as high as possible. A DFB laser can support several transverse modes. All these side modes have to be suppressed in order to avoid any crosstalk in WDM systems. Fortunately, again, modern DFB lasers have very good SMSR (40 dB is the typical value; 35 dB with an electro-absorption modulator).

   c. A laser has to operate in a single longitudinal mode. This is achieved in DFB lasers by using Bragg gratings to provide positive optical feedback.

   d. Chirp has to be eliminated. This is achieved by not going in for direct modulation in DFB laser, i.e., the laser drive current is not modulated. The laser output is modulated externally with the help of an integrated external modulator.

2. *Stability*

   a. Variations in output power result in variations in line width. Power instability also causes several hidden problems associated with SM (single mode) fibres; one such is stimulated Briillouin scattering (SBS).

   b. Variations in peak wavelength are unacceptable because the channel spacing has to be stable to prevent channels getting mixed-up, leading to information loss.

3. *Reliability*

4. *Minimal power consumption*
5. *Tuneability:* It may be noted that light sources are necessary for WDM networks not only as end transmitters but also as elements of add/drop components. Hence, the use of tuneable lasers is a better alternative to using array sources. The key features of tuneable lasers are the tuning speed and the ability to emit several wavelengths simultaneously. Some of the tuneable lasers and their relative merits/demerits are presented next.

11.5.2 Tunable laser transmitters

Temperature tuning lasers

Temperature-tuning lasers exploit the dependence of refractive index of the active area of the light source on temperature, which results in drift in frequency radiated by the laser. However, the tuning range and tuning speed of these lasers are not good; the range is only 10 nm and the tuning speed a few milliseconds.

External-cavity tunable lasers

External-cavity tuneable lasers change their wavelengths by varying the resonance condition of the lasing wavelength. This is achieved by using diffraction grating as one of the laser's reflectors. Refer to Fig. 11.2.

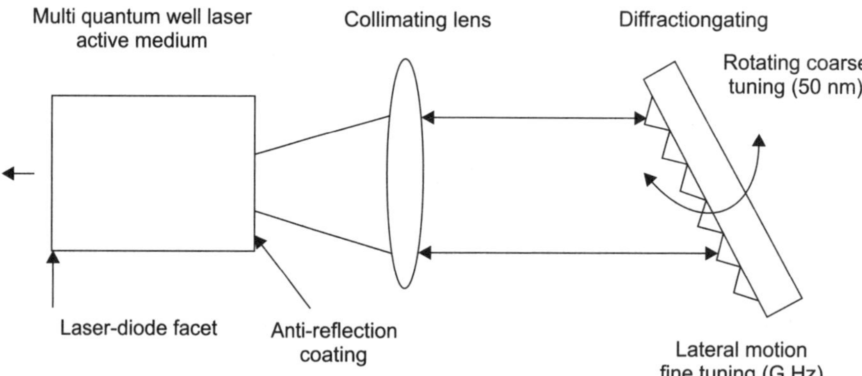

Fig. 11.2 *External cavity tuneable laser*

A diffraction grating works as a highly reflective mirror reflecting wavelength (λ_i) satisfying the condition $\delta \sin \theta = m \lambda_i$, where δ is the period of a diffraction grating, *t* the tilting angle (angle of incidence) and $m = 0, \pm1, \pm2, +3$. Only those reflected wavelengths that reproduce themselves after a round trip within the laser cavity will be radiated. This is known as resonant condition. An automatic gain controller allows the user to obtain the desired light power from the output. Since the driven current and cavity length are stable, this type of tuneable laser demonstrates exceptional characteristics: a tuneable range from 1500 nm to 1580 nm; wavelength-setting resolution of 1 pm; absolute wavelength accuracy of 0.1 nm; line-width of 700 kHz that is tightly controlled; and power stability from output 2 of ±0.01 dB/hr. The major disadvantage of an external-cavity tuneable laser is its tuning speed, which is of the order of milliseconds. Hence, external-cavity tuneable lasers are used today in WDM systems not in transmitters but in testing and measurements setups.

Sectional distributed-Bragg-reflection (DBR) tuneable laser

WDM systems need a tuneable laser with high tuning speed. Sectional distributed-Bragg-reflection (DBR) tuneable lasers are tuned at the speed of the order of tens of nanoseconds, which is much higher than the tuning speed of external-cavity lasers. Hence this type of tuneable lasers is used for transmission in WDM systems.

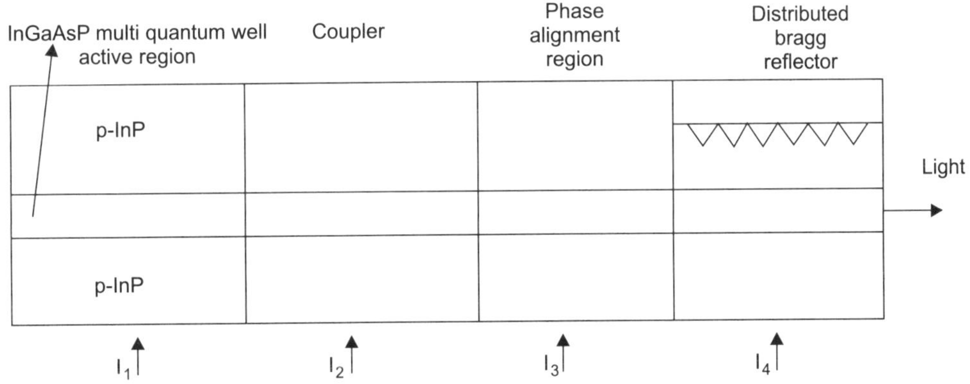

Fig. 11.3 *Four section DBR laser*

A four-section DBR laser called the grating coupler sampled reflector (GCSR) laser is the latest in WDM sources as shown in Fig. 11.3. Here, a vertical coupler section creates a set of wavelengths from which a wavelength can be chosen by tuning the phase-shift and Bragg-reflection sections. Coarse tuning is facilitated by the coupler section while fine tuning is facilitated by the phase-shift section. Some of the important parameters of GCSR lasers are:

Wavelength range	1528 to 1565 nm
Side-mode suppression ratio (SMSR)	25 dB
Laser-chip tuning time	20 ns
Output power	0 dBm
RIN (relative intensity noise)	–140 dB/Hz.

Integrated-cavity lasers

Integrated-cavity lasers offer the advantages of high-speed tuning and the ability to emit several wavelengths simultaneously, which are critical for high-speed packet switched WDM networks. Their operation is shown in Fig. 11.4. Here, a number of active media (optical amplifiers) are terminated by a common cleaved-mirror facet at one end. These amplifiers are optically connected to an optical multiplexer/filter. The multiplexer has a single output port optically connected to the second cleaved mirror facet. These facets (mirrors) constitute a laser cavity. Light amplified by the first active medium passes through the multiplexer where the individual waveguide for this medium exists. The combination of the first amplifier and the unique path of the light within the cavity generates an individual wavelength. Similarly, the different amplifier–waveguide sets generate number of wavelengths (i.e., many individual lasers in a single cavity) and the multiplexer combines all the beams into a single output. The merit of this laser lies in its tuning—no moving parts for tuning! Tuning is achieved by simply activating an individual active medium. Integrated cavity lasers have tuning speeds less than 3 ns.

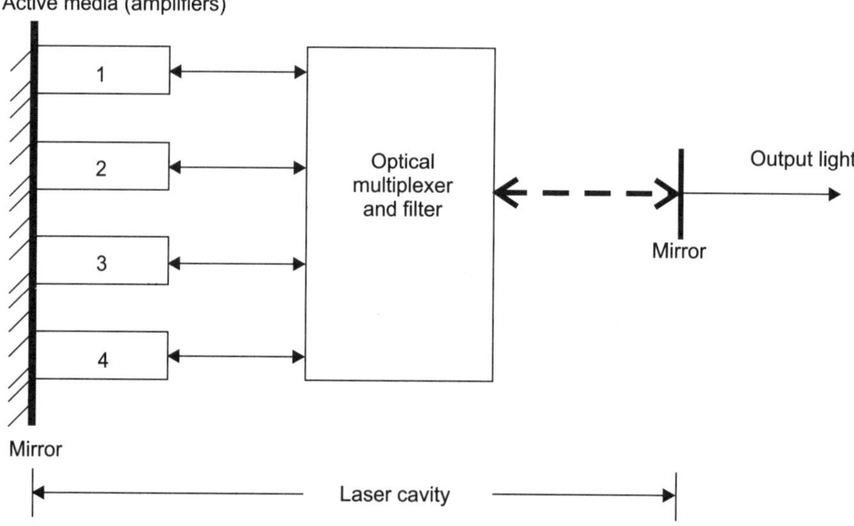

Fig. 11.4 *Integrated cavity tuneable laser*

Besides high-speed tuning, integrated cavity lasers can have channel spacing of even 50 GHz. Its only demerit is the chip size, which is larger than a DFB laser array and increases with the number of radiating wavelengths.

To sum up, we have various WDM sources: laser-array, temperature-tuning, external-cavity and integrated-cavity light sources. They differ in many ways, but in terms of the most critical characteristics for commercial application, i.e., tuning speed and the ability to emit many wavelengths simultaneously, one of the most promising transmitters is the integrated-cavity tuneable laser. As the cost of high-speed tuning lasers is high, some of the WDM networks use an array of transmitters radiating fixed wavelengths within the EDFA bandwidth. The search is still on for better WDM light sources.

11.6 Optical Amplifiers

11.6.1 Introduction to optical amplifiers

Optical signal is attenuated due to loss characteristic of fibres, splices, connectors, components or equipment ports. To improve signal strength, there are two approaches—use of regenerators or use of optical amplifiers.

A regenerator converts an optical signal into an electrical signal, decides on logic 0 or 1, generates a new electrical signal, converts it to an optical form and transmits the re-generated signal. The problems of deterioration of signal amplitude, pulse spread and fibre dispersion are eliminated by a regenerator. However, regenerator can work only on a single wavelength, bit rate and line coding format. This calls for the use of number of regenerators to suit a WDM plan. Also, if the bit rates or WDM plans are changed, re-deployment/re-engineering will be necessary to acommodate the changes. Optical amplifiers boost an optical signal without optical–electrical conversion, a capability that allows them to work with a wide range of wavelengths, bit rates and line code formats. For this reason, optical amplifiers gained popularity, especially after being available as low-noise amplifiers.

Optical amplifiers offer the twin benefits of increasing the capacity of fibre optic links with WDM (without wavelength-wise regenerators) and the capability for developing real fibre-optic (all-optical) networks, not just powerful point-to-point links.

There are three types of optical amplifiers, based on the technology used:

- Semiconductor optical amplifiers (SOA)
- Fibre optical amplifiers (FOA) working on stimulated emission
- Fibre optical amplifiers (FOA) working on stimulated scattering

11.6.2 Performance consideration of optical amplifier

Gain, bandwidth and noise figure are the important parameters related to the performance of optical amplifiers. Though the value of gain depends on the principle of operation of the amplifier, two essential factors decide the gain, viz., frequency (wavelength) of input signal and power of input signal. Gain is a constant over a finite band known as bandwidth of the amplifier. When the input power is too high, gain decreases. This phenomenon is known as gain saturation.

The noise performance of an optical amplifier is evaluated by the noise figure, which is the ratio of the signal-to-noise ratio (SNR) of an input signal to the SNR of an output signal. The noise figure accounts for the fact that an optical amplifier increases the input signal (and input noise) and generates its own noise.

An optical amplifier may function as:

a. *Booster* (post-amplifier): It is placed immediately after the transmitter and magnifies the signal before sending it down a fibre. Its main function is to produce maximum optical power.

b. *In-line amplifier*: An in-line amplifier is placed in the middle of the fibre-optic link to compensate for losses caused by fibre attenuation, connections and signal distribution in the network. It has gain flatness and stability of all characteristics over the entire WDM bandwidth.

c. *Pre-amplifier*: A preamplifier magnifies the optical signal immediately before it reaches a receiver. Hence, good sensitivity, high gain and low noise are the major requirements here.

11.6.3 Semiconductor optical amplifier (SOA)

Semiconductor optical amplifier (SOA) is an active region of a semiconductor laser with very low optical feedback. It works on the principle of stimulated emission to boost an information signal. Two main types of SOAs in use in WDM systems are:

- Fabry–Perot amplifier (FPA)
- Travelling-wave amplifier (TWA).

A Fabry-Perot laser diode is an active medium placed between two parallel mirrors operating below the radiating threshold condition. In other words, an FPA SOA is a laser diode whose gain is less than its loss. A TWA is simply an active medium without any reflective facets (mirrors). Figure 11.5 illustrates these two types of SOAs.

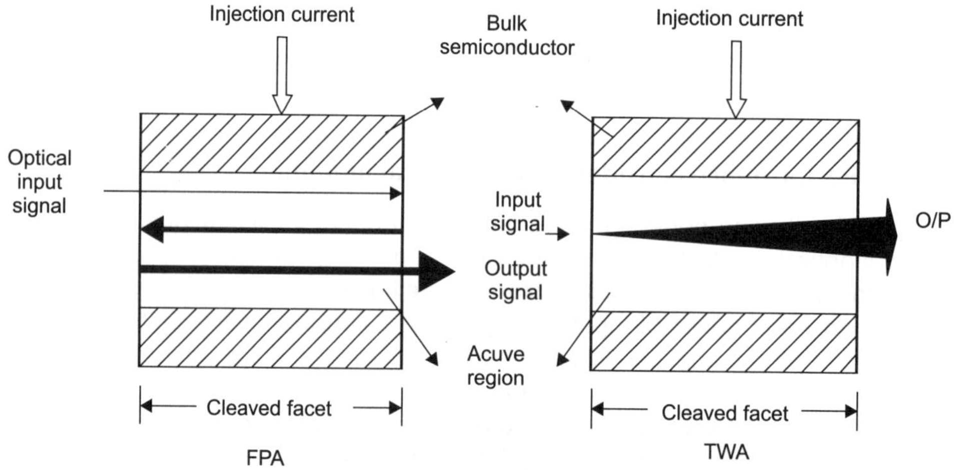

Fig. 11.5 *FPA and TWA*

The gain of a TWA is equivalent to single-pass gain of an FPA while the actual gain of an FPA exhibits ripples whose amplitude depends on the value of the mirror's reflectance. The higher the reflectance, the larger is the gain peak. If the reflectance is made equal to zero, we have a TWA.

Gain for both FPAs and TWAs depends on wavelength (frequency). A TWA has much smaller gain but covers a much wider spectral region. An FPA, in contrast, has very high gain within a narrow bandwidth. Therefore, there is a gain–bandwidth tradeoff to be considered in SOAs. Since the gain of a TWA can be increased by extending the length of the active medium while there is no means to extend the bandwidth of an FPA, the TWA is the more popular type of SOA of the two.

The advantage of SOAs is their ability to operate at 1300 nm and 1550 nm, that too simultaneously. As we will discuss later, erbium doped fibre amplifiers (EDFA) cannot work at 1300 nm, making SOAs the only optical amplifier available for both windows. Another advantage is that a bandwidth up to 100 nm can be covered by an SOA. The third advantage is the possibility of integrating SOAs along with other semiconductor and photonic devices into opto-electronic integrated circuits.

Disadvantages of SOAs are: higher crosstalk compared to EDFA, sensitivity to the state of polarisation of an incoming signal, the technological difficulties in fabricating SOAs with low reflectances, and their temperature sensitivity.

Hence, EDFAs are mostly in use as optical amplifiers on WDM networks (to some extent Raman amplifiers are in use, i.e., distributed optical amplifier) while SOAs are popular in optical fibre cable in the TV segment.

11.6.4 Fibre optical amplifiers (FOA) working on stimulated emission

Fibre optical amplifier (FOA) working on stimulated emission consists of a piece of specially fabricated fibre (heavily doped silica fibre) spliced with a transmission fibre and connected to a pump laser. The energy delivered by the pump laser excites the atoms of the doping material of the fibre at the upper energy level, where they are stimulated by photons of the information signal. In this excited condition they transfer the energy gained from the laser pump to the information signal and fall back to a lower energy level.

The doping material is erbium, hence, the FOA is known as erbium doped fibre amplifier (EDFA), which is just a silica fibre heavily doped with erbium with a coupler and laser pump input as shown in Fig. 11.6. The piece of fibre heavily doped with ions of erbium acts as an active device. To this erbium doped fibre, information signal (1550 nm) and laser pump signal (980 nm or 1480 nm) are coupled by a coupler. The information signal is amplified while the pumping signal loses its energy (gives to information signal). The pump can be co-propagating (lower noise, lower output power) or counter-propagating (higher noise, higher output power). In practice, bi-directional pump is used to get the best of both. The second coupler removes the residual pumping light and noise, else it may turn out as a laser. An isolator is also used to stop back-reflected light. Gain of EDFA is about 23 dB. EDFA's bandwidth is 30 nm, from 1530 nm and 1560 nm. To extend the bandwidth, gain equalisers are included in the structure of an EDFA. We have EDFAs with equalisers having bandwidth of 80 nm (1530 nm to 1610 nm).

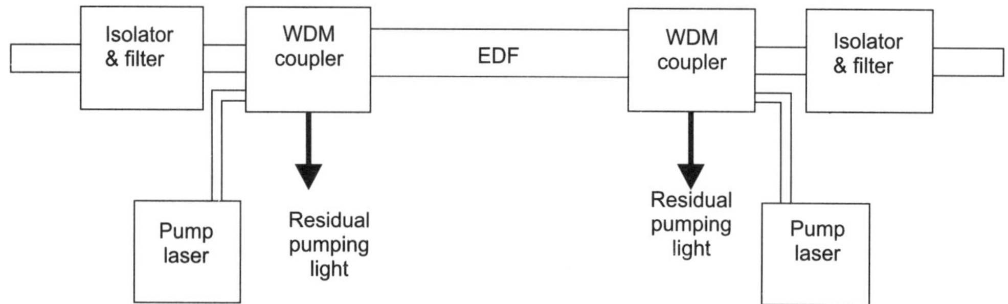

Fig. 11.6 EDFA schematic

Q 11.4 Why is EDFA is the most appealing of amplifiers?

When erbium ions are doped into silica fibre, their energy levels split into number of energy bands (each band is a set of closely spaced levels). The prominent bands are three, as shown in Fig. 11.7.

Energy from the pump laser excites erbium ions at upper energy band (level 3). The information signal (around 1550 nm) stimulates transition of the excited ions to a lower energy band. The ions come down to level 2 quickly and then to level 1. Transition between level 2 (intermediate band) and level 1 (lower energy band) results in radiation of photons with wavelength of 1550 nm. This energy is transferred to information signal.

It may be recalled from chapter 1 that attenuation of optical fibre is the lowest (0.25 dB/km) at 1550 nm. It is gift of nature that EDF gives amplification at 1550 nm. So, EDFA is the most appealing choice for fibre amplifier.

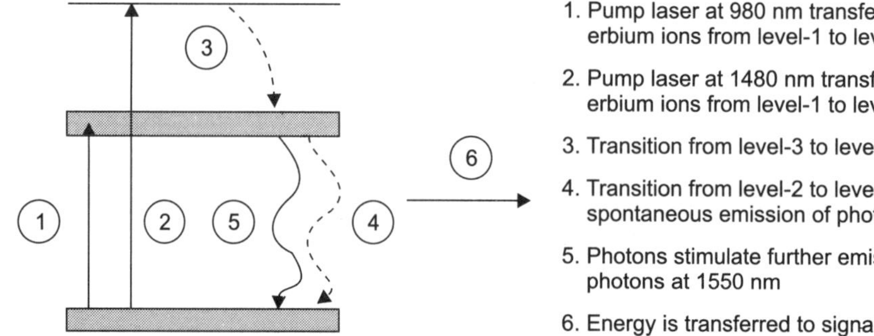

1. Pump laser at 980 nm transfers erbium ions from level-1 to level-3
2. Pump laser at 1480 nm transfers erbium ions from level-1 to level-2
3. Transition from level-3 to level-2 is fast
4. Transition from level-2 to level-1 causes spontaneous emission of photons at 1550 nm
5. Photons stimulate further emission of photons at 1550 nm
6. Energy is transferred to signal at 1550 nm.

Fig. 11.7 Energy bands and emission process in EDFA

The advantages of EDFA are large bandwidth, low crosstalk and the ability to keep its characteristics constant when one or more input channels are added or dropped. The disadvantages are slice and splice effects, i.e., the slice effect of noise and the problem of splicing EDFA to normal fibre. These are explained next.

Slice effect

The main source of noise in an EDFA is the amplified spontaneous emissions. Some of the excited erbium atoms drop to lower level spontaneously, radiating photons within the EDFA bandwidth. These spontaneous photons stimulate other transitions, which means there is amplification. Since this amplified spontaneous light is out of phase and therefore not coherent with respect to the information signal, it constitutes noise. This noise is distributed evenly along the EDFA bandwidth but for a detrimental portion called the slice, which lies within the spectral width of a specific channel (wavelength). This noise slice interferes with the information signal and the resulting beat signal is a major contributor to EDFA noise.

Splice effect

Splicing an erbium-doped fibre to the transmission fibre (single mode fibre) is a problem because of the great diameter mismatch between them. The core diameter of a single mode transmission fibre is between 8–10 microns, while the core diameter of an erbium-doped fibre varies from 2.8 to 5.2 microns. This mismatch causes losses of about 1.87 dB.

The splice joint loss on transmission fibre due to cuts or joining of fibres belonging to two drums while laying of fibres is of the order of 0.05 dB; so a loss of 1.87 dB is very high. In fact, a loss of 1.87 dB means that the light power after splicing is only 65% of the power before splicing. Obviously, something needs to be done to reduce the loss.

There are three ways of reducing this loss: (a) down-tapering, (b) up-tapering or TEC method, and (c) using interim fibre. In the down-tapering method, both regular and amplifier fibres are stretched out when melted by fusion. This process reduces the core diameter of a regular fibre more than that of an erbium-doped fibre, thus lessening the core diameter mismatch. In the up-tapering method, also called the thermally diffused expanded-core method (TEC), the core diameter of an amplifier fibre is increased by diffusing the dopants on the heated end of the fibre. Heating is done by the same fusion splicer used for splicing. The result: a reduction in core diameter mismatch. In the interim fibre method, an intermediate fibre is used between transmission fibre and EDF to reduce the core diameter mismatch at either end, but there is additional splice loss in this method.

Q 11.5 *When we talk about noise in EDFA, is it optical noise or electrical noise that we are referring to?*

Let us first understand optical noise and electrical noise of an optical amplifier. Optical noise is the noise measured in an optical field by optical measurement devices such as optical power meter and optical spectrum analyser. Electrical noise is a characteristic of an optical amplifier measured at a point just beyond the photodetector, that is, after the conversion of an optical signal into an electrical signal.

From optical considerations, the interference between two light signals—the information signal and the amplified spontaneous emission—causes intensity fluctuations in the detected light. From electrical considerations, these light fluctuations appear as photocurrent fluctuations, resulting in a certain signal-to-noise ratio. Hence, when we talk of noise in EDFA, it is the manifestation of optical noise as electrical noise plus the intrinsic electric noise in the associated electronics.

Q 11.6 *Can we really implement the three methods discussed above to reduce splice loss in EDFA?*

It is possible only if both the regular and amplifier fibres have the same cladding and coating diameters. Both the down- and up-tapering methods reduce splicing losses to 0.187 dB from 1.87 dB while the interim fibre method reduces the loss to around 0.21 dB.

Q 11.7 *EDFA being work-horse, is it possible to get all functional varieties, viz. booster amplifier, in-line amplifier and pre-amplifier?*

Yes, EDFA booster amplifiers are available with maximum output power of 16 dBm (39.8 mW) and 19 dBm (79.4 mW). EDFA in-line amplifier output power is only 14 to 18 dBm, but it has the best multi-channel gain flatness. Gain variation is only ±0.5 dB over the bands of operation. EDFA pre-amplifier, like any pre-amplifier, should have low noise and better sensitivity. The noise figure of EDFA preamplifier is 4.0 dB and sensitivity is –40 dBm, i.e., it can operate at 0.1 μW of input power.

Q 11.8 *While discussing SOA, it was mentioned that SOA has polarisation dependent characteristics whereas EDFA does not. Can it be compared quantitatively?*

Polarisation sensitivity of gain is measured by a parameter known as 'polarization-dependent loss (PDL)'. PDL is essentially the gain-polarisation sensitivity of a fibre amplifier. It is usually mentioned for an input signal of –20 dBm.

Consider an input of –20 dBm. The amplifier can have two output powers—10.2 dBm and 10.7 dBm—depending on the state of polarisation of the input signal. Then the polarisation sensitivity or PDL is 10.7 dBm –10.2 dBm = 0.5 dB.

For EDFA the PDL is 0.2 dB whereas for SOA it is 0.5 dB.

11.6.5 FOA working on stimulated scattering – Raman amplifiers

EDFA works on stimulated emission principle whereas Raman amplifier works on stimulated Raman scattering (SRS) principle. SRS is broadband in effect and hence Raman amplifiers can work in C-band (1530nm–1560 nm), L-band (1560nm–1610 nm), S-band (1480nm–1530 nm), S+ band (1420 nm–1480 nm) and 1310 nm band. EDFA can work only in C-band and L-band.

Raman amplifier can be used as a discrete amplifier or a distributed amplifier. When used as discrete amplifier, it consists of a long spool of fibre with a pump laser. In the distributed case it is used to complement EDFAs by providing additional gain in a distributed manner in ultra-long haul links. Use of Raman amplifier on a span as distributed amplifier is shown in Fig. 11.8.

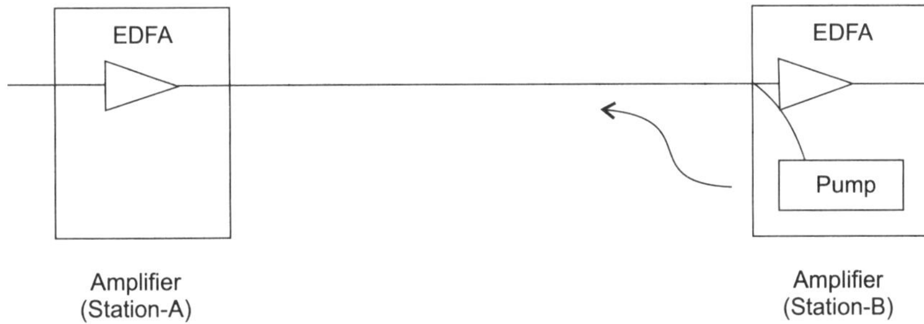

Fig. 11.8 *Deployment of Raman amplifier and pump laser*

Q 11.9 Why is the pump signal made to propagate backwards in Raman amplifier?

Raman amplifier response is instantaneous to laser pump input whereas in EDFA it is slower. Consequently, Raman amplifier shows gain variation as a follow-up of variations in laser pump input level. This causes cross-talk. To minimise this the pump signal is made to propagate backwards so that fluctuations in the pump power are averaged over the signal propagation time in the fibre.

Also, a modulating signal at the pump laser wavelength depletes the pump power (modulates the pump signal, rather) and causes gain variation. This is also smoothed out by backward propagation as described above.

11.7 WDM Receivers

11.7.1 Technical requirement of WDM receivers

WDM receivers need to operate in the WDM bandwidth of 80 nm, from 1530 nm to 1610 nm. In general, the spectral response of detector varieties (p-i-n and APD photodiodes) discussed in chapter 8, are even wider; however, their responsivity (photocurrent per unit power incident) exhibits nonlinearly within the WDM bandwidth. Hence, additional measures for equalising spectral response must be taken. Also, receiver sensitivity has to be designed considering the entrance of many channels (wavelengths) simultaneously and independently.

Another major concern in the design of receivers is inter-channel crosstalk, which degrades signal-to-noise ratio (SNR). Since transmitters radiate fixed wavelengths, WDM receivers have to be able to select the desired wavelength. This selection is done by detectors, whose tuning time realisation is a challenge in itself.

Let us first list out the technical requirements of WDM receivers.

1. *Spectral width:* The spectral width required is determined by the range of frequencies the EDFA can handle, i.e., 1530 nm to 1610 nm. p-i-n photodiodes and APDs have bandwidth from 1150 nm to 1600 nm. But the responsivity is not linear and over entire bandwidth of 1150 nm to 1600 nm, the variation is 40% whereas over the WDM bandwidth, the variation is 10%. Therefore, either the spectral response within the required range has to be flattened, or unequal amplification (equalisation technique) has to be established to keep the output signal from different wavelength channels at approximately the same power level.

2. *Receiver sensitivity:* Although a receiver's sensitivity may be good enough to handle a single channel with a given bit-error rate (BER), it does not guarantee that the receiver will work well in a WDM network. This is due to SNR degradation caused by channel cross-talk. This implies that more power is required to maintain the same BER. Furthermore, the narrower the channel spacing, the more detrimental is the effect of crosstalk and higher is the required receiver sensitivity.

3. *Tuning time:* The tuning-time for selecting a specific wavelength depends largely on the type of network. Transport networks, such as long-haul and backbone links, support connection-oriented traffic. In other words, the traffic of such networks go from the point of origin to destination without being diverted. Here, the tuning-time requirement is rather relaxed, within a few milliseconds. But in packet-switching networks, the tuning time has to be a small fraction of the packet time, of the order of nanoseconds.

4. *Temperature sensitivity:* WDM receivers have to sustain operation over a wide temperature range of –40 to +85 °C, as the networks have to work satisfactorily in different climatic conditions. The receiver module should have provision for monitoring and control of temperature, a consideration that adds to the cost of the receivers.

5. *Polarization independence:* A receiver should be able to handle signals with any type of polarization. This is a necessary requirement because the received signal is usually polarised randomly.
6. *Miscellaneous:* Requirements such as low power consumption, compactness of receiver module and associated electronics, reliability of the receiver module, immunity to electromagnetic interference, internal noise, etc., are the usual essential features of WDM receivers, similar to any receiver.

11.7.2 WDM receiver varieties

We need a WDM receiver capable of choosing an individual channel from the many transmitted over a network. Since in WDM network a channel is associated with a wavelength, the issue at hand is of wavelength division multiple access. There are two approaches to achieving this.

Tuneable transmitter and fixed receiver (TTFR)

Refer to Fig. 11.9(a). In this scheme the receiver selects only a fixed wavelength. To address a specific receiver, the transmitter has to tune its emitting wavelength. For example, transmitter Tx-l addresses receiver Rx-1 by tuning its radiating wavelength at λ_1.

Fixed transmitter and tunable receiver (FTTR)

Refer to Fig. 11.9(b). In this scheme, the receiver is tuned to pick up the desired wavelength while each transmitter emits a fixed wavelength. For example, Tx-1 emits only λ and Rx-1 picks it up by tuning to λ_1.

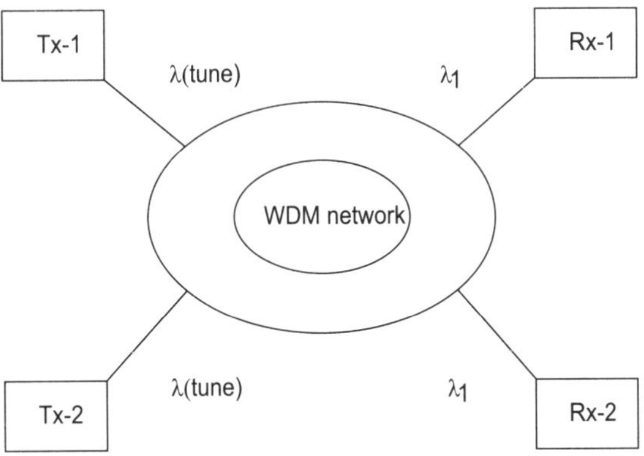

Fig. 11.9(a) *TTFR approach for WDMA*

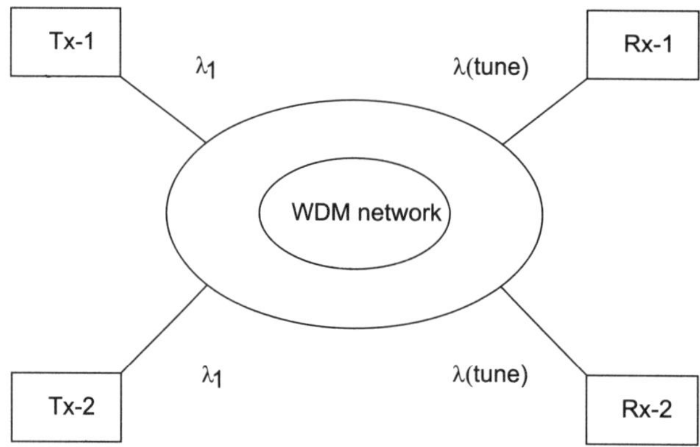

Fig. 11.9(b) *FTTR approach for WDMA*

Since most of the transmission in WDM networks is done with fixed-wavelength transmitters, the second approach, FTTR, is the one mostly in use. In FTTR, there are two methods of tuning a receiver.

Active approach

Refer to Fig. 11.10(a). In this method, a tuneable filter actively seeks a desired channel. When the desired wavelength is found, it is transmitted to a standard receiver. The search here is in the optical domain.

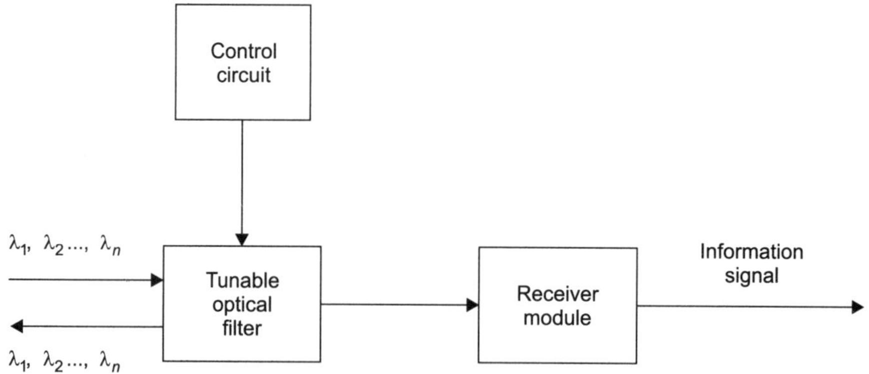

Fig. 11.10(a) *Active approach to tuning in FTTR*

Passive approach

Refer to Fig. 11.10(b). This method uses an optical de-multiplexer which separates the received wavelengths and directs each of them to an individual photodiode. All the photodiodes are placed in one array. Selection of the desired signal is done by electronic components that switch from one photodiode to another. The search in this method is in the electronic domain.

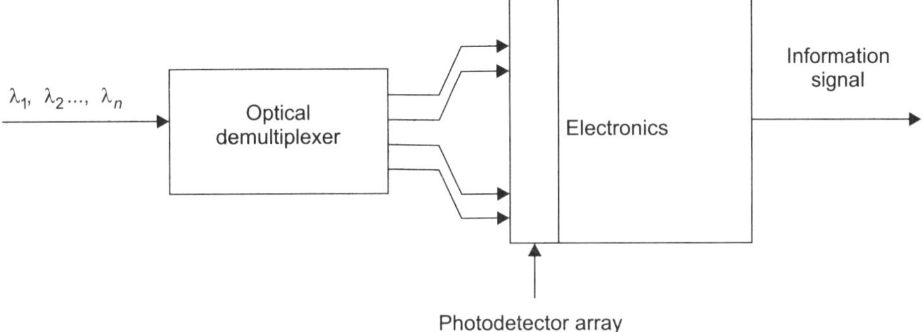

Fig. 11.10(b) *Use of de-multiplexer for tuning in FTTR*

The optical de-multiplexer can be in one stage or two stages. For example, in the 16 wavelength plan, the first de-multiplexer breaks 16 channels into 4 and then 4 de-multiplexers separate each of the 4 channels into individual wavelengths and direct them to individual photodetectors. Photodetectors are monolithically fabricated as an array in a chip. The centre-to-centre space between individual diodes can be as small as 250 nm. The modulation bandwidth of each photodiode is made to suit the channel spacing of the WDM plan.

Since the switching from one wavelength to another in the passive approach is performed electronically, the switching speed is high and selection can be done within few nanoseconds. This advantage, along with the availability and reliability of passive optical de-multiplexers, makes passive selection the method of choice in WDM networks.

The trend in WDM networks is towards opto-electronic integrated circuit (OEIC) transceivers. The implication of this is that the transceiver arrays could one day be integrated in one chip containing all the necessary components similar to the linear/logic/processor ICs.

11.8 Optical Switches

11.8.1 Functions of optical switches

Optical switches are used for the following functions.

Light path switching in WDM networks: In this application, optical switches facilitate provisioning of light paths in wavelength cross-connects, i.e., the traffic is interchanged on wavelengths in WDM network so that with the available wavelength plan of the WDM network, traffic between two nodes are carried on different wavelengths.

Protection switching between different fibres: In this application, an optical switch transfers traffic from the primary fibre to the protection fibre in case of a failure (cut) in the primary fibre.

Packet switching in optical packet switched networks: The optical switch carries out packet-by-packet switching of the incoming stream of packets.

Modulation switching: Optical switches turn-on or turn-off the data in front of external modulators.

Table 11. 3 gives the switching time and number of ports required for the different optical switch applications discussed above.

Table 11.3 *Optical switch applications – requirement of switching time and ports*

Application	Switching time required	Number of ports
Light-path switching	1–10 ms	>1000
Protection switching	1–10 ms	2 to 1000
Packet switching	1 ns	>100
Modulation switching	10 ps	1

Q 11.10 Why is switching time in case of packet switching and modulation switching of the order of nano second and pico seconds respectively?

The packet switching time has to be smaller than the packet duration. The smallest packet is an ATM cell which is 53 byte long and its highest transmission speed is 10 Gbps of STM-64. So, an ATM cell duration on STM-64 link is about 42 ns, and hence, packet switching time should be of the order of few nanoseconds. In modulation switching the switching time should be less than a fraction of rise time (100 ps) of the data signal, i.e., of the order of few picoseconds.

11.8.2 Important parameters of optical switches

Optical switches are characterised by the following parameters.

1. *Switching time*, which is the time taken to effect the switching for the intended application (Table 11.4).
2. *The number of ports* of the switch for the intended application (Table 11.4).
3. The ratio of the output power in ON state to OFF state, termed the *extinction ratio*, should be as high as possible (about 25 to 50 dB).
4. *Insertion loss* or the difference between input power and output power, should not only be low but also uniform for all ports (about 1 dB).

5. *Cross-talk*, the ratio of power at the output from the desired input to the power from all other inputs should be better than 84 dB.
6. *Polarization dependent loss* should be low, especially for switches associated with external modulators, as the polarization is to be preserved.
7. *Latching property*, implying the ability of a switch to remain connected to the path even after the switching control signal is taken-off so that traffic can pass through.
8. *Monitoring facility* to read-out current connection states (paths) of the switches.
9. *Reliability*, mentioned in two ways: (i) the number of cycles of operation (few million cycles), and (ii) the time gap (period of non-operation) after which the switch can reliably operate.

11.8.3 Large optical switches

Switches with number of ports from 100 to few thousands are made using multiple switches in arrays or similar configurations. However, it is not possible to connect any type of switches into large arrays.

A switch is said to be blocking if there are constraints in connecting any unused input port to any unused output port. A switch is non-blocking if any unused input port can be connected to any unused output port. There are three varieties in non-blocking switches.

1. Non-blocking in strict sense
2. Non-blocking in wide sense
3. Re-arrangeable non-blocking

A switch is non-blocking in the wide sense if a connection is made without re-routing any existing connection. Such switches require specific routing algorithms to route connections so that future connections are not blocked.

A switch is non-blocking in the strict sense if any unused input port is connected to any unused output port regardless of how the previous connections were made.

A non-blocking switch that requires re-routing of connections to achieve non-blocking property is known as 'rearrangeable non-blocking switch'.

To build large switches, rearrangeble non-blocking variety can be used with the advantage that fewer numbers are required but the associated control algorithms may be complex. In other varieties of control, the algorithms are simpler but the number of switches used may be large. The basic architectures used in switches are cross-bar, Clos, Spanke, Benes, Spanke–Benes.

A cross-bar switch is wide-sense non-blocking. A 4×4 ($n = 4$) cross-bar switch using sixteen 2×2 switches is shown in Fig. 11.11.

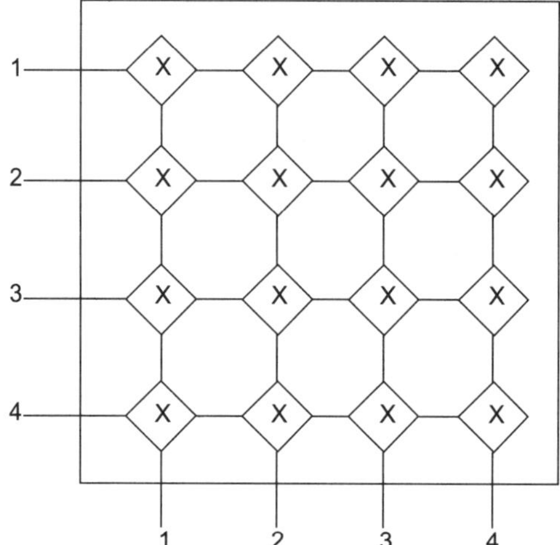

Fig. 11.11 *A 4 × 4 cross-bar switch with sixteen 2 × 2 switches*

Clos architecture is non-blocking in the strict sense. An $n \times n$ Clos switch is constructed as given below.

Let $n = mk$, where m and k are integers. The first stage and third stage consist of k number of $m \times p$ switches.

The middle stage has p number of $k \times k$ switches. To minimise cost, p is chosen as greater than equal to $2m - 1$ (to achieve non-blocking).

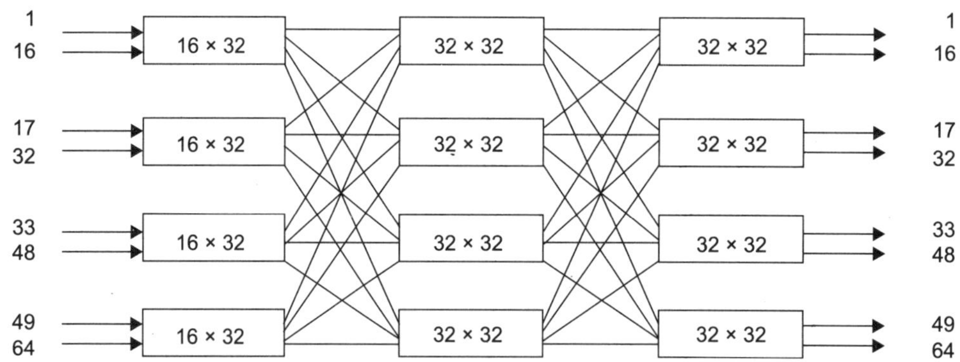

Fig. 11.12 *Three-stage Clos architecture for 64 × 64 switch (n = 64)*

Figure 11.12 shows the realisation of a 64 × 64 switch (n = 64) using three-stage Clos architecture. Note that in this example,

$$n = 64 = 16 \times 4 \qquad m = 16 \qquad k = 4 \qquad p = 32.$$

Spanke architecture is non-blocking in the strict sense. For an $n \times n$ switch we require $2n$ number of $1 \times n$ switches as shown in Fig. 11.13.

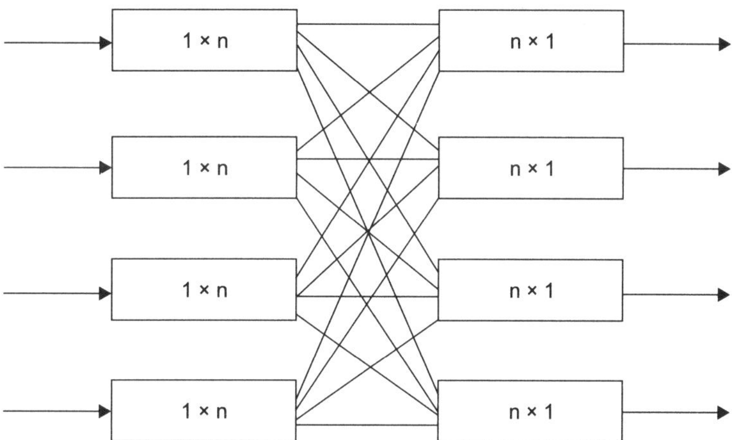

Fig. 11.13 *Spanke switch n × n using 2n numbers of 1 × n switches (n = 4)*

Benes architecture is the most efficient in terms of number of ports for realising a large $n \times n$ switch. Figure 11.14 shows Benes architecture (re-arrangeable non-blocking) implementation of 8×8 switch using twenty 2×2 switches, i.e. 80 ports. (Note that cross-bar architecture requires $8 \times 8 = 64$ numbers of 2×2 switches = 256 ports; Clos requires ninety-six 2×2 switches = 384 ports; Spanke requires sixteen 1×8 elements = 128 ports. Clearly, Benes is the most efficient of the lot.)

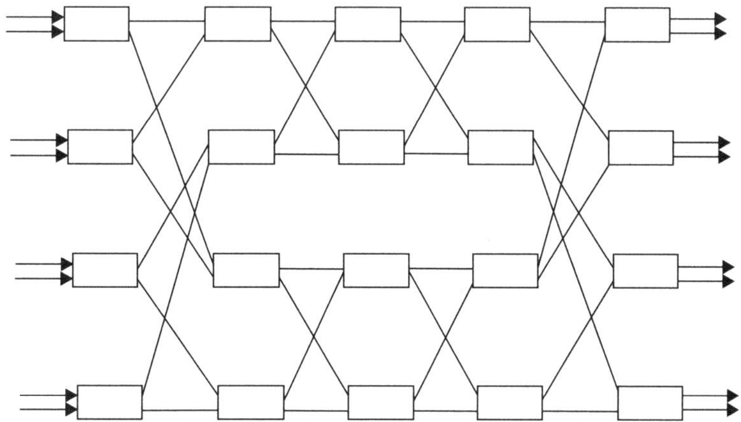

Fig. 11.14 *Benes architecture for 8 × 8 using twenty 2 × 2 switches*

Spanke–Benes architecture combines the best of cross-bar and Benes architectures. It has no waveguide cross-overs. For $n \times n$ switch, it requires $n(n-1)/2$ switches[1]. Refer to Fig. 11.15 showing Spanke–Benes architecture using twenty-eight 2 × 2 switches for an 8 × 8 switch.

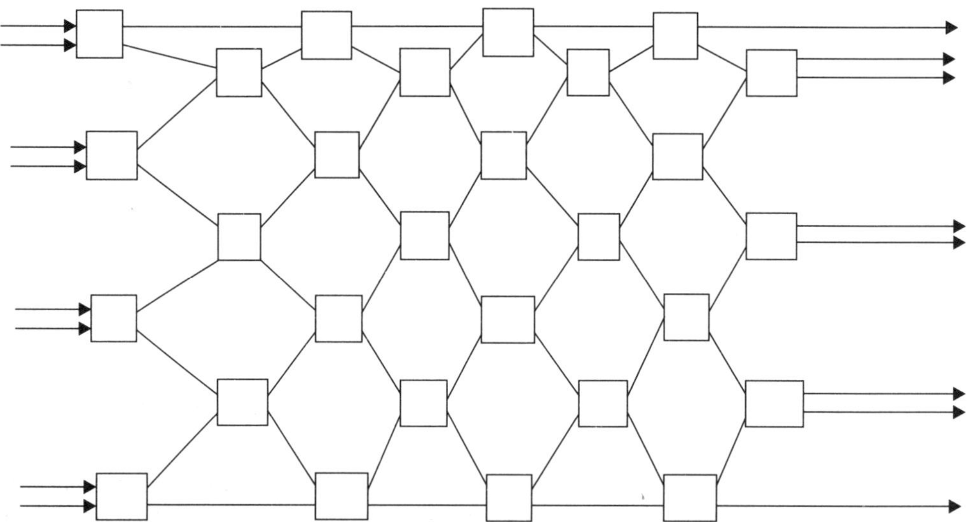

Fig. 11.15 *Spanke–Benes architecture for 8 × 8 using twenty-eight 2 × 2 switches*

Table 11.4 summarises the basic architectures of switches.

Table 11.4 *Features of various switch architectures*

Architecture	Non-blocking variety	No. of switches in the architecture of order n	Min. loss in dB considering 1 dB loss per switch (link)	Max. loss considering 1 dB loss per switch (link)
Cross-bar	Wide-sense	$n \times n$	1	$2n-1$
Clos	Strict sense	$4\sqrt{2}\, n^{1.5}$	3	$5\sqrt{(2n)}-5$
Spanke	Strict sense	$2n$	2	2
Benes	Rearrangeable	$(n/2)[2 \log_2 n(n-1)]$	$2 \log_2(n-1)$	$2 \log_2(n-1)$
Spanke–Benes	Rearrangeable	$(n/2)(n-1)$	$n/2$	n

11.8.4 Optical switch technologies

The various technologies used to realise optical switches are listed below and their parameters are compared. Since the focus of the book is on hierarchies and networks, switch fabrication technologies are covered in brief only; for a detailed discussion on the same, refer to chapter 3 of reference[2].

There are six varieties of optical switch fabrication technologies.

1. *Micro-electro-mechanical system (MEMS) switches*
   MEMS are fabricated using silicon substrates. A single wafer has a large number of miniature moveable mirrors fabricated in the wafer, with dimensions of the order microns to mm, depending on the number of mirrors to be accommodated. The mirrors are deflected by electronic actuation. Mirrors can be 2D or 3D, manufactured by surface micro-machining or bulk micro-machining, respectively. Switches with 2D MEMS are available up to 32 × 32 and with 3D MEMS are available up to 1000 × 1000 arrays.

2. *Bubble-based waveguide switches*
   This switch consists of waveguides that cross each other. At the crossover points, trenches filled with the index matching fluid are present. Under normal conditions light propagates along the trench. If the fluid is heated, it forms a bubble and no longer matches the index and reflects light. Thus, each cross-over point acts like a 2 × 2 switch.

3. *Liquid crystal switches*
   The operation of these switches is based on polarisation effect. Polarisation of light passing through liquid crystal cell can be rotated/not rotated, and thus switching can be effected.

4. *Electro-optic switches*
   The switching time of an electro-optic switch is very less, of the order of few picoseconds. Lithium niobate waveguide is subjected to external voltage to allow light or do otherwise.

5. *Thermo-optic switches*
   These switches are optic interferometers constructed on waveguide material whose refractive index is a function of temperature. The relative phase difference between two arms of propagation causes constructive interference or destructive interference of optical signal, and thus the signal is allowed or not allowed. Silica substrates or polymer substrates are used to make these switches. The waveguides in the two arms are made of lithium niobate.

   It may be noted that, in electro-optic switches, the actuation is by electric voltage whereas in thermo-optic switches, it is by temperature change. While an electro-optic switch has switching time of the order of ps, a thermo-optic switch has switching time of few ms.

6. *Semiconductor optical amplifier switches (SOA)*
   By varying the bias voltage of semiconductor optical amplifier, it can be made to work as amplifier or signal absorber.

A comparison of important parameters of various switches is presented in Table 11.5.

Table 11.5 *Comparison of important parameters of various switches*

Switch type	Size	Loss (dB)	Cross-talk (dB)	Polarisation dependent loss (dB)	Switching time
2D MEMS	32 × 32	5	55	0.2	10 ms
3D MEMS	1000 × 1000	5	55	0.2	10 ms
Bubble-based	32 × 32	7.5	50	0.3	10 ms
Liquid crystal	2 × 2	1	35	0.1	4 ms
Electro-optic	8 × 8	8	35	1	10 ps
Thermo-optic	4 × 4	8	40	Negligible	3 ms
SOA	4 × 4	0	40	Negligible	1 ns

11.9 Wavelength Converters

11.9.1 Basics of wavelength converters

Wavelength converters convert data from one (incoming) wavelength to another (outgoing) wavelength. In WDM networks, wavelength converters have the following applications.

1. *Transponders*
   Signals from a client may not come within the WDM plan wavelength. For example, a client's signal may be a 1310 nm signal whereas WDM network may be operational in C or L band. In such situations, wavelength converters are deployed as transponders which accept the client signal and offer it on a wavelength compatible with the WDM plan.

2. *Optimum utilisation of available wavelengths*
   A fixed input and fixed output converter converts the incoming signals to fixed-output signals. A variable input and fixed output converter converts a variety of incoming wavelengths to fixed output wavelengths. A variable input and variable output device in WDM network can convert any incoming wavelength to any outgoing wavelength. Deployment of suitable converters provides the required flexibility in an optical path, including protection optical paths in a WDM networks using the total available wavelengths.

3. *Network boundary devices*
   When two WDM networks of different bands or WDM plans are to work together, wavelength converters are used as boundary devices to map the wavelengths within compatible bands from one network to the other.

11.9.2 Methods of wavelength conversion

There are four methods of achieving wavelength conversion.

1. ***Optoelectronic conversion***
The input signal is converted to an electronic form, regenerated and then re-transmitted on a different wavelength. This is usually in done using a variable input and fixed output converter.
If the receiver converts the incoming photons into an electronic signal which is amplified to drive a laser operating at a desired wavelength, it is called 1R regeneration. This process is suitable for analog as well as digital data. In this process, noise gets added.
In 2R regeneration, the signal is re-shaped but not re-timed. It is suitable for digital data only. This process introduces phase jitter in the output signal.
In 3R regeneration, the effects of nonlinearities, dispersion and noise are totally reset by compensation. Also, a circuitry can be included for performance monitoring. But, optical path management overheads are to be added and processed. Refer to Fig. 11.16 showing 3R regeneration using opto-electronic conversion.

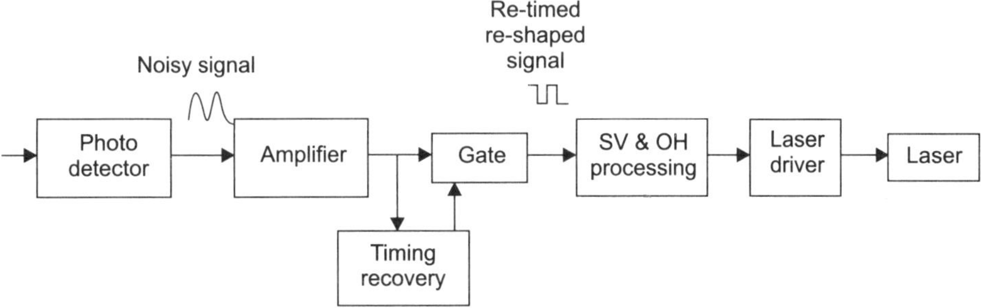

Fig. 11.16 *3R regeneration using optoelectronic wavelength conversion*

2. ***Optical gating conversion***
In this method, an optical device whose characteristics change with the intensity of input signal is used. This change is transferred to another unmodulated 'probe signal' at a different wavelength going through the device. Thus, the information envelope of the input signal is transferred to the output signal. This technique is known as cross-gain modulation (CGM).
As the gain of SOA is dependent on input power, they make good CGM devices. However, polarisation insensitive SOAs are required to be used as wavelength converters. Such an SOA can operate as wavelength converter in a wide band (100 nm) and at STM-64 rate. The only problem is the associated high cross-talk, to overcome which the input signal power requirement is high (about 0 dBm). Refer to Fig. 11.17 showing optical gating conversion using CGM device.

Fig. 11.17 *Optical gating conversion using CGM device*

3. ***Interferometric conversion***
   While optical gating technique is a CGM technique, interferometric conversion uses CPM (cross phase modulation) technique. The device used is a Mach–Zender interferometer (MZI). Refer to Fig. 11.18(a) showing an MZI. The two arms have a path difference, which if equal to an integral multiple of $\lambda/2$, can cause destructive interference and stop the signal of wavelength λ.

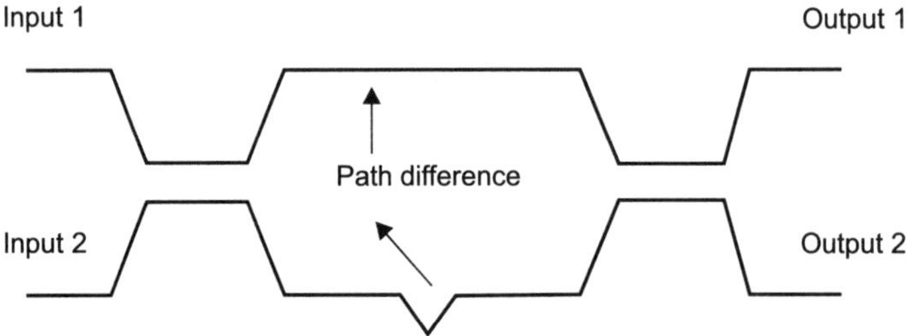

Fig. 11.18(a) *Mach–Zender interferometer*

Refer to Fig. 11.18(b) showing usage of MZI for wavelength conversion. In this method, SOAs are embedded inside the Mach–Zender Interferometer (MZI). Input 1 is signal at λ_s with data envelope and input 2 is signal at λ_p. As the path difference is integral multiple of $\lambda_s/2$, it is suppressed and the intensity modulating envelope, i.e., the data envelope is transferred to λ_p.

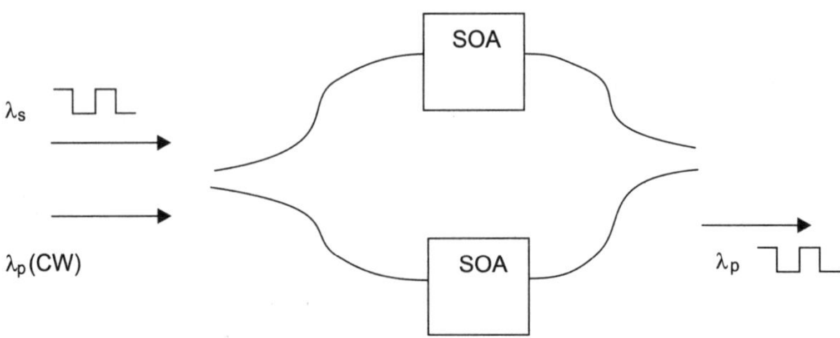

Fig. 11.18(b) *Wavelength converter using MZI*

4. ***Wave-mixing***
   If a signal is at frequency f_s and probe at frequency f_p, then four-wave mixing produces signals at frequency $2f_p - f_s$ and $2f_s - f_p$ at the output (in addition to f_s, f_p) provided these frequencies are within the bandwidth of the amplifier. In this conversion, both amplitude and phase are preserved. SOAs are used to get the signal of the desired level.

11.10 Add–Drop Approaches in WDM

Multiplexing, as we have been discussing, is the means of sharing bandwidth or medium to increase traffic flow. An increase in the volume of traffic flow by going higher-up in multiplexing hierarchy is not the end of the design objective—the network should have the ability to deliver a signal (or channel or stream) from an individual sender to an individual recipient. This ability implies that the network elements should be capable of adding and dropping any signal. In case of an E1 network, add–drop MUXes drop and add channels at the time-slot level. In case of PDH networks (E3 or E4), There is no add–drop facility as de-multiplexing down the hierarchy is needed to pull out an E1. In case of SDH networks, we have add–drop MUXes which operate at container level on the STM frame to facilitate adding/dropping of signals. This is illustrated in Fig. 11.19.

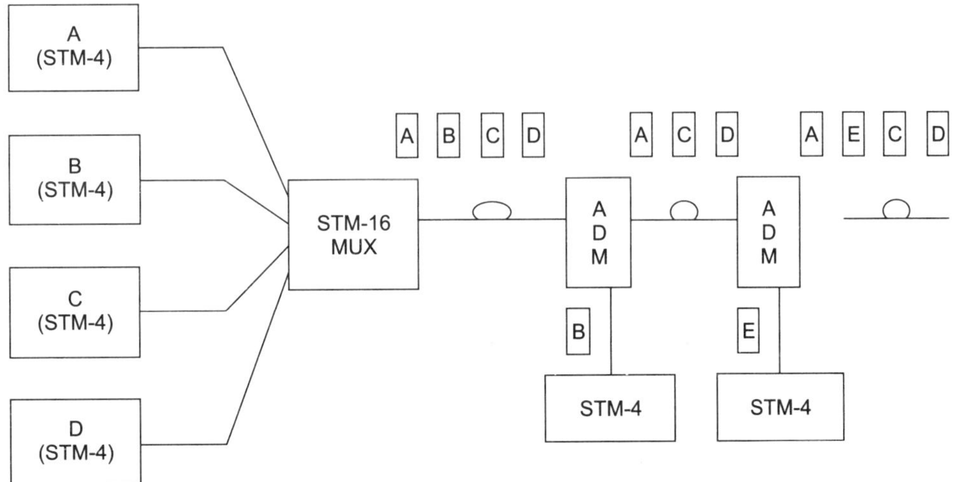

Fig. 11.19 *Add–drop approach in SDH*

Essentially, an STM add/drop NE includes receivers and transmitters because they have to convert an optical signal into an electrical signal, multiplex and de-multiplex this electrical signal, and then convert the processed signal back into an optical form.

In WDM, we have STM signals on each wavelength. There are two basic approaches to the add/drop functionality in WDM: multi-access approach and wavelength routing approach.

Multi-access approach

In multi-access approach, each node transmits signals on its wavelength and all these signals are presented to a WDM MUX. The MUX broadcasts all the signals to all the nodes. Each node selects the desired wavelength by filtering the entire signal. This approach is shown in Fig. 11.20. The WDM MUX is a passive device that does wavelength-division multiplexing and transmits multiplexed signal along the fibre. The constraint in this approach is the signal power, which is limited because the WDM MUX splits the signal power equally among all the nodes; hence traffic from only few of the nodes can be multiplexed. On the other hand, this system can be independently accessed by many nodes.

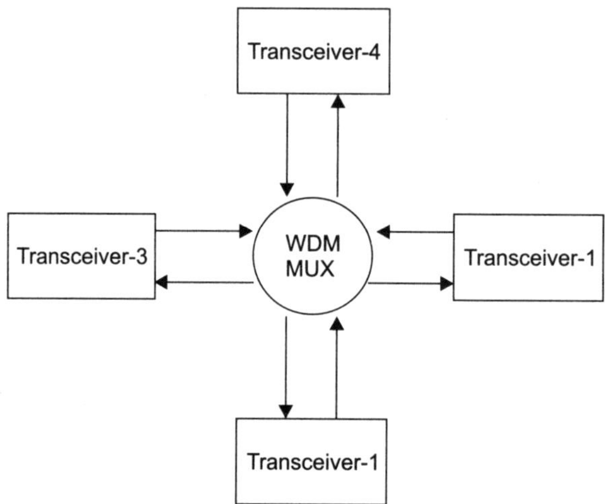

Fig. 11.20 *Multi-access approach for add/drop in WDM*

Wavelength-routing approach

In wavelength routing approach, a signal from an individual node on a specific wavelength is transmitted through a set of active switches. Any wavelength can be routed to any node, which is a great advantage. The network is however sophisticated in technology and architecture.

Wavelength-routing approach is based on the 'light path' concept. A light path is an optical channel between two NEs that is routed through a number of intermediate nodes. A light path or an optical channel segment between two nodes is identified by a wavelength. The wavelength of an optical channel is re-useable. Intermediate nodes may change or switch wavelengths. The light path is temporary, i.e., it is set up and taken off as per traffic requirement under the control of NMS. Refer to Fig. 11.21. Wavelength λ_1 (shown as dotted line) is used between NE6–NE1 via intermediate nodes A and F. It is also used between NE3–NE4 via C and D. Wavelength λ_2 (shown as normal line) is used between NE2–NE6 via intermediate nodes B, A, G and F. It is also used between NE3–NE5 via C, G, D and E.

Since there is no power splitting between the links of the network, many connections are possible.

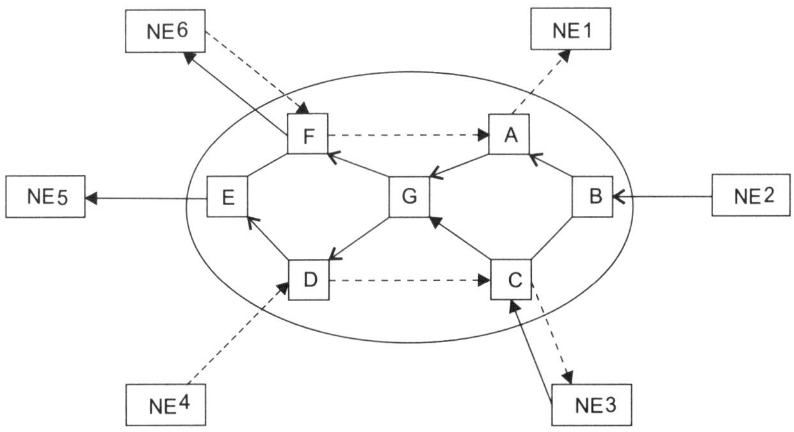

Fig. 11.21 *Wavelength-routing approach for add/drop in WDM*

There is, however, one fundamental restriction in the use of a wavelength-routing network. If two or more light paths use the same fibre, they have to use different wavelengths to avoid interfering with one another. In case of interruption on any link, alternate light path has to be set up without interfering with the wavelengths in transmission.

From the above discussion it is clear that wavelength routing networks hold great promise for transmission solutions provided the light-paths are properly managed. This concept of 'managed light path services' led to the visualisation of 'optical layer' in the NEs and new framing procedures to suit the same. Thus, an altogether new hierarchy for transmission of frames evolved, called the optical transport hierarchy (OTH).

The essentials of OTH and features of managed light path services are discussed in detail in the next chapter.

Summary

In this chapter the principles of WDM are presented. WDM varieties, viz. wide-band WDM, DWDM and CWDM are explained. ITU-T grids for WDM are presented. The types of WDM transmitters, optical amplifiers, receivers, optical switches and wavelength converters are presented. Add–drop approaches in WDM network are presented.

References

[1] Spanke, R. A. and. Benes, V. E., An n-stage planar optical permutation network, *Applied Optics*, p26, April 1987.

[2] Rajiv Ramaswami and Kumar N. Sivarajan, *Optical Networks* (Chapter 3,Optical switch technologies), Second Edition, Morgan Kaufman Publishers, 2006.

12

Essentials of OTH

12.1 Introduction

12.1.1 Why OTH and OTN?

The maximum bit rates possible with TDM have been achieved in SDH along with OAM&P (operations, administration, maintenance and provisioning) capabilities. With the advent of WDM and DWDM technologies, simultaneous transmission of a number of STMs on OFC has become a reality. WDM and TDM have become complimentary to each other and have taken transmission capacities to the order of terabits per second (Tbps) over OFC.

In this new landscape a new layer termed the 'optical layer' provides light paths to other layers such as SONET/SDH/IP/ATM which have network functional capabilities like multiplexing, add–drop and cross-connects. To effectively handle the varied traffics and network functionalities, a new transmission hierarchy called the optical transmission network hierarchy (OTN) evolved. Since the optical layer provides light path to higher layers like SONET/SDH/IP/ATM, it can be viewed as a server layer to these client layers. If the requirements of the managed light path services can be identified, the features of OTN hierarchy can be correlated to them and OTN easily understood.

12.1.2 Requirements of managed light path services

The following are the key requirements of managed light path services.

1. Setting up and withdrawing of light paths as dictated by client traffic as also for maintenance purposes.
2. Bandwidth of light path to be made available as per client traffic requirement.
3. A BER better than 1×10^{-12}, and a performance management system to be in place in the network to ensure this.

4. Client signals to be converted to optical signals and vice versa through suitable interfaces. Signal types, protocols, bit rates are to be incorporated in the optical layer.
5. Unidirectional/bi-directional light paths and associated protection mechanisms to be supported.
6. Drop and continue functions to be provided for network survivability when multiple rings are connected.
7. Multicasting to be supported for distribution of video conferencing information.
8. 3R regeneration to keep jitter within limits.
9. Propagation delays on light paths not to reduce throughput of client protocols below specified limits.
10. Fault management functions to be supported extensively.

The optical layer handles the complex tasks of multiplexing, switching and routing, management. In order to delineate management functions and to provide suitable boundaries and interfaces for equipment, sub-layers of the optical layer need to be considered.

Q 12.1 *What is meant by 1R, 2R and 3R regenerations?*

1R implies that the signal is re-amplified; 2R that the signal is re-amplified and re-shaped and 3R implies that the signal is re-amplified, re-shaped and re-timed.

12.1.3 How ITU-T recommendations for OTN incorporate the requirements

As already discussed, optical transport network (OTN) was created with the intention of combining the benefits of SONET/SDH and the OAM&P features of SONET/SDH with the bandwidth expansion capabilities offered by dense wavelength-division multiplexing (DWDM) technology. In addition, the OAM&P functions are to be enhanced to incorporate the requirements of managed light-path services mentioned in the previous section.

Hence, the purpose of the ITU-T recommendations for OTN is manifold:

1. Lay down the architecture of optical transport networks[1].
2. Specify the optical transport hierarchy(OTH) for the OTN.
3. Specify the functionality of overheads in support of multi-wavelength optical networks.
4. Define frame structures, bit rates and formats for mapping client signals.

5. Define linear protection mechanisms for OTN[2].
6. Define physical layer interfaces to optical transport network[3]

While item (1) is taken care in ITU-T Rec. G.872, items (2),(3) and (4) are taken care of in ITU-T Rec. G.709. Item (5) is covered by ITU-T Rec. G.873(1) and item (6) by G.959.1.

The above recommendations put together run into nearly a thousand pages. Obviously, the terms, definitions and descriptions thereof cannot be presented in this book. Our goal is to provide the reader a clear understanding of OTN, OTH and enhanced OAM&P features with managed light path services. Hence, we start with an overview of the transmission process using OTH across OTN. Then, elements of OTN, their interfaces and the layers or processing stages in digital and optical domain are discussed. Also, various transport entities of OTH, overheads in them for OAM&P functions and managed light path services are examined in detail. In the next chapter, mapping of various client signals onto OTH and concatenation are discussed.

12.2 Overview of Transmission Process using OTH across OTN

The transmission process using OTH across OTN is understood by knowing

- how the client traffic is encapsulated into optical channel
- the digital hierarchy or wrapping followed in digital domain
- the layered structure in optical domain for transportation of optical channels.

Let us proceed to get familiar with them.

12.2.1 Traffic encapsulation

Figure 12.1 shows the step-wise encapsulation of client traffic into optical payload unit (OPU), optical data unit (ODU) and finally into optical transport unit (OTU). The user payload or client traffic is encapsulated into optical payload unit (OPU) of order *k*. This unit is an octet based frame consisting of 4 rows and 3810 columns. The first two columns carry overhead information (OPU overhead). The remaining 3808 columns are for payload. The overhead content depends on the type of traffic described in detail in the following sections.

In the next stage of encapsulation, the optical data unit (ODU*k*) is formed by adding 14 columns of additional overhead to OPU for the rows numbered 2, 3 and 4. This overhead is known as ODU overhead.

The remaining one row, viz. the first row, is filled with FAS (7 bytes from column 1 to 7) and the optical transport unit overhead or the OTU overhead (7 bytes from column 8 to 14), making complete the 4 row × 3824 column frame. The addition of FEC (forward error correcting code) bytes (4 rows × 256 columns) at the right end makes the frame size 4 rows × 4080 columns. This then makes up the optical transport unit (OTU).

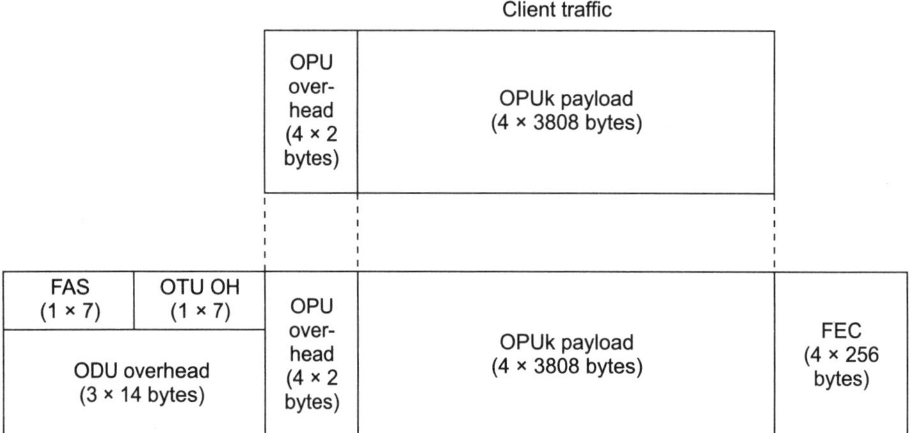

Fig. 12.1 *Step-wise encapsulation of client traffic into OTU*

Row-wise and column-wise structure of (OTU) is shown in Fig. 12.2. The order of transmission is first from left to right and then from top to bottom, i.e., row-wise left to right. Within each byte the MSB, which is the left-most is transmitted first.

Col./Row No.	1 7	8 14	15 16	17 3824	3825 4080
1	FAS	OUT OH	OPU OH	OPU payload	FEC
2	ODU overhead				
3					
4					

Fig. 12.2 *Frame structure of OTU*

The above description of OTH is an oversimplified version. The questions that could possibly arise in the minds of the reader are the following.

Q 12.2 Why are the overheads added step-wise, i.e., first the 2 columns and then 14 columns and finally the FEC?

To create an OTU frame, the client signal rate is first adapted at the OPU layer. The adaptation consists of adjusting the client signal rate to the OPU rate. Its overhead contains information for supporting the adaptation of the client signal. Once adapted, the OPU is mapped onto the ODU. In this process the overheads necessary for ensuring end-to-end supervision and tandem connection monitoring up to six levels as shown in layer-model in Fig. 12.3 are added.

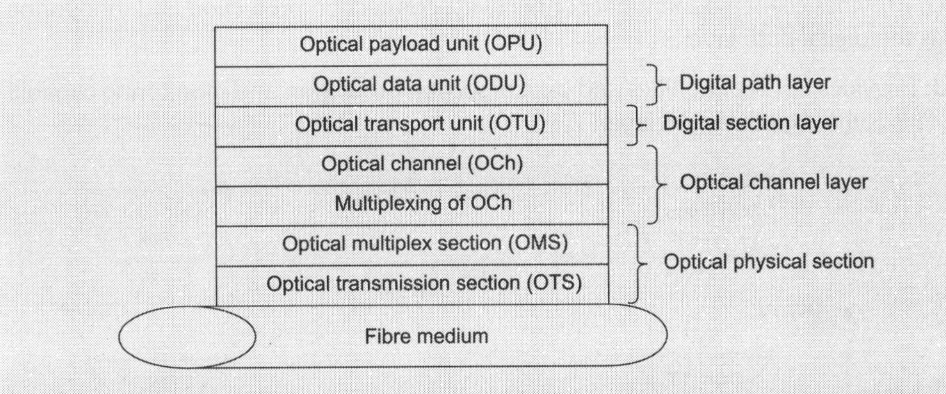

Fig. 12.3 *Layer model of OTH*

Finally, the ODU is mapped into an OTU, which provides the framing as well as section monitoring and FEC.

Q 12.3 *Why do we have such long FEC and how is it computed?*

As already discussed, an OTU frame is divided into four rows. Each row is further considered as 16 sub-rows comprising 255 bytes each. Reed–Solomon type RS(255/239) FEC is used in OTU. The FEC can correct up to eight (bytes) errors per sub-row (codeword) or detect up to 16 byte errors. Hence, we have 16 bytes of FEC for each sub-row or 16 × 16 = 256 bytes FEC for each row.

Such elaborate FEC provides a method of significantly reducing the error bursts in transmission due to noise as well as other optical phenomena that occur at high transmission speeds. Thus, FEC allows an increase in the optical link budget, i.e., it enables providers to support longer spans in between optical repeaters.

12.2.2 Digital wrapper in OTH

As can be seen in Fig. 12.3, OTH has basically two distinct domains: digital and optical. The termination of the OTS, OMS and OCh layers is performed at the optical level of the OTN. It is at the termination of the OTU layer that further functionality is added. This layer is the digital layer and offers specific overhead to manage the OTN's digital functions.

From Fig. 12.3, it can be seen that the layers upwards from OCh, i.e., OTU, ODU, OPU, provide the digital transmission hierarchy for carrying over lower optical layers. These layers, OTU, ODU and OPU together are known as 'digital wrapper'. Functions of the layers in digital wrapper are:

OPU: Provides support to map client signals, viz. STM-N, IP packets, ATM cells and Ethernet frames into structured frame. This is known as the digital-wrap outer layer.

ODU: Provides client independent connectivity, connection protection and monitoring. This is the *digital path* layer.

OTU: Provides FEC capabilities and optical section protection and monitoring capabilities. This is the *digital section* layer.

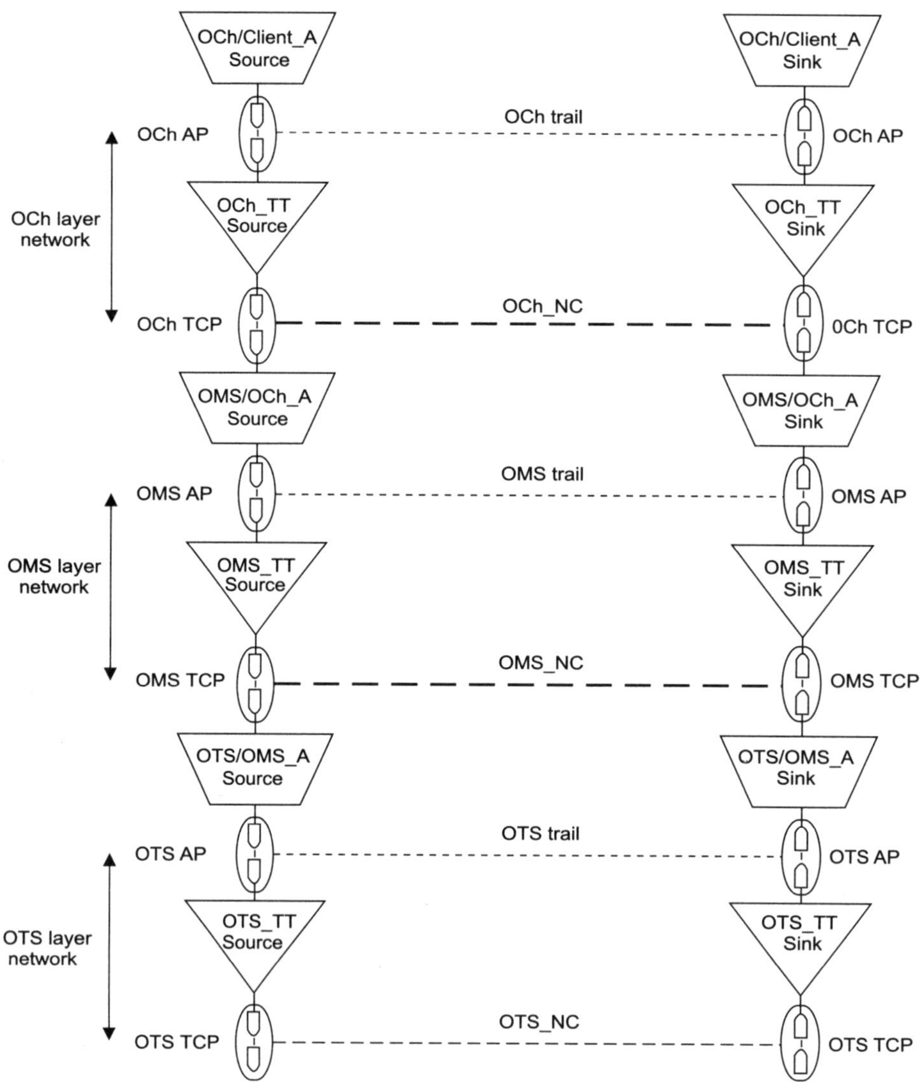

Fig. 12.4 *Layers and networks within OTN*

OTU, ODU (including the ODU tandem connection monitoring) and OPU layers can all be analysed and monitored. As per ITU G.709, the three line rates are offered:

OTU1 (255/238 × 2.488 320 Gbps ≈ 2.666057143 Gbps), also referred to as 2.7 Gbps.
OTU2 (255/237 × 9.953280 Gbps ≈ 10.709225316 Gbps), also referred to as 10.7 Gbps.
OTU3 (255/236 × 39.813120 Gbps ≈ 43.018413559 Gbps), also referred to as 43 Gbps.

Each line rate is adapted to service different client signals:

OC-48/STM-16 is transported via OTU1
OC-192/STM-64 is transported via OTU2
OC-768/STM-256 is transported via OTU3

Test equipment support the above client signals as well as the following test signals.

Null client (all 0s) is transported via OTUk ($k = 1, 2, 3$)
PRBS 2^{31}–1 is transported via OTUk ($k = 1, 2, 3$)

OTUks ($k = 1, 2, 3$) are transported using the OCh (optical chaneel), each channel being assigned a specific wavelength of the ITU grid. Several channels can be mapped into the OMS (optical multiplex section) and then transported via the OTS (optical transport section) layer. The OCh, OMS and OTS layers each have their own overhead for management purposes at the optical level. The overhead of these optical layers is transported outside of the ITU grid in an out-of-band channel called the optical supervisory channel (OSC).

When the OTU frame structure is complete (OPU, ODU and OTU), ITU G.709 provides OAM&P functions that are supported by the overhead.

12.2.3 Layered structure in optical domain for transportation of optical channels

The digital wrap received from digital domain is to be carried across OTN using WDM/DWDM techniques. To do so, OTN must have layered structure that functionally supports transport, multiplexing, routing, supervision and survivability of client signals that are processed predominantly in the photonic domain.

The current optical technology has limitations on the aspects of supervision and performance assessment capabilities in the network. Hence, digital processing is needed in order to overcome these limitations. Accordingly, ITU-T Rec.G.805 specified a layered structure for optical transport network (OTN) comprising three layers—optical channel (OCh), optical multiplex section (OMS) and optical transmission section (OTS) as illustrated in Fig. 12.4. Layers are so defined that OTN can be viewed as peer-to-peer independent network at each layer. The layer below is the server for the next higher layer peer-to-peer network. For the overall optical transmission section, physical media or the optical fibre is the server. Functions of each layer in the optical domain are elaborated below.

Optical channel layer network

This layer network provides end-to-end networking of optical channels with transparency of client traffic, which may be SDH, STM-N, PDH, cell-based ATM, etc. This layer is capable of

- rearranging connection of optical channel for flexible network routing,
- optical channel overhead processing for ensuring integrity of the optical channel adapted information,
- optical channel operations, administrations, and maintenance functions for enabling network level operation and management functions such as connection provisioning, QoS and network survivability.

Optical multiplex section layer network

This layer network is concerned with networking of a multi-wavelength optical signal. This layer is capable of

- optical multiplex section overheads processing for ensuring integrity of information adapted from the multi-wavelength optical multiplex section
- OAM functions of the optical multiplex section for enabling section level operations and management functions, section survivability.

Optical transmission section layer network

This layer network provides functionality for transmission of optical signals on optical media of various types, e.g., G.652, G.653 and G.655 fibre. This layer is capable of

- optical transmission section overhead processing for ensuring integrity of the information adapted from optical transmission section,
- OAM functions of the optical transmission section for enabling section level operations and management functions and survivability of transmission section.

12.3 Network Entities and Transport Structures in OTH

Having understood the basics of OTH, let us take a quick look at the definitions of network entities and transport structures and their meanings/functionalities in OTH. Definitions are as per ITU-T Rec. G.709, Rec.G.872 and shown in italicised font. Explanatory notes are in normal font.

12.3.1 Network entities in OTH

1. ***Optical transport network (OTN)*** *is a transport network bounded by optical channel access points.*

Note: Interconnection of nodes on an OFC does not make an OTN. The keyword is 'access point', i.e., every node on OTN must provide optical channel (OCh) access.

2. ***Optical network element (ONE)*** *is a network element belonging to OTN.*

Note: ONE can be optical cross connect (OXC) or optical add–drop MUX (OADM).

3. *The interface at an optical transport network node which is used to interconnect with another optical transport network node is known as an optical transport network node interface (ONNI).*

Note: Each network element or node on an OTN has an interface module that connects it to the fibre. This module supports a WDM/DWDM plan (specified in terms of wavelength band and separation between wavelengths) and number of carriers. ITU-T grid for WDM has been explained in section 11.4. In Table 12.1, a 32-wavelength plan supported by a manufacturer is shown as an example. All the 32 wavelengths need not be used—24, 16 or 8 of them can be used. Wavelengths out of the available 32 on the interface which are included for the plans of 8/16/24/32 are also shown in Table 12.1. It may be noted that frequency separation between each wavelength is 100 GHz. One band with the associated set of wavelengths is for forward transmission and the other band with associated set of wavelengths is for backward transmission on a span. Each band has a guard wavelength.

Spectral grids for DWDM and CWDM are specified in ITU-T G.694.1 grid plan and ITU-T G.694.2 grid plans. These are explained in chapter 13. An interface can be within one administrative domain or within the boundary of two administrative domains.

Table 12.1 *DWDM with 32-wavelength plan*

Wavelength (nm)	**Band**	**Whether included in the following plan**			
		32-wavelength	**24-wavelength**	**16-wavelength**	**8-wavelength**
1505.60	Red	**Spare**	**Spare**	**Spare**	**Spare**
1559.79	Red	Yes	-	-	-
1558.98	Red	Yes	Yes	Yes	-
1558.17	Red	Yes	-	-	-
1557.36	Red	Yes	Yes	Yes	Yes
1556.55	Red	Yes	Yes	-	-
1555.75	Red	Yes	Yes	Yes	Yes
1554.94	Red	Yes	-	-	-
1554.13	Red	Yes	Yes	Yes	-
1553.33	Red	Yes	-	-	-

(Continues...)

Table 12.1 *Continued*

Wavelength (nm)	Band	Whether included in the following plan			
		32-wavelength	24-wavelength	16-wavelength	8-wavelength
1552.52	Red	Yes	Yes	Yes	Yes
1551.72	Red	Yes	Yes	-	-
1550.92	Red	Yes	Yes	Yes	Yes
1550.12	Red	Yes	Yes	-	-
1549.32	Red	Yes	Yes	Yes	-
1548.51	Red	Yes	Yes	-	-
1547.72	Red	Yes	Yes	Yes	-
1541.30	Blue	**Spare**	**Spare**	**Spare**	**Spare**
1540.56	Blue	Yes	-	-	-
1539.77	Blue	Yes	Yes	Yes	-
1538.98	Blue	Yes	-	-	-
1538.19	Blue	Yes	Yes	Yes	-
1537.40	Blue	Yes	-	-	-
1536.61	Blue	Yes	Yes	Yes	-
1535.82	Blue	Yes	-	-	-
1535.04	Blue	Yes	Yes	Yes	-
1534.25	Blue	Yes	Yes	-	Yes
1533.47	Blue	Yes	Yes	Yes	-
1532.68	Blue	Yes	Yes	-	Yes
1531.90	Blue	Yes	Yes	Yes	-
1531.12	Blue	Yes	Yes	-	-
1530.33	Blue	Yes	Yes	Yes	-
1529.55	Blue	Yes	Yes	-	Yes
1528.77	Blue	Yes	Yes	Yes	Yes

4. ***Inter-domain interface (IrDI)*** *is a physical interface that represents the boundary between two administrative domains.*

5. ***Intra-domain interface (IaDI)*** *is a physical interface within an administrative domain.*

Note: An optical network element (ONE) which has only IaDI is an internal network element, or an internal ONE. An optical network element which has a IaDI and an IrDI can be boundary network element, or a boundary ONE. The terms ONE, internal ONE, boundary ONE, IaDI, IrDI can be appreciated while referring to Fig. 12.5.

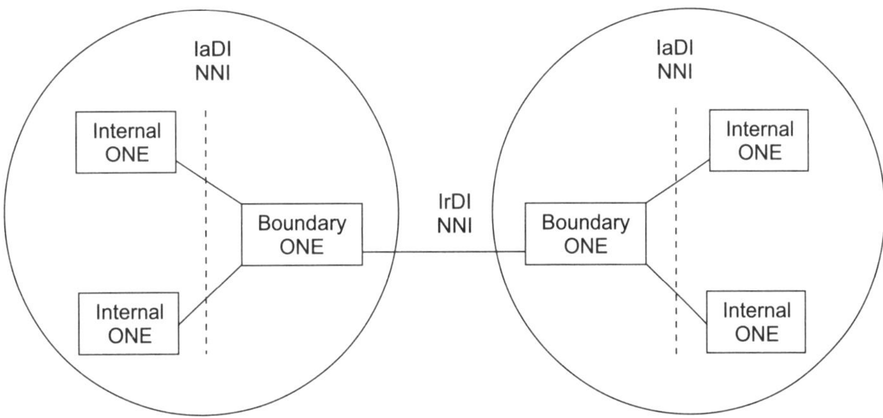

Fig. 12.5 *Illustration of internal and boundary ONEs, IaDI-NNI and IrDI-NNI*

12.3.2 Transmission structures in OTH

As explained in section 11.2, the digital hierarchical portion of OTH has a 3-stage encapsulation, namely, OPU, ODU, OTU. The OPUk is mapped into ODUk, and the ODUk mapped into OTUk. The OTUk is carried on one optical channel (OCh). So, the bit rate of an optical channel depends on OTUk being carried in it. Since more than one wavelength is used in WDM, each wavelength is one optical channel (OCh). However, no interface can support arbitrary number of wavelengths or any OTUk on each bandwidth because the source bandwidth is finite. The finite bandwidth available can be put to use either with large number of wavelengths (OChs) and lower bit rate on each such wavelength, or a smaller number of wavelengths (OChs) and higher bit rate on each such wavelength. For example, let the bandwidth available be 2400 Gbps. We can utilise it in the following forms.

(a) 24 OCh and carry one OTU3 (40 Gbps) on each

(b) 48 OCh and carry one OTU2 (10 Gbps) on each

(c) 96 OCh and carry one OTU1 (2.5 Gbps) on each

In general, an interface can support *n* OCh (wavelengths) with the lowest rate OTUk, or a number of OCh less than *n* and a higher rate OTUk on each. Thus, we have either full functionality or reduced functionality in OTH at OTM, OTS, OMS, OPS, OCG, OMU,

OCC and OCh. Certain indices are used to denote the level of functionality at each optical layer. These are *n*, *r*, *m* and *k*. Transport modules are thus:

optical transport module (OTM-n[r]m)
optical carrier group of order n (OCG-n[r]m)
optical channel carrier (OCCn[r]m)

In this section the usage of indices *n*, *r*, *m* and *k* is explained first and the transport modules later.

Indices *n*, *r*, *m* and *k*

n: The index *n* is used to represent the order of the OTM (also that of OTS, OMS, OPS, OCG, OMU as we shall see later). It represents the maximum number of wavelengths that can be supported at the lowest bit rate supported on the wavelength. It is possible that a reduced number of higher bit-rate wavelengths are supported. $n = 0$ represents the case of a single channel without a specific colour (wavelength) assigned to the channel.

r: The index *r*, if present, is used to indicate the reduced functionality of OTM (also of OCG, OCC and OCh). Note that for $n = 0$ the index *r* is not required as it always implies reduced functionality.

m: The index *m* is used to represent the bit rate or the set of bit rates supported on the interface. This is one or more of *k* digits, where each *k* represents a particular bit rate. The valid values for *m* are 1, 2, 3, 12, 23 and 123. For example, $m = 23$ implies the interface supports OTM*k* for $k = 2$ and $k = 3$, i.e., the interface supports OTM2 and OTM3.

k: The index *k* is used to represent the supported bit rate and the different versions of OPUk, ODUk and OTUk. The value $k = 1$ represents an approximate bit rate of 2.5 Gbps, $k = 2$ represents an approximate bit rate of 10 Gbps and $k = 3$ represents an approximate bit rate of 40 Gbps.

6. ***Optical transport hierarchy (OTH)*** *is a hierarchical set of digital transport structures standardised for the transport of suitably adapted payloads over optical transmission networks.*

Note: A client signal is mapped into OPU1, an OPU1 to ODU1 and an ODU1 to OTU1. Since OPU1 rate is 2.5 Gbps, it can take STM-16. The next level, STM-64, can be mapped into OPU2; then OPU2 into ODU2 and ODU2 into OTU2. Alternately, four ODU1s can be mapped into ODTUG2 which can be mapped into ODU2 followed by ODU2 into OTU2. In a similar fashion, STM-256 signal can be mapped into ODU3. Alternately, four ODU2s or 16 ODU1s can be mapped into ODTUG3, which is mapped into ODU3. ODU3 is mapped into OPU3 and, OPU3 into OTU3. The mapping process is illustrated in Fig. 12.6.

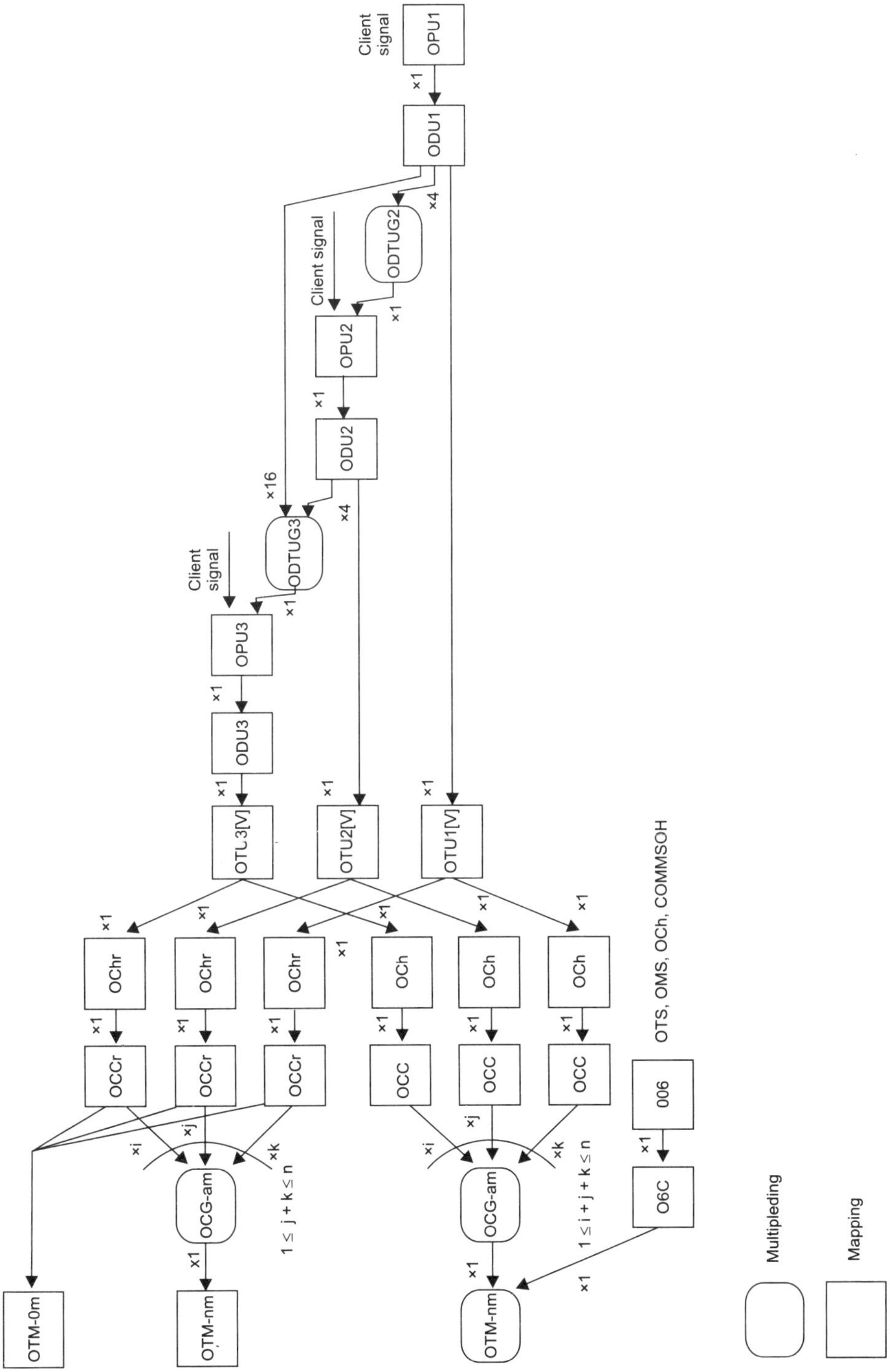

***Fig. 12.6** Illustration of OTH multiplexing*

Bit rates for OPUk, ODUk and OTUk with tolerances are given in Tables 12.2, 12.3 and 12.4. OTUk/ODUk/OPUk frame periods are given in Table 12.5.

Table 12.2 *OPUk bit rates and tolerances*

OPU type	OPU payload bit rate	Bit rate tolerance
OPU1	2488.320 Mbps	
OPU2	(238/237) × 9953.280 Mbps	±20 ppm
OPU3	(238/236) × 39813.120 Mbps	
OPU1-Xv	X × 2488.320 Mbps	
OPU2-Xv	X × (238/237) X 9953.280 Mbps	±20 ppm
OPU3-Xv	X × (238/236) X 39813.120 Mbps	

Table 12.3 *ODUk bit rates and tolerances*

ODU type	ODU payload bit rate	Bit rate tolerance
OPU1	(239/238) × 2488.320 Mbps	
OPU2	(239/237) × 9953.280 Mbps	±20 ppm
OPU3	(239/236) × 39813.120 Mbps	

Table 12.4 *OTUk bit rates and tolerances*

OTU type	OTU payload bit rate	Bit rate tolerance
OTU1	(255/238) × 2488.320 Mbps	
OTU2	(255/237) × 9953.280 Mbps	±20 ppm
OTU3	(255/236) × 39813.120 Mbps	

Table 12.5 *OPUk/ODUk/OTUk frame periods*

OTU/ODU/OPU type	Frame period
OTU1/ODU1/OPU1/OPU1-Xv	48.971 μs
OTU2/ODU2/OPU2/OPU2-Xv	12.191 μs
OTU3/ODU3/OPU3/OPU3-Xv	3.035 μs

7. ***Optical transport module (OTM-n[r]m):*** *The OTM is the information structure that is transported across an ONNI. The index n and m define the number of supported wavelengths and bit rates for OTM with full functionality/reduced functionality.*

Note: **OTM with full functionality (OTM-n.m)**

The OTM-*n.m* consists of up to *n* multiplexed optical channels and an OTM overhead signal to support the non-associated overhead. It is the information structure used to support optical transmission section (OTS) layer connections in the OTN. The characteristic information of the optical transmission section layer (OTS_CI) consists of information payload (OTS_CI_PLD) and the optical transmission section overhead information fields (OTS_CI_OH). The optical transmission section overhead (OTS_OH) information fields are contained within the OTM overhead signal (OOS) information structure. The order of an OTM-n is defined by the order of the OMU-n that it supports.

OTM with reduced functionality (OTM-0.m, OTM-nr.m)

The OTM-0 consists of a single optical channel without a specific colour assigned. The OTM-*nr.m* consists of up to *n* multiplexed optical channels. Non-associated overhead is not supported.

The OTM-nr.m/OTM-0 is the information structure used to support the optical physical section (OPS) layer connections in the OTN. The characteristic information of the optical physical section layer (OPS_CI) consists of information payload (OPS_CI_PLD). Non-associated overhead is not supported. The order of an OTM-nr is defined by the order of the OCG-nr that it supports.

Q 12.4 *Illustrate OTM-0.m with an example.*

OTM-0.m implies optical channel with an optical signal (wavelength not specified). OTM-0.1 carries OTU1, OTM-0.2 carries OTU2 and OTM-0.3 carries OTU3. The same is illustrated in Fig. 12.7(a).

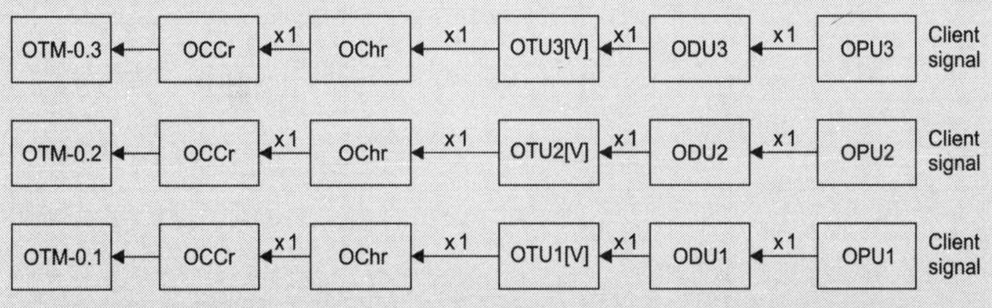

Fig. 12.7(a) *Illustration of multiplexing in OTM-0.1, OTM-0.2, OTM-0.3*

Q 12.5 *Illustrate OTM-n[r].m with an example.*

Let us take the example of OTM-16r.m. It supports 16 optical channels on a single span with 3R regeneration at each end. In other words, OTM-16r.m signal is a signal with 16 optical channel carriers (OCCr) numbered from OCCr#0 to OCCr#15.

The index *m* can be 1, 2, 3, 12, 23 or 123, i.e., *m* supports *k*=1, *k*=2, *k*=3; *k*=1 and 2, *k*=2 and 3, and *k*=1, 2 and 3. So, we have the following six OTM-16r interface signals, i.e., on each of the 16 optical channel carriers (OCCrs), any one of the following signals are supported:

- OTM-16r.1 which caries i ($i \leq 16$) OTU1 signals.
- OTM-16r.2 which carries j ($j \leq 16$) OTU2 signals.
- OTM-16r.3 can carries k ($k \leq 16$) OTU3 signals.
- OTM-16r.12 can carry i ($i \leq 16$) OTU1 signals and j ($j \leq 16$) OTU2 signals, with $i+j \leq 16$.
- OTM-16r.23 can carry j ($j \leq 16$) OTU2 signals and k ($k \leq 16$) OTU3 signals, with $j+k \leq 16$.
- OTM-16r.123 can carry i ($i \leq 16$) OTU1 signals, j ($j \leq 16$) OTU2 signals, k ($k \leq 16$) OTU3 signals with $i+j+k \leq 16$.

Multiplexing structure of OTM-16r.m is illustrated in Fig. 12.7(b).

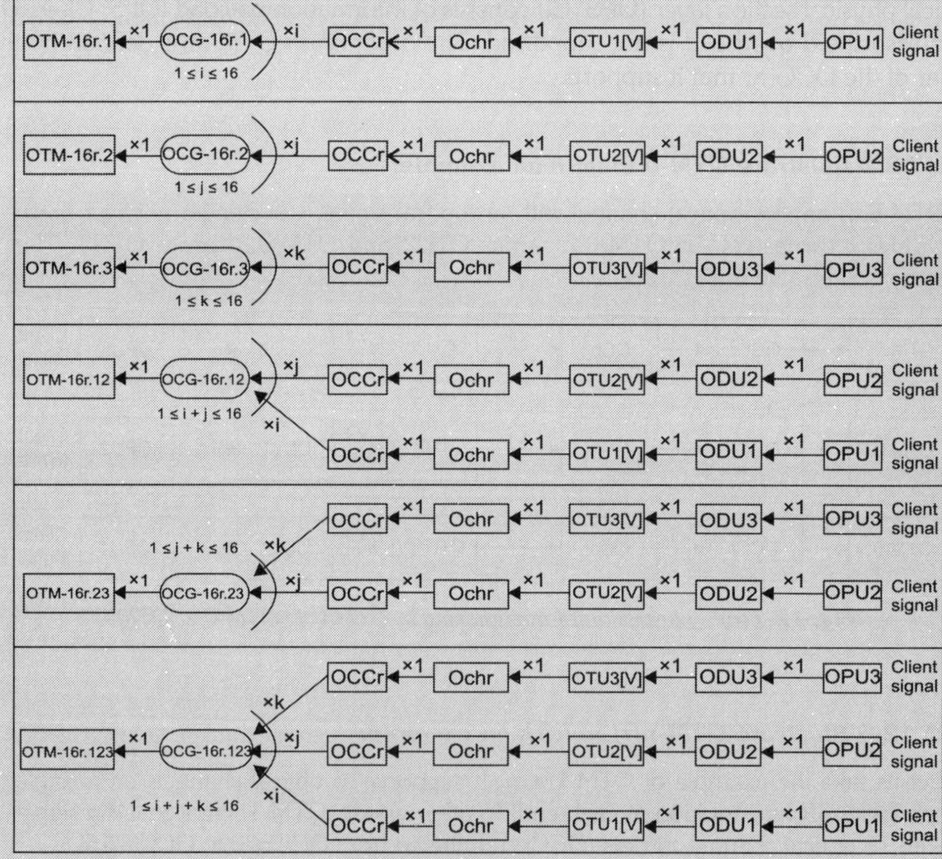

Fig. 12.7(b) *Multiplexing structure of OTM-16r.m for m=1, 2, 3, 12, 23, 123*

In the above diagram, rectangle indicates mapping and rectangle with rounded corners represents multiplexing.

8. ***Optical channel (OCh[r]):*** *The OCh is the information structure used to support the OCh trail. The OCh client signals defined in ITU-T Rec.G.709 are the OTUk signals. Other digital client signals (e.g., STM-N) are also supported by the OTM.*

Note: Two OCh structures are possible corresponding to OTM with full functionality and OTM with reduced functionality.

Optical channel with full functionality (OCh): The OCh is an information structure consisting of the information payload (OCh_PLD) with a certain bandwidth and non-associated overhead (OCh_OH) for management of the optical channel.

Optical channel with reduced functionality (OChr): The OChr is an information structure consisting of the information payload (OCh_PLD) with a certain bandwidth. Non-associated overhead is not supported.

9. ***Optical channel transport unit (OTUk[V]):*** *The OTUk is the information structure used for transport of an ODUk over one or more optical channel connections. It consists of the optical channel data unit and OTUk related overhead (FEC and overhead for management of an optical channel connection). It is characterised by its frame structure, bit rate and bandwidth. OTUk capacities for k = 1, 2, 3 are defined.*

Note: Since interfaces are intra-domain and inter-domain, we have two versions of OTUk:

(i) The completely standardised OTUk (OTUk) used on OTM IrDIs and possibly on OTM IaDIs.

(ii) Functionally standardised OTUk (OTUkV), the partly standardised OTUk used on OTM IaDIs.

10. ***Optical channel data unit (ODUk):*** *The ODUk is an information structure consisting of the information payload (OPUk) and ODUk related overhead. ODUk capacities for k = 1, 2, 3 are defined.*

11. ***ODUk path (ODUkP):*** *The optical channel data unit k path (ODUkP) is the information structure used to support the end-to-end ODUk trail.*

12. ***ODUk TCM (ODUkT):*** The optical channel data unit *k* TCM (ODUkT) is the information structure used to support the TCM trails. Up to 6 TCM sublayers are supported.

13. ***Optical channel payload unit (OPUk):*** *The OPUk is the information structure used to adapt client information for transport over an optical channel. It comprises client information together with any overhead needed to perform rate adaptation between the client signal rate and the OPUk payload rate and other OPUk overhead supporting the client signal transport. This overhead is adaptation specific. OPUk capacities for k = 1, 2, 3 are defined.*

14. ***Optical channel carrier (OCC[r]):*** *The optical channel carrier represents a tributary slot within the OTM-n.*

Note: Since OTM has 2 versions, i.e., full functionality and reduced functionality, we have two versions of OCC:

(i) **OCC with full functionality (OCC)** consisting of the OCC payload (OCCp) and OCC overhead (OCCo). The OCCp carries the OCh_CI_PLD and is assigned to a wavelength/frequency slot of the WDM group. The OCCo carries the OCh_CI_OH and is transported within the OOS information structure.

(ii) **OCC with reduced functionality (OCCr)** consisting of the OCC payload (OCCp). The OCCp carries the OCh_CI_PLD and is assigned to a wavelength/frequency slot of the WDM group. Non-associated overhead is not supported.

Q 12.6 OCC may carry OTU1 or OTU3. Why there is no such characterisation at OCC level?

Yes, it is desirable to differentiate one OCC tributary slot (e.g., one able to carry an OTU1) from another OCC tributary slot (e.g., one able to carry an OTU3). Characterisation of the OCC in this regard is under further study by ITU-T.

15. ***Optical carrier group of order n (OCG-n[r]):*** *n optical channel carriers occupying fixed, defined positions in an OTM payload are termed optical carrier group (OCG[r]).*

Note: Since we have two versions of OCh, two versions of OCG are available:

(i) OCG with full functionality (OCG-n): The OCG-n consists of up to n OCC payload (OCCp) and OCC overhead (OCCo).

(ii) OCG with reduced functionality (OCG-nr): The OCG-nr consists of up to n OCC payload (OCCp). Non-associated overhead is not supported.

16. ***Optical multiplex unit (OMU-n, n ≥ 1):*** *The OMU-n is the information structure used to support optical multiplex section (OMS) layer connections in the OTN.*

Note: The characteristic information of the optical multiplex section layer (OMS_CI) consists of information payload (OMS_CI_PLD) and optical multiplex section overhead information fields (OMS_CI_OH). The OMS_CI_PLD consists of the OCG-n payload. The OMS_CI_OH consists of the OCG-n overhead and OMS specific overhead and is transported within the OOS information structure. The order of the OMU is defined by the order of the OCG that it supports.

17. ***Optical physical section of order n (OPSn):*** *A layer network that provides functionality for transmission of a multi-wavelength optical signal on optical media of various types, e.g., G.652, G.653 and G.655 fibre.*

Note: A multi-wavelength signal includes the case of just one optical channel. It combines the transport functionality of the OMS and OTS layer networks without their supervisory information. OPSn capacities for $n = 0$ and $n = 16$ are defined.

18. ***OTM overhead signal (OOS):*** *The OOS is the information structure used for transport of OTM non-associated overhead over the optical supervisory channel.*

Note: The non-associated overhead consists of the optical transmission section overhead, optical multiplex section overhead and optical channel non-associated overhead. It is characterised by its frame structure, bit rate and bandwidth.

19. ***Optical supervisory channel (OSC):*** *It is the physical carrier outside of the amplifier band that provides transport of the OTM overhead signal.*

Note: The optical supervisory channel is an optical carrier that transfers overhead information between optical transmission section transport entities. The optical supervisory channel supports more than one type of overhead information and some of this may be used by one or more transport network layers.

Link capacity adjustment scheme (LCAS)

LCAS in the virtual concatenation source and sink adaptation functions provides a control mechanism to increase or decrease the capacity of a link in a hit-less manner to meet the bandwidth needs of the application. It also provides a means of removing member links that have experienced failure. The LCAS assumes that in cases of capacity initiation (increases or decreases) the construction or destruction of end-to-end path is the responsibility of the network and element management systems.

Overhead information

Six types of overhead information are defined.

1. *Trail termination overhead information*: It is the information generated by the trail termination source and extracted by the trail termination sink to monitor the trail. This overhead information is specific to a layer network and is independent of any client/server relationship between network layers.
2. *Client-specific overhead information*: It is associated with a particular client/server relationship and is therefore processed by a particular adaptation function.
3. *Auxiliary channel overhead information*: Auxiliary channel overhead information is the information that may be transferred by an optical network layer but which does not necessarily have to be associated with a particular connection. An example of such an auxiliary channel is a data communications channel for the purposes of transferring management data between management entities.
4. *Reserved overhead information*: Reserved bytes of overhead portion for future specification.

5. *Unassigned overhead information*: This overhead may be of types 1, 2, 3 or 4 as defined above.

6. *Network operator-specific overhead information*: Network operator-specific overhead information may be used by an operator to support unique optical networking needs and/or for service differentiation. The information content is not standardised.

12.4 Details of Framing and Overheads

In section 12.2, a brief description of transmission process and frame structures of OTH was presented. In this section we go into details of frame structure, overheads and purpose of each component in the overhead. Refer to Fig. 12.8 showing the frame structure of OCh. An OCh frame is the frame carried on each wavelength of WDM plan. The constituents are:

Frame alignment signal (FAS)
OTU overhead (OUT OH)
ODU overhead (ODU OH)
OPU Overhead (OPU OH)
OPUk payload
FEC

<table>
<tr><td>FAS
(1×7)</td><td>OTU OH
(1×7)</td><td rowspan="2">OPU
over-
head
(4×2
bytes)</td><td rowspan="2">OPUk pay-load
(4×3808 bytes)</td><td rowspan="2">FEC
(4×256
bytes)</td></tr>
<tr><td colspan="2">ODU overhead
(3×14 bytes)</td></tr>
</table>

Fig. 12.8 *Step-wise encapsulation of client traffic into OTU*

12.4.1 FAS

FAS, which is 7 bytes long, in fact has two portions: frame alignment sequence of 6 bytes and multi-frame alignment sequence of 1 byte. Refer to Fig. 12.9 showing details of FAS. The frame alignment sequence of 6 bytes, similar to SONET/SDH, is used to provide framing for the OCh signal. The sequence is 'F6 F6 F6 28 28 28'. In order to provide enough I/O transitions for synchronisation, scrambling is used over the entire OTU frame except for the FAS bytes.

The multiframe alignment signal (MFAS) byte is used to extend command and management functions over several frames. As 256 frames of OCh make a multi-frame, MFAS extends over 256 frames (MFAS counts from 0 to 255).

FAS#1	FAS#2	FAS#3	FAS#4	FAS#5	FAS#6	MFAS
F6	F6	F6	28	28	28	
						00
						..
						FF

Fig. 12.9 *Details of FAS and MFAS*

12.4.2 OTU overhead

As can be seen from Fig. 12.8, OTU overhead has 7 bytes. It consists of:

Section monitoring (SM) : 3 bytes
General communication channel (GCC0) : 2 bytes
Reserved bytes (RES) : 2 bytes

Refer to Fig. 12.10(a) showing details of OTU overhead.

1	2	3	4	5	6	7
SM			GCC0		RES	

Fig. 12.10(a) *Details of OTU overhead*

Let us examine GCC0 and RES first, and then SM. The general communication channel 0 (**GCC0**) is a channel used for transmission of information between OTU termination points. The reserved (RES) bytes are currently undefined in the standard.

Details of SM: Section monitoring (SM) bytes are used for monitoring a section between regenerators. The three bytes in it are:

Trail trace identifier (TTI) : To monitor correct pairing of regenerators
Byte parity check (BIP-8) : For parity check
Alignment indicator (AI) : To report alignment errors

Refer to Fig. 12.10(b) showing details of the three bytes of SM.

Since one byte of TTI is not sufficient to check the correct pairing of access points (which obviously require long sequence as code in a large OTN), TTI is distributed over 64 frames, i.e., TTI is in fact a 64-byte signal, with one byte going in every frame of OCh. In OCh multi-frame, which has 256 bytes, TTI is repeated four times. Out of the 64 bytes of TTI, bytes 1 to 15 are for 'source access point identifier (SAPI)' and bytes

17 to 31 are for 'destination access point identifier (DAPI)'. API has three sub-fields, namely CC (country code), ICC (ITU carrier code) and UAPC (unique access point code). Refer to Fig. 12.10(c) showing the sub-fields of API.

Alignment indicator (AI) is one-byte long. Bits 1-2-3-4 of AI constitute BEI/BIAE. Interpretation of these 4 bits is as follows.

Value 0000 to 1000	: BIP error
Value 1001 to 1111 except 1011	: BIAE, no BIP error
Value 1011	: No BIP error

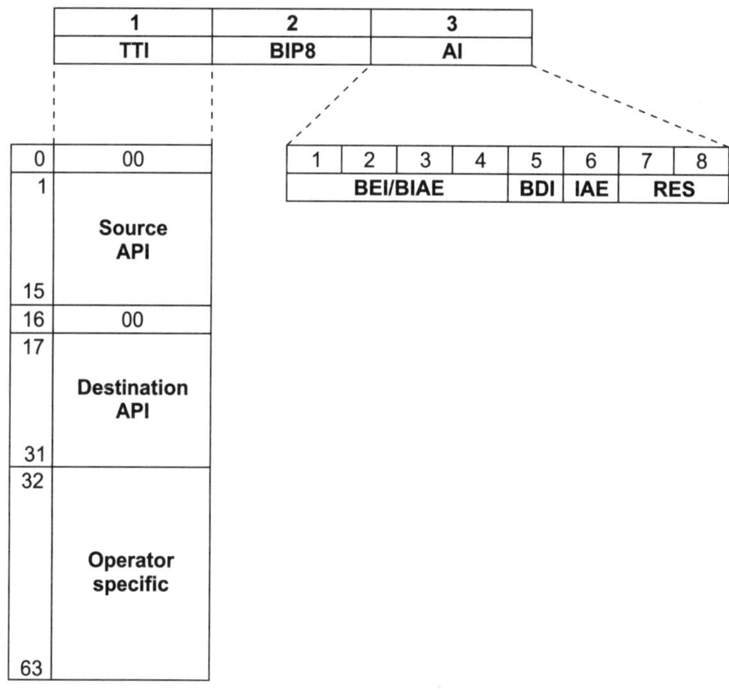

Fig. 12.10(b) *Details of SM*

<table>
<tr><td colspan="3">IS character #</td><td colspan="12">NS character #</td></tr>
<tr><td>1</td><td>2</td><td>3</td><td>4</td><td>5</td><td>6</td><td>7</td><td>8</td><td>9</td><td>10</td><td>11</td><td>12</td><td>13</td><td>14</td><td>15</td></tr>
<tr><td colspan="3">CC</td><td>ICC</td><td colspan="11">UAPC</td></tr>
<tr><td colspan="3">CC</td><td colspan="2">ICC</td><td colspan="10">UAPC</td></tr>
<tr><td colspan="3">CC</td><td colspan="3">ICC</td><td colspan="9">UAPC</td></tr>
<tr><td colspan="3">CC</td><td colspan="4">ICC</td><td colspan="8">UAPC</td></tr>
</table>

\# Note: IS stands for 'international segment' and NS stands for 'national segment'

Fig. 12.10(c) *Sub-fields of API*

The other bits of AI are: backward defect indicator (BDI) and incoming alignment error (IAE). A value of 1 in these bits indicates error and a value of 0 indicates no error.

12.4.3 Details of optical data unit overhead (ODU overhead)

The ODU overhead is broken into several fields: RES, PM, TCMi, TCM ACT, FTFL, EXP, GCC1/GCC2 and APS/PCC. Refer to Fig. 12.11 showing ODU overhead bytes.

	1	2	3	4	5	6	7	8	9	10	11	12	13	14
1	FAS (Frame alignment signal)							OTU overhead						
2	RES			TCM ACT	TCM6			TCM5			TCM4			FTFL
3	TCM3			TCM2			TCM1			PM			Experimental use	
4	GCC1		GCC2		APS/PCC				RES					

Fig. 12.11 *ODU overhead bytes*

Let us understand the purpose and interpretation of each field.

RES

This field is reserved for future international standardisation. At present, all bytes are set to 0.

TCM/ACT

This is a single byte field used for activation and de-activation of TCM fields. Each one of the fields, TCM1 to TCM6, is of 3 bytes, the purpose of which is to provide tandem connection monitoring. TCM field enables signal management across multiple networks. In each TCM field the first byte is termed TTI (trail trace identifier), the second BIP-8 (bit-interleaved parity). The third one is the TCM utility byte. Refer to Fig. 12.12 showing the details of the three bytes of TCM*i*.

TTI in TCM field: This is a 64-byte multi-frame signal, similar to path trace J0 in SDH. In the 64 bytes, the first 32 bytes are API (access point identifiers). Of these 32 bytes, the first 16 are for source (0×00 followed by 15 bytes of SAPI or source access point identifier) and the next 16 are for destination (0×00 followed by 15 bytes of DAPI or destination access point identifier). The remaining 32 bytes of TTI are operator-specific.

BIP-8 in TCM field: BIP is an 8-bit field computed over the whole OPU and payload and inserted into BIP-8 field two frames later.

Utility-byte in TCM sub-field: The third byte in the TCM field is the 'utility byte'. In this, bits 1 to 4 stand for BEI/BIAE, i.e., 'backward error indication/backward incoming alignment error'. These bits carry information on interleaved-bit blocks detected in error in the upstream direction.

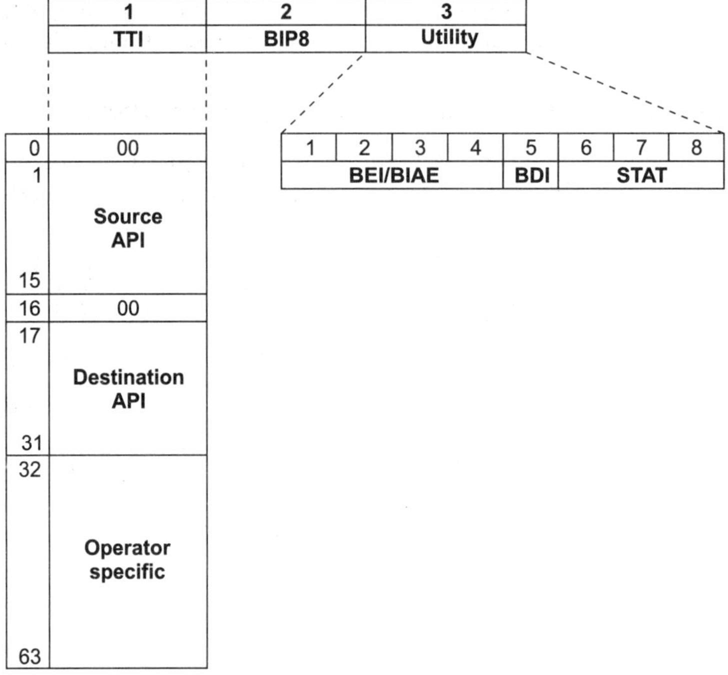

Fig. 12.12 *Details of TCMi*

Bit-5 in the TCM utility field carries BDI (backward defect indication). If BDI=0, there is no defect. IF BDI=1, there is OTUk backward defect.

Bits 6, 7 and 8 of TCM utility field (i.e. byte-3 of TCM field) convey the information shown in Table 12.6(a).

Table 12.6(a) *Interpretation of bits 6, 7and 8 of TCM byte-3*

Bits 6, 7, 8 of TCM byte 3	Interpretation
000	No source TC
001	In use without IAE
010	In use with IAE
011	Reserved
100	Reserved
101	Maintenance signal ODUk - LCK
110	Maintenance signal ODUk - OCI

Fault type and fault location (FTFL) channel

This is multi-frame signal spread over 256 frames which provides information on fault type and fault location. The interpretation of FTFL is given in Table 12.6(b). The interpretation of fault indication field given by byte 0 and byte 128 of FTFL is given in Table 12.6(c).

Table 12.6(b) *FTFL interpretation*

Byte no.	Direction	Contents
Byte 0		Fault indication field
Bytes 1 to 9	Forward	Operator identifier field
Bytes 10 to 127		Operator-specific field
Byte 128		Fault Indication field
Bytes 129 to 137	Backward	Operator identifier field
Bytes 138 to 255		Operator-specific field

Table 12.6(c) *Interpretation of byte 0 and byte 128 of FTFL*

FTFL code (Hex)	Interpretation
00	No fault
01	Signal fail
02	Signal degrade
03 to FF	Reserved

Structure of the 'operator identifier field' is shown in Table 12.6(d); the byte number in FTFL in which it is carried in the forward direction is shown in the first row. The byte number carrying the identifier in the backward direction is shown in the second row. In Table 12.6(d), CC stands for country code and ICC for ITU carrier code. NULL stands for null or padding bytes.

Table 12.6(d) *Operator identifier field of FTFL*

<table>
<tr><td>FTFL byte no. in forward direction →</td><td>1</td><td>2</td><td>3</td><td>4</td><td>5</td><td>6</td><td>7</td><td>8</td><td>9</td></tr>
<tr><td>← FTFL byte no. in reverse direction</td><td>129</td><td>130</td><td>131</td><td>132</td><td>133</td><td>134</td><td>135</td><td>136</td><td>137</td></tr>
<tr><td></td><td colspan="3">CC</td><td colspan="6">National segment code</td></tr>
<tr><td></td><td colspan="3">CC</td><td>ICC</td><td colspan="5">NULL</td></tr>
<tr><td></td><td colspan="3">CC</td><td colspan="2">ICC</td><td colspan="4">NULL</td></tr>
<tr><td></td><td colspan="3">CC</td><td colspan="3">ICC</td><td colspan="3">NULL</td></tr>
<tr><td></td><td colspan="3">CC</td><td colspan="4">ICC</td><td colspan="2">NULL</td></tr>
<tr><td></td><td colspan="3">CC</td><td colspan="6">ICC</td></tr>
<tr><td></td><td colspan="9">NULL</td></tr>
</table>

Path monitoring (PM)

There are three bytes for path monitoring (PM) located in the third row. These bytes enable monitoring of particular sections within the network and aid fault location. In each TCM field, the first byte is termed TTI (trail trace identifier), the second BIP-8 (bit-interleaved parity) and the third the PM utility byte (BEI, BDI and STAT). The path monitoring field is similar to the SM field of OTU overhead. Structurally, it is almost the same as TCM field of ODU overhead. Refer to Fig. 12.13.

The TTI in the PM field is a 64-byte multi-frame signal, similar to the path trace J0 in SDH. In the 64 bytes, first 32 bytes are API (access point identifiers), of which the first 16 are for source (0×00 followed by 15 bytes of source access point identifier or SAPI) and the next 16 are for destination (0×00 followed by 15 bytes of destination access point identifier or DAPI). The remaining 32 bytes of PM are operator-specific.

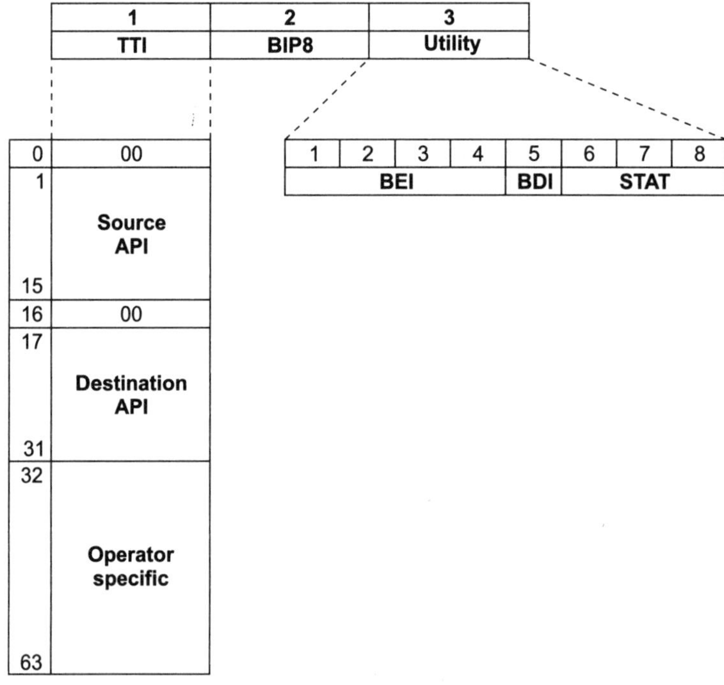

Fig. 12.13 *Details of PM*

The BIP-8 in PM field is an 8- bit field computed over the whole OPU and payload and inserted into BIP-8 field two frames later.

The third byte in PM sub-field is the *utility byte*. In this, bits 1 to 4 stand for BEI (backward error indication). These bits carry information on interleaved-bit blocks detected in error in the upstream direction. The interpretation of BEI is shown in Table 12.7. Bit-5 in the PM utility field carries the **BDI** (backward defect indication). If BDI = 0, there is no defect. IF BDI = 1, there is OTUk backward defect. **STAT** sub-field of the utility byte of PM (bits 6,7,8 of the 3rd byte of PM) provide an indication of the presence or absence of maintenance signals. Its interpretation is given in Table 12.8.

Table 12.7 *Interpretation of BEI*

ODUk PM's BEI sub-field	BIP violation
0000	0
0001	1
0010	2
0011	3
0100	4
0101	5
0110	6
0111	7
1000	8
1001 to 1111	0

Table 12.8 *Interpretation of STAT sub-field of the utility byte of PM*

PM byte 3, bits 6,7,8	Interpretation
000	Reserved
001	Normal path signal
010	Reserved
011	Reserved
100	Reserved
101	Maintenance signal ODUk- LCK
110	Maintenance signal ODUk- OCI
111	Maintenance signal ODUk- AIS

Experimental use field

This field has 2 bytes and is located next to PM in row 3. It is reserved for experimental use.

General communication channel 1 and 2 (GCC1 and GCC2)

The general communication channels (GCC1 and GCC2) allow communication between two NEs.

APS/PCC

The automatic protection switching (APS) and the protection communication channel (PCC) is a 4-byte field for switching over to protection path and related communication which supports up to eight levels of nested APS/PCC signals. These are associated to a dedicated connection monitoring level depending on the value of the multiframe.

Q 12.7 *What is meant by tandem connection monitoring (TCM)? How does it co-exist with path monitoring (PM)?*

TCM enables the user and user's signal carriers to monitor the quality of the traffic which is transported between segments or connections in the network. While SONET/SDH allowed a single level of TCM to be configured, OTH (ITU G.709) allows six levels of tandem connection monitoring to be configured. TCM supports monitoring of ODUk connections for one or more of the following network applications:

1. Optical UNI to UNI tandem connection monitoring: monitoring the ODUk connection through the public transport network (from public network ingress network termination to egress network termination).
2. Optical NNI to NNI tandem connection monitoring: monitoring the ODUk connection through the network of a network operator (from operator network ingress network termination to egress network termination).
3. Sub-layer monitoring for linear 1+1, 1:1 and 1:*n* optical channel sub-network connection protection switching to determine signal fail and signal degrade conditions.
4. Sub-layer monitoring for optical channel shared protection ring (SPring) protection switching in order to determine the signal fail and signal degrade conditions.
5. Monitoring an optical channel tandem connection for the purpose of detecting a signal fail or signal degrade condition in a switched optical channel connection in order to initiate automatic restoration of the connection during fault and error conditions in the network.
6. Monitoring an optical channel tandem connection for, say, fault localisation or verification of delivered quality of service.

As already shown in Fig. 12.11, the six TCM fields are numbered as TCM 1 to TCM6. Each TCM has 3 bytes: TTI, BIP8 and utility (BEI/BIAE, BDI, STAT). The content of TCM fields except STAT are undefined (the pattern will be all 1s, 0110 0110 or 0101 0101 repeating) during the presence of maintenance signal in PM (ODU-k AIS, ODU-k OCI, ODU-k LCK).

There are various types of monitored connection topologies: cascaded, nested and overlapping. Examples of cascading and nesting topologies are shown in Fig. 12.14.

Each of the six TCMi fields in the ODU overhead is assigned to a monitored connection. There can be from zero to six levels or connections that can be configured for each ODU trail. In Fig. 12.14, for example, monitored connections A1-A2/B1-B2/C1-C2 and A1-A2/B3-B4 are nested, while B1-B2/B3-B4 are cascaded. Corresponding TCMs in use are shaded alike. TCMs not in use are left un-shaded in the figure.

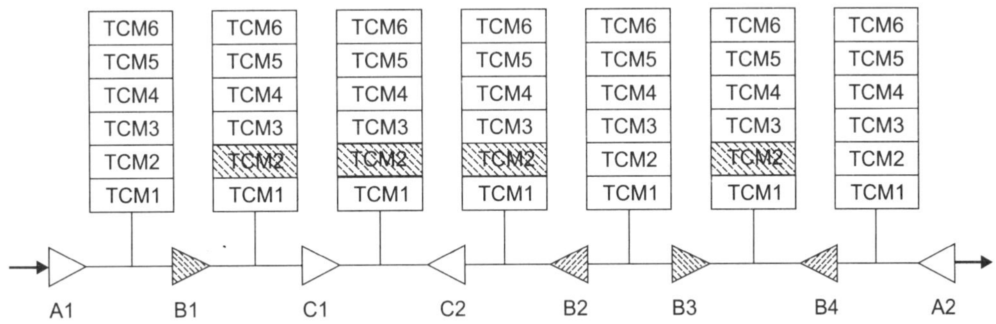

Fig. 12.14 *Nested and cascaded ODUk monitored sections*

Figure 12.15 shows the overlapping monitored conditions (B1-B2 and C1-C2). The corresponding TCMs in use are shaded alike while TCMs not in use are left un-shaded.

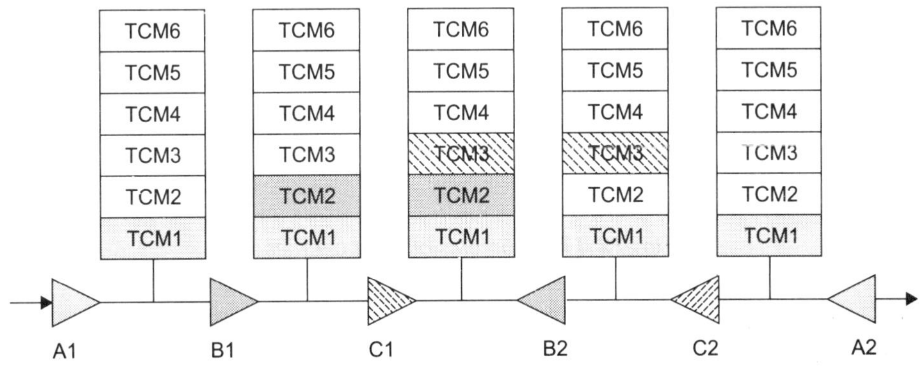

Fig. 12.15 *Nested and cascaded ODUk monitored sections*

12.4.4 Optical payload unit (OPU) overhead description

The starting point in an OTN is the mapping of client traffic to the optical channel payload unit (OPU) while duly adding the OPU overhead. The OPU overhead regulates the mapping and concatenation of the client signals and provides information on the type of signal being transported. Refer to Fig. 12.16. The OPU is a structured signal of order k ($k=1, 2, 3$) and denoted as OPUk. OPUk frame is an octet-based frame of size

4 rows × 3810 columns. Of this, two columns are for OPU overhead and 3808 columns are for client payload. The entire OPU (2 overhead + 3808 client payload = 3810 columns) is located from column 15 to 3824 in OCh. Columns 15 and 16 carry OPU overhead and columns 17 to 3824 (3808 columns) carry OPU client payload. Refer to Fig. 12.17 for the composition of OPU overhead.

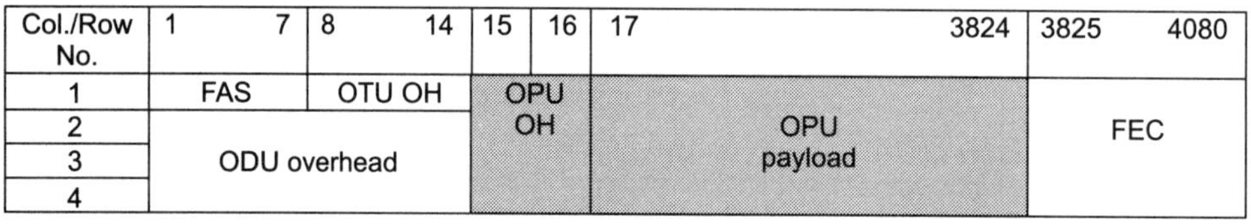

Col./Row No.	1 7	8 14	15	16	17 3824	3825 4080
1	FAS	OTU OH	OPU OH		OPU payload	FEC
2	ODU overhead					
3						
4						

Fig. 12.16 *OPU and OPU overhead in OCh (OTU frame)*

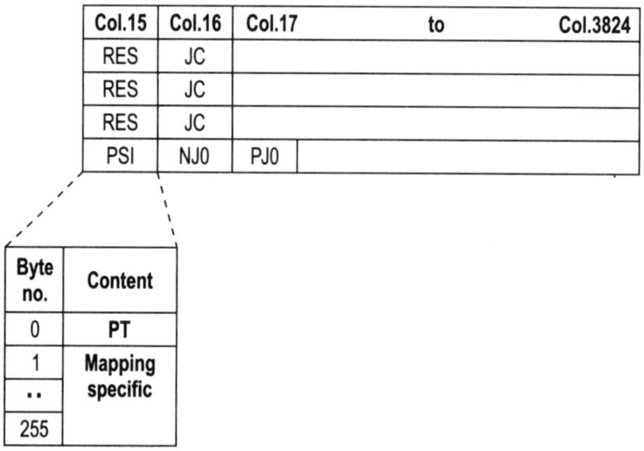

Col.15	Col.16	Col.17 to Col.3824
RES	JC	
RES	JC	
RES	JC	
PSI	NJO	PJO

Byte no.	Content
0	PT
1	Mapping specific
..	
255	

Fig. 12.17 *Contents of OPU overhead*

The primary overhead field in OPU-OH is the payload structure identifier (PSI). This is not just a single byte but a 256-byte multi-frame signal. The other fields in the OPU overhead are dependent on the mapping capabilities associated with the OPU. For an asynchronous mapping (the client signal and OPU clock are different), the justification control (JC) bytes are available to compensate for clock rate differences, while for a purely synchronous mapping (client source and OPU clock are the same), the JC bytes become reserved.

Let us summarise the purpose of each byte of OPU overhead.

1. RES is reserved for future use.
2. JC, NJO and PJO are required for asynchronous mappings. For synchronous mappings, all JCs are 0 and PJO is a data byte.

3. Interpretation of JC, NJ0 and PJ0 for asynchronous mappings is given in Table 12.9.

Table 12.9 *Interpretation of JC, NJ0, PJ0 in OPU overhead*

JC	NJ0	PJ0
00	Justification byte	Data byte
01	Data byte	Data byte
10	Not generated	Not generated
11	Justification byte	Justification byte

4. PSI is a multi-frame signal spread over 256 frames. In the 0th frame, PSI field has PT (payload type code) and the first to 255th frame, the PSI field has mapping and concatenation-specific information.
5. PT codes and interpretations are given in Table 12.10.

Table 12.10 *PT code interpretation*

PT code (Hex)	Interpretation
01	Experimental mapping
02	Asynchronous STM-N mapping
03	Bit-synchronous STM-N mapping
04	ATM mapping
05	GFP mapping
06	Virtual concatenated signal
10	Bit stream with octet timing mapping
11	Bit stream without octet timing mapping
20	ODU multiplex structure
55	Not used
66	Not used
80 to 8F	Reserved for proprietary use
FD	NULL test signal mapping
FE	PRBS test signal mapping
FF	Not used

12.4.5 Forward error correction (FEC)

The ITU G.709 standard supports forward error correction (FEC) in the OTU frame and is the last part added to the frame before the frame is scrambled. FEC provides a method for significantly reducing the number of transmitted errors arising due to noise as well as other optical phenomena that occur at high transmission speeds. This enables service providers to support longer spans in between optical repeaters.

Refer to Fig. 12.18 where the FEC portion in OCh (OTU frame) has been highlighted. FEC occupies column 3825 to 4080, i.e., 256 bytes per row of OTU. Seeing this stretch, one should not conclude that the 256-byte long FEC is computed row-wise and attached from column 3825 to 4080. In fact, each row of 4080 bytes is considered as 16 sub-rows with 255 bytes per sub-row (255 × 16 = 4080). These are byte interleaved while forming one row of OTU.

Col./Row No.	1 … 7	8 … 14	15	16	17 … 3824	3825 … 4080
1	FAS	OTU OH	OPU OH		OPU payload	FEC
2	ODU overhead					
3						
4						

Fig. 12.18 *FEC portion of OCh (OTU frame)*

In the 255 bytes of each sub-row, 239 are for information and 16 for FEC. Thus, in one row of OUT, which has 16 sub-rows, we have 16 × 16 = 256 bytes pertaining to FEC, which are placed from column 3825 to 4080 after interleaving.

The FEC uses a Reed–Solomon (RS) (255/239) coding technique. This means that 239 bytes are required to compute a 16-byte parity check. The FEC can correct up to eight (bytes) errors per sub-row (codeword) or detect up to 16 byte errors without correcting any. Combined with the byte interleaving capability included in the ITU G.709 implementation, the FEC is resilient to error bursts and up to 128 consecutive bytes can be corrected per OTU frame row. Since there are 16 sub-rows in each row, 16 × 16 = 256 errors per row or 256 × 4 = 1024 errors per OTU frame can be detected. Each code word can correct up to 8 errors, i.e., 8 errors per sub-row or 8 × 16 = 128 errors per row or 128 × 4 = 512 errors per OTU frame.

12.5 Maintenance Signals

Maintenance signals are indication bits sent through the frames on OTN to convey defect conditions. The following groups of maintenance signals are provided in OTH: OTS, OMS, OCh, OUT and ODU maintenance signals.

12.5.1 OTS, OMS and OCh maintenance signals

These are summarised in Table 12.11.

Table 12.11 *OTS, OMS and OCh maintenance signals*

Group	Maintenance signal	When generated
OTS maintenance signal	OTS payload missing indication (OTS-PMI)	OTS-PMI is generated when the OTS payload does not contain an optical signal.
OMS maintenance signal	OMS forward defect indication – Payload (OMS-FDI-P)	OMS-FDI-P is generated to indicate an OMS server layer defect in the OTS network.
	OMS forward defect indication – Overhead (OMS-FDI-O)	OMS-FDI-O is generated when the transport of OMS OH via the OOS is interrupted due to a signal fail condition in the OOS.
	OMS payload missing indication (OMS-PMI)	OMS-PMI is generated when none of the OCCs contain an optical signal.
	OCh forward defect indication – Payload (OCh-FDI-P)	OCh-FDI is generated for an OCh server layer defect in the OMS network layer. When the OTUk is terminated, the OCh-FDI is continued as an ODUk-AIS signal.
	OCh forward defect indication – Overhead (OCh-FDI-O)	OCh-FDI-O is generated when the transport of OCh OH via the OOS is interrupted due to a signal fail condition in the OOS.
	OCh open connection indication (OCh-OCI)	The OCh-OCI signal indicates to the downstream transport processing functions that the OCh connection is not bound to, or not connected (via a matrix connection) to a termination source function. This indication helps in distinguishing a missing optical channel due to a defect or due to the open connection resulting from NMS command.

12.5.2 OTUk maintenance signal

The OTUk maintenance signal is the OTUk-AIS, which is a generic-AIS signal. The generic-AIS signal is a signal with a 2047-bit polynomial number 11 (PN-11) repeating sequence. The PN-11 sequence is defined by the generating polynomial $1 + x^9 + x^{11}$ as shown in Fig. 12.19.

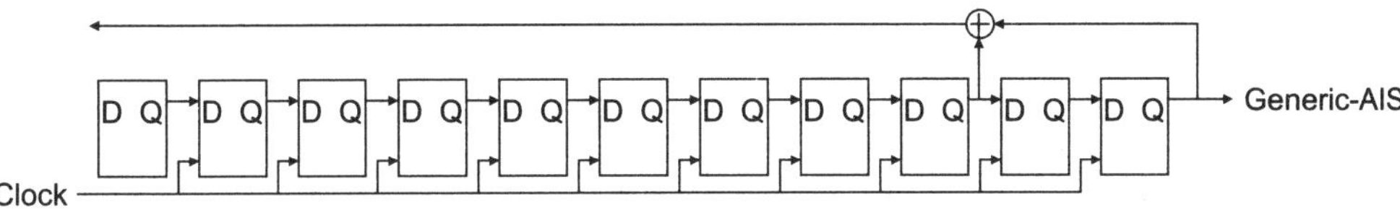

Fig. 12.19 *Generic AIS generating circuit*

12.5.3 ODUk maintenance signal

Three ODUk maintenance signals are defined in ITU-T Rec. G.709: ODUk-AIS, ODUk-OCI and ODUk-LCK.

ODUk alarm indication signal (ODUk-AIS)

The ODUk-AIS is specified as all 1s in the entire ODUk signal, excluding the frame alignment signal (FAS), OTUk overhead (OTUk OH) and ODUk FTFL. Refer to Fig. 12.20(a) showing AIS indication area of an OTU frame.

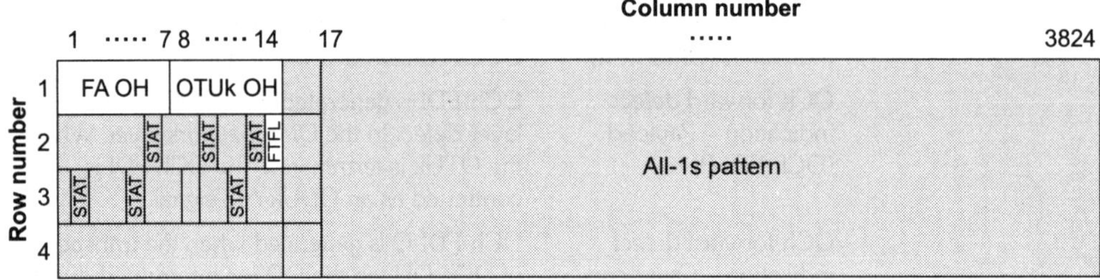

Fig. 12.20(a) *AIS indication area of OTU frame*

The presence of ODUk-AIS is detected by monitoring the ODUk STAT bits in the PM and TCMi overhead fields.

ODUk open connection indication (ODUk-OCI)

ODUk-OCI is specified as a repeating '0110 0110' pattern in the entire ODUk signal, excluding the frame alignment signal (FAS) and OTUk overhead (OTUk OH). Refer to Fig. 12.20(b) showing ODUk-OCI indication in OTU frame. The presence of ODUk-OCI is detected by monitoring the ODUk STAT bits in the PM and TCMi overhead fields.

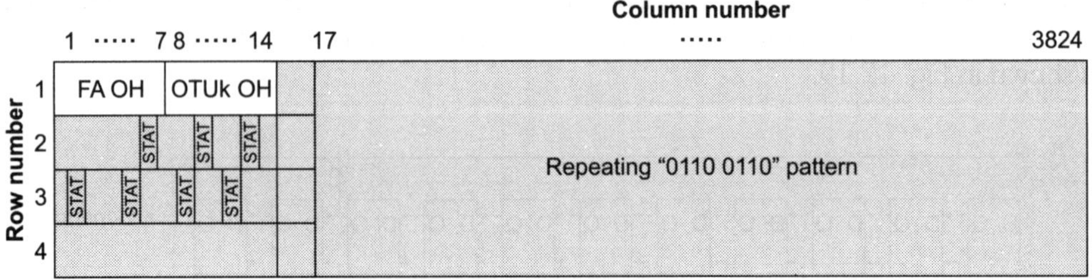

Fig. 12.20(b) *ODUk-OCI indication in OTU frame*

ODUk locked (ODUk-LCK)

ODUk-LCK is specified as a repeating '0101 0101' pattern in the entire ODUk signal, excluding the frame alignment overhead (FA OH) and OTUk overhead (OTUk OH). Refer to Fig. 12.20(c) showing ODUk-LCK indication in OTU frame.

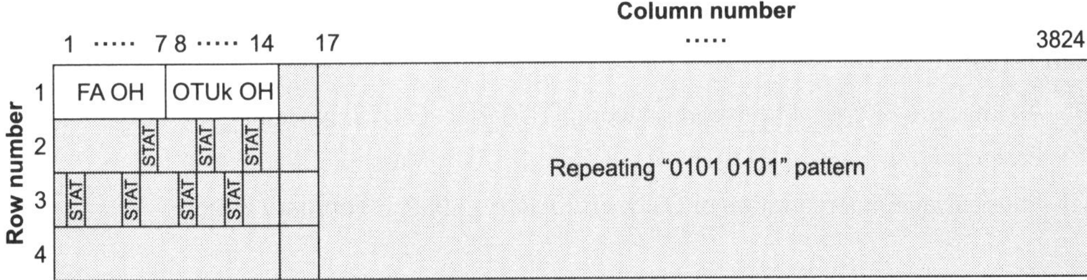

Fig. 12.20(c) *ODUk-LCK indication in OTU frame*

12.6 Scrambling

The OTUk signal must have sufficient signal transitions or timing content at the ONNI, which is possible if long sequences of 1s and 0s are avoided. Since we do not have control of the client signal, any long sequence of 1s and 0s are to be scrambled. The operation of the scrambler will be functionally identical to that of a frame synchronous scrambler of sequence length 65535 operating at the OTUk rate.

The generating polynomial is $1 + x + x^3 + x^{12} + x^{16}$. Refer to Fig. 12.21 showing the functional diagram of the frame synchronous scrambler. The scrambler will be reset to FFFF (HEX) on the most significant bit of the byte following the last framing byte in the OTUk frame, i.e., the MSB of the MFAS byte. This bit and all the subsequent bits to be scrambled are added modulo 2 to the output from the x^{16} position of the scrambler. The scrambler runs continuously throughout the complete OTUk frame. The framing bytes (FAS) of the OTUk overhead must not be scrambled. Scrambling is performed after FEC computation and its insertion into the OTUk signal.

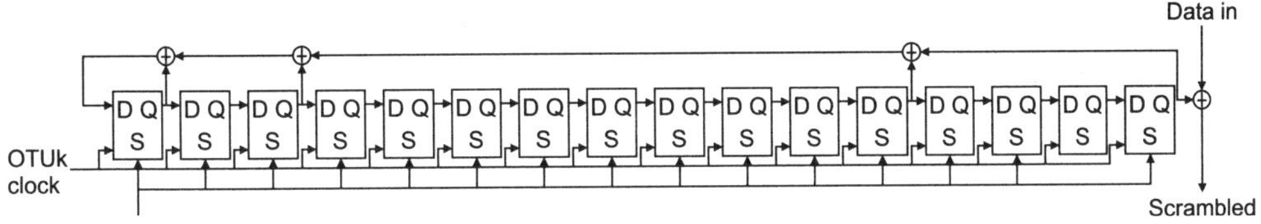

Fig. 12.21 *Frame synchronisation scrambler*

Summary

In this chapter the requirements of optical layer and managed light path services in WDM are highlighted. How these requirements resulted in evolution of OTH, its frame structure and transport entities are discussed along with the concepts of digital wrapper and optical channel. Structures of ODUk, OTUk, OPUk are thoroughly explored and the provisions for managed light path services in them are explained.

References

[1] Architecture of optical transport network, ITU-T Rec. G.B72, November 2001.

[2] Protection mechanism for OTN, ITU-T Rec. G873.1, March 2006.

[3] Physical layer interpretation of OTN, ITU-T Rec. G.959.1, February 2006.

13

Mapping Client Signals in OTH

13.1 Mapping of Client Signals – The Scope

As discussed in previous chapter, optical transport hierarchy has a payload area of 4 × 3808 bytes known as OPUk payload area. In this chapter we will discuss how OPUk payload area is filled-up or mapped with various client signals:

1. STM signals (STM-16/64/256)
2. ATM cells
3. GFP frames
4. Test signals
5. Non-specific client bit-streams

We will also discuss how virtual concatenation of OPUk is done and how client signals are mapped into concatenated OPUks.

Mapping of ODUk signals into ODTUjk signals is also discussed.

13.2 Mapping of STM Signals

13.2.1 Mapping methodology

STM-16/64/256 signals are continuous bit rate signals and hence denoted by CBR2G5, CBR10G, CBR40G respectively, where CBR stands for 'continuous bit rate' and G stands for giga bit.

The maximum bit-rate tolerance between OPUk and the client signal clock which can be accommodated by this mapping scheme is ± 65 ppm. With a bit-rate tolerance of ± 20 ppm for the OPUk clock, the client signal's bit-rate tolerance can be ± 45 ppm.

Mapping of STM signals into OPUk (k = 1,2,3) is performed in two different modes:

1. **Asynchronous mapping**
   The OPUk clock for asynchronous mapping is a locally generated clock (within the limits specified above), which is independent of the STM client signals [i.e., $4^{(k-1)} \times$ 2.488320 kbit/s (k = 1,2,3)].

2. **Bit synchronous mapping**
   The OPUk clock for bit synchronous mapping is derived from the incoming STM client signals [i.e., $4^{(k-1)} \times 2.488320$ kbit/s (k = 1,2,3)].

A generic frame structure for OPUk is followed while mapping by any of the two modes. This generic frame structure is shown in Fig. 13.1. Let us examine the overhead portion and payload portion for this mapping.

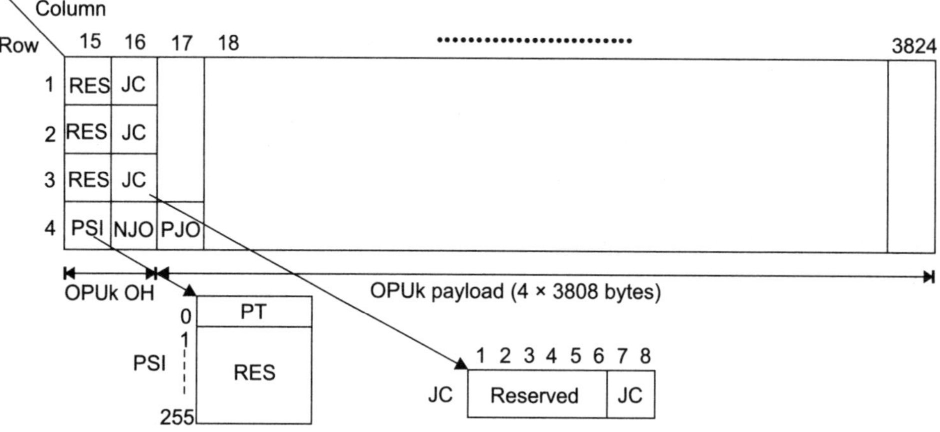

Fig. 13.1 *Generic frame structure for mapping STM signals onto OPUk*

The OPUk overhead for these mappings consists of:

1. Payload structure identifier (PSI) including the payload type (PT) and 255 bytes reserved for future international standardisation (RES)
2. Three justification control (JC) bytes
3. One negative justification opportunity (NJO) byte
4. Three bytes reserved for future international standardisation (RES).

Each JC byte consists of two bits for justification control and six bits reserved for future international standardisation.

The OPUk payload for these mappings consists of 4 × 3808 bytes, including one positive justification opportunity (PJO) byte. The justification control (JC) signal, which is located in rows 1, 2 and 3 of column 16, bits 7 and 8, is used to control the two justification opportunity bytes, NJO and PJO, that follow in row 4.

Q 13.1 *If same generic frame is used for asynchronous and bit-synchronous mapping, how are they distinguished?*

It may be recalled from section 12.4.4 describing OPU-OH that the value of PT = 02 for asynchronous mapping and 03 for bit-synchronous mapping.

Q 13.2 *How are justification bytes generated and interpreted?*

The asynchronous and bit synchronous mapping processes generate the JC, NJO and PJO according to Tables 13.1 and 13.2 respectively.

The de-mapping process interprets JC, NJO and PJO according to Table 13.3. Majority vote (two out of three) is used to make the justification decision in the de-mapping process to protect against an error in one of the three JC signals.

Table 13.1 *JC, NJO, PJO generation in asynchronous mapping of STM signals*

JC bits 7 & 8	NJO	PJO
00	Justification byte	Data byte
01	Data byte	Data byte
10	Not generated	Not generated
11	Justification byte	Justification byte

Table 13.2 *JC, NJO, PJO generation in bit-synchronous mapping of STM signals*

JC bits 7 & 8	NJO	PJO
00	Justification byte	Data byte
01	Not generated	Not generated
10	Not generated	Not generated
11	Not generated	Not generated

Table 13.3 *JC, NJO, PJO interpretation while de-mapping*

JC bits 7 & 8	NJO	PJO
00	Justification byte	Data byte
01	Data byte	Data byte
10#	Justification byte	Data byte
11	Justification byte	Justification byte

10 code is not generated by mapping process. This may be received by de-mapping circuit due to bit errors and hence PJO is treated as data byte.

The values contained in NJO and PJO when they are used as justification bytes is all-0s. The receiver is required to ignore the value contained in these bytes whenever they are used as justification bytes.

Q 13.3 *How are incoming signal failure conditions handled in mapping?*

During incoming STM signal failure condition, the failed incoming signal is replaced by the generic AIS signal as shown in Fig. 12.19 and is then mapped into the OPUk.

During signal fail condition of the incoming ODUk/OPUk signal (e.g., in the case of an ODUk-AIS, ODUk-LCK, ODUk-OCI condition) a generic AIS pattern as shown in Fig. 12.19 is generated as a replacement signal for the lost STM signal.

13.2.2 Mapping STM-16 signal into OPU1

Please refer Fig. 13.1 showing generic OPUk frame. It has bytes from column 17 to 3824 with 3808 bytes in each row. Groups of 8 successive bits (not necessarily a byte) of STM-16 signal are mapped into a data (D) byte of the OPU1 (see Fig. 13.2). Positive or a negative justification action is performed once per OPU1 frame.

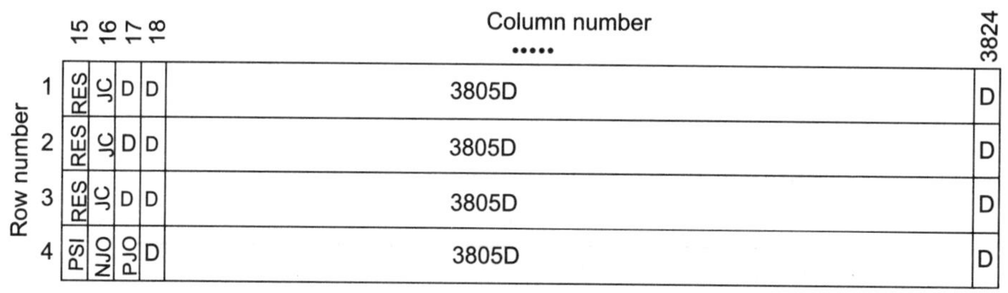

Row number	15	16	17	18	Column number •••••	3824
1	RES	JC	D	D	3805D	D
2	RES	JC	D	D	3805D	D
3	RES	JC	D	D	3805D	D
4	PSI	NJO	PJO	D	3805D	D

Fig. 13.2 *Mapping STM-16 into OPU1*

13.2.3 Mapping STM-64 signal into OPU2

In this case, the 3808 bytes in each row of OPU are considered as:

118 × 16 data bytes = 1888 bytes from column 17 to 1904
16 fixed stuff bytes (FS) from column 1905 to 1920
119 × 16 data bytes = 1904 bytes from column 1921 to 3824

Groups of 8 successive bits (not necessarily a byte) of the STM64 signal are mapped into the D-bytes in OPU2 (see Fig. 13.3). Sixty-four fixed stuff (FS) bytes (16 per row × 4 rows) are added from columns 1905 to 1920. It is possible to perform either a positive or a negative justification once per OPU2 frame.

Column number

Row number	15	16	17 1904	1905 1920	1921 3824
1	RES	JC	118 × 16D	16FS	119 × 16D
2	RES	JC	118 × 16D	16FS	119 × 16D
3	RES	JC	118 × 16D	16FS	119 × 16D
4	PSI	NJO	PJO 15D+ 117×16D	16FS	119 × 16D

Fig. 13.3 *Mapping STM-64 signal into OPU2*

13.2.4 Mapping STM-256 signal into OPU3

In this case, 3808 bytes in each row of OPU are considered as:

78 × 16 data bytes = 1248 bytes from column 17 to 1264
16 fixed stuff bytes (FS) from column 1265 to 1280
79 ×16 data bytes = 1264 bytes from column 1281 to 2544
16 fixed stuff bytes (FS) from column 2545 to 2560
79 × 16 data bytes from column 2561 to 3824.

Groups of 8 successive bits (not necessarily a byte) of the STM-256 signal are mapped into D-bytes OPU3 (see Fig. 13.4). 32 fixed stuff bytes (FS) bytes are added per row (in two instalments of 16 each, from column 1265 to 1280 and 2545 to 2560, making a total of 128 fixed stuff (FS) bytes per OPU3 frame. Either a positive or a negative justification action is possible once per frame.

Column number

Row number	15	16	17 1264	1265 1280	1281 2544	2545 2560	2561 3824
1	RES	JC	78 ×16D	16FS	79 × 16D	16FS	79 × 16D
2	RES	JC	78 ×16D	16FS	79 × 16D	16FS	79 × 16D
3	RES	JC	78 ×16D	16FS	79 × 16D	16FS	79 × 16D
4	PSI	NJO	PJO 15D+77×16D	16FS	79 × 16D	16FS	79 × 16D

Fig. 13.4 *Mapping of STM-256 signal into OPU3*

13.3 Mapping of ATM Cell Stream into OPUk

The OPUk overhead for the ATM mapping consists of:

1. Payload structure identifier (PSI), which is a multi-frame signal over 256 frames having the payload type (PT) and 255 reserved bytes (RES) for future international standardisation.
2. Seven bytes reserved for future international standardisation (RES).

It may be recalled from the description OUT-OH in section 12.4.4 that the value of PT = 04 for ATM cell mapping. A constant bit rate ATM cell stream, when mapped into OPUk payload area, creates an ATM virtual path (ATM VP) signal. Rate adaptation is accomplished in this process either by inserting idle cells or by discarding cells.

Refer to Fig. 13.5. The ATM cell stream is mapped into the OPUk payload area with byte alignment between the two. The OPUk payload area is 3808 × 4 = 15232 bytes whereas ATM cell size is 53 octets (bytes). The OPUk payload area is not an integer multiple of ATM cell size. So, ATM cell may cross the OPUk frame boundary.

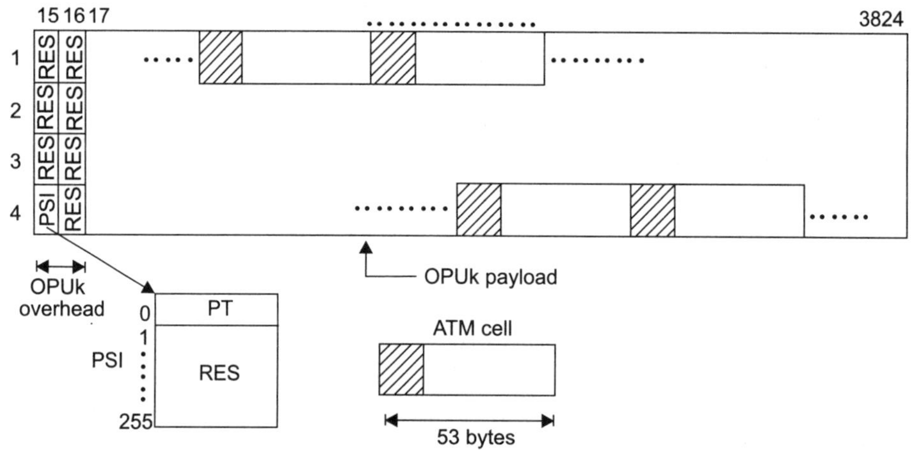

Fig. 13.5 *Mapping ATM cells into OPUk*

Each ATM cell has 53 octets, of which 48 are information bytes and 5 are header bytes. The ATM cell header contains a header error control (HEC) field, which may be used in a way similar to a frame alignment word to achieve cell delineation. It may be noted that the frame alignment signal is not a constant but varies from cell to cell as explained below.

The HEC method uses the correlation between the header bits to be protected by the HEC (32 bits) and the control bit of the HEC (8 bits) introduced in the header after computation with a shortened cyclic code with generating polynomial $g(x) = x^8 + x^2 + x + 1$. The remainder from this polynomial is then added to the fixed pattern '01010101' in order to improve the cell delineation performance. This method is similar to the conventional frame alignment recovery the where the alignment signal is not fixed but varies from cell to cell.

The ATM cell information field (48 bytes) is scrambled before mapping into the OPUk. A self-synchronizing scrambler with generator polynomial, $x^{43} + 1$, is used (as per ITU-T Rec. I.432.1). The scrambler operates for the duration of the cell information field. During the 5-byte header, the scrambler operation is suspended and the scrambler state is retained. At the receiving end, after termination of the OPUk signal, the ATM cell information field is descrambled before being passed to the ATM layer.

Cell information field scrambling is required to provide security against false cell delineation and the cell information field replicating the OTUk and ODUk frame alignment signal.

For more information on HEC cell delineation, readers may please go through reference [1].

13.4 Mapping of GFP Frames into OPUk

The OPUk payload for the GFP mapping consists of 4 × 3808 bytes. The mapping of generic framing procedure (GFP) frames is performed by aligning the byte structure of every GFP frame with the byte structure of the OPUk payload. Please refer Fig. 13.6.

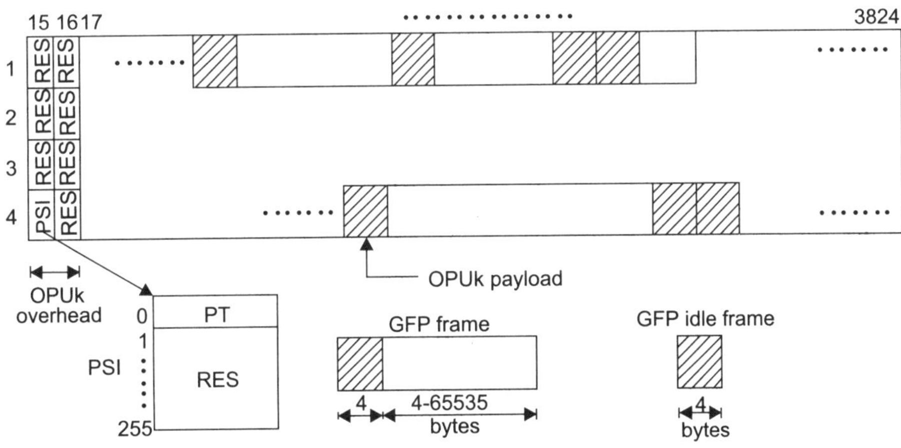

Fig. 13.6 *GFP mapping into OPUk*

Since the GFP frames are of variable length (4 bytes to 65535 bytes), a frame may cross the OPUk frame boundary. When there is no data to be sent, an idle GFP frame (4 bytes) is inserted. Thus, despite discrete data flow and variable frame length, continuity is maintained at OPUk. The GFP frame stream is scrambled during encapsulation to avoid replication of OTUk and ODUk FAS.

The OPUk overhead for the GFP mapping consists of:

1. Payload structure identifier (PSI), which is a 256 byte multi-frame signal with the first byte being the payload type (PT) and the rest of 255 bytes reserved for future international standardisation (RES).
2. Seven bytes reserved for future international standardisation (RES).

Value of PT = 05 for GFP mapping, as mentioned in section 12.4.4.

13.5 Mapping of Test Signal into OPUk

Test signals are (i) all-0s pattern, also known as NULL client signal, (ii) 2147483647-bit pseudo random test sequence or $(2^{11} - 1)$ PRBS sequence as per ITU-T Rec. O.150. We discuss mapping of these test signals into OPUk.

13.5.1 Mapping of NULL client signal

OPUk payload area for NULL client signal is 4 rows × 3808 columns in which all-0s pattern is mapped. The OPUk overhead for the NULL mapping consists of:

1. Payload structure identifier (PSI), which is a 256 byte multi-frame signal having payload type (PT) as its first byte and the rest of the 255 bytes reserved (RES) for future international standardisation.
2. Seven bytes reserved (RES) for future international standardisation.

It may please be recalled that the value of PT = FD for mapping NULL test signal. The mapping of NULL test signal is shown in Fig. 13.7.

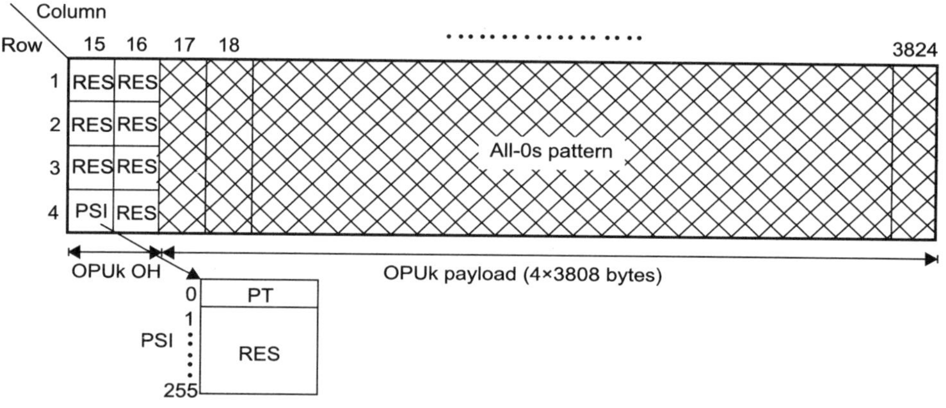

Fig. 13.7 *Mapping of NULL client signal into OPUk*

13.5.2 Mapping of $(2^{31}–1)$ PRBS sequence

To map PRBS test sequence into OPUk, 3808 bytes in each row of OPUk are considered as successive bytes with group of 8 successive bits of PRBS $(2^{31} - 1)$, which is a 2147483647-bit sequence, being mapped into each such byte. The OPUk overhead for the PRBS mapping consists of:

1. Payload structure identifier (PSI), which is a multi-frame signal with the payload type (PT) in first byte and the rest (255 bytes) reserved (RES) for future international standardisation.
2. Seven bytes reserved (RES) for future international standardisation.

It may be recalled that value of PT = FE for mapping PRBS test signal. The mapping of PRBS test pattern is shown in Fig. 13.8.

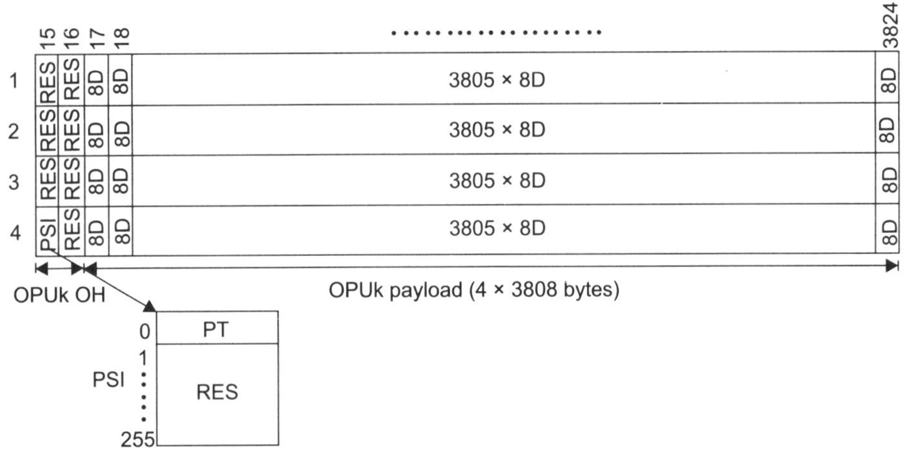

Fig. 13.8 *Mapping PRBS test signal into OPUk*

13.6 Mapping of Non-Specific Client Bit Stream into OPUk

Client signals other than the specific client signals (STM, ATM, GFP, test signals) are known as non-specific client signals. Any non-specific client signals can be mapped into OPUk. The seven reserved bytes of OPUk-OH can be used to identify these bit streams. The payload area for mapping is the 4 rows × 3808 columns area of OPUk (Fig. 13.9). The seven RES bytes are marked CS to denote that these are 'customer specific' and not belonging to the types STM, ATM, GFP, etc.

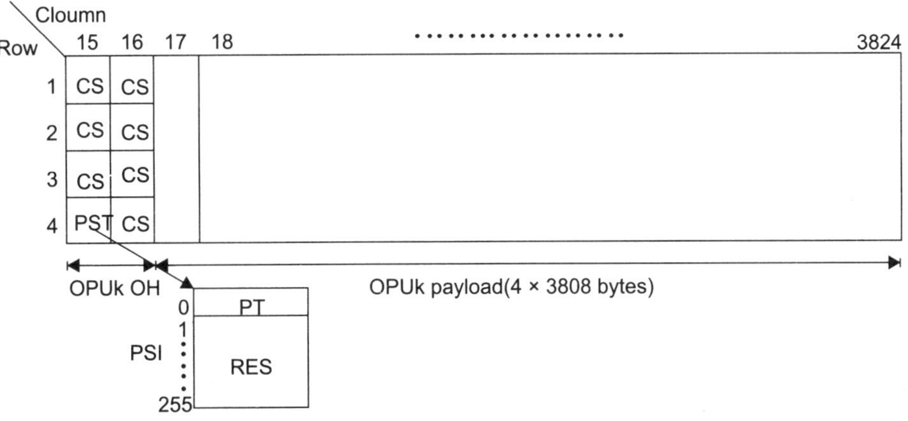

Fig. 13.9 *Mapping customer specific signals (other than STM, ATM, GFP, etc.) into OPUk*

13.7 Concatenation of OPUk

13.7.1 Essentials of concatenation

In OTN, virtual concatenation of OPUk is resorted to for enhanced container (payload) capacity. Virtual concatenated OPUk is denoted by OPUk-Xv, $k = 1\ldots3$, $X = 1\ldots256$). Let us examine the payload area and overhead area of OPUk-Xv. Refer to Fig. 13.10 showing the frame structure of OPUk-Xv.

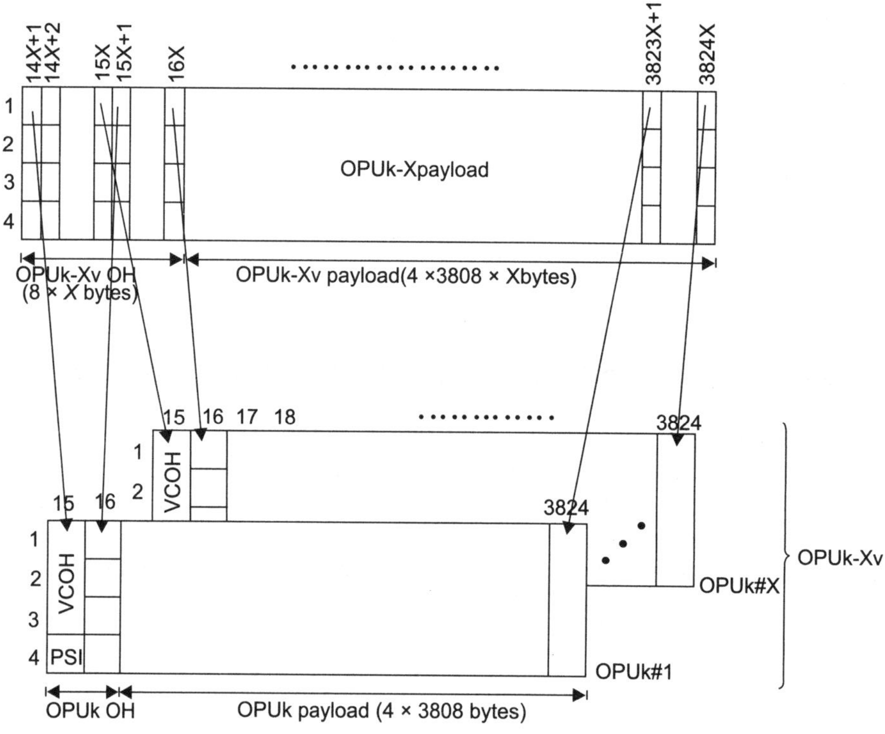

Fig. 13.10 *OPUk-Xv frame structure*

The OPUk-Xv payload area is from columns $16X + 1$ to $3824X$ and the OPUk-Xv overhead area is from column $14X + 1$ to $16X$.

Q 13.4 What is the payload capacity of OPUk-Xv?

The payload capacity of OPUk-Xv is 'X' *times* OPUk payload capacity, i.e.

$X \,.\, [238/(239-k)] \,.\, 4^{(k-1)} \,.\, 2488320$ kbps.

Q 13.5 *How is OPUk-Xv transported across OTN?*

Each OPUk of the OPUk-Xv is transported in an ODUk and X ODUks form an ODUk-Xv. Each ODUk of the ODUk-Xv is transported individually through the network. Due to different propagation delays of the ODUks, differential delays crop up between ODUks and also OPUks. These are compensated and individual OPUks are realigned at the receiving end.

13.7.2 OPUk-Xv overhead description

Please refer to Fig. 13.11 showing OPUk-Xv overhead.

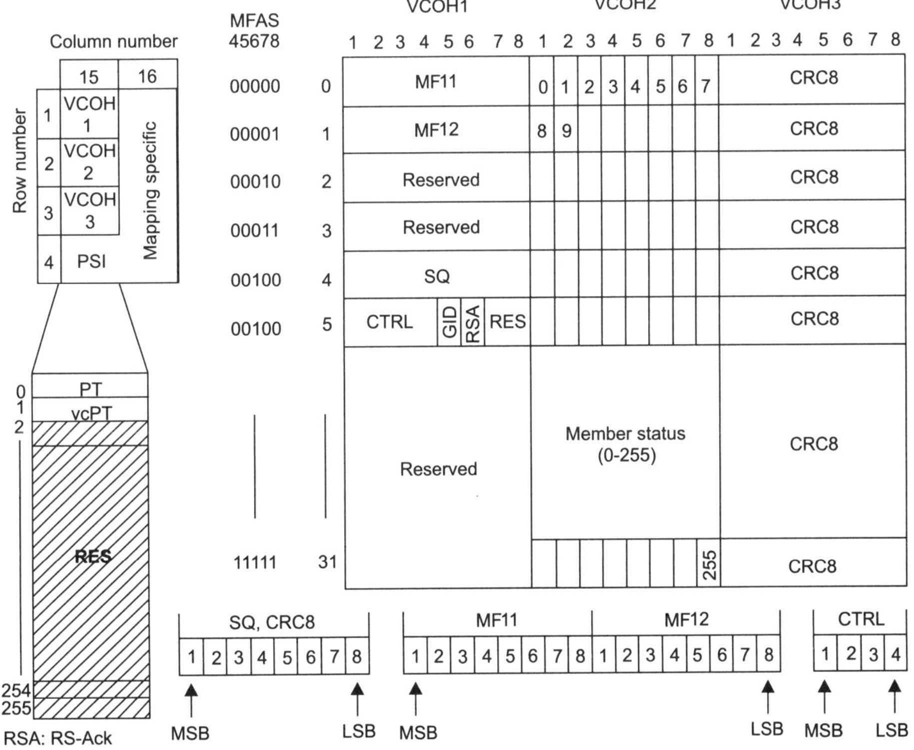

Fig. 13.11 *OPUk-Xv overhead*

It can be observed that –

1. The OPUk-Xv overhead consists of:

   a. *X* times PSI

   b. X times virtual concatenation overhead used for a virtual concatenation-specific sequence and multi-frame indication and overhead (i.e. justification control and opportunity bits associated with the mapping of client signals into OPUk).

2. PSI and VCOH are for individual OPUk of OPUk-Xv, while the mapping specific overhead is for the concatenated signal.
3. PSI content is identical for each OPUk of the OPUk-Xv. It is a 256 byte multi-frame signal of which the first byte is payload type (PT) and second byte for indicating the presence of virtual concatenation (vcPT). Rest of the 254 bytes are reserved (RES).
4. vcPT values are defined as per Table 13.4.
5. VCOH is 3 bytes per OPUk. These 3 bytes are VCOH 1/2/3. These are located in rows 1, 2 and 3 of column 15.
6. Each of the three VCOH bytes, i.e. VCOH1/2/3 is a 32 byte multi-frame signal, and hence repeated 8 times over 256 multi-frame of PSI.
7. The sub-fields in VCOH1 are MF11, MF12, SQ and one byte consisting of bit fields CTRL,GID, RSA, RES.
8. VCOH2 has member status field (MST).
9. VCOH3 has 8 CRC bytes, numbered CRC-1 to CRC-8.

Table 13.4 *vcPT values and interpretation*

vcPT in HEX code	Interpretation
01	Experimental mapping
02	Asynchronous mapping of STM signals
03	Bit-synchronous mapping of STM signals
04	ATM mapping
05	GFP mapping
10	Bit-stream with octet timing mapping
11	Bit-stream without octet timing mapping
55	ODUk maintenance signals
66	ODUk maintenance signals
80–8F	Reserved for proprietary use
FD	NULL test signal mapping
FE	PRBS test signal mapping
FF	ODUk maintenance signals
Others	The rest of 228 spare codes left for future international standardisation

Let us understand the purpose of the above sub-fields of VCOH 1/2/3.

OPUk-Xv virtual concatenation multi-frame indicator (MFI1, MFI2)

The differential delay that crops up between OPUks of a concatenated group due to transportation of different optical paths is to be compensated. To do so, a two-stage multi-frame logic is followed. The first stage uses MFAS in the frame alignment overhead area for the 8-bit multi-frame indicator. MFAS is incremented every ODUk frame and counts from 0 to 255.

The second stage uses the MFI1 and MFI2 overhead bytes in the VCOH. They form a 16-bit multi-frame counter with the MSBs (most significant bits) in MFI1 and the LSBs (least significant bits) in MFI2. MFI1 is located in the zero-th byte of VCOH1 and MFI2 in first byte of VCOH1.

The multi-frame counter of the second stage counts from 0 to 65535 and is incremented at the start of each multi-frame of the first stage (MFAS = 0). Note that $2^{16} = 65536$, which is the range of count (0 to 655535) of the second stage multi-frame counter.

The resulting overall multi-frame (a combination of the first multi-frame and second multi-frame counter) is 16 777 216 ODUk frames long (range of count of first stage × range of count of second stage = 256 × 65536 = 16777216). At the start of the OPUk-Xv, the multi-frame sequence of all individual OPUks of the OPUk-Xv is identical. The realignment process compensates a differential delay of at least 125 μs.

OPUk-Xv sequence indicator (SQ)

The purpose of SQ is to identify the sequence/order in which the individual OPUks of the OPUk-Xv are combined to form the contiguous OPUk-X-PLD. It is an 8-bit sequence number (hence has values $X = 2^8 = 256$). SQ is transported in the fourth byte of VCOH1, with bit-1 being the MSB and bit-8 the LSB. The fourth byte of VCOH1 can be written as VCOH1[4].

Each OPUk of an OPUk-Xv has a fixed unique sequence number in the range of 0 to (X–1). The OPUk carrying the first time slot of the OPUk-Xv has SQ = 0; the OPUk carrying the second time slot has SQ = 1, and so on up to SQ = 255.

Q 13.6 *How is SQ assigned in case of applications requiring fixed bandwidth?*

For applications requiring fixed bandwidth, SQ is assigned once in the beginning and remains fixed (not configurable). This allows the constitution of the OPUk-Xv either to be checked without using the trace, or be transported via a number of ODUk signals which have their trail termination functions as part of an ODUk trail termination group.

OPUk-Xv LCAS control words (CTRL)

The link capacity adjustment scheme (LCAS) control word (CTRL) is located in bits 1 to 4 in fifth byte of VCOH1, i.e., VCOH1 [5]. Bit 1 of VCOH1 [5] is the MSB, bit 4 is the LSB. Readers may go through reference[2] to find out more about LCAS control commands, coding and operation.

OPUk-Xv LCAS member status field (MST)

The LCAS member status field (MST) reports the status of the individual OPUks of the OPUk-Xv. One bit is used per OPUk to report the status from destination to source. VCOH2 [0] to VCOH2 [31] are used to carry MST. Readers may go through reference [2] to learn about MST control commands, coding and operation.

> ***Q 13.7 What is the time taken to carry MST of all OPUks?***
>
> MST of all 256 OPUks is transferred in 1567 μs for $k = 1$, 390 μs for $k = 2$ and 97 μs for $k = 3$.

OPUk-Xv LCAS group identification (GID)

The LCAS group identification (GID) facilitates verification at the receiver end that all the arriving channels have originated from the same transmitter. Bit 5 of VCOH1 [5] is used for the GID.

OPUk-Xv LCAS re-sequence acknowledge (RS-Ack)

Bit 6 of VCOH1 [5] is used for 're-sequence acknowledge (RS-Ack)'. Re-sequence acknowledge is an indication from the destination to the source that a re-sequence, a sequence increase or a sequence decrease has been detected.

OPUk-Xv LCAS cyclic redundancy check (CRC)

OPUk-Xv LCAS cyclic redundancy check (CRC) is a CRC-8 check on virtual concatenation on LCAS OH. The CRC-8 is calculated over VCOH1 and VCOH2 on a frame-by-frame basis and then inserted into VCOH3. The CRC-8 polynomial is $x^8 + x^2 + x + 1$.

OPUk-Xv VCOH reserved overhead

The reserved VCOH is set to all 0s.

OPUk mapping specific overhead

These bytes are located in columns $15X + 1$ to $16X$, i.e., X times four bytes in the OPUk overhead for mapping specific overhead (Fig. 13.11).

13.8 Mapping of Client Signals into OPUk-Xv

13.8.1 Mapping of STM64/256 into OPUk-4v

Mapping of STM-64/256 into OPUk-4v is performed in two modes: asynchronous and bit-synchronous. The generic frame OPUk-4v used in this mapping is shown in Fig. 13.12.

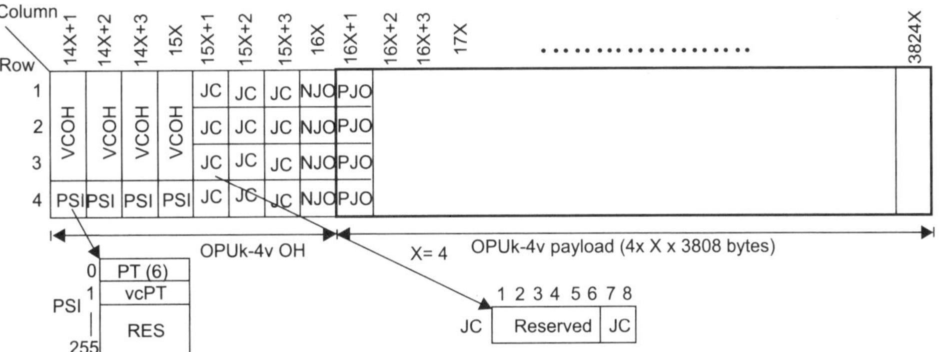

Fig. 13.12 *Generic frame OPUk-4v for mapping STM-64/256*

The mapping process is described below.

1. The OPUk-4v overhead consists of:
   a. 4 times PSI (which has PT, vcPT and 254 RES bytes).
   b. 4 times VCOH.
   c. Three justification control bytes (JC).
   d. One NJO (negative justification opportunity) byte.
2. OPUk-4v payload has 4 × 3808 bytes, which includes one positive justification opportunity (PJO) byte per row.
3. Justification control signals are bits 7 and 8 of JC bytes located in columns 61, 62 and 63 ($15X + 1$, $15X + 2$, $15X + 3$ with $X = 4$) and are used to control the justification opportunity bytes (NJO, PJO) located in columns 64, 65 ($16X + 1$, $16X + 2$ with $X = 4$).
4. Usage of JC, NJO, PJO is governed by the following rules:
   a. JC, NJO, PJO are generated for asynchronous and bit-synchronous mapping in accordance with Tables 13.1 and 13.2 respectively.
   b. De-mapping is done in accordance with Table 13.3.

c. Majority vote is followed in interpreting the three JC signals.

d. NJO and PJO have all-0s when used as justification bytes and are to be ignored.

5. If the incoming STM signal fails, generic-AIS signal is mapped into OPUk-4v.

6. If the incoming OPUk-4v fails (i.e., ODUk-AIS, ODUk-LCK, ODUk-OCI condition), the STM signal supposed to be received is lost. In such situations, a generic AIS pattern is generated as replacement for the lost STM signal.

7. For asynchronous mapping, OPUk-4v is created from a locally generated clock, which is independent of STM client signal. Mapping is by positive/negative/zero justification.

8. In case of bit-synchronous mapping, the OPUk-4v clock is derived from the STM client signal. The mapping does not use justification (PJO contains data bytes and NJO has justification bytes with justification bits, i.e., JC set to 00).

9. STM-64 is mapped into OPU1-4v as follows:

a. Set of 8 successive bits of STM-64 (not necessarily a byte) are mapped into D bytes of OPU1-4v as shown in Fig. 13.13 (a).

b. There is scope for positive or negative justification once per OPU1-4v row (i.e., 4 times per OPU1-4v frame).

	14X+1	14X+2	14X+3	15X	15X+1	15X+2	15X+3	16X	16X+1	16X+2 16X+3 17X ... X = 4 ... 3824X
1					JC	JC	JC	NJO	PJO	4 x 3808D – 1
2					JC	JC	JC	NJO	PJO	4 x 3808D – 1
3					JC	JC	JC	NJO	PJO	4 x 3808D – 1
4	PSI	PSI	PSI	PSI	JC	JC	JC	NJO	PJO	4 x 3808D – 1

(Rows 1–3 of columns 14X+1 to 15X: VCOH, VCOH, VCOH, VCOH)

Fig. 13.13(a) *Mapping STM-64 signal into OPU1-4v*

10. STM-256 is mapped into OPU2-4v in the following manner

a. Sets of 8 successive bits of STM-256 signal (not necessarily same byte) are mapped into D bytes of OPU2-4v as shown in Fig. 13.13(b).

b. 4 × 64 FS bytes are added in columns 7617, 7618 (1904X + 1 to 1904X + 2).

c. There is scope to perform positive or negative justification once per OPU2-4v row (i.e., 4 times per OPU2-4v frame).

Row	14X+1	14X+2	14X+3	15X	15X+1	15X+2	15X+3	16X	16X+1	16X+3 16X+2 17X …………	1904X 1904X+1 ….. 1920X	1920X+1 ……… 3824X
1	VCOH	VCOH	VCOH	VCOH	JC	JC	JC	NJO	PJO	4 x 118 x 16D –1	4 x 16FS	4 x 119 x 16D
2	VCOH	VCOH	VCOH	VCOH	JC	JC	JC	NJO	PJO	4 x 118 x 16D –1	4 x 16FS	4 x 119 x 16D
3	VCOH	VCOH	VCOH	VCOH	JC	JC	JC	NJO	PJO	4 x 118 x 16D –1	4 x 16FS	4 x 119 x 16D
4	PSI	PSI	PSI	PSI	JC	JC	JC	NJO	PJO	4 x 118 x 16D –1	4 x 16FS	4 x 119 x 16D

Fig. 13.13(b) *Mapping STM-64 signal into OPU2-4v*

13.8.2 Mapping of STM-256 signal into OPUk-16v

Mapping of STM-256 into OPUk-16v is performed in two modes: asynchronous and bit-synchronous. The generic frame OPUk-16v used in this mapping is shown in Fig. 13.14.

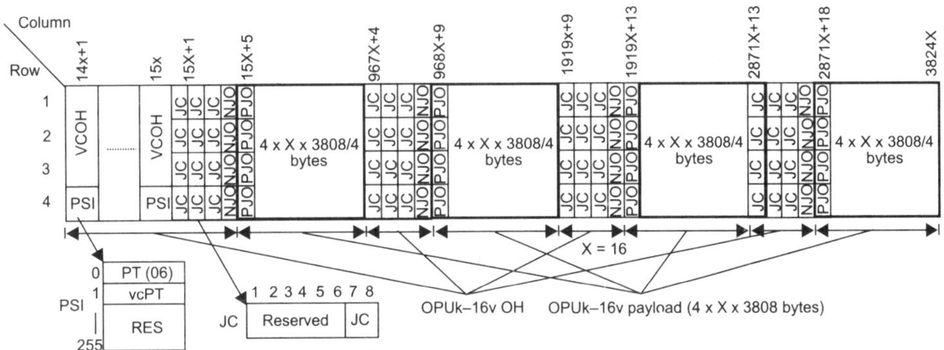

Fig. 13.14 *Generic frame OPUk-16v for mapping STM-256*

The mapping process is described below:

1. OPUk-16v overhead consists of:
   a. 16 times PSI (which has PT, vcPT and 254 RES bytes).
   b. 16 times VCOH.
   c. 4×3 justification control bytes (JC).
   d. 4×1 NJO (negative justification opportunity) byte.
2. The OPUk-16v payload has 4 blocks of 4×15232 bytes (please note $15232 = 4 \times 3808$), including 4×1 positive justification opportunity (PJO) bytes per row.

3. Justification control signals are bits 7 and 8 of JC bytes located in columns 241, 242, 243 ($15X + 1$, $15X + 2$, $15X + 3$ with $X = 16$), used to control the justification opportunity bytes (NJO, PJO) located in columns 244, 245 ($16X + 1$, $16X + 2$ with $X = 16$).
4. Usage of JC, NJO, PJO is governed by the following rules:
   a. JC, NJO, PJO are generated for asynchronous and bit-synchronous mapping in accordance with Tables 13.1 and 13.2 respectively.
   b. De-mapping is done in accordance with Table 13.3.
   c. Majority vote is followed in interpreting the three JC signals.
   d. NJO, PJO have all-0s when used as justification bytes and are to be ignored.
5. If the incoming STM signal fails, a generic AIS signal is mapped into OPUk-16v.
6. If the incoming OPUk-4v fails (i.e., ODUk-AIS, ODUk-LCK and ODUk-OCI condition), the STM signal supposed to be received is lost. In such a situation, a generic AIS pattern is generated as replacement for the lost STM signal.
7. For asynchronous mapping, OPUk-16v is created from a locally generated clock, which is independent of the STM client signal. Mapping is by positive/negative/zero justification.
8. In case of bit-synchronous mapping, OPUk-16v clock is derived from STM client signal.. Mapping does not use justification (PJO contains data bytes and the NJO justification bytes containing justification bits JC set to 00).
9. STM-256 is mapped into OPU1-16v as follows:
   a. Sets of 8 successive bits of STM-256 (not necessarily a byte) are mapped into the D bytes of OPU1-16v as shown in Fig. 13.15.
   b. There is scope for positive or negative justification four times per OPU1-16v row (i.e., 16 times per OPU1-16v frame).

Column number

Row number	14X+1	.....	15X	15X+1				15X+5	X=16	967X+4				968X+9		1919X+9				1919X+13		2871X+13				2871X+18	3824X
1	VCOH	.....	VCOH	JC	JC	JC	NJO	PJO	15231D	JC	JC	JC	NJO	PJO	15231D	JC	JC	JC	NJO	PJO	15231D	JC	JC	JC	NJO	PJO	15231D
2	VCOH	.....	VCOH	JC	JC	JC	NJO	PJO	15231D	JC	JC	JC	NJO	PJO	15231D	JC	JC	JC	NJO	PJO	15231D	JC	JC	JC	NJO	PJO	15231D
3	VCOH	.....	VCOH	JC	JC	JC	NJO	PJO	15231D	JC	JC	JC	NJO	PJO	15231D	JC	JC	JC	NJO	PJO	15231D	JC	JC	JC	NJO	PJO	15231D
4	PSI	.....	PSI	JC	JC	JC	NJO	PJO	15231D	JC	JC	JC	NJO	PJO	15231D	JC	JC	JC	NJO	PJO	15231D	JC	JC	JC	NJO	PJO	15231D

Fig. 13.15 *Mapping STM-256 signal into OPU1-16v*

13.8.3 Mapping of ATM cells into OPUk-Xv

The OPUk-Xv overhead for the ATM mapping consists of:

1. X times PSI (which has PT, vcPT and 254 RES bytes).
2. X times 3 VCOH bytes.
3. X times 4 bytes reserved for future international standardisation.

It may be recalled from the description OUT-OH in section 12.4.4 that the value of PT = 04 is for ATM cell mapping. A constant bit rate ATM cell stream, when mapped into OPUk-Xv payload area, creates an ATM virtual path (ATM VP) signal. Rate adaptation is accomplished in this process either by inserting idle cells or by discarding cells.

Figure 13.16 shows the ATM cell stream mapped into the OPUk payload area with byte alignment between the two. The OPUk-Xv payload area is 3808 × 4X = 15232X bytes; whereas the ATM cell size is 53 octets (bytes), i.e. OPUk-Xv payload area is not an integer multiple of ATM cell size. So, ATM cell may cross the OPUk-Xv frame boundary.

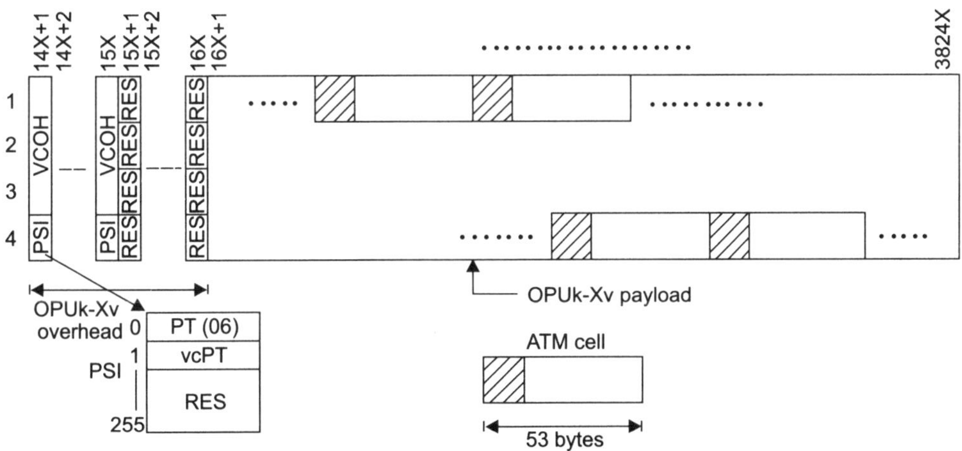

Fig. 13.16 *Mapping ATM cells into OPUk-Xv*

Each ATM cell has 53 octets, of which 48 are information bytes and 5 are header bytes. The ATM cell header contains a header error control (HEC) field, which may be used in a similar way to a frame alignment word to achieve cell delineation. It may be noted that frame alignment signal is not a constant but varies from cell to cell as explained below.

In HEC method, uses the correlation between the header bits to be protected by the HEC (32 bits) and the control bit of the HEC (8 bits) introduced in the header after computation with a shortened cyclic code with generating polynomial

$g(x) = x^8 + x^2 + x + 1$. The remainder from this polynomial is then added to the fixed pattern '01010101' in order to improve the cell delineation performance. This method is similar to conventional frame alignment recovery procedure where the alignment signal is not fixed but varies from cell to cell.

The ATM cell information field (48 bytes) is scrambled before mapping into the OPUk. A self-synchronising scrambler with generator polynomial $x^{43} + 1$ is used (as per ITU-T Rec. I.432.1). The scrambler operates for the duration of the cell information field. During the 5-byte header, the scrambler operation is suspended and the scrambler state is retained. At the receiving end, after termination of the OPUk-Xv signal, the ATM cell information field is descrambled before being passed on to the ATM layer.

Cell information field scrambling is required to provide security against false cell delineation and against cell information field replicating the OUTk and ODUk frame alignment signal.

13.8.4 Mapping of GFP frames into OPUk-Xv

The OPUk-Xv payload for the GFP mapping consists of $4X \times 3808$ bytes. The mapping of generic framing procedure (GFP) frames is performed by aligning the byte structure of every GFP frame with the byte structure of the OPUk payload. Refer to Fig. 13.17. Since the GFP frames are of variable length (4 bytes to 65535 bytes), a frame may cross the OPUk frame boundary. When there is no data to be sent, an idle GFP frame (4 bytes) is inserted. Thus, despite discrete data flow and variable frame length, continuity is maintained at OPUk-Xv. The GFP frame stream is scrambled during encapsulation to avoid replication of OTUk and ODUk FAS.

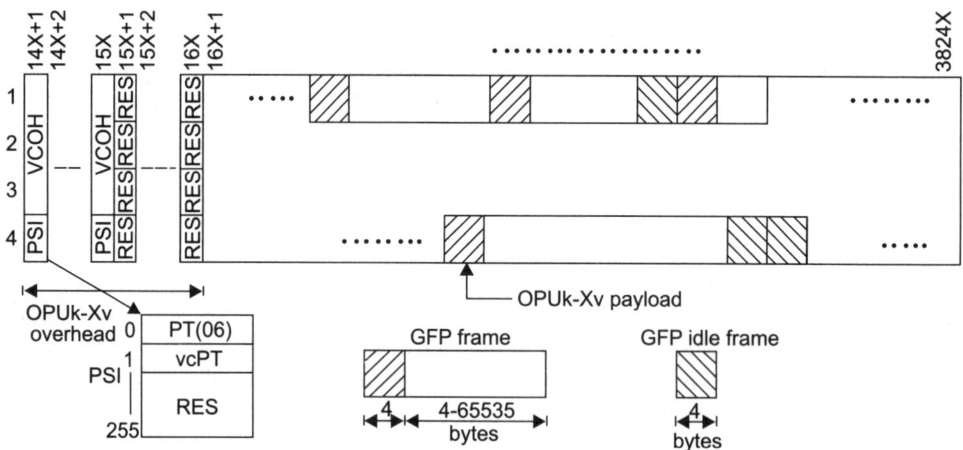

Fig. 13.17 *GFP mapping into OPUk-Xv*

The OPUk-Xv overhead for the GFP mapping consists of:

1. *X* times PSI (which has PT, vcPT and 254 RES bytes).
2. *X* times 3 VCOH bytes.
3. *X* times 4 bytes reserved for future international standardisation.

The value of PT = 05 for GFP mapping, as mentioned in section 12.4.4.

13.8.5 Mapping of NULL test signal into OPUk-Xv

OPUk-Xv payload area for NULL client signal is 4X × 3808 bytes in which an all-0s pattern is mapped. The OPUk-Xv overhead for the NULL mapping consists of:

1. *X* times PSI (which has PT, vcPT and 254 RES bytes).
2. *X* times 3 VCOH bytes.
3. *X* times 4 bytes reserved for future international standardisation.

It may please be recalled that value of PT = FD for mapping NULL test signal. Mapping of NULL test signal into OPUk-Xv is shown in Fig. 13.18.

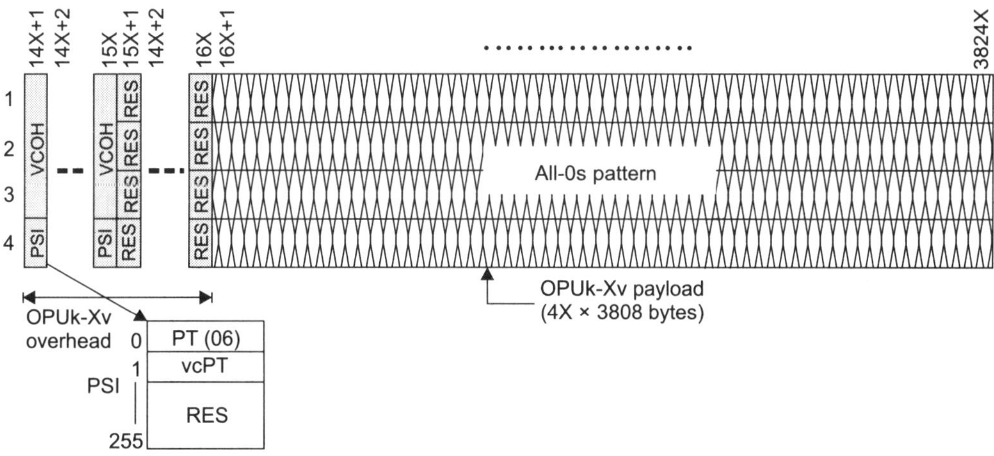

Fig. 13.18 Mapping of NULL test signal into OPUk-Xv

13.8.6 Mapping of PRBS test signal into OPUk-Xv

The PRBS test sequence for OPUk-Xv is ($2^{31} - 1$), i.e., a 2147483647-bit sequence. Sets of 8-bits of this sequence are mapped into D bytes of OPUk-Xv as shown in Fig. 13.19. The OPUk-Xv overhead for the PRBS mapping consists of:

1. X times PSI (which has PT, vcPT and 254 RES bytes).
2. X times 3 VCOH bytes.
3. X times 4 bytes reserved for future international standardisation.

It may be recalled that the value of PT = FE for mapping PRBS test signal.

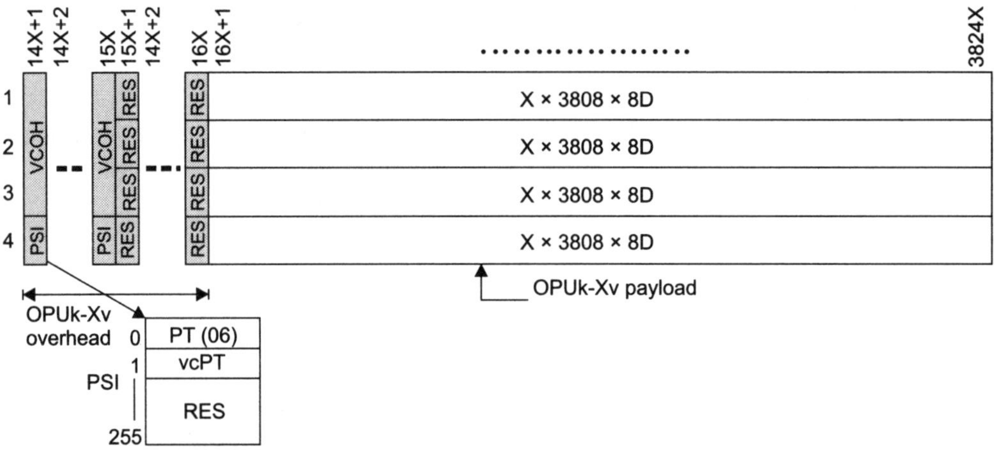

Fig. 13.19 Mapping of PRBS test sequence into OPUk-Xv

13.8.7 Mapping non-specific client signals into OPUk-Xv

Client signals other than the specific client signals (STM, ATM, GFP, test signals), which are known as non-specific client signals, can be mapped into OPUk-Xv by bit-synchronisation method. Justification bytes can be added as per requirement. Please refer Fig. 13.20.

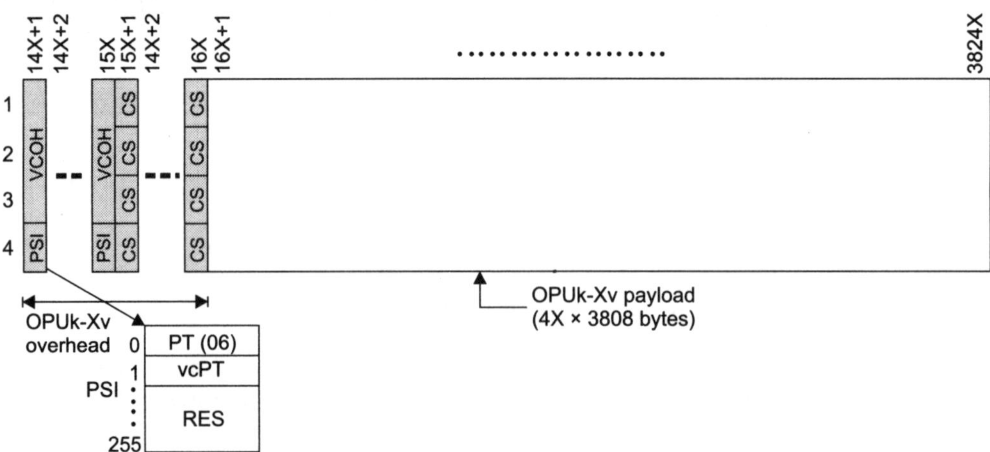

Fig. 13.20 Mapping non-specific client signals into OPUk-Xv

The OPUk-Xv overhead for customer-specific signal mapping (other than STM, ATM, GFP, etc.) consists of:

1. *X* times PSI (which has PT, vcPT and 254 RES bytes).
2. *X* times 3 VCOH bytes.
3. *X* times 4 bytes used for client specific purpose (marked CS in Fig. 13.20).

The payload consists of 4X × 3808 bytes.

13.9 Mapping of ODUk Signals into ODTUGjk

ODTUGjk stands for optical data tributary unit group of order *j*,*k*. This is similar to TUG in SDH. At the macro-level, OPUk has 4 × 3808 bytes. Let us examine how OPUk is organised as tributary-units. OPUk is, in fact, divided into number of *tributary slots* (*TS*) which are interleaved within OPUk. Each such tributary slot has a part in OPUk-OH also, meant for justification. The tributary slot, together with the associated justification (JOH), form the *tributary unit*.

So, at the tributary-unit level, bytes of ODUj frame are mapped into OPUk payload area of tributary slots to form ODTUjk. The associated justification overhead bytes are mapped into the OPUk-OH. Thus, OUTUjk is formed. Let us examine TS allocation in case of OPU2 and OPU3, and also see the associated TU structures.

13.9.1 Optical payload unit tributary slot allocation

OPU2 tributary slot allocation

Figure 13.21 shows OPU2 tributary slot (TS) or tributary unit allocation. One single OPU2 tributary slot occupies one-fourth the OPU2 payload area. An OPU2 TS has 4 rows × 952 columns. It may be noted that:

1. The justification overhead is located in column 16 of rows 1 to 4. It consists of JC and NJO signals of the four OPU2 TSs.
2. Justification overhead is assigned to TSs on a per-frame basis. It is available for a TS once every four frames.
3. A 4-frame multi-frame structure is used for justification overhead assignment. This multi-frame is locked to bits 7 and 8 of the MFAS byte as shown in Table 13.5.

Table 13.5 *JOH TS and MFAS bits correlation*

MFAS bits 7 and 8	Justification overhead for TS
00	1
01	2
10	3
11	4

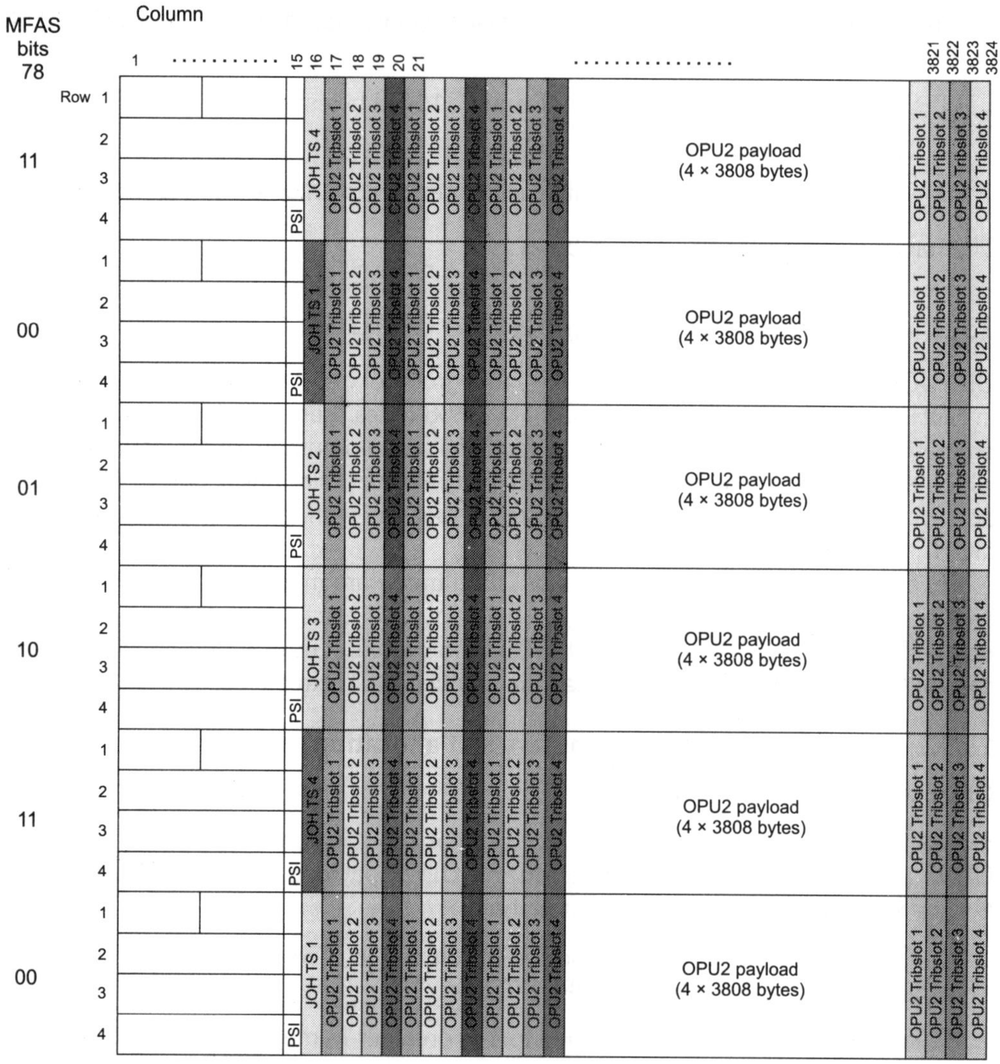

Fig. 13.21 *OPU2 tributary-slot allocation*

OPU3 Tributary slot allocation

Figure 13.22 shows OPU3 TS allocation. An OPU3 tributary slot occupies one-sixteenth of the OPU3 payload area, i.e., there are 16 tributary-units or slots in OPU3. Each OPU3 TS has 4 rows × 238 columns. The 16 OPU3 TSs are byte-interleaved in the OPU3 payload area. It may be noted that:

1. Justification overhead (JOH) of the 16 OPU3 TSs are located in column 16 of rows 1 to 4. JOH for each OPU3 TS consists of JC and NJO signals.

2. Justification overhead is assigned to TSs on a per-frame basis. It is available for a TS once every 16 frames.
3. A 16-frame multi-frame structure is used for justification overhead assignment. This multi-frame is locked to bits 5, 6, 7 and 8 of the MFAS byte as shown in Table 13.6.

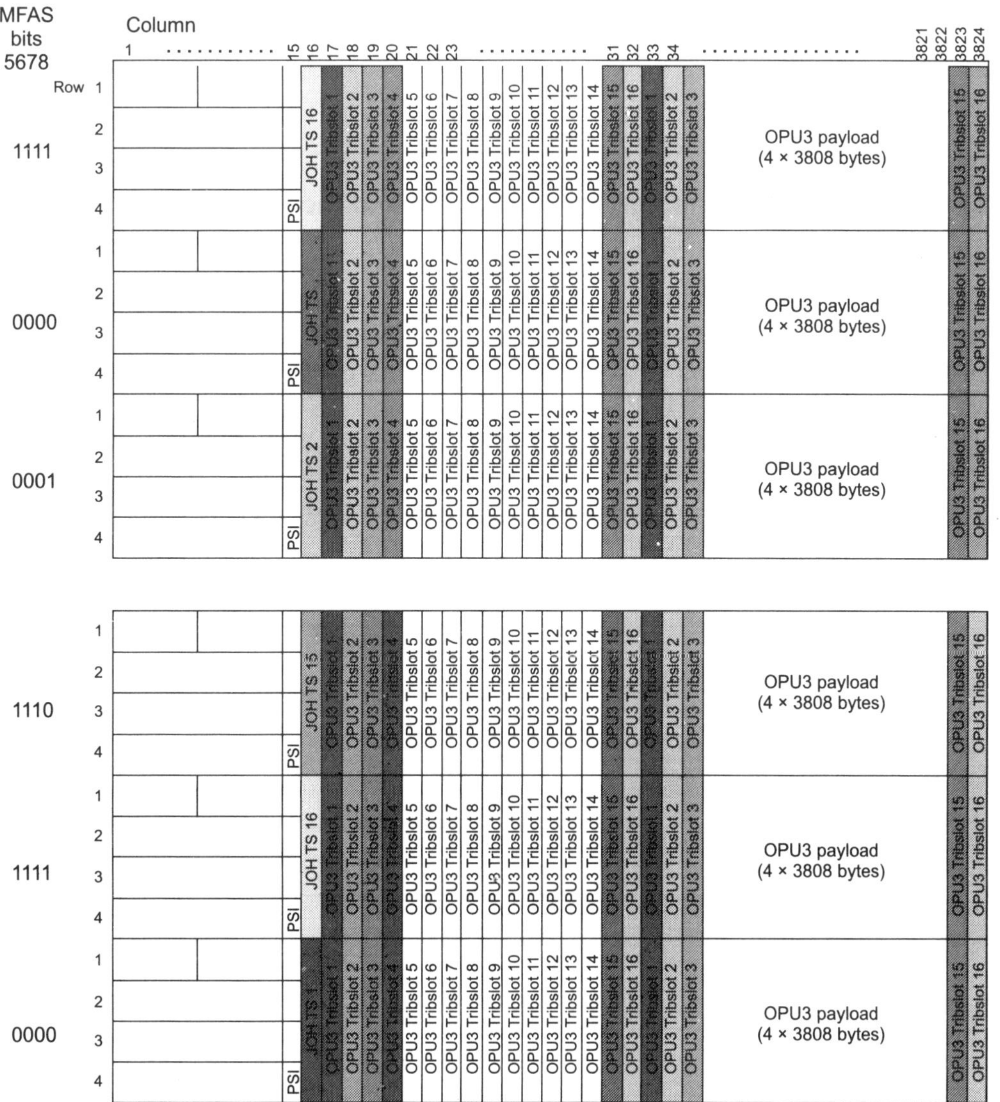

Fig. 13.22 *OPU3 Tributary slot allocation*

Table 13.6 *JOH TS and MFAS bits correlation*

MFAS bits 5, 6, 7, 8	JOH TS	MFAS bits 5, 6, 7, 8	JOH TS
0000	1	1000	9
0001	2	1001	10
0010	3	1010	11
0011	4	1011	12
0100	5	1100	13
0101	6	1101	14
0110	7	1110	15
0111	8	1111	16

13.9.2 Optical data tributary units (ODTUjk) and their mapping

ODTU12

The optical channel data tributary unit 12 (ODTU12) is a structure with 952 columns by 16 (4 × 4) rows plus 1 column of justification overhead (JOH). It carries a justified ODU1 signal. The location of the JOH column depends on the OPU2 tributary slot used when multiplexing the ODTU12 in the OPU2.

The ODTU12 structure is illustrated in Fig. 13.23(a). It also shows the mapping of ODU1 into ODTU12. The ODTU12 is mapped into OPU2 in one of the four OPU2-TS as shown in Fig. 13.23(b). ODTU12 JOH bytes are mapped into OPU2 OH bytes allotted to the OPU2 TS of concern.

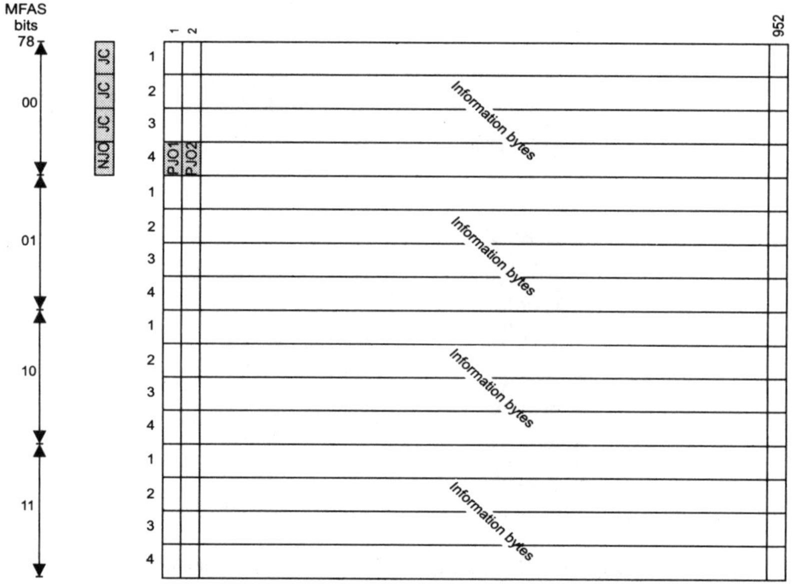

Fig. 13.23(a) *ODTU1 frame format and mapping of ODU1*

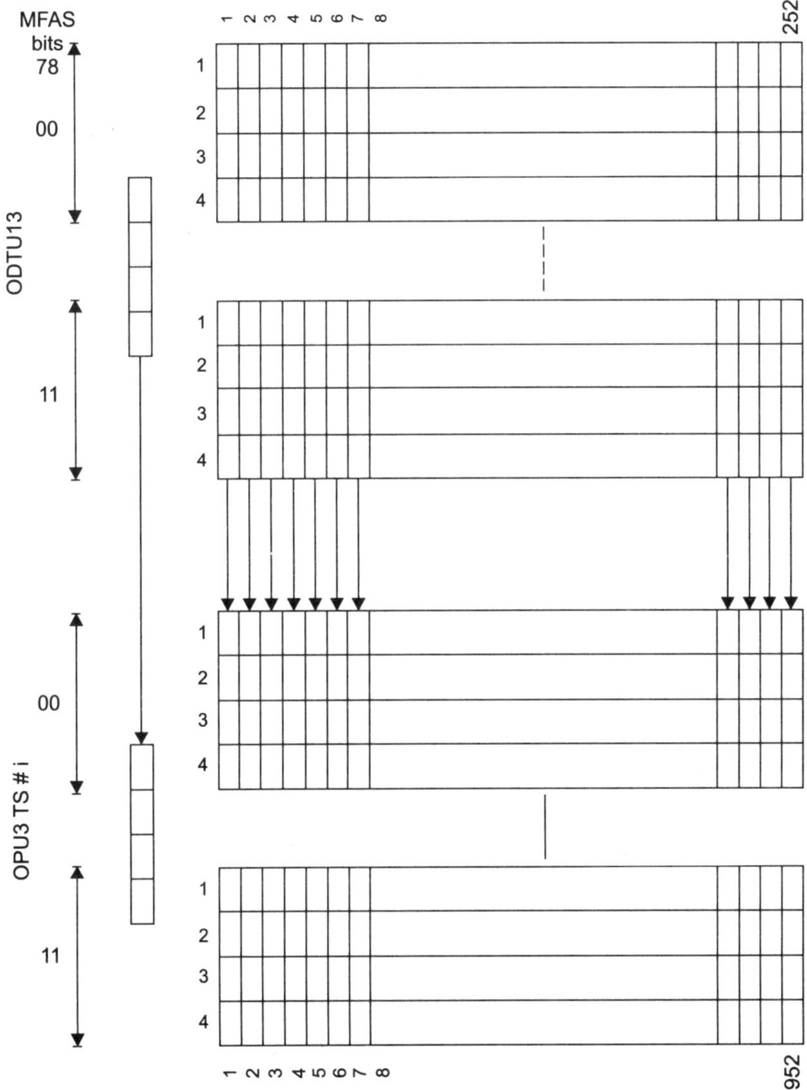

Fig. 13.23(b) *Mapping of ODTU12 into OPU2-TS*

ODTU13

The optical channel data tributary unit 13 (ODTU13) is a structure with 238 columns by 64 (16 × 4) rows plus 1 column of justification overhead (JOH). It carries a justified ODU1 signal. The location of the JOH column depends on the OPU3 tributary slot used when multiplexing the ODTU13 in the OPU3.

The ODTU13 structure is shown in Fig. 13.24(a), which also shows the mapping of ODU1 into ODTU13. The ODTU13 is mapped into OPU3 in one of the 16 OPU3-TS as shown in Fig. 13.24(b). ODTU12 JOH bytes are mapped into OPU3 OH bytes allotted to the OPU3 TS of concern.

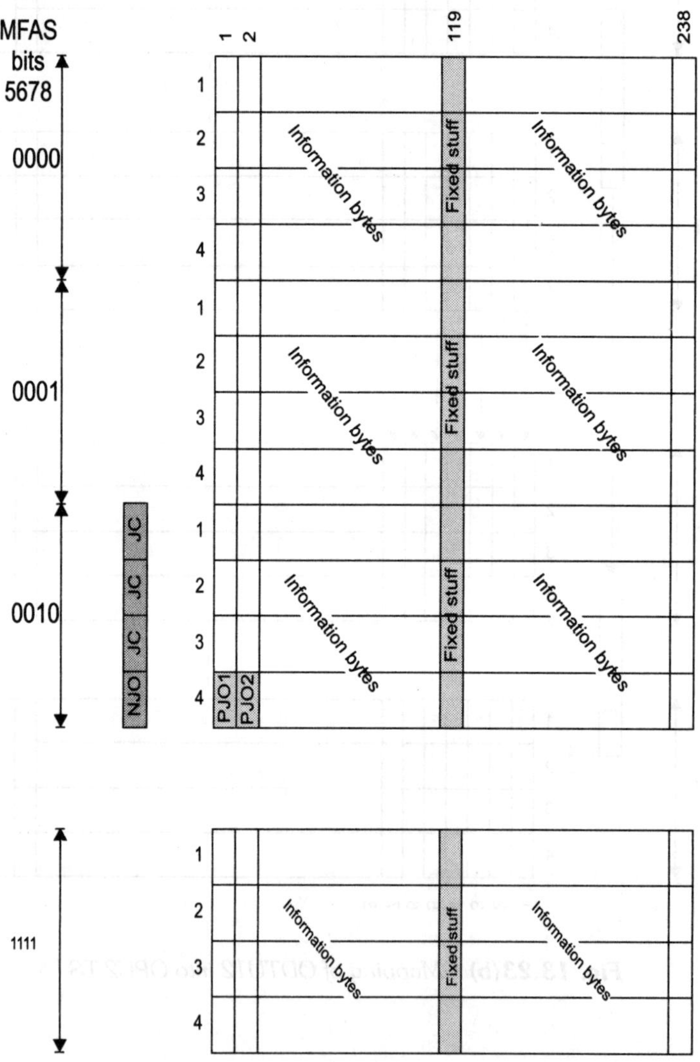

Fig. 13.24(a) *ODTU13 frame format and mapping of ODU1*

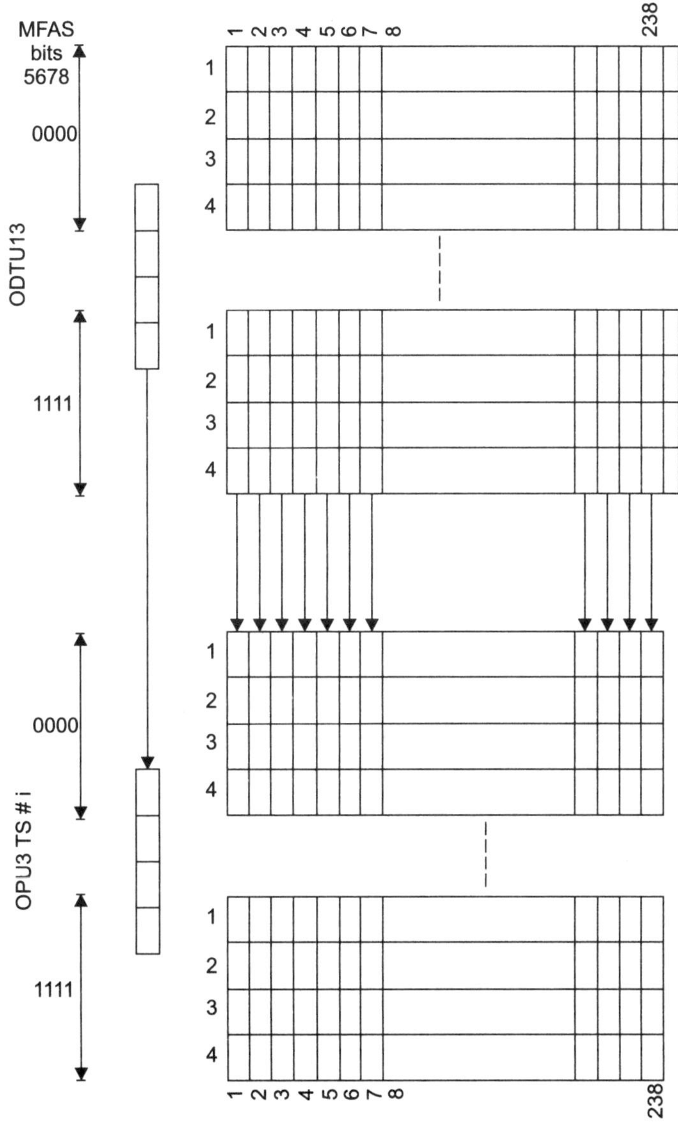

Fig. 13.24(b) *Mapping of ODTU13 into OPU3 TS*

ODTU23

The optical channel data tributary unit 23 (ODTU23) is a structure with 952 columns by 64 (16 × 4) rows plus 4 times 1 column of justification overhead (JOH). It carries a justified ODU2 signal. The location of the JOH column depends on the OPU3 tributary slot used when multiplexing the ODTU23 in the OPU3.They may not be equally distributed.

The ODTU23 structure is shown in Fig. 13.25(a), which also shows the mapping of ODU2 into ODTU23. The ODTU23 is mapped into OPU3 in 4 of the 16 OPU3 TS as shown in Fig. 13.25(b).

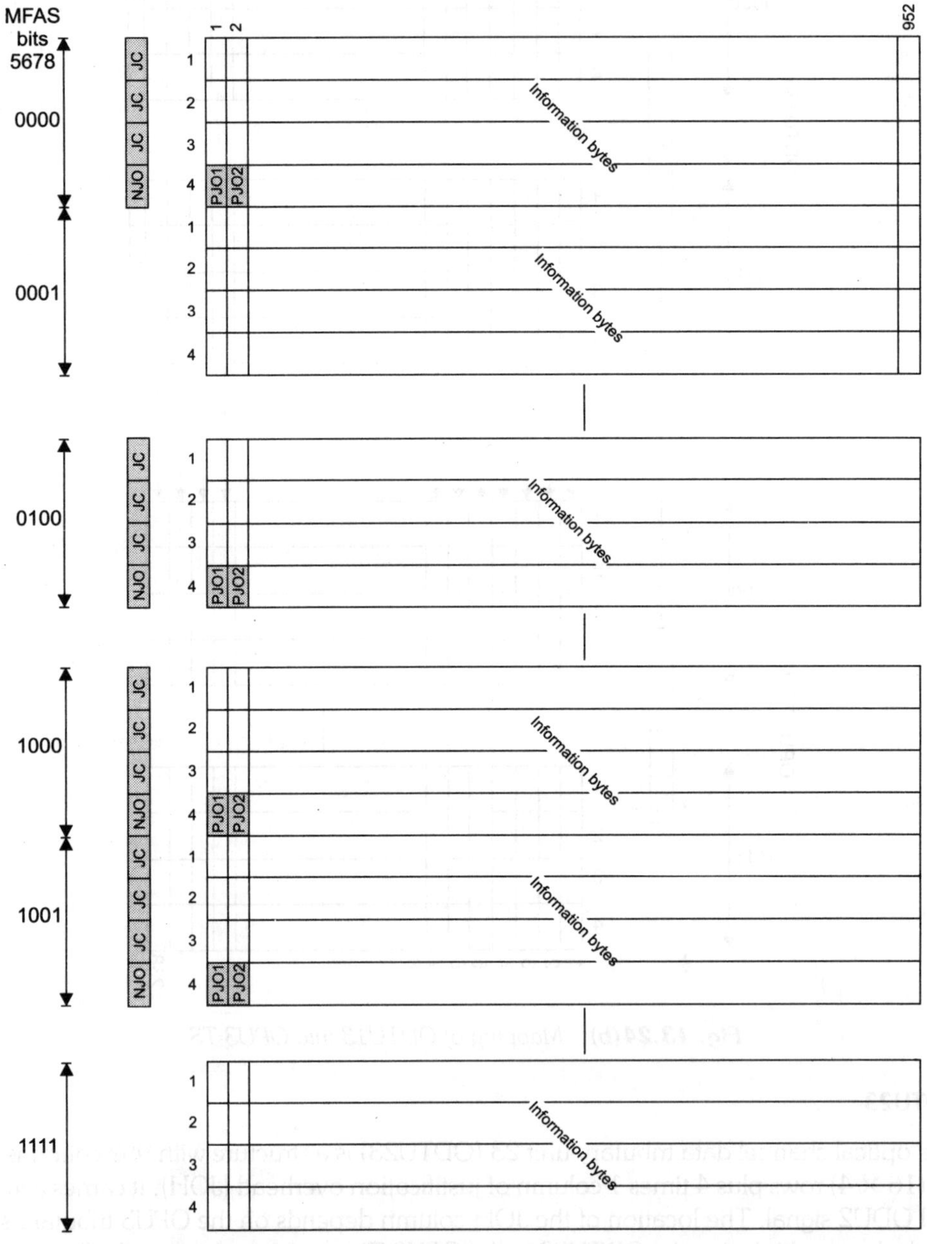

Fig. 13.25(a) *ODTU23 Frame format and mapping of ODU2 into ODTU23*

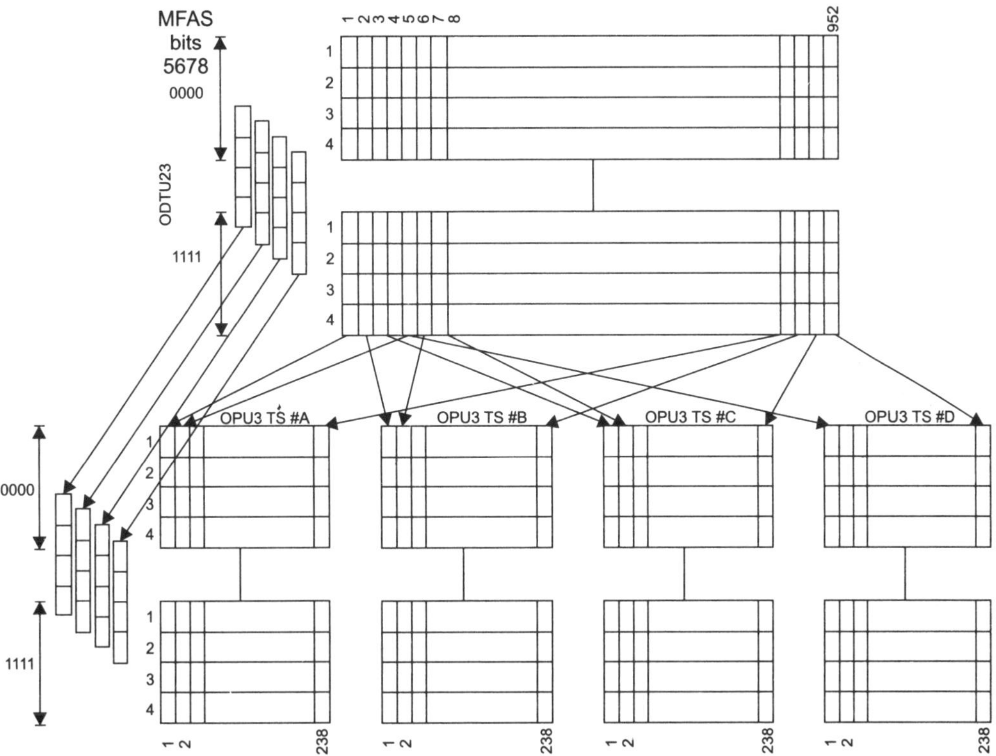

Fig. 13.25(b) *Mapping of ODTU23 into OPU3 in 4 of the 16 OPU3 TS*

Q 13.8 What are the multiplex overheads in the mapping of ODU1 to ODTU and how are they mapped?

The OPUk multiplex overhead consists of

1. Payload structure identifier (PSI), which is a 256-byte multi-frame signal with 16-byte long multiplex structure identifier (MSI) located in the mapping specific area (byte no. 2 to 17) in PSI.
2. Justification control (JC) bytes.
3. NJO, PJO1 and PJO2.

Figure 13.26 shows the location of these bytes. Let us examine the sub-fields of OPUk multiplex overhead.

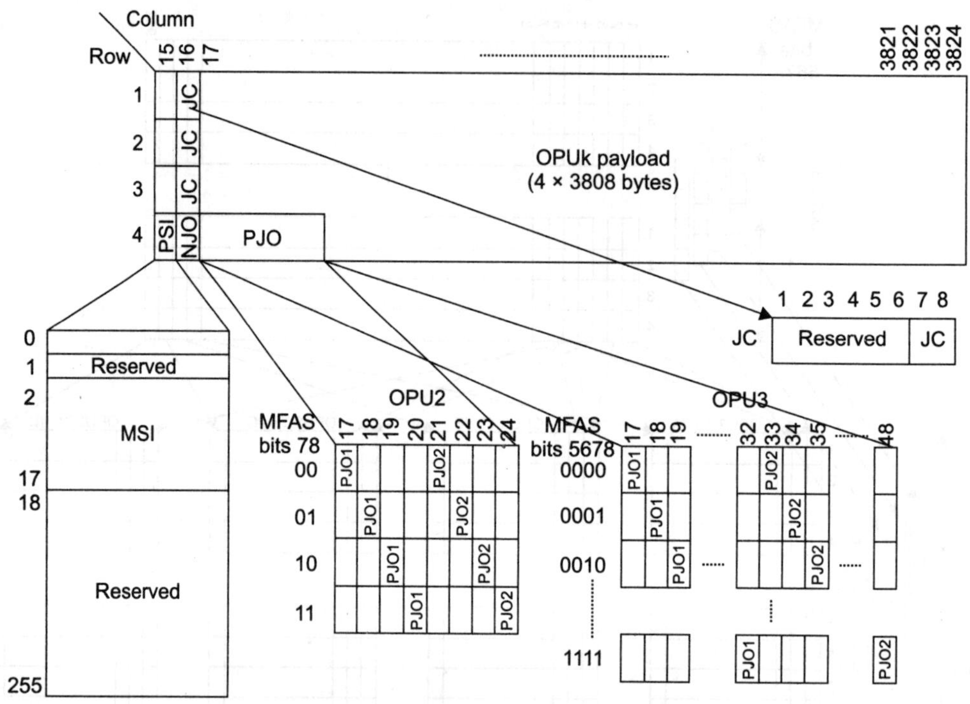

***Fig. 13.26** OPUk Multiplex overhead*

OPUk multiplex structure identifier (MSI)

As already mentioned, MSI is a 16 byte signal carried in PSI. Its framing rules and utilities are listed below.

1. MSI located in the mapping specific area of PSI (byte no. 2 to 17) to indicate TS of OPU.
2. The number of TS bytes in MSI, depending on the OPUk being handled. For OPU2, there are 4 TS bytes in MSI. For OPU3, there are 16 TS bytes in MSI. The other bytes of MSI which do not carry TS indication are all set to 0.
3. General format of TS byte of MSI is shown in Fig. 13.27. This byte has 2 bits (bits 1 and 2) to indicate ODU type and 6 bits (bits 3 to 8) to indicate the tributary port number. The tributary port number field facilitates flexible assignment of ODUs to tributary slots (TS).

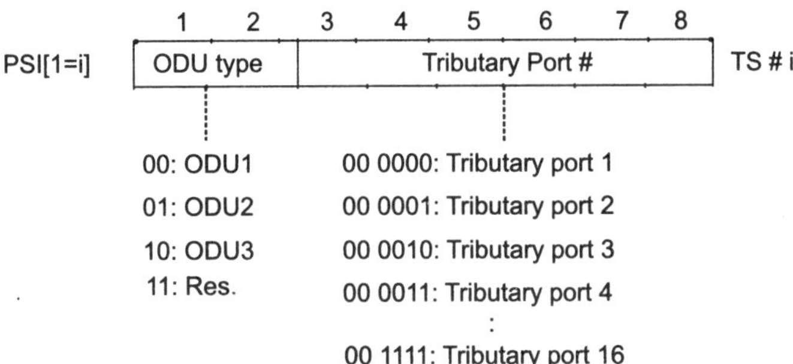

Fig. 13.27 *General format of TS byte of MSI*

4. TS bytes for OPU2 tributary slots are shown in Fig. 13.28(a).

	1 2	3 4 5 6 7 8	
PSI[2]	00	00 0000	TS1
PSI[3]	00	00 0001	TS2
PSI[4]	00	00 0010	TS3
PSI[5]	00	00 0011	TS4

Fig. 13.28(a) *TS bytes of OPU2-MSI*

5. TS bytes for OPU3 are shown in Fig. 13.28(b).

	1 2	3 4 5 6 7 8	
PSI[2]	ODU type	Tributary port #	TS1
PSI[3]	ODU type	Tributary port #	TS2
PSI[4]	ODU type	Tributary port #	TS3
PSI[5]	ODU type	Tributary port #	TS4
PSI[6]	ODU type	Tributary port #	TS5
PSI[7]	ODU type	Tributary port #	TS6
PSI[8]	ODU type	Tributary port #	TS7
PSI[9]	ODU type	Tributary port #	TS8
PSI[10]	ODU type	Tributary port #	TS9
PSI[11]	ODU type	Tributary port #	TS10
PSI[12]	ODU type	Tributary port #	TS11
PSI[13]	ODU type	Tributary port #	Ts12
PSI[14]	ODU type	Tributary port #	TS13
PSI[15]	ODU type	Tributary port #	TS14
PSI[16]	ODU type	Tributary port #	TS15
PSI[17]	ODU type	Tributary port #	TS16

Fig. 13.28(b) *TS bytes of OPU3- MSI*

RES field in OPUk-PSI

239 bytes (bytes 18 to 255) of the PSI of OPUk are reserved for future international standardisation. These are all set to 0s.

Multiplex justification overhead (JOH)

Refer to Fig. 13.26 again. The JOH bytes are located in column 16 of OPUk and consist of

1. Three JC bytes, one in each of the rows 1, 2, 3.
2. One NJO byte in row 4.

Bits 7 and 8 of each JC byte are used for justification control. The other six bits are reserved for future international standardisation.

Mapping ODUj into ODTUjk

The mapping process is described below.

1. ODUj signal mapping into ODTUjk is in asynchronous mode.
2. ODUj is attached with FAS and fixed stuff bits of OTUj overhead as shown in Fig. 13.29.

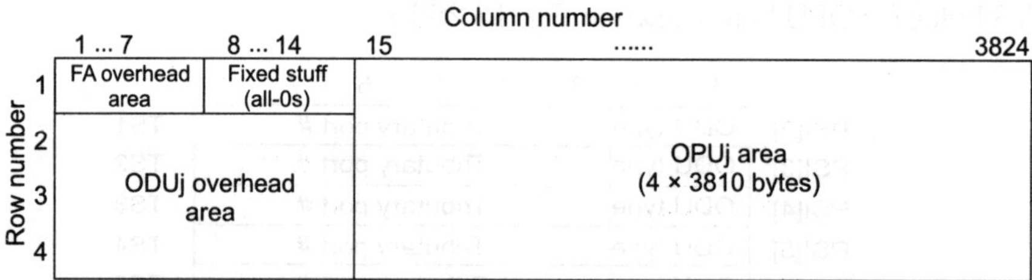

Fig. 13.29 *Extension of ODUj with FAS, fixed stuff of OTUj overhead*

3. OPUk for carrying ODUj structure is created from locally generated clock and is independent of ODUj.
4. Extended ODUj is mapped into locally generated ODUk by asynchronous mapping with –1/0/+1/+2 positive/zero/negative justification.
5. Then, the ODUj bytes are mapped into ODTUjk bytes.

The generation and interpretation of NJO, PJO1 and PJO2 is shown in Table 13.7.

Table 13.7 *Generation and interpretation of NJO, PJO1 and PJO2*

JC 7 & 8	NJO	PJ01	PJ02	Interpretation
00	Justification byte	Data byte	Data byte	No justification (0)
01	Data byte	Data byte	Data byte	Negative justification (–1)
10	Justification byte	Justification byte	Justification byte	Double positive justification (+2)
11	Justification byte	Justification byte	Data byte	Positive justification (+1)

6. NJO, PJO1 and PJO2 have all-0s when used as justification bytes. In such a case, the receiver ignores them.
7. If the incoming ODUj fails, ODUj-AIS is mapped into ODTUjk.
8. If the incoming signal contains ODUj-OCI (open connection indication), ODUj-OCI is mapped into ODTUjk. The point to note is that OPUk and ODTUjk (k = 2, 3) are created from a locally generated clock, which is independent of ODUj (j = 1, 2) signal.
9. ODUj (j = 1, 2) is mapped into ODTUjk (k = 2, 3) by using –1/0/+1/+2 positive/zero/negative justification.
10. De-mapping of ODUj signals from ODTUjk signal (j = 1, 2; k = 2, 3) is done by using extended ODUj signal from OPUk under control of JOH (JC, NJO, PJO1 and PJO2)
11. During failure condition (ODUk-AIS, ODUk-LCK, ODUk-OCI), an ODUj-AIS pattern is generated as replacement for the lost ODUj signal.

Mapping of ODU1 into ODTU12 and ODTU13 are shown in Fig. 13.23(a) and 13.24(a) respectively, along with ODTU frame formats. Mapping of ODU2 into ODTU23 is shown in Fig. 13.25(a) along with ODTU23 frame format.

Q 13.9 *Can we trace the steps in multiplexing four ODU1s into ODU2?*

Multiplexing of four ODU1s into ODU2 is shown in Fig. 13.30. The ODU1 signals including FAS are adapted to ODU2 clock via justification. These adpted ODU1 signals are byte interleaved into OPU2 payload area and their JC and NJO bytes are frame-interleaved into OPU2 overhead area.

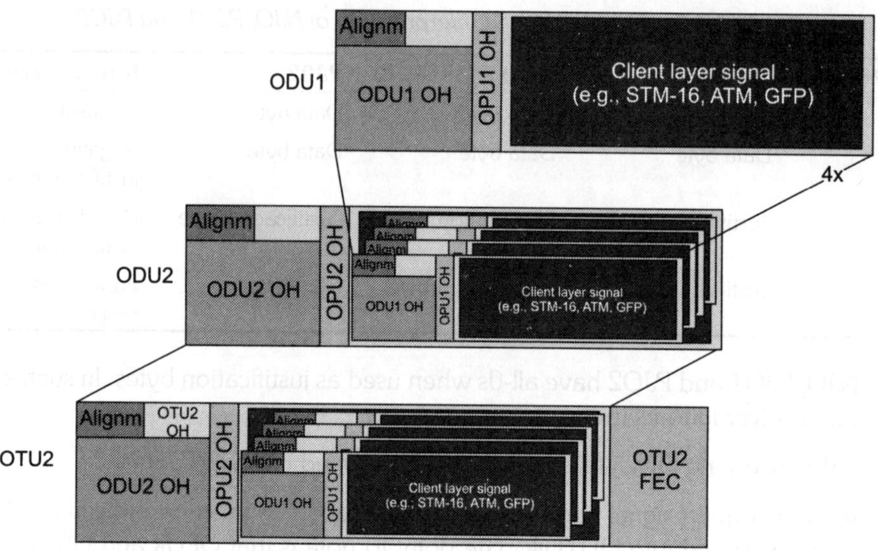

Fig. 13.30 *Steps in Multiplexing 4 ODU1 s into ODU2*

The ODU2 overhead is added and then ODU2 mapped into OTU2. The OTU2 overhead and frame alignment overhead are added to make the signal ready for transmission via an OTM signal after adding the OPU2 –OH to OPU2 payload.

Summary

In this chapter the multiplexing process in OTH, i.e., mapping of various payload entities in OTH is explored and the associated structure identifiers are discussed. Concatenation and multiplexing of lower order OTH entities into higher order entities are explored.

References

[1] ITU-T Rec. I.432.1.

[2] ITU-T Rec. G.7042/Y.1305

14

Optical Transport Networks and Applications

14.1 Introduction

We have discussed essentials of WDM and OTH in the previous chapters. In this chapter the architectural features of optical transport networks (OTN), NEs of OTN and applications of OTN are presented. OTN consists of the following elements:

Optical line terminals (OLTs)
Optical add–drop multiplexers (OADMs)
Optical cross-connects(OXCs)
Optical amplifiers (OAs)

Optical amplifiers have been covered in detail in chapter 11. Let us turn our attention to OLT, OADM and OXC.

An OLT multiplexes multiple wavelengths into a single fibre and de-multiplexes a composite WDM signal into individual wavelengths. OLTs are used at either end of a point-to-point link. An OADM is deployed in linear or ring topology at locations where some fraction of wavelengths needs to be dropped (terminated) locally and others need to be added (routed to other destinations). An OXC performs similar functions of OADM but on a larger number of ports; the wavelengths involved are also on a larger scale. OXCs are deployed in mesh topologies.

The architectural features of OTN are explained in section 14.2. OLTs, OAs, OADMs and OXCs are explained in sections 14.3 to 14.6. Applications of OTN viz. MPλS (multi protocol lambda switching), GMPLS (generalised MPLS), PON (passive optical networks) and NGN are presented in subsequent sections.

14.2 Architectural Features of OTN

An OTN is illustrated in Fig. 14.1. The users or clients or customers utilising the bandwidth of OTN are connected to OLTs, OADMs and OXCs. The clientele include IP routers, ATM switches, SONET terminals and ADMs. The number of wavelengths supported on a segment between nodes (adjacent) depends on the interface used and the currently operational traffic. The architectural features of OTN are listed below.

1. *Wavelength re-use*: A wavelength can be used on multiple segments but on the same segment there should not be any overlap. This feature enables the OTN to support a large number of light paths with limited number of wavelengths.
2. *Wavelength conversion*: Since only few wavelengths are supporting multiple light paths, to avoid overlap of wavelengths, it may be necessary to switch wavelengths. Wavelength conversion may also be necessary at the boundary of the network for inter-working with an adjacent network.
3. *Transparency*: Light paths across OTN are protocol-insensitive. They support a variety of protocols so that data streams of various protocols or origins get carried simultaneously in the aggregate OTH frame on the light path. In other words, the optical layer can support a variety of higher layers concurrently.
4. *Circuit switching*: Switching is not performed at the optical layer but at the higher application layer (IP or ATM). Circuit switching feature of OTN implies the ability of optical layer to set-up or take-down light-path.
5. *Survivability*: In the event of failures, the network is configurable to route optical paths over alternate paths.
6. *Light path topology*: The topology seen by higher layers utilising the services of the optical layer is the light-path topology. For example, for an IP application running on OTN the light path appears as a set of links between IP routers. The light path topology can be set up as per traffic requirement.

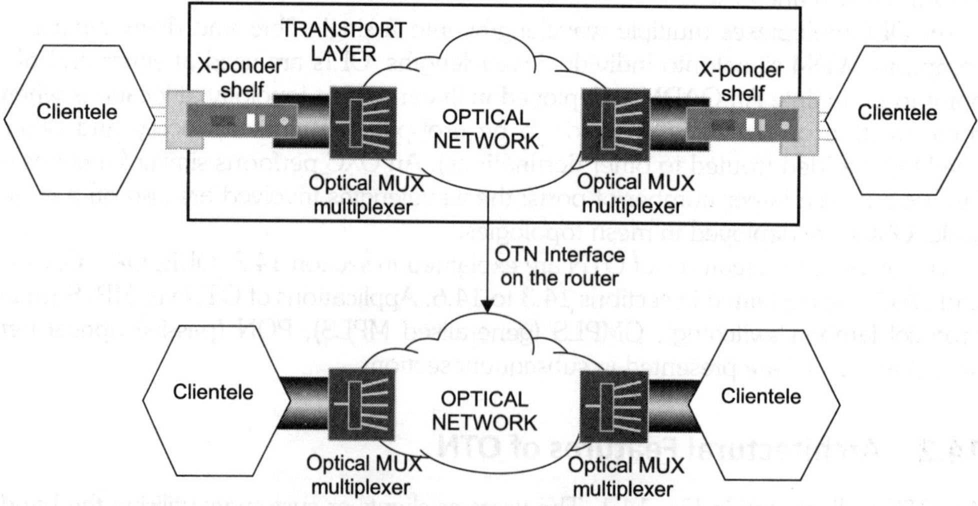

Fig. 14.1 *OTN architecture*

14.3 Optical Line Terminal (OLT)

An OLT is used at either ends of a point-to-point link. It consists of transponders, wavelength MUX/DEMUX and optical amplifier and termination for the optical supervisory channel (OSC). Refer to Fig. 14.2 showing the functional block diagram of an OLT.

The transponders convert the incoming signal from a client to a wavelength suitable for transmission over WDM network. Non-ITU wavelengths are also accepted from clients but converted to ITU wavelengths. If client signals are already on ITU wavelengths, transponders are not needed; this reduces the cost, size and power consumption of an OLT.

Signals coming out of the transponders are multiplexed using arrayed waveguide gratings or dielectric thin film filters or fibre Bragg gratings (which were described in chapter 11). After multiplexing, an optical amplifier is used to improve the signal level.

In the receive direction, the received signal is amplified and sent through de-multiplexer to extract individual wavelengths which are sent through transponders to get client signals directly to the client if the client link is working on an ITU wavelength.

The optical supervisory channel (OSC) is carried on a separate wavelength, different from the one carrying the traffic. The OSC facilitates OAM&P functions.

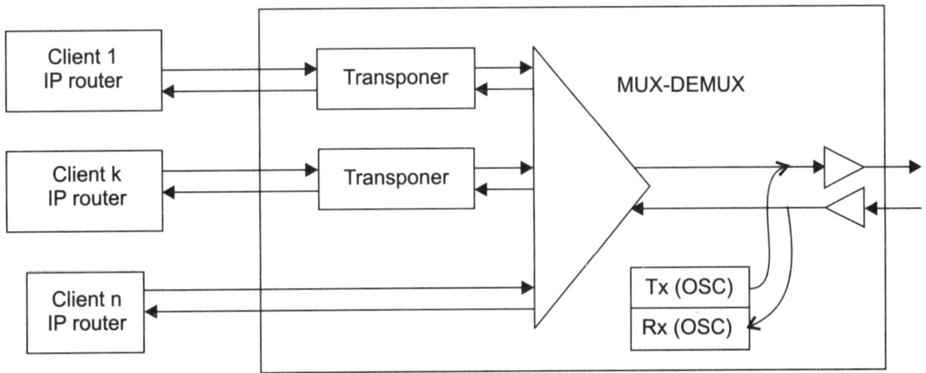

Fig. 14.2 *Functional block diagram of an OLT*

14.4 Optical Amplifier

The principles of working of optical amplifiers and types of optical amplifiers have already been discussed in chapter 11. The functional block diagram of a multi-stage gain amplifier with dispersion compensation is shown in Fig. 14.3.

The optical service channel (OSC) is extracted at the input and inserted at output. Amplifiers are provided at two stages. Dispersion compensator fibre is also in cascade. Raman laser pump provides additional gain over the fibre span. If the WDM plan has both C and L bands, these are separated at the input and separate amplifiers are used for each band.

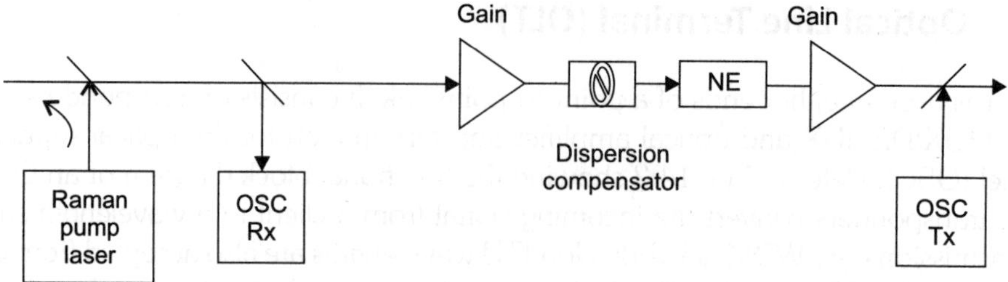

Fig. 14.3 Multi-stage optical amplifier arrangement

14.5 OADM

14.5.1 OADM basics

In metropolitan area networks as well as in long-haul networks, OADM facilitates cost-effective means of handling traffic by way of effective utilisation of available wavelengths in an add–drop MUX at each OADM. For example, consider a network consisting of three nodes, A, B and C. The traffic between A and C is more than A–B or B–C. If we have a 4 wavelength WDM plan, we use three wavelengths on A–C and one wavelength (the same wavelength) on A–B and B–C. We need to deploy the OLT at A, C and an OADM at B. Otherwise, we have to use point-to point links, A–B, B–C, A–C and deploy one OLT each at A, C and a pair of OLTs at B.

Some of the important considerations in selecting OADM are listed below. The SEARCH:

S – Support required for how many wavelengths and add/drops?
E – Ease of add/drop, i.e., any disruption of traffic to add/drop a wavelength
A – Architectural support for add/drop, i.e., use of one card to drop *n* wavelengths
R – Re-configurability of OADM under NMS
C – Constraints in add/drop for specific wavelengths/group of wavelengths/bands
H – Hierarchical impact of add/drop of additional channels on OADM and NW

14.5.2 OADM architecture

OADM has three different architectures: parallel, serial and band-drop.

Parallel architecture

Refer to Fig. 14.4. In this architecture, all the incoming channels are de-multiplexed and channels other than the dropped ones are passed through. The advantage of this architecture is that any desired number of wavelengths can be dropped/passed through. So, the design constraints are minimum. However, this architecture is not cost-effective as de-multiplexing is to be done for all channels even if the requirement is to drop a few.

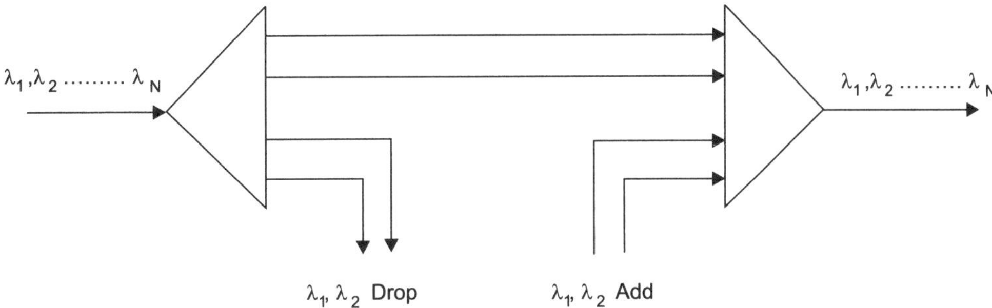

Fig. 14.4 *Parallel architecture of OADM*

The parallel architecture is realised using dielectric thin-film filters and arrayed waveguide gratings. If a large number of channels are to be dropped, modularity in add/drop as shown in Fig. 14.5 is followed. In this method, interleave type filters are used where group-wise separation of wavelengths is employed in the first stage of filtering followed by separation of individual wavelengths.

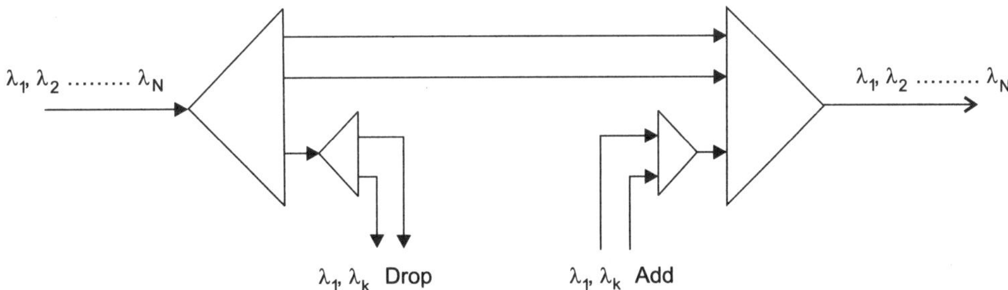

Fig. 14.5 *Parallel architecture of OADM – band drop or modular version*

Serial architecture

In this architecture (Fig. 14.6), a single channel is dropped and added at a time. Single channel OADM (SC-OADM) is used in series to drop as many wavelengths as derived. It works on Bragg-gratings or dielectric thin-film filters. The cost is low, proportional to the number of channels, but may require additional amplifiers if more channels are dropped (each SC-OADM introduces approximately a 3 dB loss).

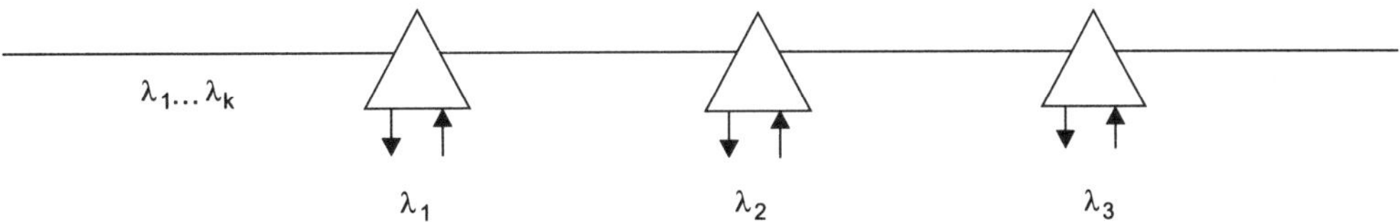

Fig. 14.6 *Serial architecture with single wavelength add–drop at a time*

Band-drop approach

In this architecture (Fig. 14.7), a fixed group of channels is dropped and added from an aggregate set of channels. For example, a set of four adjacent channels is dropped as a band and added from a set of 32 wavelengths. In this architecture, flexibility in network planning is at stake.

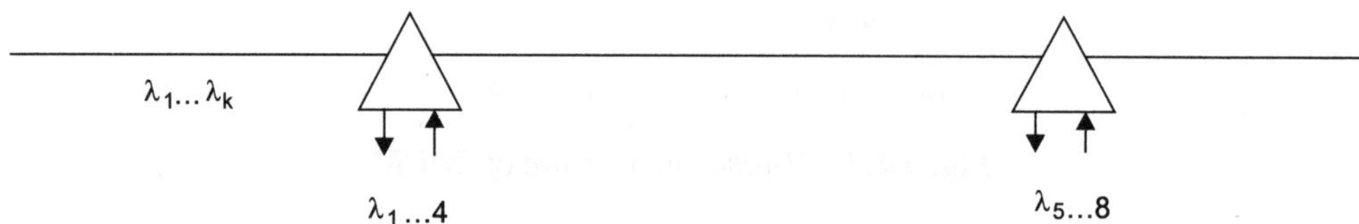

Fig. 14.7 *Serial architecture with a band of wavelengths add–drop at a time*

14.5.3 Re-configurability of OADM

For flexible traffic management on OTN, it will be possible to select the channels (wavelengths) to be dropped/added at OADM under the control of NMS rather than deciding on add/drop in advance. With a variation of parallel architecture of OADM, re-configurability can be achieved, as shown in Fig. 14.8. Optical switches facilitate add/drop of specific wavelengths. By making SC-OADM a tuneable device, i.e., capable of adding/dropping a specific wavelength as shown in Fig. 14.9, it is possible to achieve re-configuration under the control of NMS.

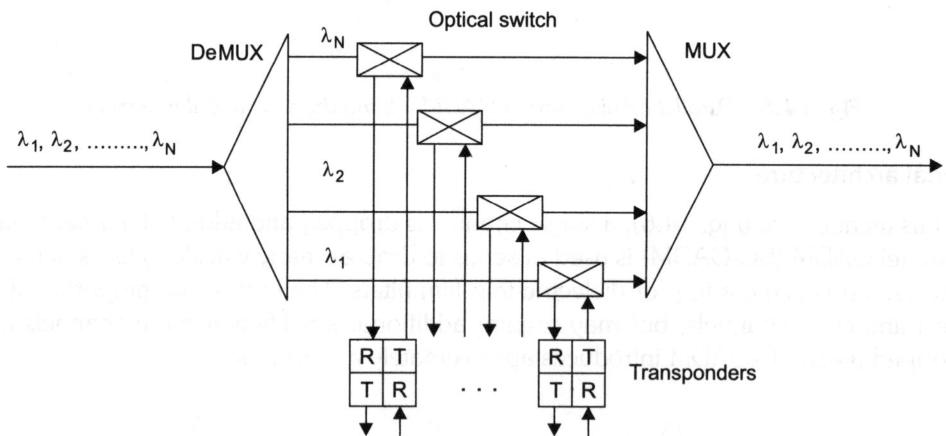

Fig. 14.8 *Partially re-configurable OADM with add/drop switches and fixed wavelength transponders*

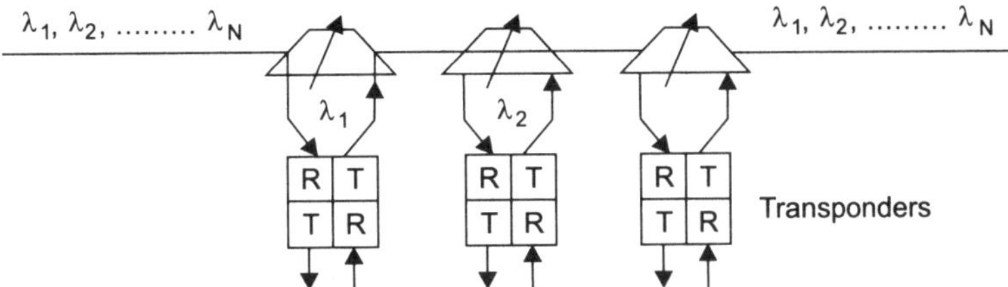

Fig. 14.9 *Partially re-configurable OADM in serial architecture with fixed wavelength transponders*

The above two approaches make OADM only partially re-configurable. To achieve full re-configurability, tuneable transponder is to be used as shown in Fig. 14.10(a) for parallel architecture and 14.10(b) for serial architecture.

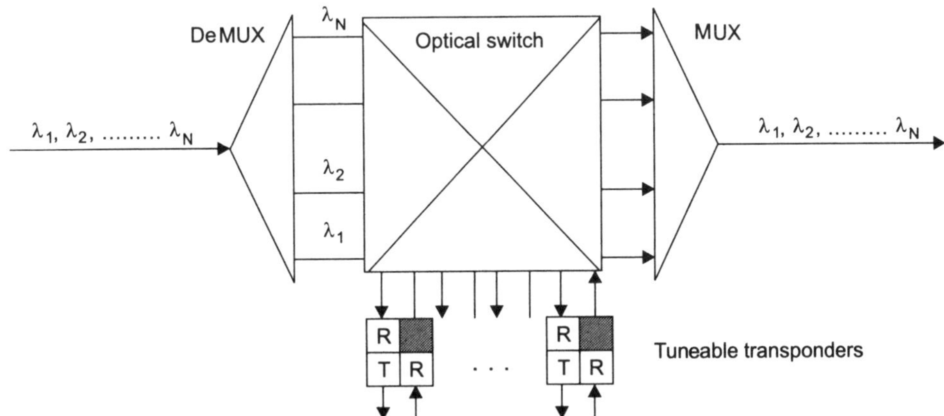

Fig. 14.10(a) *Fully tuneable OADM in parallel architecture with tuneable transponders*

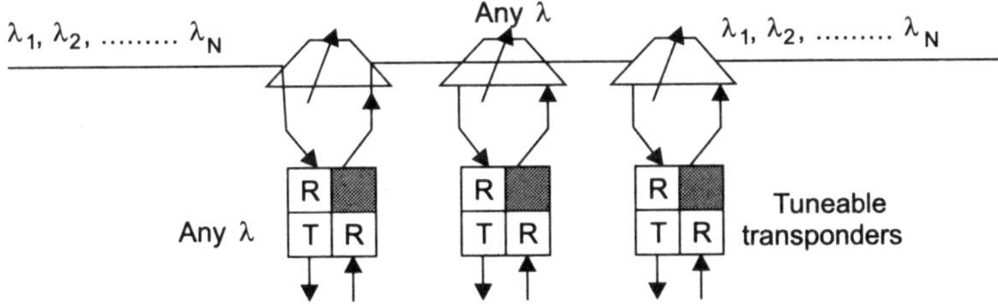

Fig. 14.10(b) *Fully tuneable OADM in serial architecture with tuneable transponders*

14.6 Optical Cross-Connect (OXC)

14.6.1 Role of OXC in OTN

To state that OXC is an NE that can cross-connect traffic among different directions is to understate the functionalities of OXC. An OXC is a switch with several ports, some of them connected to WDM equipment and others to terminating devices like SONET/ SDH ADMs, IP routers and ATM switches such that a pass-through for express traffic as well as pick-up of traffic from attached equipment onto network is realised in a cost-effective manner.

Refer to Fig. 14.11. The OLTs should not be considered as a part of OXC. Service providers on OTN consider OXC and OLT as separate NEs. OXC and OLT can be from different manufacturers as well.

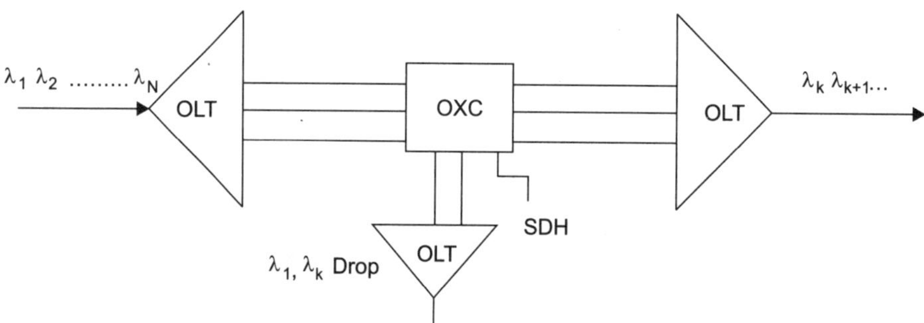

Fig. 14.11 *OXC and OLTs in OTN*

OXC has two parts: switch core and port complex. Switch core performs cross-connect functions while port-complex has interface cards which are optical-to-electrical (O/E) or optical-electrical-optical (O/E/O) converters.

OXC performs the following functions in a large network.

1. *Automatic provision of light paths*: Light-paths in a large network can be provided in automatic manner by controlling the switching at OXC through NMS to facilitate smooth handling of services
2. *Protection of light-paths*: OXC is an intelligent NE and can detect fibre cuts and equipment failures and re-route paths quickly.
3. *Bit rate and protocol transparency*: OXC supports all standard bit rates and NW transmission protocols (frame formats)
4. *Non-intrusive fault localisation*: A test equipment can be put on a dedicated port and the NW can be monitored, tested and the fault localised. Also, loop-back tests can be done if needed.

5. *Wavelength conversion*: OXC can do switching as well as wavelength conversion.
6. *Multiplexing and granularity adoption*: Multiplexing at optical layer as well as rationalisation of granularity can be done. If many under-rate signals are being multiplexed, a suitable SDH or OTH granularity can be chosen to 'pack' such signals.

14.6.2 Deployment of OXC in OTN

When an OXC is deployed in an OTN, it works together with OLTs, and OADMs. The configuration is illustrated in Fig. 14.12(a).

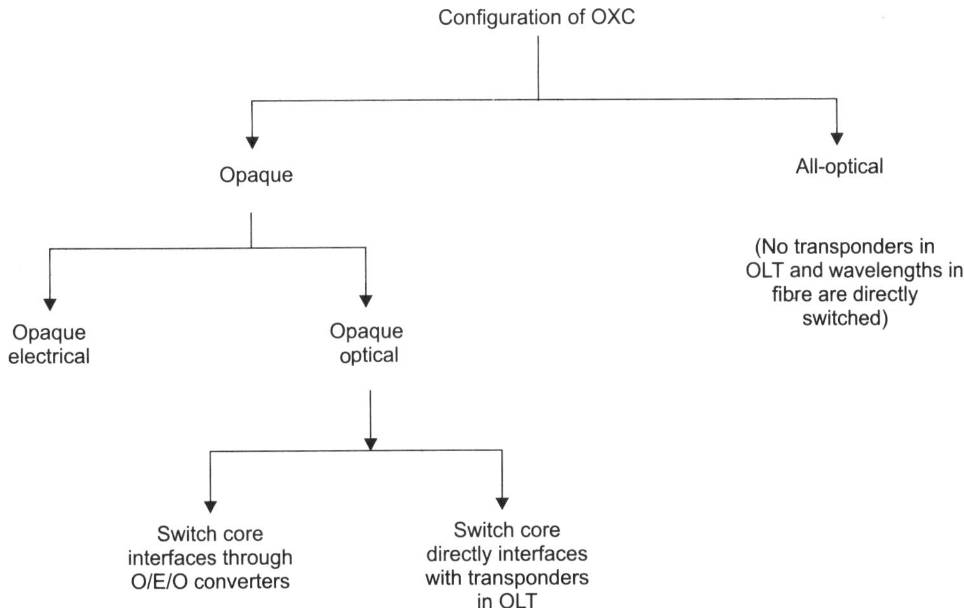

Fig. 14.12(a) *Classification of configuration of OXC*

In opaque configuration, an optical signal is converted to electrical domain as it passes through the NEs. In all-optical configuration, optical signal remains in the optical domain as it passes through the NEs. In opaque configuration, the switch core can be electrical or optical, with switching taking place in the electrical domain or the optical domain.

In opaque electrical configuration, an OXC is connected to an OLT. Within OXC, there is first an O/E conversion, followed by switching and finally, E/O conversion. This configuration is presented in Fig. 14.12(b).

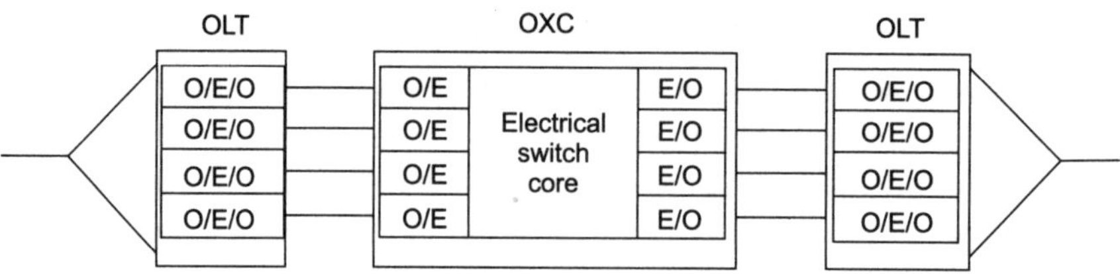

Fig. 14.12(b) *Opaque electrical*

In opaque optical configuration, we have two varieties:

i. Additional O/E/O converters with short reach interfaces between OXC and OLT deployed to suit WDM plan, as shown in Fig. 14.12(c).

ii. Optical switch core directly interfaces with transponders in OLT, as shown in Fig. 14.12(d).

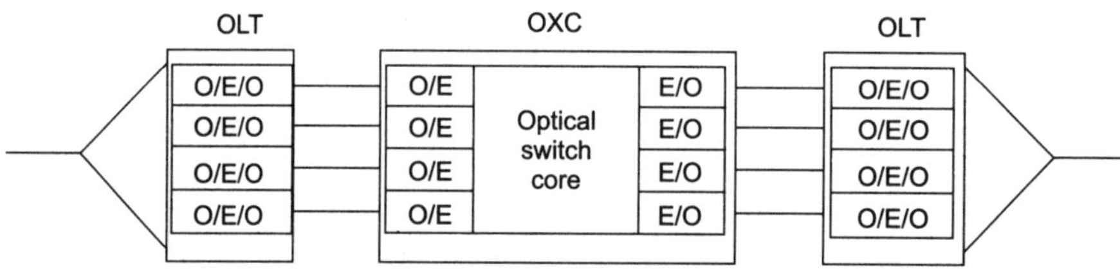

Fig. 14.12(c) *Opaque optical with additional O/E/O between OXC and OLT*

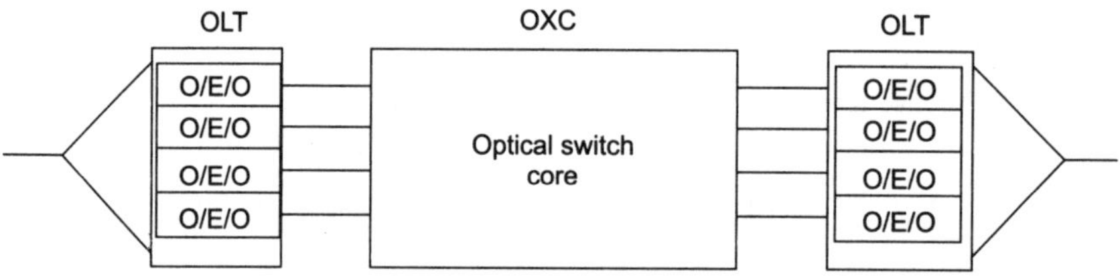

Fig. 14.12(d) *Opaque optical with direct connection to transponders of OLT*

A comparison of electrical switch core and optical switch core is presented in Table 14.1A.

An all-optical configuration is shown in Fig. 14.12(e). In this the signals are in the optical domain throughout and are switched optically. It has an optical switch core which is directly connected to a multiplexer/de-multiplexer in the OLT. Interoperability of such OXCs and OLTs is to be confirmed before planning, and it is preferable to obtain both from the same vendor.

Optical cross-connect is a versatile NE making possible the support of several applications on OTN. We will discuss these applications in the following sections.

Table 14.1A *Comparison of electrical and optical switch cores*

Electrical switch core	Optical switch core
Switching takes place in the electrical domain i.e., O/E conversion, switching and E/O conversion takes place.	Switching takes place in the optical domain. It can work with O/E/O conversion to suit WDM plan or is directly connected to transponders.
Traffic can be taken in finer granularities, i.e., at lower data rates like STM-1, STM-4.	Traffic cannot be taken in finer granularities, i.e., at lower data rates.
Electrical switch core is designed to a total capacity. If the total capacity is M Gbps, it can switch n signals of rate R Gbps, where $M = n.R$. For example, 1.28 Tbps switch core capacity implies switching of 512 OC-48 (2.5 Gbps) or 128 OC-192 (10 Gbps).	Optical switch core is bit-rate independent. For example, a 100 port can switch 100 OC-48 or 100 OC-192 streams. Thus, it is scalable and future-proof.
As the bit rate increase, cost per port of the electrical switch increases. Hence, at lower bit rates, electrical switch is cost effective.	Cost per port of optical switch is independent of bit rate. Hence, at higher bit rates, optical switch is cost effective.
Electrical switch core is not transparent: it requires separate interface card for each input type to convert into a suitable format.	Optical switch core is transparent – it can switch any traffic (Ethernet, SONET, STM). It does not require separate interface.

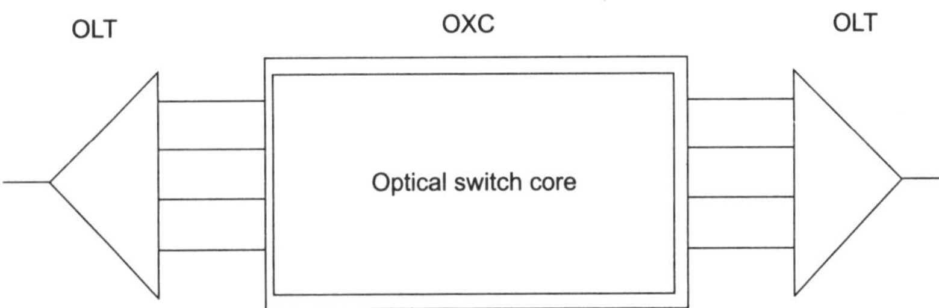

Fig. 14.12(e) *All-optical cross-connect*

Q 14.1 *What are the reconfigurable OADMs used in today's OTN?*

Today's dynamic OTN environment is characterised by high bandwidth connectivity to existing and new sites, complex traffic routing patterns , implementation of business continuity, expansion and disaster recovery (BCEDR) strategies. Hence, operators are looking for a flexible and cost-effective solution to dynamically provide 'on the fly' high-bandwidth connectivity to any network site without interrupting the existing services or re-engineering the network. While the discussion on OADM in section 14. 5

provides the conceptual framework, the technology in use is ROADM (re-configurable optical ADM). It does away with 'physical' change of filters, optical interfaces and adjusting link engineering equipment (amplifiers and regenerators) and allows on-demand connectivity.

ROADM comes in three generations:

1. *Wavelength blocker ROADM architecture (WB ROADM)*
   In this technology, wavelengths are sent on line to the sites where the WB (wavelength blocker) in conjunction with a set of optical filters, blocks the selected wavelengths from passing through the site and directs them to filters for extraction. Similarly, wavelengths can be inserted past the WB using optical combiners, and be added to the line.
2. *PLC (planar light-wave circuit) based ROADM architecture*
   The second generation of ROADMs are based on PLC (planar light-wave circuit) technology where NMS facilitates remote insertion/termination (add/drop) or routing of any particular wavelength through wavelength directors. This approach has the advantage of reduced networking cost over WB ROADM but suffers from the disadvantages of poorer optical performance and the limitation to 100-GHz spaced wavelengths. Many vendors have packaged PLC ROADMs and wavelength multiplexers/ de-multiplexers into their existing DWDM platforms in the form of a circuit pack allowing leverage of ROADM features on installed base.
3. *Enhanced ROADM (eROADM) or micro electro-mechanical system (MEMS) ROADM*
   In WB ROADM and PLC RAODM, optical branching remains a complex task. A new disruptive ROADM architecture has been recently introduced to the market, called enhanced ROADM (eROADM). It leverages the simplicity, fine precision and unmatched flexibility of micro electro-mechanical systems (MEMS) technology to allow unrestricted add/drop and redirection of any wavelength at any port (as the ports are not assigned to particular wavelengths as in previous generations), thus enabling network operators to remotely perform 'on the fly' remote and unrestricted wavelength re-routing for true network agility.

   No wonder then that in any OTN, eROADMs are provided at all junction sites as part of BCEDR strategy. Table 14.1B gives the comparison of the three generations of OADMs.

Table 14.1B *Comparison of WB ROADM, PLC ROADM and eROADM*

Parameter	WB-ROADM	PLC-ROADM	eROADM(MEMS)
Period	Year 2002–2004	Year 2004–2006	Year 2006 onwards
Expandability	All wavelengths to be installed on day 1	All wavelengths to be installed on day 1	Modular and add-on

(Continues…)

Table 14.1B *Continued*

Parameter	WB-ROADM	PLC-ROADM	eROADM(MEMS)
Optical branching	No	Yes, complex	Yes, simple
Wavelength unassigned ports (colourless ports)	No	No	Yes
Max. wavelengths	72	36	96 (more and more as per ITU-T's grid)
Patch cords' usage	Moderate	High	Low to moderate

Q 14.2 *What are the features of the OXCs used in today's OTN?*

OXCs used in OTN are designed to provide seamless integration of SDH, CWDM, DWDM on a single platform, known as MSPP (multi service provisioning platform). It means that DXCs and OXCs are designed on the same architectural plane so as to facilitate future migration. As an example, ECI's XDM platform is presented in brief next.

XDM family's 30G to 240G LO cross-connect switching matrix facilitates connectivity between various clients and also grooming to utilise backbone resources. Its OTN-based flexible optical capabilities ensure smooth migration to higher rates.

XDM platform supports SDH interfaces (STM-1 to STM-64), PDH interfaces (E1, E3, DS3), ETH/FE/GE/ATM/IP/MPLS over STM, 40/10/2.5 G, 10 GE optical interfaces. XDM platform supports BLSR, MS-SPRing, SNCP and ASON protections. XDM equipment can be configured to support CWDM wavelengths, 16/32/40/80 DWDM wavelengths.

14.7 Multi Protocol Lambda Switching (MPλS)

14.7.1 What is multi protocol lambda switching (MPλS)?

In multi protocol label switching discussed in section 9.3, the control plane dealt with with label distribution and binding end-to-end LSPs (label switched paths). In an optical network, the control plane is concerned with setting up of wavelengths (light paths), optical transport hierarchical coding, transfer rates and protection switching. If we can interwork these control planes, we can interwork MPLS and optical networks. Optical networks on which MPLS is overlaid through a standardised interworking of control planes is known as multi protocol lambda switching (MPλS).

Figure 14.13(a) illustrates the interworking of control planes of MPLS and OTN.

In MPLS, the data plane of LSRs (label switched router or the provider edge equipment discussed in section 9.3) uses label to transfer packets from input port to output port. In case of OTN, the OXC's data plane refers to the switching matrix required to connect the optical channel from an input port to an output port.

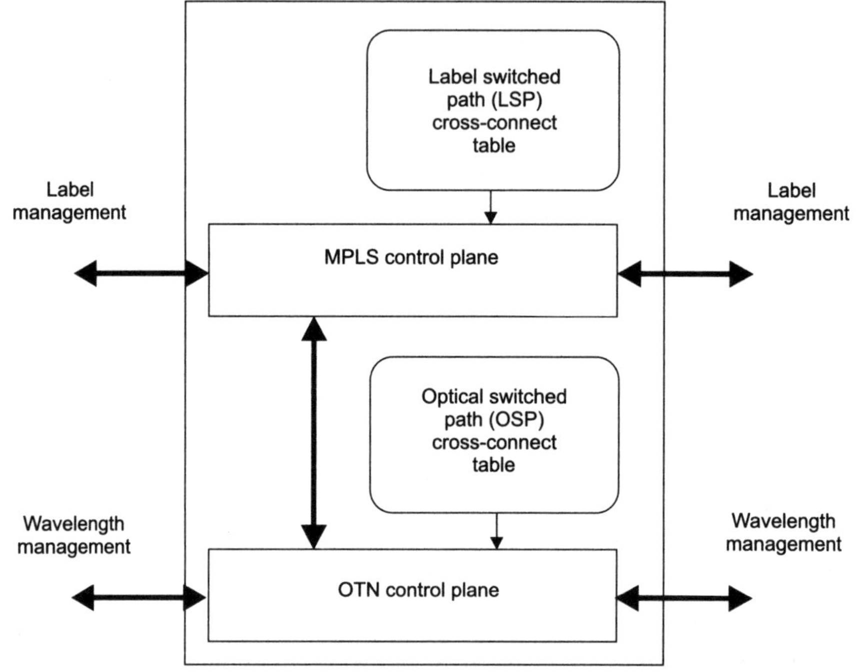

Fig. 14.13(a) *Concept of interworking of control planes of MPLS and OTN at an NE*

Figure 14.3(b) shows the protocol layers in MPLS. MPLS is in layers 2 and 3, as discussed in section 9.3. Control messages from MPLS layer move down to optical layer to get carried in optical data channels or control channels.

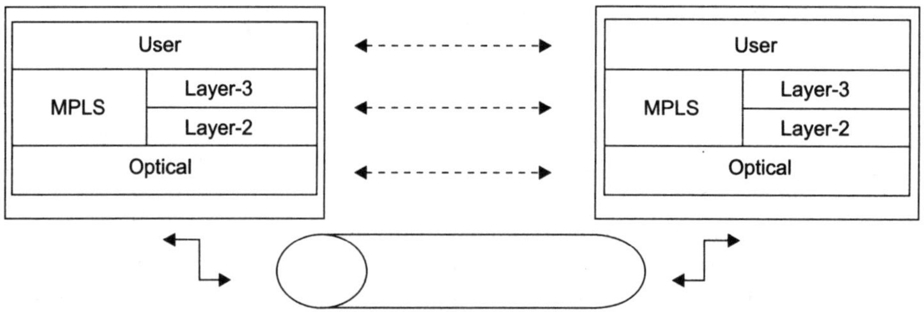

Fig. 14.13(b) *Layers in MPLS and OTN interworking*

The MPLS control plane is concerned with the discovery, distribution and maintenance of state information of MPLS network and management of LSPs.

On close observation, it is realised that the control plane of OXC is concerned with the discovery, distribution and maintenance of state information of OTN and the establishment/maintenance of optical channel trails under varying traffic conditions. While this is the similarity it shares with MPLS, there is a subtle difference: in MPLS, the forwarding information is carried explicitly as part of labels (VC labels appended to data packets as discussed in section 9.3), while in OTN, the switching information is implied through the wavelength or the optical channel.

14.7.2 Implementation of MPλS by interworking MPLS and OTN

From the discussion in the previous section, it is clear that the key to successful implementation of MPλS by interworking MPLS and OTN is to correlate MPLS labels to optical wavelengths. Refer to Fig. 14.14 showing the processing of MPLS traffic at OTN nodes.

Consider the packet flow from site A to site D under MPLS. At the ingress node or PE attached to site A, the MPLS label is correlated to appropriate wavelength i.e., channel in the OTN to each egress node or PE attached to site D. The intermediate nodes or core routers are configured to process wavelengths rather than labels to make routing decisions. Hence, we call them photonic cross-connects (PXCs). The payload carried by LSPs and optical trails is transparent to the intermediate nodes along their respective paths so that the performance as well as protection requirements are taken care of.

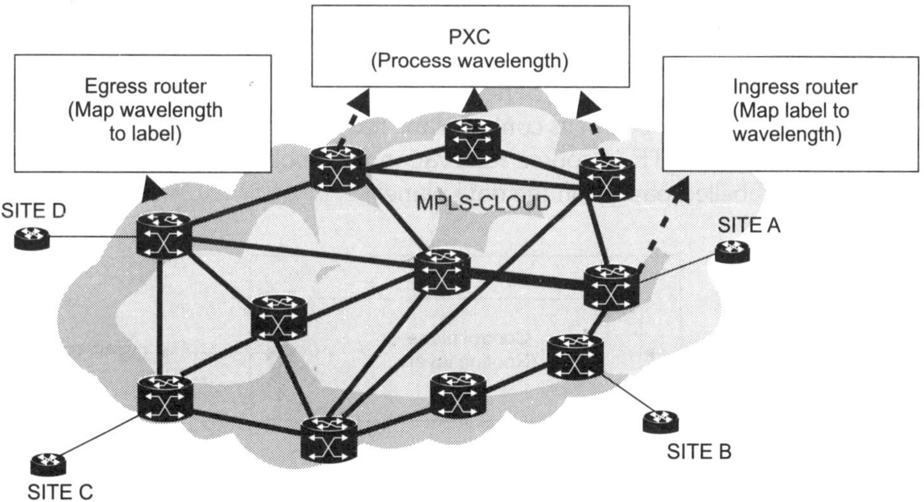

Fig. 14.14 *Processing MPLS traffic at OTN nodes*

To configure the mapping and switching functions, the PXCs should be able to exchange control information. To facilitate the same, in general, dedicated 'control wavelengths' (not data but supervisory channels) are pre-configured between adjacent PXCs or between the PXC and the router. These supervisory channels are used for exchange of signalling and OAM&P traffic.

In this approach, a PXC maintains a WFIB i.e., (wavelength forwarding information base) or an optical switching path cross-connect table per interface, or fibre. In MPLS networks, it is possible to merge labels, i.e., multiple incoming labels for the same destination can be given a single outgoing label so as to reduce the number of labels processed by the LSR. Similar merging of traffic onto single wavelength (optical channel through suitable OPU) can be done in OTN. The failure of a fibre or node invokes the protection mechanism to arrange for an alternate route.

Thus, MPλS implementation requires a close coordination between the optical control plane and the label control plane.

14.7.3 IP and MPLS over optical transport networks

Issues concerned with IP and MPLS over optical transport networks

In the previous sections we had discussed MPλS implementation by interworking MPLS and OTN. In section 9.3, MPLS was presented as a solution for quick transport of IP packets across large-meshed IP networks to avoid routing decisions on a packet-by-packet basis. So, when we implement MPLS over OTN, it would seem that we have IP through MPLS over OTN. But, in reality, a lot more needs to be done for enabling IP over OTN. Let us to take a closer look at the control and data planes of IP, MPLS and OTN, shown in Fig. 14.15(a), (b) and (c) respectively.

In the case of a node implementing IP, the control plane consists of routing protocols like OSPF (open shortest path first), IS-IS (intermediate system to intermediate system) and BGP (border gateway protocol). The data plane comprises forwarding protocol IP.

In the case of MPLS, a node in its control plane has protocols like LDP (label distribution protocol) and RSVP-TE (resource reservation protocol–traffic Engineering) which deal with MPLS labelled packets in the data plane.

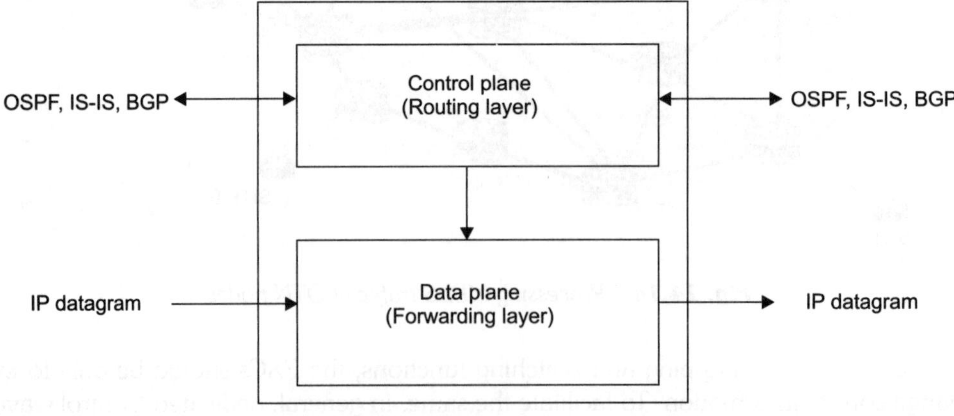

Fig. 14.15(a) *Control and data planes of IP*

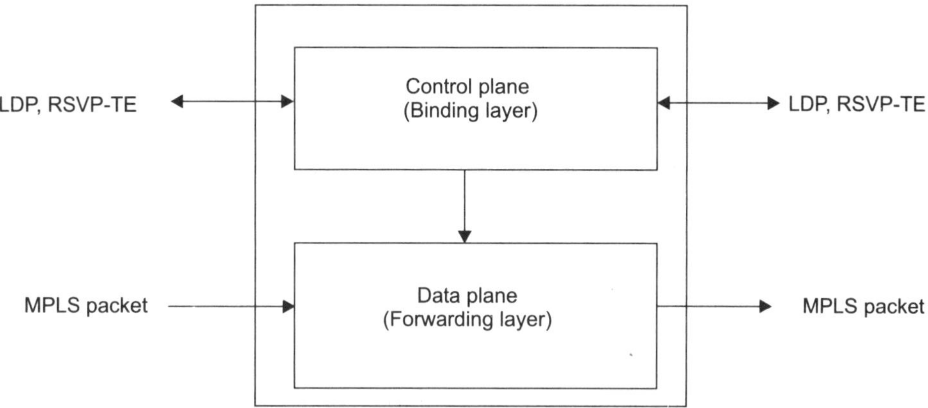

Fig. 14.15(b) *Control and data planes of MPLS*

In case of OTN, the control plane of a node has LMP (link management protocol) or a protocol to handle MPLS, which will be discussed later. The data plane deals with wavelengths.

By interworking IP, MPLS and OTN, we should get advantages of all the three viz.

- route discovery capabilities of IP control plane,
- traffic engineering capabilities of MPLS control plane,
- forwarding (switching) speed of optical data plane.

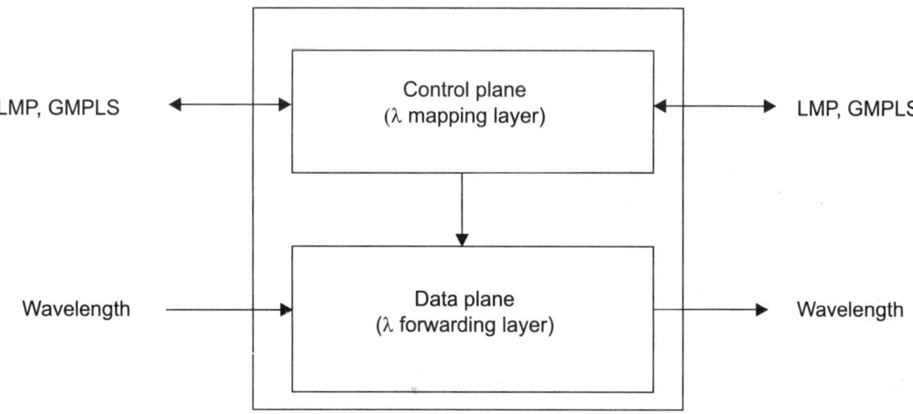

Fig. 14.15(c) *Control and data planes of OTN*

There are two methods or models to interwork IP, MPLS and OTN to realise these objectives.

1. *Domain services model*
   In this model the basic philosophy is that the control planes of PXCs and LSRs are de-coupled with little or no interaction between control plane instances.
   The OTN offers high bandwidth connectivity in the form of light-paths. Standard signalling across a UNI (user network interface) is used to invoke the services associated with light paths viz. light path creation, light path deletion, light path modification and light path status enquiry.
   Before obtaining services through UNI, a precursor program or module termed 'Service Discovery Procedure' is run. By running this, the client is allowed to determine static parameters of interconnection with OTN and the signalling protocol to be followed through the UNI.
   It may be noted that this model does not deal with and the type and nature of routing protocols.

2. *Unified services model*
   In this model the basic philosophy is to use a single traffic engineering control plane or, couple the control planes of IP, MPLS and OTN, i.e., use a single traffic engineering control plane. The OXC can be treated as any other router for control plane operations. There is no distinction between UNI or NNI. Also, there is no distinction between any router–router interface from the point of view of routing and signalling. Refer to Fig. 14.16 showing the implementation of this model in a simplified way.

    1. IP routing protocols discover the addresses and routes to the nodes with those addresses.
    2. MPLS label distribution protocols distribute labels associated with IP addresses.
    3. MPLS labels are mapped to specific wavelengths between adjacent nodes so that PXC based O/O/O operations take place (no MPLS label swapping and O/E/O operations)

Activity 1 and 2 pertain to IP and MPLS protocols and are beyond the scope of this book. Activity 3 pertains to OTN and is discussed next.

Solution to interwork IP, MPLS and OTN: GMPLS

As can be seen, the key to the implementation of IP and MPLS over OTN lies in the mapping of MPLS labels to wavelengths to achieve O/O/O operations. In normal MPLS implementation, LSRs make use of a forwarding protocol which is capable of processing and routing signals containing packet, frame or cell boundaries. So, such LSRs in OTN have to 'know' the signal in the electrical domain, i.e., they have to be O/E/O devices. Normal OXCs would do the job.

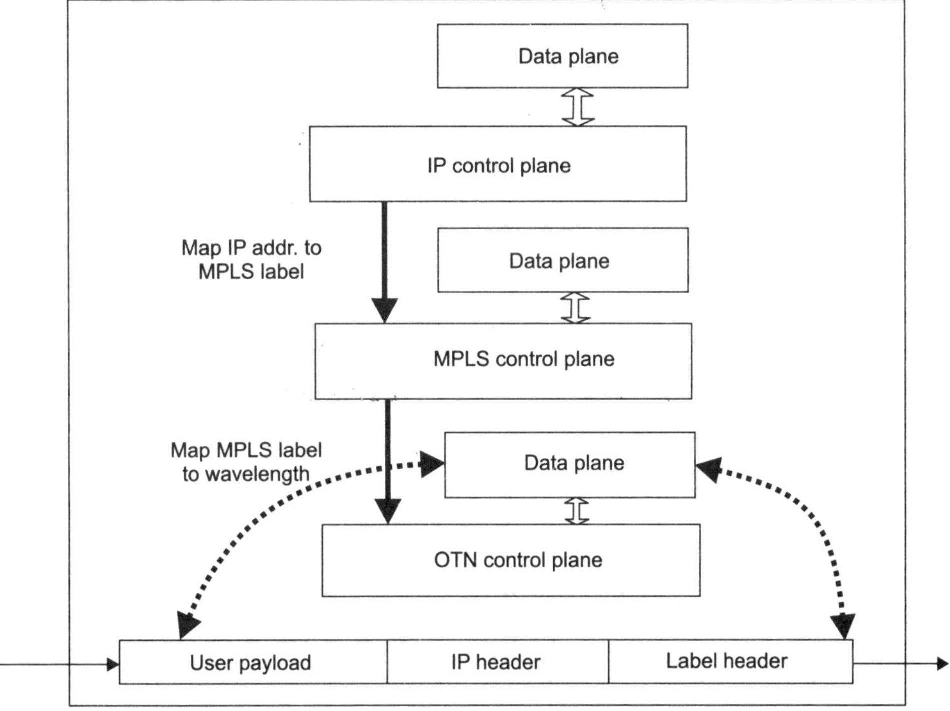

Fig. 14.16 *Interworking IP, MPLS and OTN control planes*

Suppose we have LSRs that recognise neither the cell nor the packet boundaries. Then the forwarding decision of such LSRs have to be based on one of the following:

- time-slots,
- wavelengths or physical ports i.e., label switching is time-division switching (ex. OADMs in OTH),
- wavelength switching (λ-switching) and
- spatial switching (incoming port/fibre to out-going port/fibre). Such LSRs are not entering the electrical domain; they are O/O/O devices. Thus, with such implementation we get the advantage of a total optical-domain operation. The cross-connects serving this purpose are photonic cross-connects.

To support MPLS and IP operations over OTN by way of optical time-division switching, wavelength switching spatial switching and GMPLS (generalised MPLS) extensions have been developed.

It may be noted that MPλS provides only wavelength switching, whereas GMPLS supports TDM switching, wavelength switching and spatial (fibre/port) switching. We now look at the implementation of GMPLS over OTN.

Q 14.3 *How does restriction of labels selectable by down-stream node help?*

In the optical domain there are several instances where wavelengths used by the path needs to be restricted either to a small subset of possible wavelengths or to one specific wavelength. This requirement emanates from the fact that some NEs may only be able to generate a small set of the wavelengths that the intermediate equipment may be able to switch, or because the intermediate equipment may not be able to switch a wavelength at all.

Q 14.4 *How do formalised separation of control and data channels help?*

Such separation supports technologies where control traffic cannot be sent in-band with the data traffic.

14.8 GMPLS over OTN

14.8.1 GMPLS vs. MPLS

The distinctions between GMPLS and MPLS are listed in Table 14.2.

Table 14.2 *A comparison between MPLS and GMPLS*

MPLS	GMPLS
Conventional MPLS supports switching based on packet frame cell boundaries	Generalised MPLS differs from traditional MPLS in that it supports multiple types of switching, including, lambda, and fibre (port) switching.
No additional functionalities and properties for LSPs	The support for the additional types of switching adds additional functionalities and properties to LSPs with regard to the way in which labels are requested and communicated, directional nature of LSPs; how errors are propagated and information provided for synchronising the ingress and egress NEs.
In MPLS traffic engineering, links traversed by an LSP can include an intermix of links with heterogeneous label encodings. Ex.: An LSP may span links between routers, links between routers and ATM-LSRs, and links between ATM-LSRs.	Generalised MPLS extends this by including links where the label is encoded as a time slot, or a wavelength, or a position in physical space (fibre/port).
In MPLS traffic engineering, not all LSRs are capable of recognising (IP) packet boundaries (e.g., an ATM-LSR) in their forwarding plane. An LSP that carries IP has to start and end on a router.	Generalised MPLS extends this by requiring an LSP to start and end on similar type of LSRs.

(Continues...)

Table 14.2 *Continued*

MPLS	GMPLS
MPLS cannot support signals of SDH/SONET, Gbps/10 Gbps Ethernet.	Generalised MPLS allows signals of SDH/SONET, Gbps/10 Gbps Ethernet. Label processing is taken up accordingly.
Bandwidth allocation is not in discrete units; the number of labels on LSP is more.	Bandwidth allocation on LSP can be performed in discrete units. Consequently, the number of labels on each LSP are few.
Conventional MPLS LSPs are uni-directional.	Generalised MPLS LSPs are bi-directional. The benefits are: no resource contention when allocating reciprocal LSPs via separate signalling sessions; easier failure restoration procedures; lower setup latency and lower number of messages required during setup.
No label suggestion by any NE.	Generalised MPLS allows for a label to be suggested by an upstream node. This suggestion may be overridden by a downstream node but, in some cases, at the cost of higher LSP setup time.
No technology specific parameters are included in label (signalling).	Generalised MPLS also allows for the inclusion of technology specific parameters in signalling.
Restriction on the labels that may be selected by down-stream node is possible.	In generalised MPLS, an ingress or an other upstream node may restrict the labels that may be used by an LSP along either a single hop or along the whole LSP path.
MPLS does not formally separate control and data channels.	GMPLS formalises separation of control and data channels

14.8.2 OTH support for GMPLS switching

For time-domain switching, the signal needs to be examined in the time-domain. From our discussion of OTH and G.709 in chapter 12, it may be recalled that ITUT-G.709 defines the following layers in OTH:

- Optical transmission section (OTS)
- Optical multiplex section (OMS)
- Optical channel (OCh)
- Optical channel data unit (OTUk)
- Optical channel data unit (ODUk)

Among these, only the OCh and the ODUk layers can be time-domain switching layers because they include overhead for supervision and management.

The ODUk overhead is transported in an associated manner as part of the digital ODUk frame. Recall from section 12.3 that GCC1 and GCC2 bytes provide connectivity channel between NEs through digital path layer (ODUk). Hence, these bytes can be used to send GMPLS labels.

The OCh overhead is transported in a non-associated manner (also referred to as the non-associated or NAOH) in the optical transport module (OTM) overhead signal (OOS), together with the OTS and OMS non-associated overhead. In fact, the OOS is transported via a dedicated wavelength referred to as the optical supervisory channel (OSC). It may be recalled from the description of OTM overhead that the non-associated overhead is only functionally specified and as such open to vendor specific solutions. This feature can be used to define label format to be sent through OOS of OCh.

Let us first discuss the GMPLS label format. Then, we shall deliberate the specifics for various cases of LSPs through ODUk and OCh.

14.8.3 GMPLS label format

To deal with the widened scope of MPLS in the optical and time domain, new forms of 'label' are required, which are collectively referred to as 'generalised label'. A generalised label contains enough information to allow the receiving node to program its cross-connect regardless of the type of cross-connect such that the ingress segments of the path are properly joined.

It may be noted that as the nodes are aware of the kinds of links being used, there is no need for type field in the generalised label.

Generalised requests are sent from one node to another (one cross-connect to the other) and are intended to provide sufficient information to set up resources for the connection of an LSP between the nodes. The generalised label request includes a common part (used for any switching technology) and a technology-dependent part (the traffic parameters).

The common part has 32 bits with 3 sub-fields. Refer to Fig. 14.17(a).

1. LSP encoding type (8 bits)
2. Switching type (8 bits)
3. Generalised protocol ID (G-PID) (16 bits)

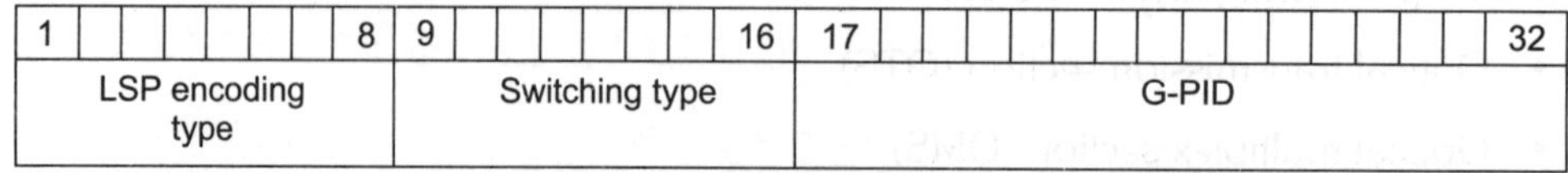

Fig. 14.17(a) *GMPLS label for OTN – common part*

When G.709 digital path layer or G.709 optical channel layer is specified in the LSP encoding type field, the technology dependent part (or traffic parameters) is carried in addition to the common part shown. The format of technology dependent part of GMPLS label for LSP of ODUk/OCh is shown in Fig. 14.17(b). It has 5 sub-fields:

1. Signal type (ST)
2. Number of multiplexed channels (NMC)
3. Number of virtual components (NVC)
4. Multiplier (MT)
5. Reserved

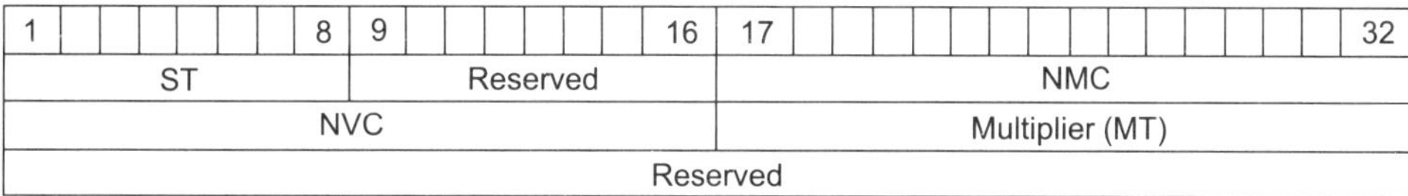

Fig. 14.17(b) *GMPLS label for OTN – traffic parameters part*

Common part or technology independent part of generalised label

A brief on the three sub-fields in the common part of GMPLS label for LSP of ODUk/OCh is presented here.

1. *LSP encoding type sub-field*
   The LSP encoding type conveys the encoding to be used with the data associated with LSP. It tells the receiving node about the framing format. It represents the nature of the LSP, and not the nature of the links that the LSP traverses. A link may carry and switch a signal of one or more of the encoding formats depending on the resource availability and capacity of the link. There are 12 LSP encoding types as shown in Table 14.2.

Table 14.2 *LSP encoding sub-field interpretation in GMPLS label*

Value	Format of traffic
1	Conventional IP format
2	Ethernet (original V2 of XEROX)
3	PDH payloads (T hierarchy)
4	PDH payloads (E hierarchy)
5	SONET ITU-T G.707
6	SONET ANSI T1.105
7	Digital wrapper
8	Lambda (photonic)
9	Fibre
10	Ethernet (IEEE 802.3)
11	SDH ITU-T G.707
12	ODUk of OTH
13	OCh of OTH

Q 14.5 *What are ANSI PDH, ETSI PDH, lambda and fibre formats?*

The ANSI PDH and ETSI PDH types are T and E hierarchy PDH. DS1 and DS3 are examples of ANSI PDH LSPs. The lambda encoding type refers to an LSP that encompasses a whole set of WDM wavelengths. The fibre encoding type refers to an LSP that encomposses a set of fibre ports.

2. *Switching type sub-field*
   This sub-field informs the cross-connect about the type of switching to be performed on a link – TDM/lambda/spatial or ATM/frame-relay, etc. This sub-field is needed for links that advertise more than one type of switching capability. Interpretation of this sub-field is given in Table 14.3.

Table 14.3 *Switching type sub-field*

Value	Type of switching
1 to 4	Packet-switch capable-1 to 4 (PSC-1 to PSC-4)
51	Layer-2 switch capable (L2SC)
100	Time-division-multiplex Capable (TDM)
150	Lambda-switch capable (LSC)
200	Fibre (Spatial) switch capable (FSC)

3. *Generalised protocol ID(G-PID) sub-field*
   This sub-field identifies the payload carried on the LSP, i.e., traffic from client for specific LSP. Their values and interpretations are shown in Table 14.4.

Table 14.4 *G-PID sub-field values in GMPLS messages and interpretation*

Value of G-PID sub-field	Traffic carried	Technology
0	Unknown	All
1–4	ANSI PDH signals (T hierarchy)	ANSI-PDH
5	Asynchronous mapping of E4	SDH
6	Asynchronous mapping of T3	SDH
7	Asynchronous mapping of E3	SDH
8	Bit synchronous mapping of E3	SDH
9	Byte synchronous mapping of E3	SDH
10	Asynchronous mapping of T2	SDH
11	Bit synchronous mapping of T2	SDH
12	Byte synchronous mapping of T2	SDH

(Continues...)

Table 14.4 *Continued*

Value of G-PID sub-field	Traffic carried	Technology
13	Asynchronous mapping of E1	SDH
14	Byte synchronous mapping of E1	SDH
15	Byte synchronous mapping of 31*DS0	SDH
16	Asynchronous mapping of DS1/T1	SDH
17	Bit synchronous mapping of DS1/T1	SDH
18	Byte synchronous mapping of DS1/T1	SDH
19	Same as 12, but in a VC-12	SDH
20	Same as 13, but in a VC-12	SDH
21	Same as 14, but in a VC-12	SDH
22–25	ANSI-PDH mapping in SONET	SONET
26	VT mapping in SONET	SONET
27	STS	SONET
28	POS-no scrambling, 16 bit CRC	
29	POS-no scrambling, 32 bit CRC	
30	POS-scrambling, 16 bit CRC	
31	POS-scrambling, 32 bit CRC	
32	ATM mapping	SONET/SDH
33	Ethernet	Lambda, Fibre
34	Ethernet	Lambda
35	SONET	Lambda
36	Digital wrapper	Lambda
37	Lambda	Fibre

Bandwidth encoding

Bandwidth encodings are carried in 32-bit number in IEEE floating point format (the unit is bytes per second). Details of the same are given in Table 14.4.

Table 14.4 *Bandwidth value sub-field in GMPLS label*

Signal	Bandwidth	Value
DS0	64 kbps	0x45FA0000
DS1	1.544 Mbps	0x483C7A00
E1	2.048 Mbps	0x487A0000
DS2	6.312 Mbps	0x4940A080

(Continues...)

Table 14.4 *Continued*

Signal	Bandwidth	Value
E2	8.448 Mbps	0x4980E800
Ethernet	10 Mbps	0x49989680
E3	44.736 Mbps	0x4A831A80
DS3	44.736 Mbps	0x4AAAA780
STS-1	51.84 Mbps	0x4AC5C100
Fast Ethernet	100 Mbps	0x4B3EBC20
E4	139.264 Mbps	0x4B84D000
OC-3/STM-1	155.52 Mbps	0x4B9450C0
OC-12/STM-4	622.08 Mbps	0x4C9450C0
Gigabit Ethernet	1 Gbps	0x4CEE6B28
OC-48/STM-16	2.5 Gbps	0x4D9450C0
OC-192/STM-64	10 Gbps	0x4E9450C0
10 Gbps Ethernet	10 Gbps	0x4E9502F9
OC-768/STM-256	40 Gbps	0x4F9450C0

Technology dependent or traffic parameters part of generalised label

A brief on the three sub-fields in traffic parameters part of GMPLS label for LSP of ODUk/OCh is presented here.

1. *Signal type (ST)*
   This field (8 bits) indicates the type of G.709 signal that comprises the requested LSP. The permitted values are given in Table 14.5.

Table 14.5 *ST sub-field interpretation in traffic parameters part*

ST Value	Type
0	Not significant
1	ODU1
2	ODU2
3	ODU3
4	Reserved
5	Reserved
6	OCh at 2.5 Gbps
7	OCh at 10 Gbps
8	OCh at 40 Gbps
9–255	Reserved

It may be noted that the value of the 'signal type' sub-field is correlated to the 'LSP encoding type' sub-field value:

a. If the LSP encoding type value is G.709 digital path layer, then the values of ST are 1 or 2 or 3 (for ODUk, with $k = 1, 2$ or 3).

b. If the LSP encoding type value is G.709 optical channel layer, then the values of ST are 1 or 2 or 3 (for OCh at 2.5, 10 or 40 Gbps).

c. If the LSP encoding type is lambda (which includes the pre-OTN optical channel layer), then ST = 0.

d. If the LSP encoding type is digital wrapper, then ST = 0.

It may also be noted that several transforms can be sequentially applied on the elementary signal to build the final signal being actually requested for the LSP. Transforms must be applied strictly in the following order:

First, a virtual concatenation (by using the NVC field) can be optionally applied directly on the elementary signal to form a composed signal.

Secondly, a multiplication (by using the multiplier field) can be applied optionally either directly on the elementary signal, or on the virtually concatenated signal obtained from the first phase.

The resulting signal is referred to as the final signal.

2. *Number of multiplexed channels (NMC)*
The NMC field (16 bits) indicates the number of ODU tributary slots used by an ODUj when multiplexed into an ODUk ($k > j$) for the requested LSP. This field is not applicable when an ODUk is mapped into an OTUk and is irrelevant at the optical channel layer. In both cases, it *must* be set to zero (NMC = 0) when sent and should be ignored when received.

   When applied at the digital path layer, in particular for ODU2 connections multiplexed into one ODU3 payload, the NMC field specifies the number of individual tributary slots (NMC = 4) constituting the requested connection. These components are still processed within the context of a single connection entity. For all other currently defined multiplexing cases, the NMC field is set to 1.

3. *Number of virtual components (NVC)*
The NVC field (16 bits) is dedicated to ODUk virtual concatenation (ODUk inverse multiplexing) purposes. It indicates the number of ODU1, ODU2 or ODU3 elementary signals that are requested to be virtually concatenated to form an ODUk-Xv signal. By definition, these signals *must* be of the same type.

   This field is set to 0 (default value) to indicate that no virtual concatenation is requested. Note that the current usage of this field only applies for G.709 ODUk LSPs i.e., values greater than zero are only acceptable for ODUk signal types.

Therefore, it *must* be set to zero (NVC = 0) and should be ignored when received i.e., when a G.709 OCh LSP is requested.

4. *Multiplier (MT)* -
   The multiplier field (16 bits) indicates the number of identical elementary signals or composed signals requested for the LSP, i.e., those which form the final signal. A composed signal is the resulting signal from the application of the NMC and NVC fields to an elementary signal type. GMPLS signalling currently implies that all the composed signals must be part of the same LSP.
   This field is set to one (default value) to indicate that exactly one instance of a signal is being requested. Intermediate and egress nodes *must* verify that the node itself and the interfaces on which the LSP will be established can support the requested multiplier value. If the requested values cannot be supported, the receiver node *must* generate a path-error message.
   Zero is an invalid value for the MT field. If received, the node *must* generate a path-error message.
5. *Reserved*
   The reserved fields (8 bits in row 1 and 32 bits in row 3) are dedicated for future use. Reserved bits *should* be set to zero when sent and *must* be ignored when received.

Having discussed the format of GMPLS labels for OTN, let us turn our attention to their handling in OTN.

14.8.4 Handling labels in OTN (OTH)

ODUk label space

It was discussed in section 12.3 that ODUk multiplexing supported by G.709 are the multiplexing of ODUj ($j = 1, 2$) into an ODUk ($k > j$) signal, i.e.,

- ODU1 into ODU2 multiplexing
- ODU1 into ODU3 multiplexing
- ODU2 into ODU3 multiplexing
- ODU1 and ODU2 into ODU3 multiplexing.

Therefore, the label space structure is a tree whose root is an OTUk signal and leaves are the ODUj signals ($k >= j$) that can be transported via the tributary slots and switched between these slots.

The digital path layer label carried in G.709 frame identifies the exact position of a particular ODUj signal in an ODUk multiplexing structure. This label has the format shown in Fig. 14.17(c). The description of the same is given below.

1. The value space for the t1, t2 and t3 sub-fields are defined as follows.
   t1 (1-bit):

- t1 = 1 indicates an ODU1 signal.
- t1 is not significant for the other ODUk signal types (i.e., the t1 value *must* be set to 0 and ignored).

t2 (3-bit):

- t2 = 1 indicates an ODU2 signal that is not further sub-divided.
- t2 = 2 to 5 indicates the tributary slot used by the ODU1 in an ODTUG2 mapped into an ODU2 (via OPU2).
- t2 is not significant for an ODU3 (i.e., t2 value *must* be set to 0 and ignored).

t3 (6-bit):

- t3 = 1 indicates an ODU3 signal that is not further sub-divided.
- t3 = 2 to 17 indicates the tributary slot (t3th–1) used by the ODU1 in an ODTUG3 mapped into an ODU3 (via OPU3).
- t3 = 18 to 33 indicates the tributary slot (t3th–17) used by the ODU2 in and ODTUG3 mapped into an ODU3 (via OPU3).

2. Reserved bits *must* be set to zero when sent and are to be ignored when received.
3. In case of ODU2 into ODU3 multiplexing, four labels are required to identify the four tributary slots used by the ODU2; these tributary time slots have to be allocated in an ascending order. If the label sub-field value t[i] = 1 (i, j = 1, 2 or 3) and t[j] = 0 ($j > i$), the corresponding ODUk signal ODU[i] is directly mapped into the corresponding OTUk signal ($k = i$). This is referred to as the mapping of an ODUk signal into an OTUk of the same order. Therefore, the numbering starts at 1; zero is used to indicate a non-significant field. A label field equal to zero is an invalid value.

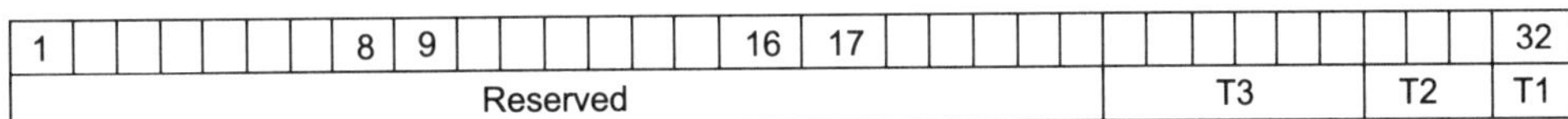

Fig. 14.17(c) *ODUk label format*

Label distribution rules

1. In case of ODUk in OTUk mapping, only one label can appear in the generalised label. The unique label is encoded as a single 32-bit label value (as defined in previous sub-section).
2. In case of ODUj in ODUk ($k > j$) multiplexing, the explicit ordered list of the labels in the multiplex is given (this list can be restricted to only one label when NMC = 1). Each label indicates a component (ODUj tributary slot) of the

multiplexed signal. The order of the labels must reflect the order of the ODUj into the multiplexer (not the physical order of tributary slots). This ordered list of labels is encoded as a sequence of 32-bit label values (as defined in previous sub-section).

3. In case of ODUk virtual concatenation, the explicit ordered list of all labels in the concatenation is given. Each label indicates a component of the virtually concatenated signal. The order of the labels must reflect the order of the ODUk to concatenate (not the physical order of time-slots). This representation limits virtual concatenation to remain within a single (component) link.
4. In case of multiplexed virtually concatenated signals, the first set of labels indicates the components (ODUj tributary slots) of the first virtually concatenated signal, the second set of labels indicates the components (ODUj tributary slots) of the second virtually concatenated signal, and so on. This ordered list of labels is encoded as a sequence of 32-bit label values (as defined in previous sub-section).
5. In case of ODUk virtual concatenation, the number of label values is determined by the NVC value. Multiplexed ODUk virtual concatenation additionally uses the NMC value to determine the number of labels per set (equal in size).
6. In case of multiplication (when using the MT field), the explicit ordered list of all labels taking part in the composed signal is given. The above representation limits the multiplication to remain within a single (component) link. In case of multiplication of multiplexed virtually concatenated signals, the first set of labels indicates the components of the first multiplexed virtually concatenated signal, the second set of labels indicates components of the second multiplexed virtually concatenated signal, and so on.
7. This ordered list of labels is encoded as a sequence of 32-bit label values (as defined in previous sub-section). In case of multiplication of (equal) ODUk virtual concatenated signals, the number of label values per signal is determined by the NVC value. Multiplication of multiplexed (equal) ODUk virtual concatenation additionally uses the NMC value to determine the number of labels per set (equal in size).

Optical channel label space

At the optical channel layer, the label space must be consistently defined as a flat space whose values reflect the local assignment of OCh identifiers corresponding to the OTM-n.m sub-interface signals ($m = 1, 2$ or 3). Note that these identifiers do not cover OChr since the corresponding connection function (OChr-CF) between OTM-nr.m/OTM-0r.m is not defined in ITU-T G.798.

The OCh label space values are defined by either absolute values (i.e., channel identifiers or channel ID, also referred to as wavelength identifiers) or relative values (channel spacing also referred to as inter-wavelength spacing). The latter is strictly confined to a per-port label space while the former could be defined as a local or a global (per node) label space. Such an OCh label space is applicable to both OTN optical channel layer and pre-OTN optical channel layer.

Examples of label processing in ODUk and OCh

The following examples are given in order to illustrate the processing of labels described so far.

Example 1: ODUk in OTUk mapping

When one ODU1 (ODU2 or ODU3) signal is directly transported in an OTU1 (OTU2 or OTU3), the upstream node requests results simply in an ODU1 (ODU2 or ODU3) signal request. In such conditions, the downstream node has to return a unique label since the ODU1 (ODU2 or ODU3) is directly mapped into the corresponding OTU1 (OTU2 or OTU3). Since a single ODUk signal is requested (signal type = 1, 2 or 3), the downstream node has to return a single ODUk label which can be, for instance, one of the following when the signal type = 1:

t3 = 0, t2 = 0, t1 = 1 indicating a single ODU1 mapped into an OTU1.
t3 = 0, t2 = 1, t1 = 0 indicating a single ODU2 mapped into an OTU2.
t3 = 1, t2 = 0, t1 = 0 indicating a single ODU3 mapped into an OTU3.

Example 2: ODU1 into ODUk multiplexing (*k* > 1)

When one ODU1 is multiplexed into the payload of a structured ODU2 (or ODU3), the upstream node requests results simply in a ODU1 signal request. In such conditions, the downstream node has to return a unique label since the ODU1 is multiplexed into one ODTUG2 (or ODTUG3).

The latter is then mapped into the ODU2 (or ODU3) via OPU2 (or OPU3) and then mapped into the corresponding OTU2 (or OTU3). Since a single ODU1 multiplexed signal is requested (signal type = 1 and NMC = 1), the downstream node has to return a single ODU1 label which can take for instance one of the following values:

t3 = 0, t2 = 4, t1 = 0 indicates the ODU1 in the third TS of the ODTUG2.
t3 = 2, t2 = 0, t1 = 0 indicates the ODU1 in the first TS of the ODTUG3.
t3 = 7, t2 = 0, t1 = 0 indicates the ODU1 in the sixth TS of the ODTUG3.

Example 3: ODU2 into ODU3 multiplexing

When one unstructured ODU2 is multiplexed into the payload of a structured ODU3, the upstream node requests results simply in a ODU2 signal request. In such conditions, the downstream node has to return four labels since the ODU2 is multiplexed into one ODTUG3. The latter is mapped into an ODU3 (via OPU3) and then mapped into an OTU3. Since an ODU2 multiplexed signal is requested (signal type = 2 and NMC = 4), the downstream node has to return four ODU labels which can take for instance the following values:

t3 = 18, t2 = 0, t1 = 0 (first part of ODU2 in first TS of ODTUG3).
t3 = 22, t2 = 0, t1 = 0 (second part of ODU2 in fifth TS of ODTUG3).
t3 = 23, t2 = 0, t1 = 0 (third part of ODU2 in sixth TS of ODTUG3).
t3 = 26, t2 = 0, t1 = 0 (fourth part of ODU2 in ninth TS of ODTUG3).

Example 4: When a single OCh signal of 40 Gbps is requested

When a single OCh signal of 40 Gbps is requested (signal type = 8), the downstream node must return a single wavelength label as specified in previous sub-section.

Example 5: When requesting multiple ODUk LSP

When requesting multiple ODUk LSP, the a multiplier (MT) (value > 1) and an explicit list of labels is returned to the requestor node.

When the downstream node receives a request for a 4 × ODU1 signal (signal type = 1, NMC = 1 and MT = 4) multiplexed into a ODU3, it returns an ordered list of four labels to the upstream node: the first ODU1 label corresponding to the first signal of the LSP, the second ODU1 label corresponding to the second signal of the LSP, etc. For instance, the corresponding labels can take the following values:

First ODU1: t3 = 2, t2 = 0, t1 = 0 (in first TS of ODTUG3).
Second ODU1: t3 = 10, t2 = 0, t1 = 0 (in ninth TS of ODTUG3).
Third ODU1: t3 = 7, t2 = 0, t1 = 0 (in sixth TS of ODTUG3).
Fourth ODU1: t3 = 6, t2 = 0, t1 = 0 (in fifth TS of ODTUG3).

14.8.5 Waveband switching

A special case of lambda switching is the waveband switching. A waveband represents a set of contiguous wavelengths which can be switched together to a new waveband. This is resorted for optimisation purpose at optical cross-connect the to reduce the distortion on the individual wavelengths and to provide better separation of the individual wavelengths.

Waveband switching introduces another level of label hierarchy, so a waveband label is required to support the same. As far as the MPLS protocols are concerned, there is little difference between a waveband label and a wavelength label except that semantically the waveband can be subdivided into wavelengths whereas the wavelength can only be subdivided into time or statistically multiplexed labels.

A waveband label has the format shown in Fig. 14.18. The sub-fields are:

1. *Waveband ID (32 bits)*: A waveband identifier. The value is selected by the sender and reused in all subsequent related messages.
2. *Start label (32 bits)*: Indicates the channel identifier of the lowest value wavelength making up the waveband from the object/ sender's perspective.
3. *End label (32 bits)*: Indicates the channel identifier of the highest value wavelength making up the waveband from the object/sender's perspective.

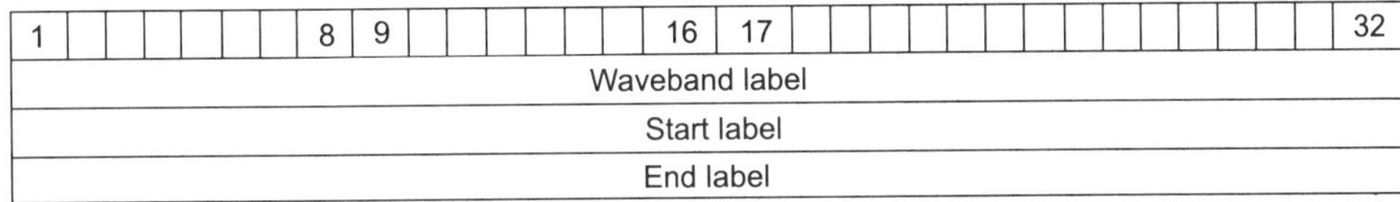

Fig. 14.18 *Waveband label format*

14.9 Fibre Access Network and OTN Support

Our focus so far has been on long-haul networks with interconnection of several NEs and on their applications. Let u turn our attention to the issue of fibre access to the subscriber and explore the support OTN gives for the same.

Subscriber access has traditionally been on copper in PSTN. With the development of ADSL, high speed internet access has also been made possible on copper. As discussed in chapter 1, the latest in DSL technology viz. ADSL2+ extends the frequency band to 2.2 MHz on copper and transmission rate to 20 Mbps over 1.2 km. One such channel can support one HDTV channel, two channels of standard definition TV and high-speed internet access.

Another solution for subscriber access is WiMAX. The IEEE 802.16 series of WMAN standard specifies operation in the 2 GHz to 66 GHz band with coverage of 50 km and transmission speed of 70 Mbps for the same. In some of the countries in North America, hybrid fibre coaxial (HFC) has been tried. In HFC, data access is provided on fibre and on coaxial cable on a bandwidth of 50–550 MHz, 78 AM-VSB (amplitude modulated vestigial sideband) television channels with 6 MHz separation.

A simpler solution to the providing access provision would be an OFC based 'passive optical network (PON)'. In fact, PON has the potential to eliminate copper totally in access network. With this, the hassles of wireless access can be avoided. Refer to Fig. 14.19.

PON is the network from CO (central office of a telecom company) to the fibre terminating nodes called optical network units (ONUs). Each ONU serves a number of subscribers.

When the fibre termination or ONU is at about 1 km from the home or the network interface unit (NIU), the PON is known as FTT cab (fibre to the cabinet). From cabinet, the distribution is on copper. When the fibre termination or ONU is within 100 m of a building, each ONU serving few homes to tens of homes, the PON is known as FTTC (fibre to the curb) or fibre to the building (FTTB). When fibre reaches home, i.e., ONU and NIU merge, it is FTTH (fibre to the home). In this case the PON is right up to the home.

Whatever be the case, the CO to ONU link which constitutes the PON has several alternatives for architecture. It can be, for instance, star (each ONU connected on exclusive fibre pair to CO), ring (few ONUs are on fibre pair ring). The most common architecture of PON are TPON (originally called PON for telephony and now TDM PON) and WPON (WDM PON). Refer to TPON shown in Fig. 14.20.

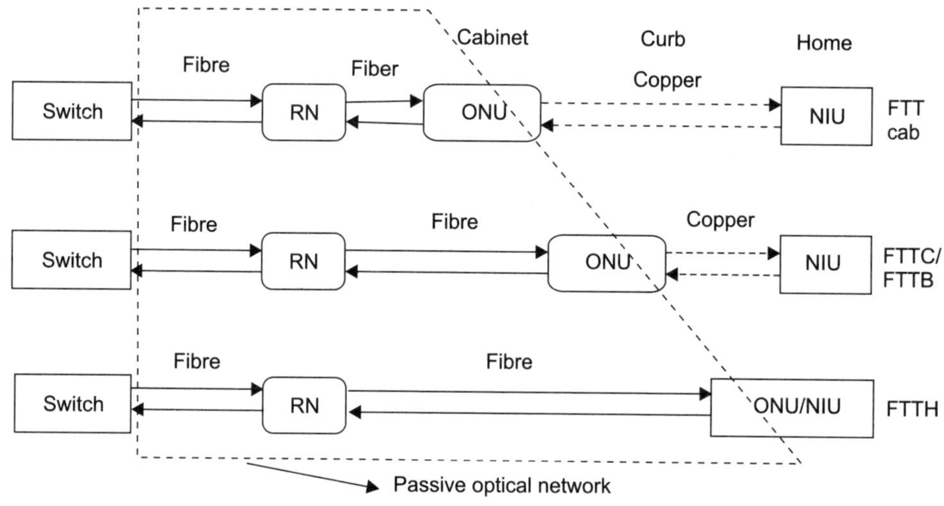

Fig. 14.19 *Different cases of PON reach: FTT cab, FTTC/FTTB, FTTH*

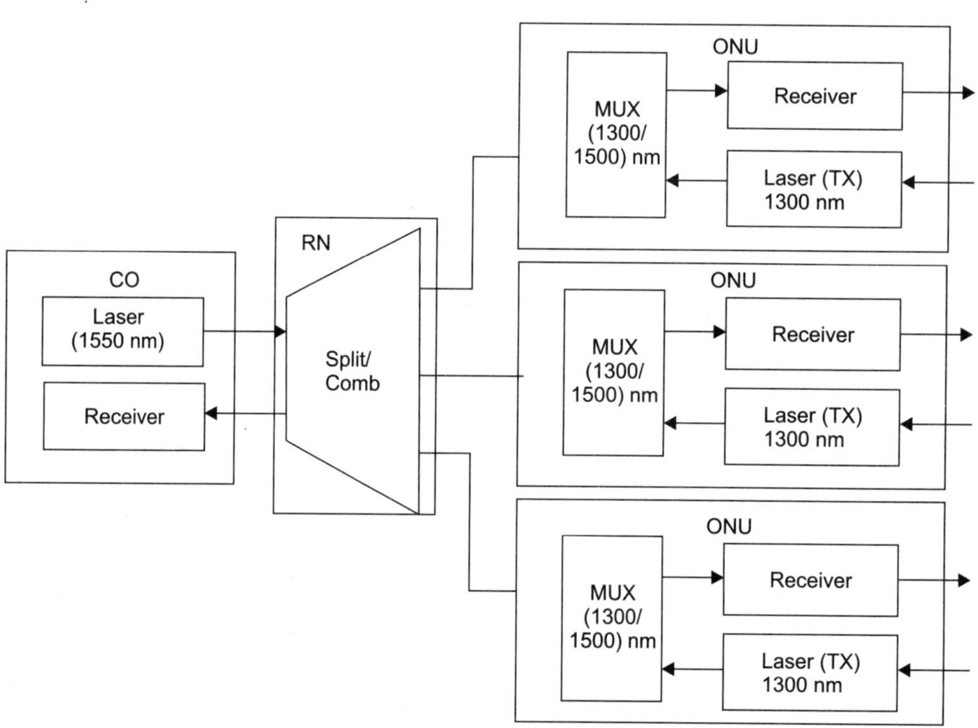

Fig. 14.20 *TPON (CO to ONU – minimum STM4 and ONU to CO – minimum STM-1)*

The CO sends downstream traffic to all ONUs through a passive star coupler. Specific time slots are assigned to individual ONUs to facilitate switching. TPON makes use of SDH (on down link a minimum of STM-4 and on up-link a minimum of STM-1).

Refer to Fig. 14.21 showing WPON. In this, the CO broadcasts multiple wavelengths to all ONUs and each ONU selects a particular wavelength. ONU and CO form an OTN. The transceiver at CO is now an array of WDM transceivers. But, the loss in splitter/combiner is quite significant. To avoid loss in splitter/combiner, WPRON (wavelength routing architecture) is resorted to, which is shown in Fig. 14.22. We can also have point-to-point dedicated services to ONUs in this architecture. ONUs and CO are NEs of OTN. Passive arrayed waveguides (AWG) are used for routing different wavelengths to the desired ONUs.

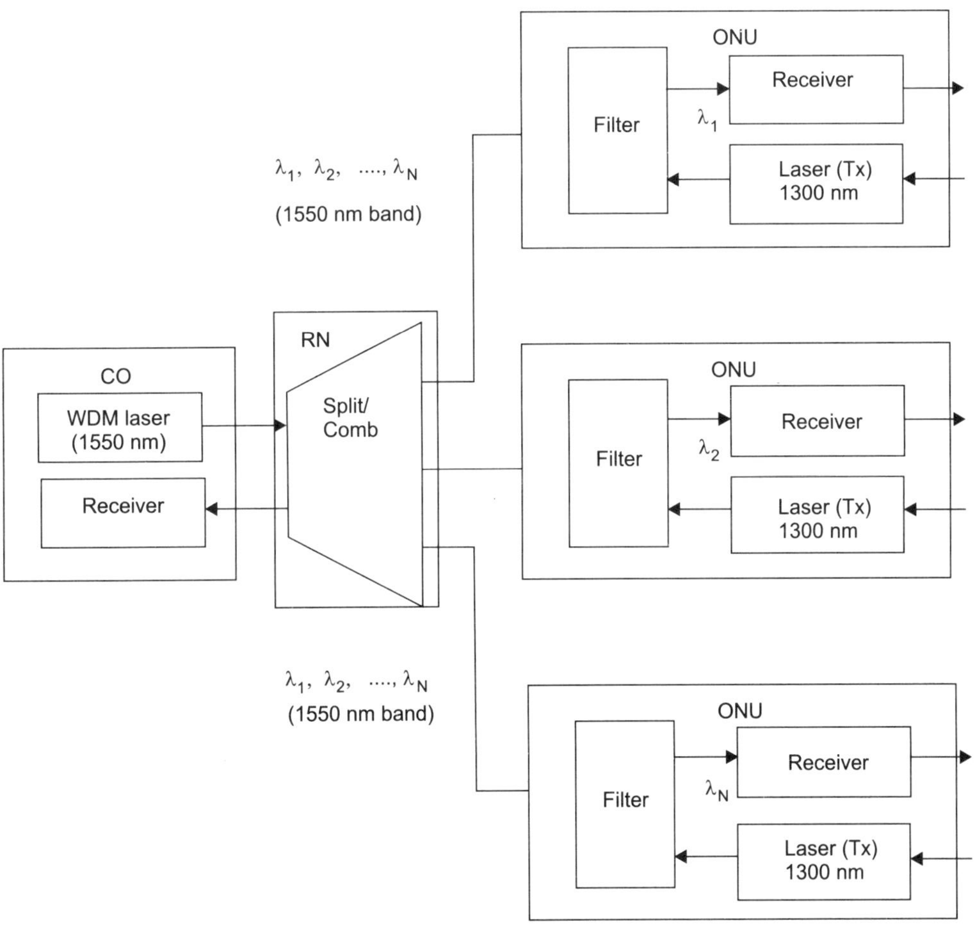

Fig. 14.21 *WPON (ONUs have electronics working at desired wavelength)*

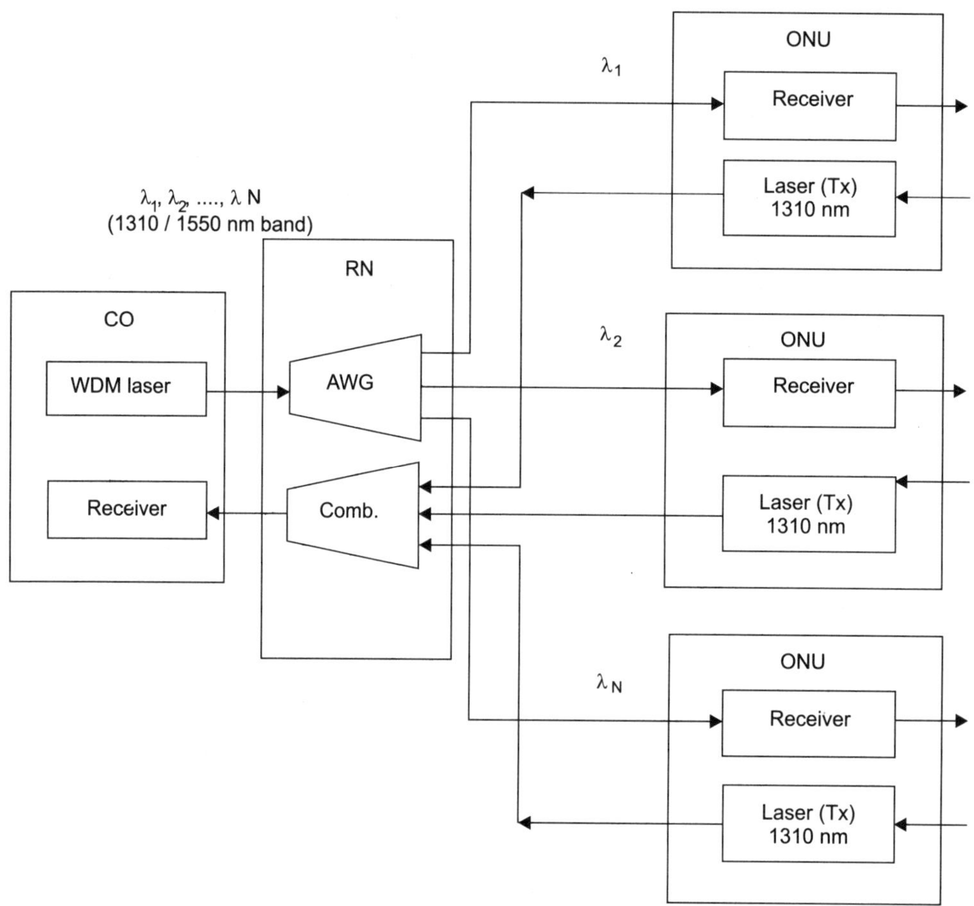

Fig. 14.22 *Wavelength routing PON (WPRON)*

14.10 OTN and NGN

14.10.1 Overview of NGN

The use of IP flow format for VoIP , SDH network support for transportation of the same and the evolution of VoIP to carrier grade have been discussed in section 9.6. Further, it has been discussed in section 9.7 that the need to go beyond the precincts of VoIP to support multimedia and innovative value added services in a cost-effective manner, with implementation strategies of scalable gateways and integration with existing PSTN switches rather than their expansion led to the evolution of next generation networking switch or NGN switch. While features of NGN switch and SDH network support for connectivity have been discussed in section 9.7, a full-fledged discussion on NGN itself has been deferred to this section. In fact, in NGN, IP flow is not just for VoIP or multimedia; it is an omnipresent phenomenon. Let us now move on to full-fledged discussion on NGN to understand how it is the convergence platform for many with its all pervasive IP flow.

Dramatic changes have taken place in telecom sector in the recent past, both on technology front as well as the services front. On the technology front these changes are – availability of quick updates of any technology or version, decrease in cost of network, and increase in sophistication of the equipment. On the services sector, fixed line telephony has been overtaken by mobile telephony and IP telephony. Again, mobile telephony is being overtaken by mobile access for V+D+video. Though IP is expanding with its migration from IPv4 to IPv6, the IP business model has changed from one of low-risk high-profit model to a high-risk, low-profit one.

All these developments call for emergence of a new network which would converge these evolutions towards:

1. Seamless multiple access
2. Next generation soft switch
3. Next generation IPv6 based network
4. Next generation of cellular networks 3G and beyond
5. Next generation automatic switching optical network.

According to ITU-T, NGN is a packet based network able to provide telecom services and able to make use of multiple broadband, QoS enabled technologies and in which service-related functions are independent from the underlying transport-related technologies. It offers unfettered access by users to different service providers. It allows generalised mobility which will allow consistent and ubiquitous provision of services to users.

According to ITU-T Rec. Y.2001[1], the NGN is characterised by the following fundamental aspects:

- Packet-based transfer
- Separation of control functions among bearer capabilities, call/session and application/service
- Decoupling of service provision from transport, and provision of open interfaces
- Support for a wide range of services, applications and mechanisms based on service building blocks (including real time/streaming/non-real time services and multi-media)
- Broadband capabilities with end-to-end QoS and transparency
- Interworking with legacy networks via open interfaces
- Generalised mobility
- Unfettered access by users to different service providers
- A variety of identification schemes which can be resolved to IP addresses for the purposes of routing in IP networks

- Unified service characteristics for the same service as perceived by the user
- Converged services between fixed and mobile networks
- Independence of service-related functions from underlying transport technologies
- Support of multiple last mile technologies
- Compliance with all regulatory requirements, for example, concerning emergency communications and security/privacy, etc.

At the core of NGN is the IP technology which allows service providers to converge technologies, services, networks, business prospects, terminals and finally regulation policies.

With IP services at the core, many functions of significance in voice networks become meaningless and consequently many layers in hierarchical architecture in various entities disappear. The network converges to simple architecture, low-cost deployment offering unified services and products to subscribers through single access across fixed/wire-less/internet with a motto of 'one bill – all services'.

This convergence towards NGN cannot take place overnight. It is estimated that it would take around a decade to fifteen years to realise. The right approach to know the role of OTN in such convergence would be to discuss the five evolutions listed earlier and analyse how OTN will support them.

14.10.2 Evolution towards seamless multiple broadband access – Role of OTN

As discussed in section 14.8, the alternatives for subscriber access for high speed access are ADSL variants, WiMAX, HFC and PON. PON appears to be the ideal solution for the following reasons:

1. Active equipment is not in the access network (no EMI)
2. Failure rate of line and equipment is less (if fibre is properly laid)
3. PON bandwidth is the highest among all the alternatives
4. PON service is transparent
5. No spectrum related problems

The disadvantage is that the cost of WDM electronics associated with WPON is high. It is expected to reduce in few years. Also, sharing of ONU by number of users reduces the cost per subscriber.

ITU-T has issued recommendation for gigabit PON (GPON) which has the following features:

1. Full service support including voice, TDM, Ethernet, ATM, leased lines, wireless extension.

2. Physical reach of at least 20 km with a logical reach support within the protocol of 60 km.
3. Support for various bit rate options using the same protocol, including symmetrical 622 Mbit/s, symmetrical 1.25 Gbit/s, asymmetrical (2.5 Gbit/s downstream and 1.25 Gbit/s upstream) and more.
4. Operation, administration, maintenance and provisioning (OAM&P) capabilities offering end-to-end service management.
5. Security at the protocol level for downstream traffic due to the multicast nature of PON.

The ITU-T recommendations concerned with GPON are as follows.

1. **G.984.1: General characteristics of GPON**
   This recommendation provides examples of services, user network interfaces (UNI) and service node interfaces (SNI) that are required by network operators. In addition, it shows the principal deployment configuration.
   Wherever possible, this recommendation maintains characteristics from the ITU-T G.982 and G.983.x series of recommendations in order to promote backward compatibility with the existing optical distribution networks (ODN) that comply with these recommendations.
2. **G.984.2: Physical media dependent (PMD) layer specification of GPON**
   This recommendation specifies the physical layer requirements and specifications for the physical media dependent (PMD) layer. It covers systems with nominal line rates of:
   a. 1244.160 Mbit/s (1.25 Gbps) and 2488.320 Mbit/s (2.5Gbps) in the downstream direction
   b. 155.520 Mbit/s, 622.080 Mbit/s, 1244.160 Mbit/s (1.25 Gbps) and 2488.320 Mbit/s (2.5Gbps) in the upstream direction. Both symmetrical and asymmetrical (upstream/downstream) gigabit-capable passive optical network (G-PON) systems are described.
3. **G.984.3: Transmission convergence layer specification**
   This recommendation specifies the frame format, media access control method, ranging method, OAM functionality and security in G-PON networks.

As can be seen, PON discussed in section 14.8 and GPON discussed above require high-end STM rates. As fibres cannot be laid exclusively for each customer right from the switch, there would be aggregation up to an intermediate point from the switch, after which there would be distribution to individual premises. When several such streams of PON/GPON are aggregated, the data streams will necessarily have to be in OTH, i.e., the access network for fixed line part of NGN is optical in core. In other words, PON/GPON are feasible through OTN, appropriately laid for the aggregated data rates, with survivability incorporated into it.

Q 14.6 2.5 Gbps is the STM-16 rate and 622.08 Mbps is the STM-4 rate. What is the significant of 1.25 Gbps?

When PONs are in a ring with shared protection bandwidth, the STM-16 capacity is divided for working and protection traffics, making it 1.25 Gbps.

14.10.3 Evolution towards next generation soft switch – support from OTN

OTN is concerned with transmission network whereas soft switch is concerned with switching. They may appear to be uncorrelated but are not so.

Soft-switch has a structure different from the traditional close-switch structure. It is based on distributed control, open interfaces and generic protocols. Services in soft-switch are controlled in a centralised way. The soft-switch requires OAM&P of real-time services, QoS guarantees, security management and protection mechanisms suitable for distributed processing. Ultimately, it also requires a convergent platform with mobile services.

The IMS (IP multimedia subsystem) of 3GPP (3G partnership program) is the convergent platform. IMS is an IP multimedia core network system with IP as basis which has no relevance to access modes and access technology. It supports VoIP, data and multimedia services. The core network, subscriber database, charging system and service development platform are shared by all services (fixed line, mobile and IP).

So, OTN provides the connectivity (other than the air-interfaces) for a soft switch architecture requiring the features mentioned above with ultimate convergence with IMS.

14.10.4 Evolution towards IPv6 – support from OTN

Around 1.2 billion IPv4 address were left by the end of year 2007, which got exhausted by the end of year 2008. IPv4 has limitations on QoS and security. So, it is time to migrate from IPv4 (32 bit address) to IPv6 (128 bit address). There are several advantages of IPv6 the enumeration of which would require at least one full chapter to do justice to the task. As our focus is on transmission hierarchies and networks, we will go through the prominent five features of IPv6 only.

1. Because of 128-bit address in IPv6, addressing problems on network layer are eliminated. Internet can be expanded to enterprise network and home network
2. Built-in mobile IPv6 protocol makes terminal movement and access across different media feasible. Seamless usage of fixed line access (fibre/ADSL), WLAN, 3G, WiMAX is feasible.
3. Permanent IPv6 address can be given to transmitter and end-to-end encryption will be easier.
4. The automatic discovery and configuration function of IPv6 makes NE management easier. Plug and play function involving mobile nodes, household appliances and communication devices can be easily realised.

5. Hierarchical addressing structure improves routing efficiency. Also, priority levels can be given to outgoing packets by source node (a boon for real time applications).

IPv6 with MPLS (and variants) will provide the bearer layer for NGN. Support for MPLS (and variants) by OTN have already been discussed in the previous sections.

Furthermore, OTN helps the cause of migration from IPv4 to IPv6. Optical cross-connects act as optical routers, GMPLS enabled, and provide IPv6 links. This is elaborated in Appendix-A.

14.10.5 Evolution towards next generation cellular networks

2G mobile communications systems offer data transfer rates of the order of tens of kbps. GPRS, which can be called as 2.5G system, supports data transfer rate of 115 kbps. The focus of 3G systems was to provide data transfer rates of 2 Mbps over air-interface. 3G systems are distinguished from the previous generations by data transfer rate to mobiles over air interface at 2 Mbps in stationary condition, 384 kbps in slow movement (up to 50 kmph) and 144 kbps in fast movement (> 50 kmph).

The main challenge in mobile communications is in the access method available over air interface. As different countries embarked on different initiatives for realizing 3G, initiatives were taken to create a single forum for the development of a common radio interface called the universal terrestrial radio access (UTRA). The result was the formation of the 'Third Generation Partnership Program (3GPP)' which came into existence in December 1998. The 3GPP adopted frequency division duplex (FDD) and time division duplex (TDD) modes of WCDMA for its air interface.

In North America, the cellular operators formed another group by the name 3GPP2. This group chose the multi-carrier CDMA-2000 radio system. Universal Wireless Communications Consortium (UWCC) chose the UWC136 (also known as the IMT-single carrier or IMT-SC) system. Another system, termed digital enhanced cordless telecommunications (DECT) system, was chosen by the ETSI DECT project.

All these standards, viz. W-CDMA, CDMA-2000, UWC136, DECT for access over air interface were adopted for third generation (3G) networks. These standards fall under the umbrella of International Mobile Telecommunications 2000 (IMT-2000) systems. Of these, W-CDMA of 3GPP and CDMA2000 of 3GPP2 are the most popular ones.

In WCDMA (wideband code division multiple access) network, the requirements of transport network are:

1. Connection to core network (circuit switched)
2. Interconnection between switched networks across cities
3. Transmission between radio network controller and base station, with ATM as transmission bearer at STM-1.

1 and 2 above require protected paths for every pair of switch-nodes which forms a large meshed network. Also, the ATM bearers have to go through pseudo wires in tunnel LSPs, i.e., layer-2 MPLS or layer-3 MPLS, to have seamless connection across autonomous boundaries (refer to section 9.3.3 and 9.3.4). As the network grows, we will require more transmission capacities between nodes and will need to achieve low latency. Optical cross-connects with GMPLS functionalities are thus needed, which in turn implies that OTN is again the core.

14.10.6 NGN – A complete picture

A complete picture of NGN is presented in Fig. 14.23. Whatever be the aspect, connectivity for the transceivers/switches/controllers is through an OFC back-bone in an OTN.

End-to end auto-provisioning is possible through ASON (automatically switched optical network) which uses GMPLS and OTN control planes. ITU-T Rec. G.8080 for ASON makes physical layer VPN feasible in OTN. Thus, if NGN be the converging platform for all next generation services, OTN is the support structure and IP the flow format.

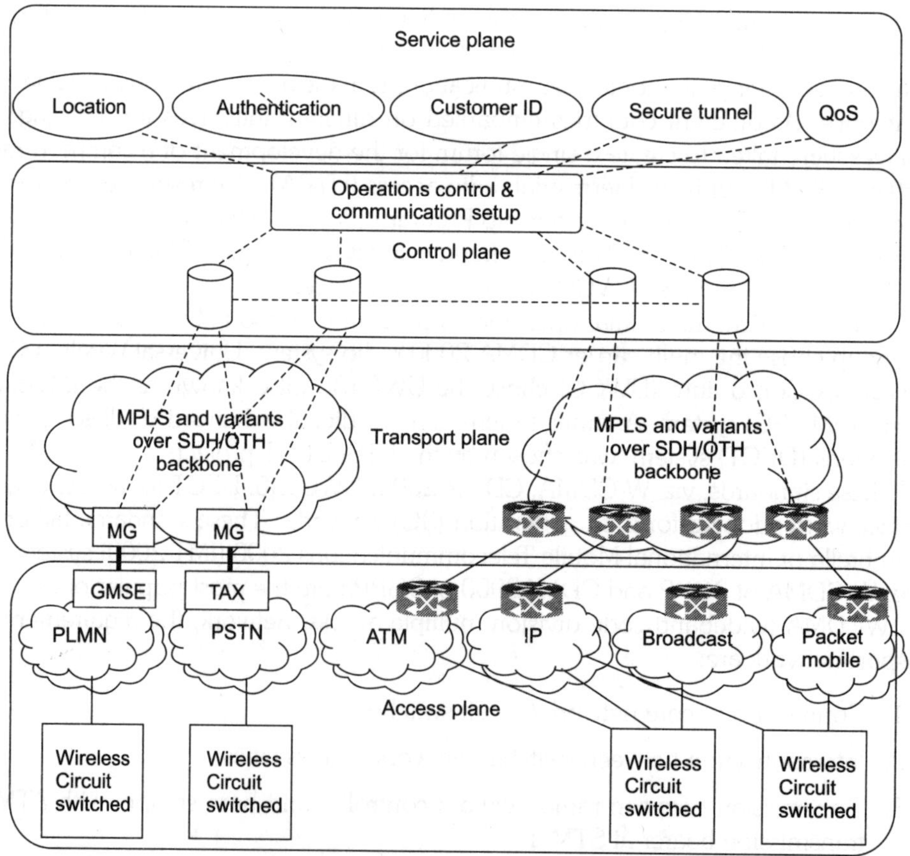

Fig. 14.23 *NGN architecture*

Summary

In this chapter, the NEs of OTN, viz., OLT, OADM and OXC are explained. Applications of OTNs in IP over optical networks, MPλS and GMPLS are discussed. The support of OTH for NGN components, viz., PON, soft-switch, IPv6 migration, 3G cellular network connectivity are explained.

References

[1] General overview of NGN, ITU-T Recommendation Y.2001, December 2004.

15

Reliability and Survivability Issues

15.1 Introduction

The control of jitter in the first and second generation hierarchies (PDH and SDH networks respectively) were discussed in chapters 5 and 10. In this chapter, control of jitter in the third generation hierarchy or OTH networks (OTN) will be discussed.

In an OTN, jitter and wander accumulate on transmission paths according to the jitter and wander generation and transfer characteristics of the ONEs (optical network elements) on the path. The optical network element may be 3R regenerators, client mappers/demapper (OADM/OXC) or client de-synchronisers. An excessive amount of jitter and wander can adversely affect both digital signals (by generation of bit errors, frame slips and other abnormalities) and analog baseband signals (by unwanted phase modulation of the transmitted optical signal). The consequences of such impairment depend on the particulars of the service being carried on the network and the on the terminating or adaptation equipment involved. Hence, the control of jitter in OTN involves the study of:

1. Network limits for the maximum output jitter and wander at an OTUk interface.
2. Jitter and wander tolerance of network interfaces (OTUk) and CBR2G5, CBR10G, CBR40G client interfaces.
3. Intrinsic jitter, i.e., output jitter in the absence of input jitter.
4. Jitter transfer function at ODUk, OTUk.
5. Effect of OTN on the distribution of synchronisation via STM-N clients.

The ITU-T recommendations G.8251 deal with the above aspects of control of jitter. Once the reliability issues mentioned above are addressed, the survivability issues, viz. the need for optical layer protection over and above the SDH protection mechanisms and their implementation will be taken up for discussion in this chapter.

15.2 Network Limits for the Maximum Output Jitter and Wander at an OTUk Interface

15.2.1 Network limits for jitter at OTUk interface

As studied in section 12.3, OTUk is the digital signal that is mapped into the optical channel (OCh). The OTUk bit rate is essentially the line rate associated with the OCh and the respective wavelength the OCh is assigned to.

The maximum permissible levels of jitter at OTUk interfaces are as per Table 15.1, as recommended by ITU-T Rec. G.8251. The jitter measured over a 60-second interval should not exceed the limits given in Table 15.1 when using the specified measurement filters. The limits specified in this table are to be met for all operating conditions, regardless of the amount of equipment that precedes the interface.

Table 15.1 *Network limits for jitter at the OTUk interfaces*

Interface	Measurement bandwidth (–3dB frequencies)	Peak-to-peak amplitude (UI_{pp})
OTU1	5 kHz to 20 MHz	1.5
	1 MHz to 20 MHz	0.15
OTU2	20 kHz to 80 MHz	1.5
	4 MHz to 80 MHz	0.15
OTU3	20 kHz to 320 MHz	6.0
	16 MHz to 320 MHz	0.15

The high-pass measurement filters of Table 15.1 have a first-order characteristic and a roll-off of 20 dB/decade. The low-pass measurement filters have a maximally-flat Butterworth characteristic and a roll-off of −60 dB/decade.

Q 15.1 *What is UI for OTU1, OTU2 and OTU3?*

For OTU1: 1 UI = [238/(255)(2.48832)] ns = 375.1 ps
For OTU2: 1 UI = [237/(255)(9.95328] ns = 98.38 ps
For OTU3: 1 UI = [236/(255)(39.81312)] ns = 23.25 ps

15.2.2 Network limits for wander

OTUk interfaces are not synchronisation interfaces. None of the ODUk clocks are significant sources of wander, and hence, there is no need for specification of network wander limits.

15.3 Jitter and Wander Tolerance of OTUk Interface

Jitter and wander tolerance of an OTUk interface indicates the minimum level of phase noise that the input port should accommodate whilst

- not causing any alarms,
- not causing any slips or loss of lock in the clock recovery phase-locked loop,
- not causing any bit errors in excess of those allowed by an equivalent 1 dB optical power penalty.

15.3.1 Measuring tolerable input jitter and wander

There are two ways of measuring the tolerable input jitter and wander.

1. **1 dB optical power penalty method:** In this method, the optical input power is reduced until a BER of 10^{-10} is reached. Then the power is increased by 1 dB and jitter is applied to the input signal. The amount of jitter that results in a BER of 10^{-10} is the limit of tolerable input jitter.

   It may be noted that as the system provides for FEC, this will be disabled for the measurement when using an external bit error detector.

2. **Error count method:** Another method is to measure BER, leaving the FEC enabled, by counting the number of corrected bit errors per unit time when an input signal with jitter is applied.

Whatever be the method adopted, the signal used for test purpose should be a jitter and wander modulated structured test sequence as defined in ITU-T Rec. G.709.

15.3.2 How is tolerance specified?

All OTUk input ports of an equipment should be able to tolerate an OCh adapted information and data signal which has

a. optical characteristics of ITU-T Rec. G.959.1,

b. ODUk clock not deviating worse than 20 ppm with respect to PRC,

c. sinusoidal phase deviation with its amplitude–frequency relationship defined in the following tables/figures OTUk, signal-wise.

Jitter tolerance of OTU1 interface

The level of sinusoidal jitter that must be accommodated by OTU1 interfaces is specified in Table 15.2 and illustrated in Fig. 15.1.

Table 15.2 *Jitter tolerance for OTU1 interface*

Frequency f (Hz)	Peak-to-peak amplitude (UI_{pp})
$500 < f \le 5$ k	$7500/f$
$5\text{ k} < f \le 100$ k	1.5
$100\text{ k} < f \le 1$ M	$150000/f$
$1\text{ M} < f \le 20$ M	0.15

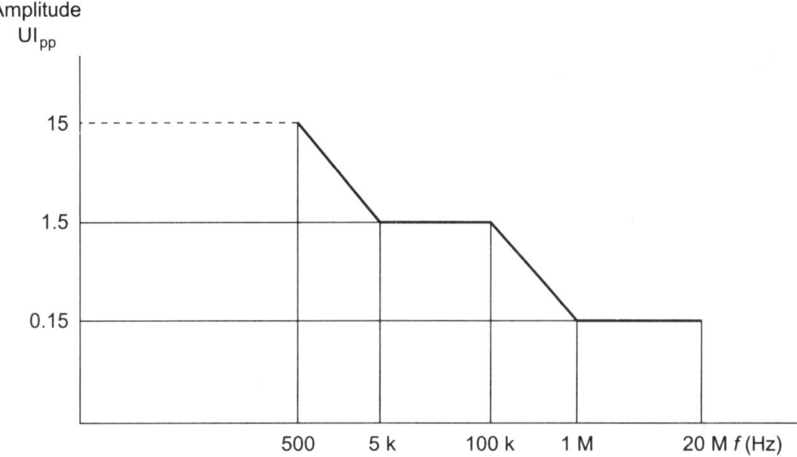

Fig. 15.1 *Graphical representation of jitter tolerance of OTU1 interface*

Jitter tolerance of OTU2 interface

The level of sinusoidal jitter that must be accommodated by OTU2 interfaces is specified in Table 15.3 and illustrated in Fig. 15.2.

Table 15.3 *Jitter tolerance for OTU2 interface*

Frequency f (Hz)	Peak-to-peak amplitude (UI_{pp})
2 k < f ≤ 20 k	30000/f
20 k < f ≤ 400 k	1.5
400 k < f ≤ 4 M	600000/f
4 M < f ≤ 80 M	0.15

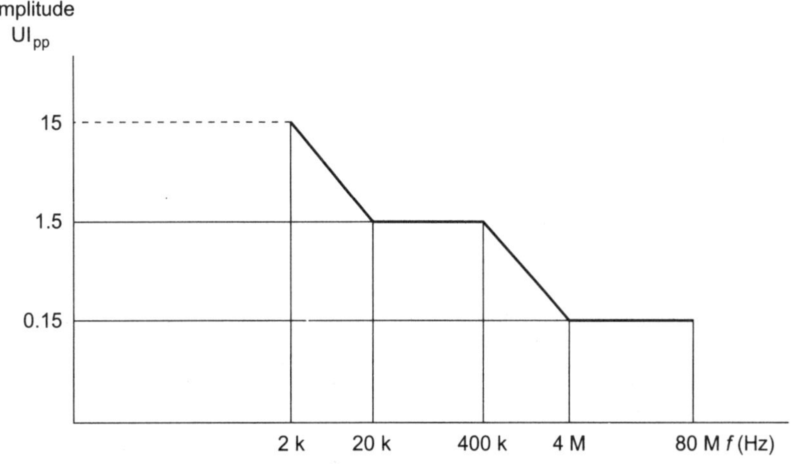

Fig. 15.2 *Graphical representation of jitter tolerance of OTU2 interface*

Jitter tolerance of OTU3 interface

The level of sinusoidal jitter that must be accommodated by OTU2 interfaces is specified in Table 15.4 and illustrated in Fig. 15.3.

Table 15.4 *Jitter tolerance for OTU3 interface*

Frequency f (Hz)	Peak-to-peak amplitude (UI_{pp})
8 k < f ≤ 20 k	120000/f
20 k < f ≤ 400 k	6.0
400 k < f ≤ 16 M	2400000/f
16 M < f ≤ 320 M	0.15

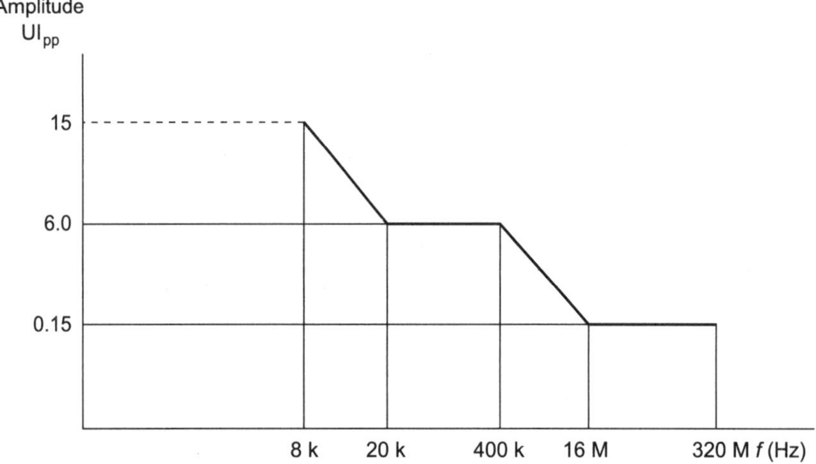

Fig. 15.3 *Graphical representation of jitter tolerance of OTU3 interface*

Q 15.2 Why do we not measure jitter and wander tolerance below 500 Hz in the case of OTU1 and below 8 kHz in the case of OTU2 and OTU3?

A phase-locked loop with corner frequency ≥ 5 kHz that can tolerate jitter and wander indicated by the 20 dB/decade sloped portion of the mask between 500 Hz and 5 kHz, as shown in Fig. 15.1 will also tolerate jitter and wander indicated by an extension of this sloped region to lower frequencies because such jitter and wander is within the bandwidth of the phase-locked loop. It means that OTU1 interfaces tolerate this jitter and wander as well below 500 Hz. Hence, there is no need to measure jitter and wander tolerance below 500 Hz. Also, for practical reasons, it is not possible to measure them below 500 Hz. For similar reasons we do not measure jitter and wander tolerance below 8 kHz in the case of OTU2 and OTU3.

15.3.3 Jitter and wander tolerance of CBR2G5, CBR10G, and CBR40G client interfaces

Network limits for jitter and wander and tolerance requirements for CBR2G5, CBR10G are derived from the corresponding requirements for STM-16 and STM-64 signals, respectively, given in ITU-T Rec. G.825. These are already covered in chapter 10. For CBR40G, i.e. STM-256 signal, these are yet to be specified by ITU-T. (It may be available in future editions of ITU-T Rec. G.825.)

15.4 Intrinsic Jitter

15.4.1 Intrinsic jitter at ODUk/OTUk interface

Intrinsic jitter at ODUk/OTUk in the absence of input jitter should be within the limits prescribed in Table 15.5 when measured over a 60-second interval.

Table 15.5 *Intrinsic jitter at ODUk/OTUk*

Interface	Measurement bandwidth (–3dB frequencies)	Peak-to-peak amplitude (UI_{pp})
ODU1, OTU1	5 kHz to 20 MHz	0.3
	1 MHz to 20 MHz	0.1
ODU2, OTU2	20 kHz to 80 MHz	0.3
	4 MHz to 80 MHz	0.1
ODU3, OTU3	20 kHz to 320 MHz	1.2
	16 MHz to 320 MHz	0.1

15.4.2 Intrinsic jitter in STM signal derived from ODUk signal

In the absence of input jitter, jitter in STM signal derived from ODUk should not exceed the values specified in Table 15.6 when measured over a 60-second interval with the specified measurement filters.

Table 15.6 *Intrinsic jitter in STM signal derived from ODUk signal*

Interface	Measurement bandwidth (–3dB frequencies)	Peak-to-peak amplitude (UI_{pp})
STM-16	5 kHz to 20 MHz	1.0
	1 MHz to 20 MHz	0.1
STM-64	20 kHz to 80 MHz	1.0
	4 MHz to 80 MHz	0.1
STM-256	80 kHz to 320 MHz	1.0
	16 MHz to 320 MHz	0.1

Q 15.3 Why is no tolerance specified for wander while jitter tolerance has been discussed in this section?

Intrinsic wander, if any, is expected to be negligible compared to the wander on the input CBR (i.e. SDH) client signal because the ODCp (ODUk clock PLL) bandwidth is relatively wider. Hence, no wander tolerance requirements need to be specified. Also, the intrinsic wander generation of the ODCp is negligible compared to the wander generated by the de-mapping process.

15.5 Jitter Transfer Requirement of ODUk, OTUk

The jitter transfer function of ODUk, OTUk is given in Fig. 15.4. The values of A, f_L, f_c and f_H for ODUk are specified in Table 15.7 and in Table 15.8 for OTUk.

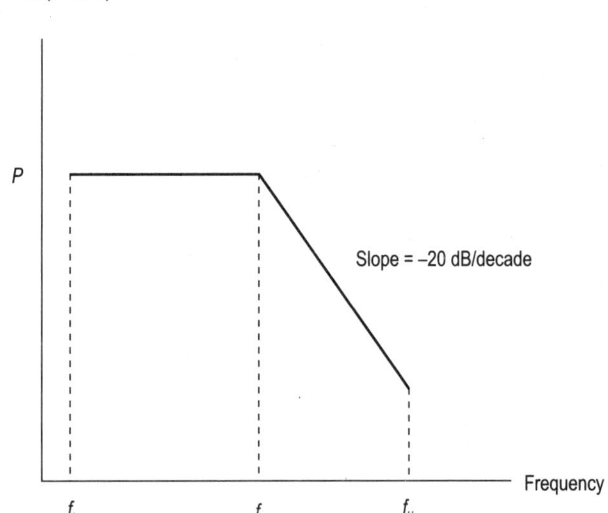

Fig. 15.4 *Jitter transfer function for ODUk*

Table 15.7 *Parameters of jitter transfer curve of ODUk*

ODUk level	f_L (Hz)	f_c (k Hz)	f_H (k Hz)	*P* (dB)
ODU1	10	1	100	0.1
ODU2	40	4	400	0.1
ODU3	160	1600	1600	0.1

Table 15.8 *Parameters of Jitter transfer curve of OTUk*

OTU_k level	f_L (Hz)	f_c (k Hz)	f_H (k Hz)	P (dB)
OTU1	2.5	250	20	0.1
OTU2	10	1000	80	0.1
OTU3	40	4000	320	0.1

Q 15.4 Why is wander transfer requirement not specified?

Since the ODUk clock PLL bandwidth is much larger than 10 Hz (i.e., the upper limit of the wander region), the regenerators transfer wander without attenuation. Hence there is no need of wander transfer specification.

15.6 Effect of OTN on the Distribution of Synchronisation via STM-N Clients

The OTN physical layer is not subjected to transport synchronisation—this is done over SDH. However, the introduction of the optical transport network (OTN) makes the STM-N signal a client signal within the OTN layer network. This might affect the synchronisation network architecture, since the STM-N signal is used as a carrier for synchronisation information (through S1 byte, pointer bytes). Hence, the effect of OTN on the distribution of synchronisation via STM-N signal is to be analysed.

Refer to Fig. 15.5 showing OTN island and SDH NEs in a synchronisation chain (trail). The OTN island in this model is a conglomerate of OTN equipment that performs mapping of an STM-N into its corresponding ODUk – multiple NEs that perform de-multiplexing and cross-connecting of ODUk's and transport over optical channels (including optical multiplexing and cross-connecting), and finally de-map STM-N.

The composition of each OTN island in this model consists of 1 OTN network element that performs the mapping operation and nine other OTN network elements that perform multiplexing operations of ODUks[1].

There are 10 OTN islands in the synchronisation reference chain (one island between each pair of SSU makes it 10 islands over the trail having 10 SSUs and a PRC), each with 10 mapper/multiplexer NEs. Hence there are 100 of those NEs altogether in the reference chain. In fact, the distribution of OTN islands over the adapted synchronization can be different; more are to be 100 NE, altogether in the reference chain.

The network limit for wander at the end of synchronisation reference chain, as defined in ITU-T Rec. G.803 (also discussed in chapter 10) is 5 μs over 24 hours. Out of this 5 μs wander budget, 10% (500 ns) is allocated for combined OTN islands. Since there are 100 NEs altogether in the 10 OTN islands, wander budget per NE is 5 ns.

Note that wander budget is not a specification for the sake of specifying something on wander but has design implications. It determines the buffer hysteresis for client signals (in alignment functions for mapping and multiplexing operations).

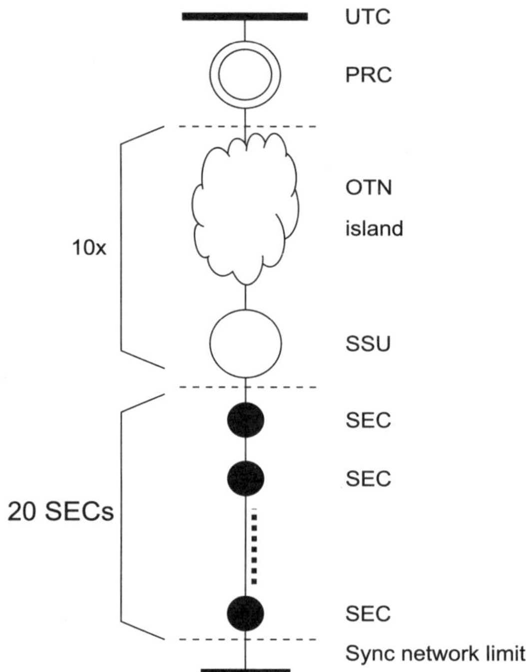

Fig. 15.5 *Synchronisation reference chain for wander budget*

Q 15.5 ITU-T Rec. G.798 specifies that the maximum allowable buffer hysteresis in alignment functions for mapping and multiplexing operations is 2 bytes for STM-16, 8 bytes for STM-64, and 32 bytes for STM-256, per OTN NE. Why is this so?

A wander budget of 5 ns for each OTN NE implies each OTN network element is allowed to contribute 5 ns to long-term wander accumulation. This 5 ns duration is approximately 12.5 UI = 1.5625 bytes for ODU1, 50 UI = 6.25 bytes for ODU2 and 200 UI = 25 bytes for ODU3. Hence, buffer hysteresis in alignment functions of mapping and de-multiplexing operations has to be 1.5625 bytes for ODU1, 6.25 bytes for ODU2 and 200 UI = 25 bytes for ODU3. Because the implementation of buffer is by whole numbers, the ODU1 buffer is 2 bytes. The ODU2 buffer value is set to 4 times ODU1, i.e., 8 bytes while the ODU3 value is set to 16 times ODU1 value, i.e., 32 bytes.

Q 15.6 *When we consider such buffer hysteresis, what is the effect on overall wander budget?*

In units of phase time, the buffer hysteresis is approximately 6.4 ns. The total long-term wander budget for the synchronisation reference chain for a 100 OTN network elements becomes 640 ns.

15.7 Network Survivability through Optical Layer Protection

Having addressed reliability issues in the previous sections, let us turn our attention to survivability of OTN in operational condition.

15.7.1 Need for optical layer protection

Optical layer provides light paths which are used by clients like SONET/SDH, IP, ATM. SONET/SDH have protection mechanisms which have been already discussed in chapter 8. The other clientele, viz. IP, ATM, etc., do not provide the same level of protection. Their motto is 'best-effort' delivery. We can improve their protection mechanism using MPLS, i.e., by setting up LSPs between nodes as discussed in section 8.6.

Another way to provide protection mechanism is to develop one within the optical layer. This can be a cost-effective solution and is presented next.

Consider a WDM ring with SDH TMs and IP routers as shown in Fig. 15.6. The working light-path from SDH TM-A to TM-B is the shortest path (clockwise on wavelength λ_1). Similarly, the working light-path from IP router C to D is on wavelength λ_1. The protection path from A to B is on longer path in anticlockwise direction, on wavelength λ_1. But, protection path from C to D, also in anticlockwise direction, cannot be on wavelength λ_1 but on a different wavelength, λ_2.

Refer to the arrangement shown in Fig. 15.7 where optical layer protection is incorporated. Here, the working path from A to B or C to D is on wavelength λ_1. Protection path, which is in anti-clockwise direction, is also on wavelength λ_1 from A to B or C to D. But, the protection path is shared so that we get the benefit of using less number of wavelengths in the optical layer protection.

From the above discussion it is clear that if only SDH protection is used, we would require two wavelengths. In general, for *N* wavelengths we require *N* protection wavelengths. If we use optical layer protection, the protection wavelengths can be saved and we can have 1 or 2 or more protection wavelengths as per value of *N*.

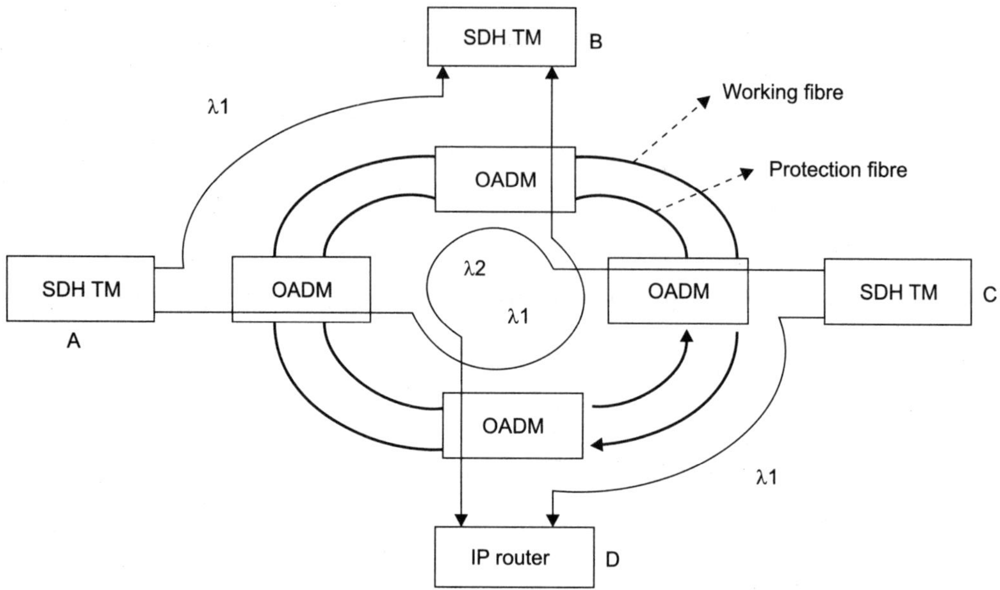

Fig. 15.6 *WDM ring without optical layer protection (no sharing of protection bandwidth)*

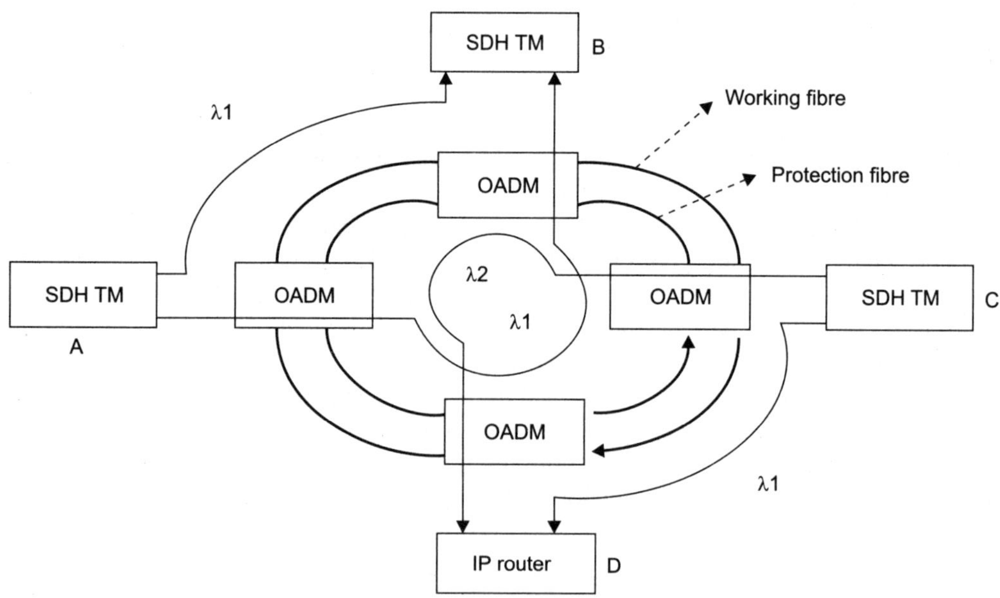

Fig. 15.7 *WDM ring with optical layer protection (sharing of protection bandwidth)*

Q 15.7 How can you say that the optical layer protection is cheaper when it involves cost per port of optical layer equipment?

If optical layer protection were not used, we would require to use IP protection which involves more ports of routers, which is a costlier proposition than using additional ports of optical layer equipment.

Q 15.8 Which is more efficient – handling client layer protection or optical layer protection?

In case of client layer protection, although the traffic is restored independently, several alarms and bit errors occur due to jitter associated with pointer movement while the NEs are adjusting to new clock references. If traffic is restored by optical layer protection, restoration is faster and simpler as it involves wavelength switching.

Q 15.9 Which is more resilient – client layer protection or optical layer protection?

Consider SDH- BLSR discussed in chapter 8. In the event of a fibre cut or equipment failure, ring switching and re-routing of traffic takes place around the ring. In this transient condition, if another cut or equipment failure takes place, it cannot be handled. If optical layer protection is available, traffic is re-routed on protection wavelength (light path) and the network is ready to handle another failure. Thus, optical layer protection makes the network more resilient.

The above three questions and their answers serve to convey the need for optical layer protection.

15.7.2 Optical layer protection schemes

Within the optical layer, there are number of layers (refer to Fig. 12.3). We will discuss where the optical layer protection is operational and then go on to the protection scheme.

SDH protection schemes are at line level (BLSR) or path level (UPSR). Likewise, optical layer protection schemes are at optical channel (path) layer, i.e., OCh layer or optical multiplex (OMS) layer. OCh protection restores one light path at a time whereas OMS protection restores a group of light-paths at a time but not an individual light path.

Protection schemes in OCh layer

Protection schemes in OCh layer are:

1. **1+1 OCh protection**
   Figure 15.8 depicts the 1+1 OCh protection scheme. In this scheme, two light-paths on disjoint routes are to be setup for each client connection. This protection

requires four WDM terminals, a splitter and a switch per wavelength. The equipment required gets multiplied by the number of channels if all channels are to be protected. If all channels are not to be protected, the number of equipment and cost can both be reduced. No sharing of bandwidth is possible in this scheme. This scheme is suitable for point-to-point, ring and mesh configurations. In case of ring, it is called OCh-DPRing (OCh dedicated protection ring) and is similar to UPSR of SDH. (OCh-DPRing is also called optical UPSR.)

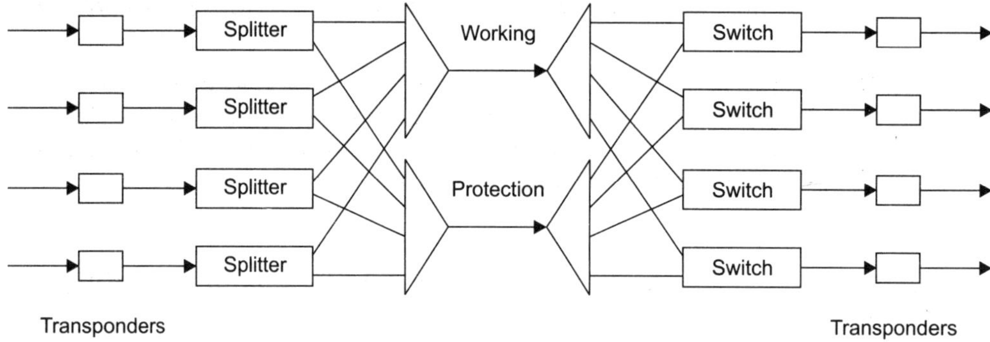

Fig. 15.8 *1+1 Ch protection scheme*

1+1 OCh protection can be implemented using separate set of transponders working on diverse paths as shown in Fig. 15.9.

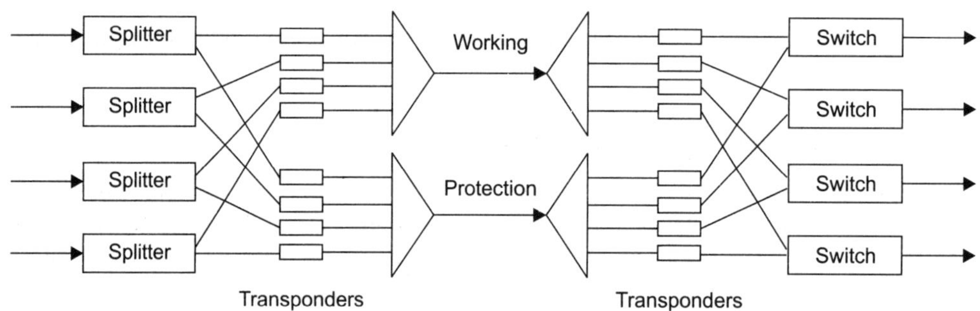

Fig. 15.9 *1+1 Ch protection scheme using multiple sets of transponders*

2. **OCh-SPRing**

OCh-SPRing (optical channel shared protection ring) is similar to SDH BLSR/4. It is operational at optical channel level, whereas BLSR/4 is operational at line level. Please refer to Fig. 15.10. When the working light-path fails, restoration is carried out using span switch or ring switch. Re-use of wavelengths in a non-overlapping manner makes OCh-SPRing more efficient (in terms of bandwidth) than OCh-DPRing.

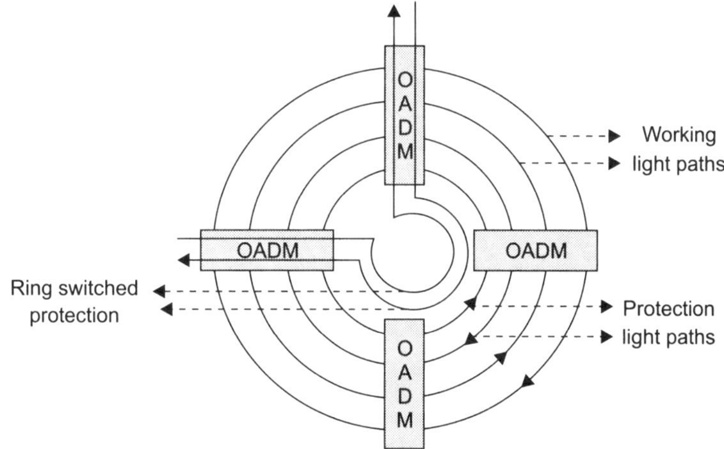

Fig. 15.10 *Ring switched protection in OCh-SPRing*

3. **OCh-mesh**

   Most of the long-haul networks are meshed. This makes implementation of OCh-SPRing tricky, as we have to ensure that no overlap of wavelengths on the rings exists within the mesh. Consider, for example, Fig. 15.11(a) showing protection by OCh-SPRing. Light-paths L1 and L2 share protection ring-1. Light path L3 uses protection ring-2. Eight wavelengths are required for dedicated protection paths. If light-paths L1 and L2 share the same path (wavelength), six wavelengths are required for protection.

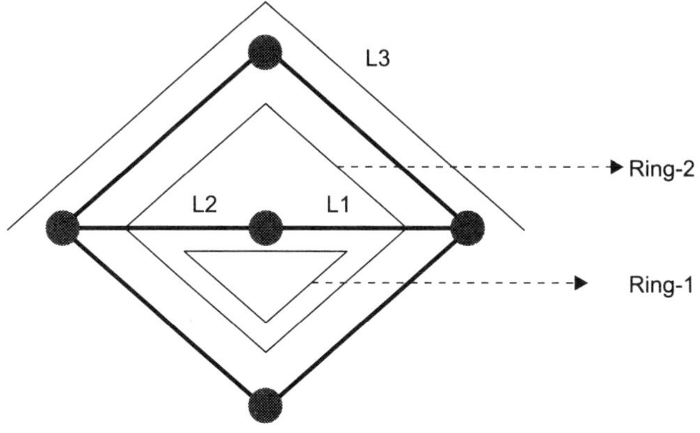

Fig. 15.11(a) *Protection in OCh-SPRing*

Refer to Fig. 15.11(b) showing mesh-protection. In this scheme the protection paths are defined in advance, but set up only when there are failures. In this case, four wavelengths are sufficient.

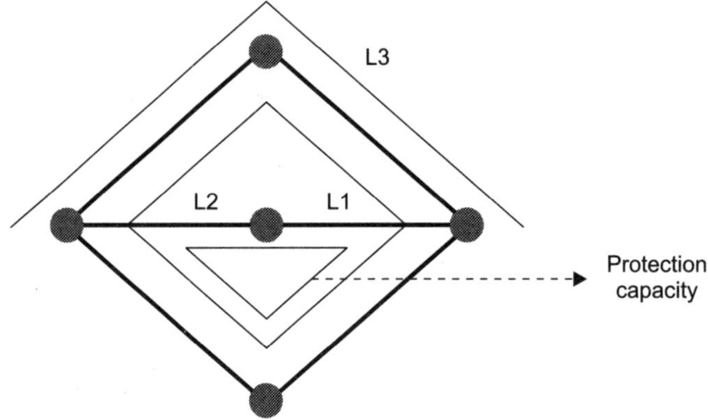

Fig. 15.11(b) *Protection in OCh-mesh*

OCh-Mesh protection scheme may require routing of protection channels. OCh mesh protection scheme is the preferred scheme due to the following reasons:

a. Networks can afford to spare some bandwidth exclusively for control traffic.
b. Protocols and signalling for routing, already developed for IP and ATM can be used for routing in the optical layer.
c. ITU-T's recommended 60 ms limit for protection time is not difficult to achieve with OCh-Mesh protection scheme.

A comparison of OCh protection schemes is presented in Table 15.9.

Table 15.9 *Comparision of OCh protection schemes*

	Protection scheme		
	1+1	**OCh-SPRing**	**OCh-Mesh**
Type	Dedicated	Shared	Shared
Topology	Mesh	Ring	Ring

Protection schemes in OMS layer

1. **1+1 OMS protection**
Figure 15.12(a) shows 1+1 OMS protection. In this scheme, the composite WDM signal is split onto two different optical paths using an optical splitter. At the receiving end the optical switch selects the better of the two. Optical splitter introduces a loss of 3 dB and the switch a loss of 1 dB. 1+1 OMS protection can be implemented using optical amplifier on each fibre and a passive combiner to

combine signals received from both directions at the receiver. At any time, one amplifier is turned on. This arrangement is shown in Fig. 15.12(b). This arrangement is more expensive to implement but has the advantage that a system failure due to switch failure can be avoided.

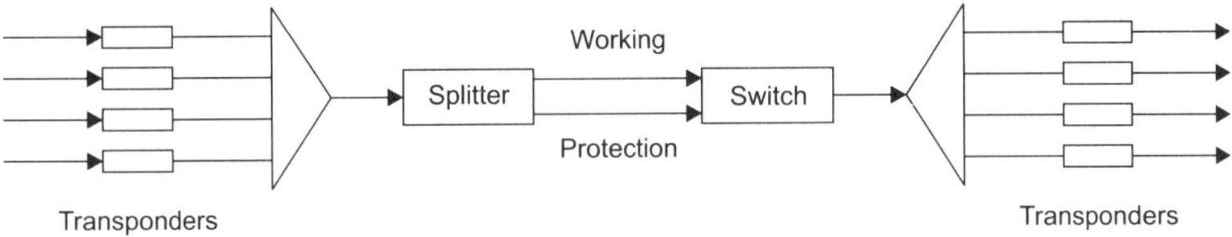

Fig. 15.12(a) *1+1 OMS protection scheme using multiple sets of transponders*

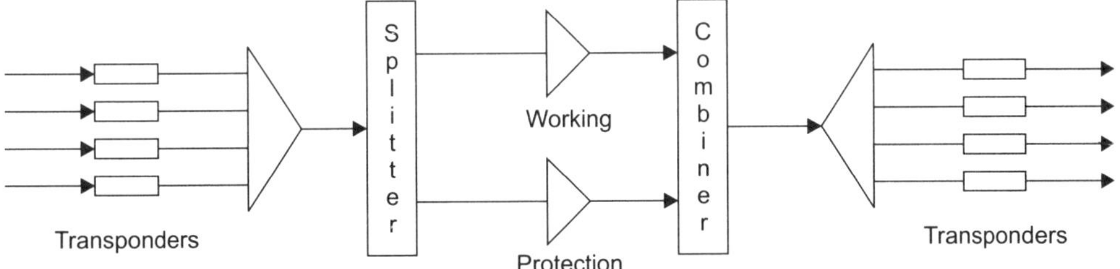

Fig. 15.12(b) *1+1 OMS protection scheme using amplifiers and combiners*

2. **1 : N OMS protection**
   In this scheme, *N* working optical paths share a common protection path. Instead of a splitter, a switch is used at the transmitter. The switch is already associated with the receiver (refer to Fig. 15.13). Switches at transmitter and receiver have coordinated working through APS protocol.

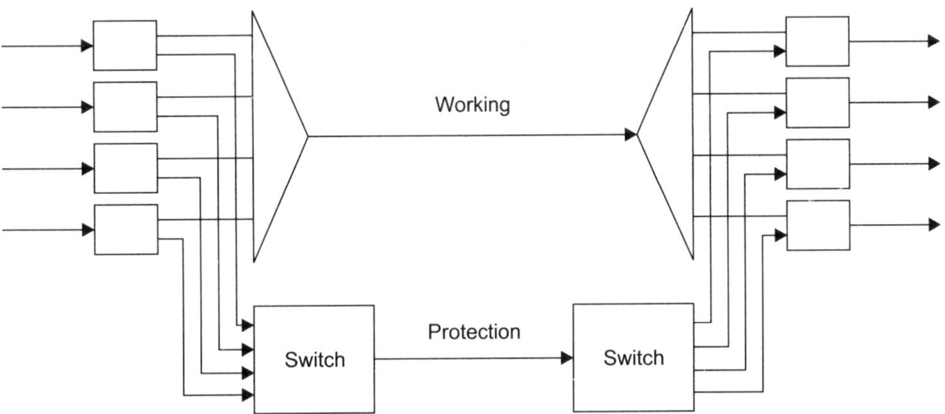

Fig. 15.13 *1 : N OMS protection scheme using multiple sets of transponders*

3. **OMS-DPRing**

   It is similar to SDH-UPSR. UPSR is operational in path layer whereas OMS-SPRing is operational in optical multiplexer layer. For this reason, it is also called optical ULSR (optical uni-directional line switched ring). Figure 15.14(a) shows the normal operation of OMS-DPRing. One pair of optical amplifier (at NE-A) are turned off and all amplifier pairs (second pair at NE-A and both pairs at other nodes in the ring) are turned on.

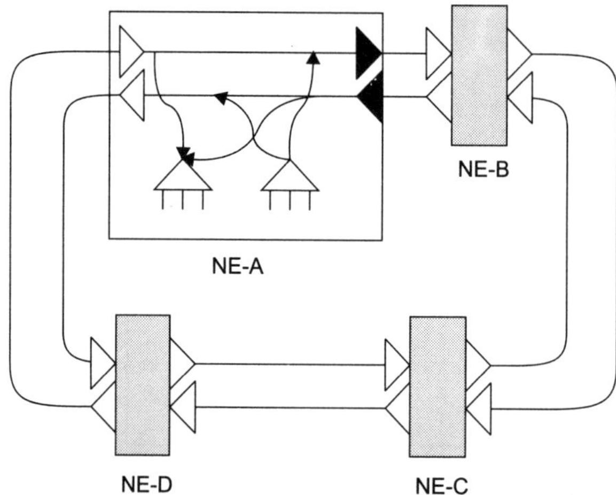

Fig. 15.14(a) *OMS-DPRing – normal operation (inactive amplifiers are shaded dark)*

Refer to Fig. 15.14(b) showing the operation of the ring in the event of a failure. The currently inactive pair (at NE-A) is turned on and a pair in a node adjacent to the cut (NE-C) is turned off.

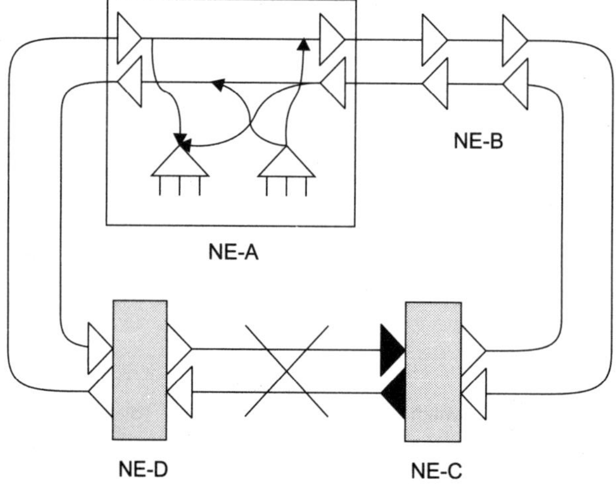

Fig. 15.14(b) *OMS-DPRing – after failure (inactive amplifiers are shaded dark)*

4. **OMS-SPRing**
   This is similar to SONET/SDH BLSR/4. Refer to Fig. 15.15, where of the four fibres shown—only two are connected to NEs. The other two fibres just form a ring.

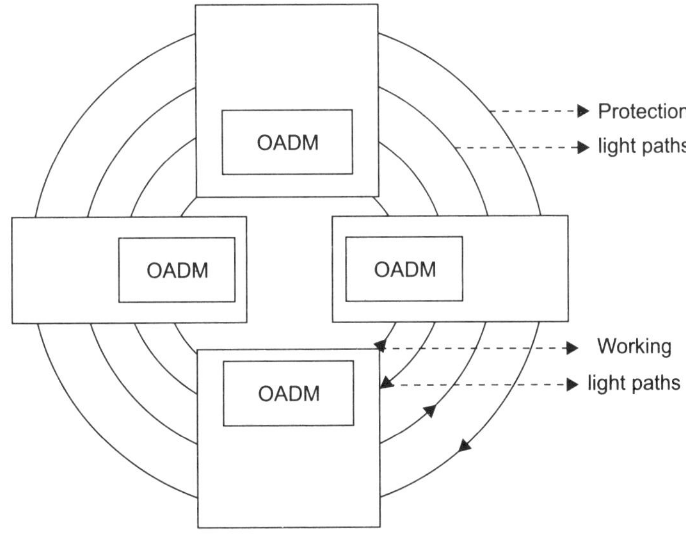

Fig. 15.15 *OMS-SPRing – normal operation*

In the event of a cut, the traffic is switched onto protection fibres span-wise or through the ring. This is a cost-effective solution as no additional equipment need to be deployed.

Q 15.10 Can we implement OMS-SPRing using two fibres?

Yes, but half the wavelengths on each fibre are to be reserved for protection purpose. In the event of failure, signals are re-routed through wavelength conversion; the operation is not at OMS layer.

15.7.3 Choice of protection scheme

Simply stated, if the network planner (operator) wants protection for all light paths together, the choice is clearly OMS layer protection. If protection is required on light-paths, the choice is OCh protection.

Further choice depends on the service class, as discussed below.

1. *Platinum class*: In this class, the highest level of availability and the fastest restoration (within 60 ms) are to be guaranteed. So, a dedicated 1+1 protection scheme can be the choice.

2. *Gold class*: In this class, high availability and quick restoration (100 to 300 ms) are required. Shared mesh-protection scheme can be the choice.
3. *Silver class*: In this class, it is the best-effort restoration. Shared protection can be the choice.
4. *Bronze class*: Only unprotected light-paths are provided. In the event of failure, the connection is lost.
5. *Lead class*: Here, the availability is the lowest. In fact, this class of traffic is carried on protection bandwidth as long as it is available. Once the protection bandwidth is allocated for protection purpose of high priority traffic, lead class traffic is interrupted.

A comparison of OMS protection schemes is presented in Table 15.10.

Table 15.10 *Comparison of OMS protection schemes*

	Protection scheme			
	1+1	**N:1**	**OMS-DPRing**	**OMS-SPRing**
Type	Dedicated	Shared	Dedicated	Shared
Topology	Point-point	Point-point	Ring	Ring

15.7.4 Limitations of optical layer protection

Optical layer protection is not the solution for all types of failures. It has the following limitations.

1. It cannot handle client equipment failures.
2. Usually, the presence or absence of optical signal (or optical signal-to-noise ratio) can be monitored, which can help to invoke protection switching. Monitoring BER and triggering protection switching is a better option but in 'optically' transparent networks, it may not be possible to measure BER.
3. Traffic is protected in units of light-paths (wavelengths), but not in parts. This has to be handled by clients (SONET/SDH, etc.).
4. Many a time, protection paths are longer than the regular path, so link budget constraints are to be checked and amplifiers to be used if required.
5. Interworking of protection schemes between different layers is to be properly scripted:
   a. Server layer protection does not depend on client layer. If the client layer is also trying to handle some protection in its domain at the same time as server layer, it may not find a path. So, the client layer protections are to be made 'revertive', i.e., client layer has to repeatedly try switching to the other path if the current

path fails. In other words, client layer protection switching has to wait till server layer converges on the protection path.

b. The above arrangement is all right if the same node is handling protection at both the layers. If different nodes are handling these protections (it may happen in any network), additional hold-off time is to be provided for the higher layer before it attempts restoration.

In general, the layer that provides fastest restoration is given the first priority and the other layer protection switching follows in the additional hold-off time while interworking protection schemes.

15.8 Dynamic Control and Management of Optical Network through ASON

In an optical network based on SDH or OTH, whenever a user requires more bandwidth, a new connection has to be requested and the service provider has to manually plan and configure the route in the network through NMS. This is not only time consuming, but also wasteful of bandwidth when the user sparingly uses the connection. Bandwidth is increasingly becoming a precious resource; for optical networks, it should be sufficient to efficiently handle resources as quickly as possible. ASON will be able to fulfil the requirements for the newer generation of optical networks, namely:

1. Fast and automatic end-to-end provisioning
2. Fast and efficient re-routing
3. Support of different clients, but optimised for IP
4. Dynamic set up of connections
5. Support of optical virtual private networks (OVPNs)
6. Support of different levels of quality of service.

ASON is a dynamic signalling-based policy-driven SDH or an OTHJ network via a distributed (or partially distributed) control plane that provides auto-recovery and dynamic connection setup. With ASON, the following OAM&P improvements are possible.

1. Fast end-to-end provisioning
2. Fast re-routing and restoration
3. New transport services such as bandwidth on demand
4. Rapid service restoration for disaster recovery
5. Switched connections within a private network

6. Support for a wide range of narrowband and broadband clients signals such as SDH/SONET, IP, Ethernet, ATM, frame relay, fibre channel (end-to-end fibre).

Architecture of ASON

ASON can be divided into three planes:

- Transport plane
- Control plane
- Management plane

The transport plane contains a number switches (optical or otherwise) responsible for transporting user data via connections. These switches are connected to each other via PI (physical interface).

The control plane takes care of the resource and connection management within an ASON network. It consists of a series of OCC (optical connection controllers) inter-connected via NNIs (network-to-network interfaces). These OCCs have the following functions:

- Network topology discovery (resource discovery)
- Signalling, routing, address assignment
- Connection set-up/tear-down
- Connection protection/restoration
- Traffic engineering
- Wavelength assignment

The management plane manages the control plane. It takes care of the configuration management of:

1. Control plane resources
2. Routing areas
3. Transport resources in control plane
4. Policy.

It also provides fault management, performance management, accounting and security management functions. Refer to Fig. 15.16 for ASON architecture and interfaces. The management plane contains the network management entity which is connected to an OCC in control plane via the NMI-A (network management interface for ASON control plane) and to one of the switches via NMI-T (network management interface for the transport network).

The traffic from a user connected to an ASON network contains data for both transport and control planes. The user is connected to transport plane via a physical interface (PI), while it communicates with the control plane via a user network interface (UNI).

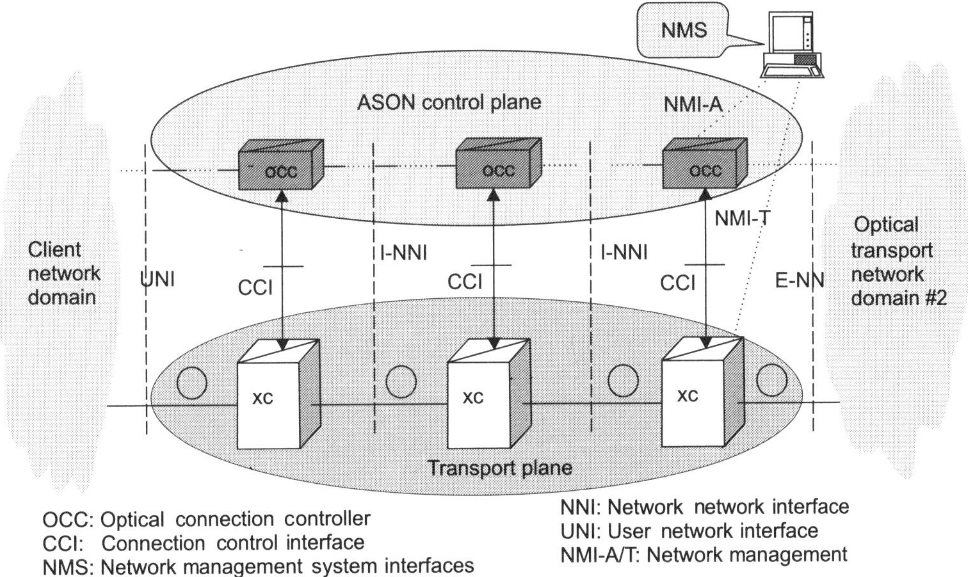

Fig. 15.16 ASON architecture and interfaces

15.9 Open Fibre Control Protocol for Optical Safety and Automatic Restart

Lasers used in optical transmitters, though operating at low power, can cause permanent damage to the human eye (blindness or black-spots on the retina). Classification of laser sources is shown in Table 15.11. In WDM systems, while checking out the level of hazard, it is not the power per channel that is considered but the total power. The simplest safety mechanism is shuttered optical connectors which shuts off power if the connectors are removed. But, this cannot help in case of a fibre cut. There are two methods to shut-off in case of a fibre cut:

1. In case of a fibre cut, one end finds a loss of incoming optical signal. So it turns off its transmitter towards the failure direction, which in turn causes loss of optical signal at the other end as well. This leads to turning-off of the transmitter.
2. In case of fibre cut, the back-reflected signal level increases which triggers a shut-down.

It is equally important to restart the system automatically after the cut is rectified. 'Optical fibre control protocol' is the standard followed for shutting down for optical safety as well as automatic restart after the rectification. The same is presented in Fig. 15.17.

Table 15.11 *Safety hazard classes of laser sources*

Class	Power limit	Application	Hazard
Class I	10 dBm at 1550 nm 0 dBm at 1310 nm	CD player	No hazard since it is enclosed
Class IIIa	17 dBm at 1550 nm	STM & WDM	Safe, but only trained personnel are allowed access
Class IIIb	>17 dBm at 1550 nm	WDM after optical amplifiers	Hazardous; Requires automatic shut-down

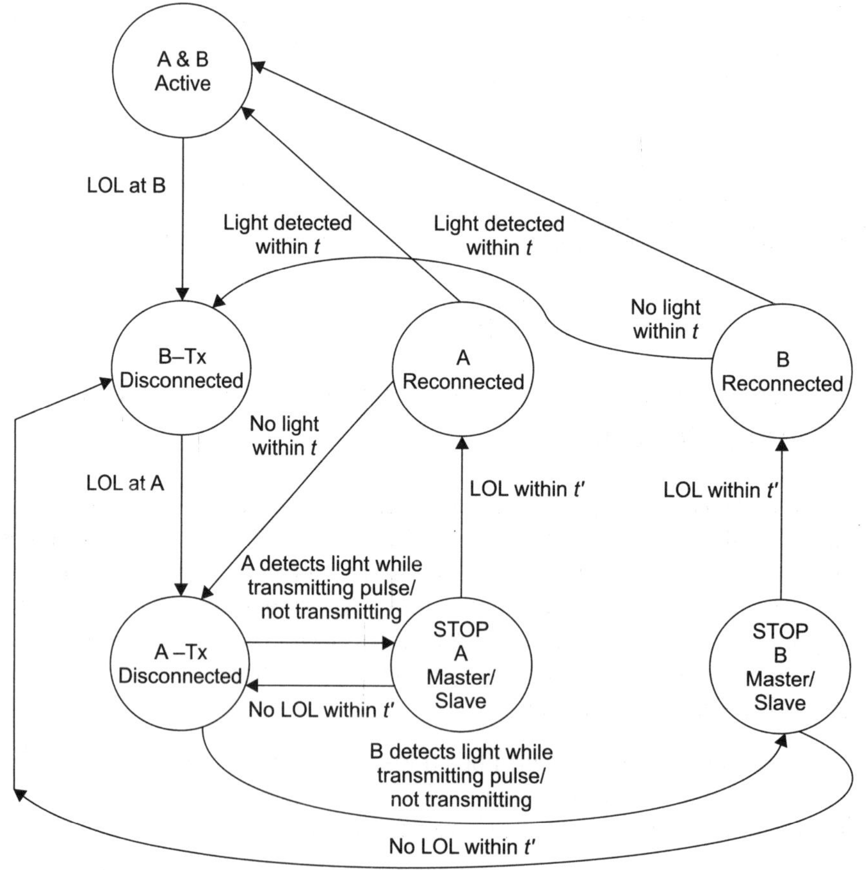

Fig. 15.17 *Open fibre control protocol*

Open fibre control protocol

Refer to Fig. 15.17.

1. A and B are in active state under working conditions. Failure of link A to B causes LOL (loss of light) at B. B turns off laser and enters 'disconnect' state.
2. Consequently, A also detects LOL and turns off laser and enters disconnect state. (If A to B link and B to A link fail simultaneously, both enter disconnect state simultaneously.)
3. While in disconnect state, A transmits a pulse of duration *t* every *T* seconds. B also does the same. If A detects light while transmitting pulse, it enters 'stop' state and becomes 'master'. If A detects light while not transmitting pulse, it enters stop state and becomes 'slave'.
4. Upon entering stop state, A turns off its laser source for a period of t' seconds. It continues to do so till a LOL is detected on the incoming link. If LOL is detected within t' seconds, A moves into reconnect state. Else, A moves to disconnect state.
5. After entering reconnect state, if A happens to be master, it sends out a pulse of duration t. If light is detected on the incoming link within the same duration, A enters active state. Else, it enters disconnect state.
6. If A happens to be slave, it monitors the incoming link for *t* seconds, and if light is detected within *t* second, it enters active state. Else, it enters disconnect state.

Steps 4 to 6 are to be taken in a similar manner for B.

Values for *t*, t' and *T* are to be chosen appropriately, considering the maximum propagation delay. Also, the average transmitted power in a disconnect state ($= tP/T$, where *P* is the laser's transmitted power on turn-on) should be below the permitted safety level.

Summary

In this chapter the reliability issues of OTN, viz. control of jitter as per ITU-T recommendations are discussed. Survivability issues of ring protection are discussed. Types of optical layer protection schemes are explored. Hazard levels in the case of open fibre for normal and WDM sources, access control/shut-down requirements and open fibre control protocol (OFCP) are explained.

References

[1] ITU-T Rec. G.8251

Appendix A

The architectural and functional features of STM-1 to STM-64 equipment from Tejas and Alcatel, along with the interfaces supported by these equipment are discussed here in brief. The information provided is not exhaustive (readers may refer to the equipment manual for complete details), nor is it in any way meant for making a product comparison. The intention here is to merely highlight the approach of a typical manufacturer in providing the aggregates, interfaces, etc.

A.1 Tejas Equipment

A 1.1 Introduction

The STM equipment of Tejas make are available as STM-1, STM-4, STM-16 and STM-64 ADMs/cross-connect equipment. While STM-1 equipment is optimised for thin-route applications, STM-4 and above are designed to support fully non-blocking cross-connects to interconnect multiple rings and linear segments with full range of redundancy options.

A.1.2 Tejas STM-1 equipment

TJ100 LT, TJ100 MC1, TJ1100 and TJ100CPE/TJ1210 are Tejas STM-1 equipment. Of these, TJ100LT is obsolete, and hence not discussed. The TJ100MC-1 is MSPP (multi-service provisioning platform) equipment supporting practical bandwidth provisioning to meet low or medium capacity bandwidth service demands. The TJ100MC-1 can be configured as terminal multiplexers (TMUX) or add–drop multiplexers (ADM), with mix-and-match tributary interfaces at E1, E3, DS3 or 10/100 Mbps Ethernet service interfaces. It has in-built non-blocking cross-connect at VC-3 and VC-12 granularity for efficient traffic grooming. It also supports FE interface.

The TJ1100 service access node is an ultra-compact STM-1 equipment meant for customer premises in access/edge networks. TJ1100 can be configured as a terminal multiplexer (TMUX) or as an add–drop multiplexer (ADM) with E1 and 10/100 Ethernet service interfaces and trunk interfaces at STM-1 rate. The product has a built-in non-blocking cross-connect at VC-3 or VC-12 granularity for efficient traffic grooming.

The TJ100CPE/TJ1210 is an ultra-compact and cost-effective customer premises bandwidth provisioning equipment designed to meet low or medium capacity bandwidth service demands. It is also an MSPP product.

In this section, the architectural features of TJ100 MC1 are presented in detail. This gives general insight into STM-1 product features. Later, a comparative study of features of TJ100 MC1, TJ1100,TJ100CPE/TJ1210 is presented.

Architectural features of TJ100MC1

TJ100 MC1's Lite tributary card (LTC) provides aggregate interfaces, clocks, processing and monitoring capability to the system. TJ100 MC1 supports tributary cards of different E1 capacities, E3/DS3 tributary card and Ethernet tributary card. On the aggregate side, it has STM-1 optical and STM-1 electrical cards. The equipment has redundant PSUs (power supply units) to generate the required DC voltages from the – 48V DC input.

The LTC provides the following interfaces:

a. Processor bus for control path communication to the other cards in the equipment sub-rack via back plane
b. Two telecom buses (STM frames of 270 rows × 9 columns of bytes, transmitted row-wise, serially) for the data path interface to the tributary cards via back plane
c. System clocks and timing signals via back plane
d. Two STM-1 optical interfaces, SC-PC type
e. Two STM-1 electrical interfaces, BNC type
f. 10/100BaseT Ethernet interface (RJ45) for communication with the network management system via LAN
g. RS232C port for local craft terminal.

Clock and data inputs and outputs, external alarm inputs and alarm outputs, diagnostic port, modem port and 2-wire order-wire for communication are also supported on this card.

LTC of TJ100 MC1 has three functional blocks.

1. *Processor subsystem (PSS)*: The processor subsystem performs the configuration, control and processing of all the other subsystems on the LTC. It consists of a 32-bit microcontroller and onboard external memory to implement the required functions. The software residing in this block controls the overall management of the system. It has watch-dog timer, RTC (real time clock) with backup battery.
2. *Cross-connect*: The LTC has a 4×4 STM-1 level cross-connect capable of switching VC-12s from any input port to any output port. It is implemented as a column-level time slot interchanger (TSI) and a space switch to achieve full 1080 × 1080 column cross connections.
3. *Timing module*: The timing module operates to a stable Stratum-3 compliant 20 MHz oscillator. This unit can synchronise to a variety of timing sources, selected

via the user interface. A digital and an analog PLL are used to synchronise the node clock to the selected reference. The selected reference is divided to yield 8 kHz and provided to the digital PLL.

Q A.1 ***What are STM-1 optical and electrical interfaces?***

In any STM equipment, the optical interfaces are through optical connectors in the form of an optical signal. In the case of TJ100 MC1, the interfaces are at STM-1 data rate, through SC-PC type optical connectors to the optical signal specification of S1.1, L1.1 or L1.2.

Electrical interfaces are through electrical connectors in the form of electrical signal. In case of TJ100 MC1, the interfaces are at STM-1 data rate, through BNC type connector (75 ohm).

It may be noted that TJ100MC1 can be configured for 2 STM-1 optical interfaces or 2 STM-1 electrical interfaces.

Note: Reference[1] presents parameters of the optical interfaces viz. output power (maximum, minimum), receiver sensitivity, receiver overload, optical path penalty, spectral range, etc.

Q A.2 ***What are the functions of watch-dog timer? In case of restart, how are the cross-connects preserved?***

Watch-dog timer is a 1024 second timer implemented on the LTC to check for software errors. If this timer expires, the operating system (OS) and the application software are restarted. This restart is service non-affecting, hence cross-connects are preserved, except for loss of management connectivity during the restart time. The time required for restarting the OS and the application software is around 3 minutes.

Q A.3 ***How are loss of synchronisation and other alarms in timing module handled?***

Synchronisation status messaging (SSM), i.e., interpretation of S-byte (sync status byte) of STM frame is implemented in the LTC software to handle loss of synchronisation signal and any alarms reported on it. This ensures that the network runs synchronously and does not create timing loops that are catastrophic to stable operation of networks.

Q A.4 ***Are there any hardware diagnostics?***

In the software, hardware errors are checked every 10 seconds. Minor, major, critical or deferred LED indications are given on the front panel according to the severity of error detected.

Q A.5 ***What are the functions of the RTC?***

A real time clock (RTC) is available onboard for time stamping of alarms and events. The RTC maintains time in the event of power failure (or powering off). The backup battery for this is a capacitor of value 1 F. This capacitor will maintain the RTC time and calendar functions for approximately four weeks after the LTC is powered off.

Q A.6 ***Why are 1080 × 1080 cross-connects required?***

As already mentioned, each telecom bus is STM-1 frame, organised as two hundred and seventy columns and nine rows of bytes, transmitted row-wise, serially. Since all the telecom busses received by the cross-connect are multi-frame aligned and no AU pointer movements are allowed, each column in any row is equivalent to the corresponding column in any other row. Hence, a 270 column cross-connect is sufficient for the whole telecom bus.

The cross-connect is memory based and uses two banks of data memory and one bank of connection memory. The incoming data is written into one bank of the data memory and read out (in the order specified in the connection memory) from the other data memory. Hence, for each telecom bus, 540 × 40 columns are needed.

Since TJ100MC1 supports two STM-1 electrical / optical interfaces or two telecom buses, cross-connects are 1080 × 1080. However, latency through the cross-connect is only 270 clocks of 19.44 MHz or 13.88 microseconds for any telecom bus.

Q A.7 ***How does PLL ensure phase lock to reference?***

The digital PLL implements a controlled rate of phase change to lock to reference. It also has time interval error (TIE) correction circuitry to ensure that phase errors are minimised. The digital PLL is composed of two PLLs – the acquisition PLL and the core PLL. The acquisition PLL does coarse tracking of the incoming reference and provides an out-of-range (OOR) alarm. The OOR alarm is raised when the incoming reference has shift beyond +/– 12 ppm with respect to the local oscillator. The OOR alarm is used to report an alarm on the user interface, whence the PLL will be moved to the 'holdover' state.

The core PLL does fine acquisition of the reference and ensures that phase error accumulation is maintained within the ITU-T G.812 standards.

Q A.8 ***Why is the analog PLL needed in addition to digital PLL?***

The digital PLL by itself is not sufficient to generate the required quality of clocks for the system. It generates too much jitter on its output. So, it is followed by an analog PLL that acts as a jitter attenuator. The analog PLL uses a parallel resonant crystal at 19.44 MHz for the jitter reduction operation.

Q A.9 What are PLL states or equipment states regarding clock reference?

The synchronisation of PLL is done in following states:

a. *Free run*: in this state, the NE is not synchronised to an external clock and its output clock will meet the Stratum-3 requirements for free-run accuracy, i.e. +/– 4.6 ppm. The NE outputs will also meet all jitter requirements on its outputs.

b. *Locked*: in this state, the node is synchronised to a selected external source and tracks the frequency changes on the reference. The NE outputs will meet Stratum-3 requirements on phase error accumulation, phase noise, jitter and timing deviation.

c. *Holdover*: in this state the NE has lost lock to the selected reference and is maintaining its last known input frequency. The synchronisation PLL could have lost lock because of an alarm on the selected reference (LOS, AIS or DoNotUse synchronisation status message) or because of an unacceptable phase or frequency variation on the reference not confirming to ITU-T Rec. G.813.

Tributary interface modules of TJ100 MC1

As already mentioned, TJ100 MC1 supports following the tributary interface modules.

1. *E1 tributary interface modules*
   These modules are designated TET16, TET21 and TET28 and provide line interfaces to 16 E1, 21 E1, and 28 E1 channels respectively, in both add and drop directions. These cards map and de-map the E1 channels into SDH frame (at programmed slots) for the tributary card to make cross connects. The interface to the LTC is through 19.44 MHz telecom bus for the SDH data path. The software control is with the LTC through the processor bus.

2. *E3/DS3 tributary interface module*
   The 1 port E3/DS3 card provides line interface to an E3/DS3 channel in both add and drop directions. The port is software configurable to support E3/DS3 rates correspondingly. The card maps E3/DS3 tributaries into an AU-3/AU-4. This card comes with 1 port or 3 ports (TE31 or TE33).

3. *Ethernet tributary module*
   TP01 Ethernet tributary interface card provides line interfaces to eight 10/100 Mbps Tx port. TP01FT tributary interface card provides four 10-base T/100 base-T Ethernet ports and four 100BASE-FX Ethernet ports.

   These cards map and de-map the Ethernet data into the virtual containers of different granularity (VC 12/VC 3/VC 4) of the SDH frame. The VC in which the data is mapped/de-mapped is programmable through software. Mapping is by GFP as discussed in section 9.2.3.

Aggregate cards of TJ100 MC1

The STM-1 aggregate card, A011, supports 1 port STM-1 optical interface. The STM-1e aggregate card supports 1 port electrical STM-1/E4 interface. LTC card of TJ100 MC1 can support 2 STM-1 optical (AO11) or 2 STM-1 electrical (STM-1e/E4) aggregate cards. More details on cards and card population rules are available in[1].

A comparative study of the essential features of TJ100 MC1, TJ1100,TJ100CPE/ TJ1210 is presented in Table A.1.

Table A.1 *Features of TJ100 MC1, TJ1100,TJ100CPE/TJ1210*

Feature	TJ100 MC1	TJ1100	TJ100 CPE/1210
Cross-connect	4 × 4 STM-1 252 × 252 VC-12 granularity Fully non-blocking	4 × 4 STM-1 Fully non-blocking	252 × 252 VC-12 Fully non-blocking
Network topology	Linear, ring	Linear, ring, bus and mesh	Linear, ring
NE configuration	ADM, TM	ADM, TM	ADM, TM
Aggregate interfaces	2 × STM-1o/e	2 × STM-1o/e	2 × STM-1o/e
Optical interface	S1.1, L1.1,L1.2	S1.1, L1.1,L1.2	S1.1, L1.1,L1.2
Electrical interface	PDH: E1, E3 SDH: STM-1 e ETH: 8 port FE card	PDH: E1 SDH: STM-1 e ETH: 4 port FE	PDH: E1, E3 SDH: STM-1 e ETH: 8 port FE card
Management	EMS with full FCAPS, FE management interface, E1 management channel with drop facility, RS232 interface for LCT	EMS with full FCAPS, RJ 45 port for LCT	EMS with full FCAPS, RJ 45 port for LCT, E1 management channel with drop facility

A.1.3 Tejas STM-4 equipment

STM-4 equipment from Tejas are TJ100 MC4/ MC4-L/XPE4, TJ1310, TJ1260/1250, TJ1270. While TJ100 MC4/ MC4-L/XPE4 are older versions and are being phased out, TJ1310, TJ1260/1250, TJ1270 are products of MSPP (multi service provisioning platform). MSPP products are compact nodes supporting bandwidth provisioning for multiple types of services and provide competitive edge in implementation. However, to start with, TJ100 MC4/ MC4-L/XPE4 are discussed so that the design approach and features are clear. Later, features of the MSPP products viz. TJ1310, TJ1260/1250, TJ1270 are presented to facilitate easy appreciation of the same.

TJ100 MC4/ MC4-L/XPE4 equipment have redundant PSUs like TJ100 MC1. All the tributary cards of TJ100 MC1, viz., TET16, TET21, TET28, E3/DS3, TP01, TP01FT are supported by these equipment.

In addition, TJ100 MC4/ MC4-L/XPE4 support the following tributary cards:

Tributary card TR01: Supports four STM-1 bandwidth of Ethernet traffic via Electrical or optical interface. This can provide line interface to eight 10/100 Mbps (FE) and/or two 1000Mbps (GE) Ethernet ports.

E1 tributary protection card (E1-TPC): It is used to provide 1:3 protection to the E1 tributary cards on the TJ100MC-4. The TPC can provide protection of up to 28 E1s per card.

E3 tributary protection card (E3-TPC): It is used to provide 1:3 protection to the E3/DS3 tributary cards on the TJ100MC-4/MC-4L/XPE4.

On the aggregate side, as the nomenclature suggests, TJ100 MC4/ MC4-L/XPE4 support STM-4 aggregate cards. These are A041 for normal reach and A041VLR for very long reach, both being 1 port STM-4 aggregate cards.

TJ100MC-4/MC-4L/XPE4 equipment also support all aggregate cards of TJ100 MC1, viz. AO11, STM-1e/E4, which are 1-port STM-1 cards. In addition, TJ100MC-4/MC-4L/XPE4 equipment support 2-port STM-1 cards – AO12 (optical) and AO12E (electrical).

Unlike TJ100 MC1, TJ100MC-4/MC-4L/XPE4 equipment have processor sub-system (PSS), cross-connect (XC) and management interface (MIC) as separate cards. PSS and XC have 1+1 hot stand-by configuration.

PSS card has 128 MB compact flash disk to hold the operating system and 64 MB SDRAM. MIC has Ethernet port for remote management, RS-232 port for local management, clock inputs/outputs and 2W order-wire interface and potential-free contacts to extend alarms.

Cross-connect (XC) is a single stage non-blocking fabric of 2.5 Gbps capacity that can connect multiple rings and linear segments. Cross-connect capacity can be used as 16 STM-1 or 4 STM-4, as shown in Fig. A.1. All types of cross-connects are supported, viz. aggregate to aggregate, aggregate to tributary, tributary to aggregate and tributary to tributary. Loop-backs and local cross-connects are also possible.

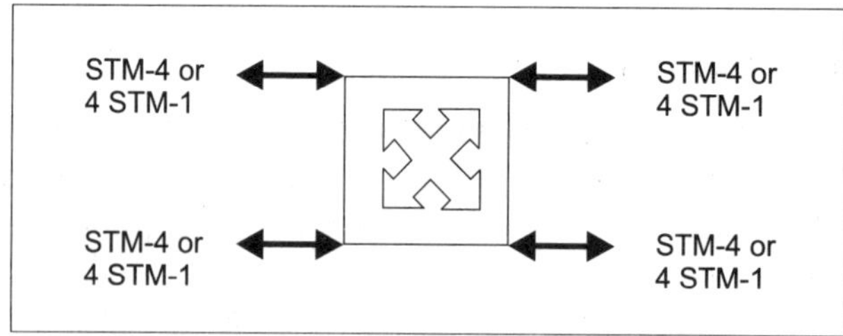

Fig. A.1 *Cross-connect in Tejas STM-4 equipment*

The general rules in cross-connects utilisation in any STM equipment are:

1. If the equipment is STM-N, the total of aggregates that can be provided in any direction cannot exceed STM-N.

2. In any direction, the total BW derived from tributary interfaces cannot exceed the aggregates in that direction.
3. Total of aggregates in all directions cannot exceed the cross-connect capacity.

In the equipment, slots are nominated for PSS, XC, PSU, MIC cards. Provision of tributary cards is governed by card population rules which are specific to the type of equipment, viz. TJ100 MC4/MC-4L/XPE4, and laid down in accordance with the above general rules in the manuals. Details can be found in[2].

Q A.10 *With PSS being separate, what happens in the case of module failure?*

PSS is non-trafic affecting. Even if both modules fail, traffic remains unaffected.

Q A.11 *Is the switching between cross-connect modules hitless?*

Yes, one XC is master and the other slave, which can be set through NMS or LCT. As slave XC is always in sync with master, switching is hit-less. Any such switching forced through NMS or LCT results in soft reboot. Cross-connects provisioned are not affected.

However, the case of forcing a working XC off by the front panel switch is known as hard reboot. Cross-connects are unavailable for few minutes and after reboot, they are available for retrieval from memory storage.

Q A.12 *With higher BW links of 2.5G and above (STM-16 and above) and with DWDM equipment taking minimum STM-16 on each wavelength, are STM-4 still in use?*

It is true that equipment now being manufactured are mostly STM-16 and upwards. STM-4 equipment in use will continue to be operational. If any link requires only STM-4, it is provided through STM-4 interface on STM-16 equipment. However, the concept of segregating PSS, XC, MIC which started with STM-4 continues with higher order equipment.

Features of MSPP STM-4 products: TJ1310, TJ1260/1250, TJ1270

The TJ1310 is designed to meet low or medium capacity bandwidth service demands. It supports end-to-end provisioning and management of voice and data services across all the segments of the optical network—from the customer premises to the core.

The TJ1260 service access node is an ultra-compact and cost-effective customer premises bandwidth-provisioning equipment designed to meet low or medium capacity bandwidth service demands. (TJ1250 is the non-redundant version of TJ1260.)

TJ1260/TJ1250 can be configured as a TM or ADM, with E1, E3, DS1, DS3 or Ethernet service interfaces and trunk interfaces at STM-4 rate. It has in-built non-blocking cross-connect at VC-3 or VC-12 granularity for efficient traffic grooming.

The TJ1270 service access node is also ultra-compact equipment but is meant for provisioning dense customer premises bandwidth. It is an ideal platform for evolving networks. TJ1270 also supports a variety of data interfaces, with carrier Ethernet capabilities required to deliver carrier-grade services such as voice over IP (VoIP) and enterprise services with stringent SLAs (service level agreements).

A comparative study of essential features of TJ1310, TJ1260/1250 and TJ1270 from application point of view is presented in Table A.2.

Table A.2 *Essential features of TJ1310, TJ1260/1250, TJ1270*

Feature	TJ1310	TJ1260/1250	TJ1270
Cross-connect	10 × 10 STM-1 (VC-12 granularity) fully non-blocking	10 × 10 STM-1 (VC-12 granularity) fully non-blocking	1008 × 1008 VC-12 fully non-blocking
NE configuration	ADM, TM	ADM, TM, RG	ADM, TM, RG
Network topology	Linear, ring	Linear, ring	Linear, ring, mesh
Aggregate interfaces	2 × STM-4	1 or 2 × STM-4	1 or 2 × STM-1 1 or 2 × STM-4
Optical interface	S4.1, L4.1, L4.2	S4.1, L4.1, L4.2	S4.1, L4.1, L4.2 8 x STM-1 o 2 x STM-4 o
Electrical interface	PDH: E1, E3 SDH: STM-1 e ETH: 8 port FE, 8 port FE switching, 2 port GE	PDH: E1, E3 SDH: 2 x STM-1 e ETH: 8 port FE	PDH: E1, E3 SDH: 8 x STM-1 e ETH: 4 FE on base 8 FE, 12 FE and 2 GE interface cards
Management	EMS with full FCAPS, FE management interface, E1 management channel with drop facility, RS232 interface for LCT	EMS with full FCAPS, FE management interface, E1 management channel with drop facility	EMS with full FCAPS, FE management interface, E1 management channel with drop facility

A.1.4 Tejas STM-16 equipment

TJ100 MC-16/MC-16X, TJ1500 are the STM-16 equipment from Tejas. In all these equipment, PSS and XC functions are brought into one card designated as cross-connect and controller card (XCC). Multifunction interface card (MIC) is separate as in STM-4 equipment.

Two types of cross-connect and controller cards are available in TJ100 MC-16/MC-16X:

1. *XCC64L*: '64' implies 64 numbers of STM-1 links are supported, i.e., 64 × 155 Mbps, which equals 10G. L stands for lower order, i.e., VC-12. In other words, XCC64L is cross-connect controller card designed to implement 10G VC-12 granularity cross-connect. It has 20G space switching capabilities and thus provides 10G strict-sense non-blocking switch capacity. With XCC64L card, the system can also support 16 telecom buses, each with STM-4 capacity.
2. *XCC128L*: '128' implies that 128 numbers of STM-1 links are supported, i.e., 128 × 155 Mbps which equals 20G. L stands for lower order (VC-12). In other words, XCC128L is a cross-connect controller card designed to implement 20G VC-12 granularity cross-connect. It has 40G space-switching capability and thus provides 20G strict-sense non-blocking switch capacity. With XCC128L card, the system supports 32 telecom buses, each with STM-4 capacity.

In the equipment chassis, two slots are allocated for XCC cards to facilitate 1+1 redundancy and both are to be fitted with the same type of XCC card. There is a hot swap switch on each XCC card. Master and slave card configuration of the redundant XCC card can be forced using the hot swap switch. When the hot swap switch is activated, the respective card renders the peer card as the master and takes over as slave. The switch is functional only on the card that is active, provided another card is available in the standby configuration.

TJ100 MC-16/MC-16X support the following tributary cards supported by TJ100 MC4:

a. 3 port E3 /DS3 card (TE33)
b. Ethernet tributary card (TP01)
c. Ethernet tributary card (TP01FT)

In addition, these equipment support following interfaces and aggregates:

a. 2 port 1000 base LX card (LQ02) for 2 GE interfaces
b. 84 port E1/T1 card (LB84) for 84 E1 interfaces
c. 8 port E4 /STM-1E interface card (PC1L8SA)
d. 4 port E4/STM-1 electrical and 8 port STM-1optical interface card (LC1L12) to provide 4 electrical ports of E4/STM-1 and 8 optical ports of STM-1
e. 16 port STM-1 optical interface (LC1L16FP)
f. 4 port STM-4 SFF line interface card (LC4L4FF) to provide STM-4 optical interface (S4.1, L4.1 and L4.2 types)
g. 1 port STM-16 MSA optics interface card (LC16L1N) to provide 1 STM-16 optical port (S16.1, L16.1, L16.2 and L16.2JE types)
h. 1 port STM-16 SFF line interface card (LC16L1FF) to provide 1 STM-16 optical port (S16.1, L16.1 and L16.2 types)

TJ1500 equipment

The TJ1500 is a STM-1/4/16 product supporting redundancy of cross-connect fabric, timing/synchronisation subsystem and control processor subsystem. It also has redundant power supply modules. In this equipment, NMS and other miscellaneous interfaces (viz. LCT interface, order-wire, alarm input/output interfaces and indication LEDs' drive) are implemented in the multifunction interface card (MFC).

Cross-connect and control functions are implemented in the XCC card. Two types of XCCs are available in TJ1500 - XCC02 card (12.5 cross-connect capacity) and XCC03 card (20G cross-connect capacity). TJ1500 has 2 slots for XCC cards to provide 1+1 redundancy, both slots being equipped with either XCC 02 or XCC 03.

The following tributary cards are supported by TJ1500:

a. 63 × E1 tributary card (TET63ME)
b. 63 × E1 tributary card (TET63)
c. 3 × E3 tributary card (TE33)
d. Ethernet tributary card (TP01)
e. Gigabit Ethernet transport card (TR03)
f. Gigabit Ethernet transport card (TR04)
g. ELAN01 tributary interface card

The following aggregate cards are supported by TJ1500:

a. Aggregate optical card (A012) which provides 2 optical STM-1 interfaces
b. Aggregate card (A018) which can be field configured to operate as either 2 × STM-1optical/electrical or 2 × STM-4 line card
c. Aggregate card (A161) which provides 1 optical STM-16 interface
d. Aggregate card (A162) which provides 1 optical STM-16 interface (for Release-3 of TJ1500 only)
e. Aggregate card (A164T) which can be field configured to operate as either 1 × STM-16 or 4 × STM-4 line card.

Q A.13 *What is meant by SFP, SFF, MSA?*

MSA means multi-source agreement between competing manufacturers to support interworking of their transceivers.

SFF transceiver stands for 'small form factor' transceiver, which supports greater *port density* (number of transceivers per inch along the edge of motherboard) than conventional optical transceiver modules. In SFF, transceiver is soldered to the host board as a pin-through-hole device.

SFP transceiver stands for 'small form factor pluggable'. It also supports greater port density like SFF and has the added advantage of being pluggable on the host board socket.

However, whatever be the optics, i.e., SFP/SFF under MSA, the optical interfaces supported are ITU-T G.957 and equivalent Bellcore GR-253-CORE compliant.

Q A.14 *What are the differences between TET63ME card and TET63 card?*

TET63ME offers 63 × E1 interface through a set of 6 pins of the 48-pin metallic connectors on the front panel of the card. TET63 offers 63 × E1 interface through a set of 4 pins of the 32-pin EURO connectors.

Q A.15 *What are the differences between TR03 and TR04 cards?*

TR03 supports 2 GE ports whereas TR04 supports 2 GE ports plus 8 FE ports. TR03 card carries GE traffic over serial 2 × STM-4 uplink using virtual concatenation (VCAT) and link capacity adjustment scheme (LCAS), i.e., if one of the two STM-4 paths has a problem, the traffic will continue with the available single STM-4 bandwidth with necessary flow-control mechanisms in the GFP mapping of Ethernet traffic onto containers.

TR04 card carries GE traffic over STM8 pipe using virtual concatenation (VCAT) and there is no link capacity adjustment scheme (LCAS). It may please be noted that both TR03 and TR04 provide virtual concatenation (VCAT) with complete 63 ms differential delay supported through external memory.

TJ100ME

There is yet another STM-16 equipment from Tejas – TJ100ME equipment. It is an ultra-compact, carrier class and cost effective bandwidth-provisioning equipment designed to manage and deliver services from the optical core to the access. TJ100ME is one of the most compact of products with only 4U height, suitable for installation in a 19 or 23 inch-high rack; it has a dense port configuration and can be configured as RG, ADM, TM, DXC and in-line amplifier in linear and ring topology networks. It supports a variety of service interfaces such as E1/DS1, E3/DS3, STM-1e/o and 10/100/1000 Mbps Ethernet tributary interfaces and trunk interfaces at STM-1/4/16 rates. It features non-blocking cross-connect at VC-3, VC-4 and VC-12 granularity.

It supports 2.5G XC as STM-4 and 7.5G XC as STM-16. On the aggregate side it supports 1 / 2 STM-1/4/16 interfaces, S/L 1.1, L 1.2, S/L 4.1, L 4.2 and S/L 16.1, L 16.2 interfaces. On the tributary side it supports E1/E3, STM-1 o/e interfaces. TJ100ME supports 4-port FE card, 8-port FE card, 12-port FE card and 2-port GE card. For management purposes, it has V.24/V.28 modem interface for remote management and FE port.

A.1.5 Tejas STM-64 equipment

TJ1600, TJ1610 and TJ1700 are STM-64 equipment from Tejas. These are all multi-service provisioning platform (MSPP) products.

TJ1600 and TJ1610 are STM-1/4/16/64 SDH multiplexer equipment designed for optical backbone, which can be configured as TM, ADM, DXC, RG with E1/T1, E3/DS3, STM-1e/1o, STM-4, STM-16 and 10/100 Mbps Ethernet tributary interfaces and trunk interfaces at STM-64 rates. These equipment have in-built non-blocking cross-connects at VC-3, VC-4 and VC-12 granularity and support drop-and-continue functionality. While TJ1600 is optimised for high-speed point-to-point as well as high density cross-connects, TJ1610 is optimised for carrier Ethernet switching through FE/GE interface cards.

TJ1700 is a compact, carrier-class, STM-64 MADM (multiple add–drop multiplexer) optimised for high-bandwidth metro and core networks. At 8 rack units in height and with a scalable 80G lower-order cross-connect, the TJ1700 is one of the densest STM-64 (MADM) available. It is power-efficient as well. It provides support for a large number of gigabit and 10-gigabit Ethernet interfaces and serves as a converged platform that can aggregate several GigE links into 10-GigE interfaces in addition to aggregating STM-16 signals to STM-64. Besides being a dense STM-64 MADM, it also supports a highly scalable Ethernet switching fabric.

Table A.3 summarises the features of TJ1600, TJ1610 and TJ1700.

Table A.3 *Comparison of features of TJ1600, TJ1610, TJ1700*

Feature	TJ1600	TJ1610	TJ1700
Cross-connects	80×80 STM-1 or 128×128 STM-1 Fully non-blocking at VC-4,VC-3, VC-12 granularity	12.5 G or 40G ×C Fully non-blocking at VC-4,VC-3, VC-12 granularity	80G LO/HO 160G HO/20G LO 320 G LO Fully non-blocking at VC-4,VC-3, VC-12 granularity
Network topology	Linear, ring, mesh and bus	Linear, ring, mesh and bus	Linear, ring, mesh and bus
NE configuration	RG,TM,ADM,DXC	RG,TM,ADM,DXC	TM,ADM,DXC
Aggregate interface	STM-16, STM-64	STM-16, STM-64	STM-64
Optical interfaces	STM-16: DWDM, S16.1, L16.1,L16.2, STM-4: S4.1, L4.1, L4.2 STM-1: S1.1, L1.1, L1.2 ETH: FE,GE	STM-16: DWDM, S16.1, L16.1,L16.2, STM-4: S4.1, L4.1, L4.2 STM-1: S1.1,L1.1,L1.2 ETH: FE,GE	STM-64: DWDM, S64.1, L64.1, L64.2 STM-16: DWDM, S16.1, L16.1, L16.2, STM-4: S4.1, L4.1, L4.2 STM-1: S1.1, L1.1,L1.2 ETH: FE,GE,10Gig

(Continues...)

Table A.3 *Continued*

Feature	TJ1600	TJ1610	TJ1700
Electrical interfaces	PDH: E1,E3 SDH: STM-1e ETH: FE,GE	PDH: E1,E3 SDH: STM-1e ETH: FE,GE	PDH: E1 interface SDH: STM-1 e ETH: FE,GE,10Gig
Management	EMS with full FCAPS, FE management interface, RS232 interface for LCT V.24/V.28 modem interface for remote management	EMS with full FCAPS, FE management interface, RS232 interface for LCT V.24/V.28 modem interface for remote management	EMS with full FCAPS, FE management interface, RS232 interface for LCT V.24/V.28 modem interface for remote management
Redundancy	Power supply unit, cross-connect, timing and control system	Power supply unit, cross-connect, timing and control system	Power supply unit, cross-connect, timing and control system

A.2 Alcatel Equipment

A.2.1 Introduction

Alcatel's SDH equipment is based on OMSN (optical multi service node) concept. In OMSN concept, an NE can operate as an STM node and also on a CWDM grid. As STM node it supports transceivers operating at STM-1, STM-4, STM-16 and STM-64 rates. It can be configured as multiple line TM, ADM and cross-connect for connecting to multiple line segments, rings and meshed rings. As CWDM NE, it can operate on ITU –T G.694.2 wavelength grid 1470 – 1490 – 1510 – 1530 – 1550 – 1570 – 1590 – 1610 nm.

A wide range of ports can be added/dropped/connected to the NE as per the needs of traffic: E1s, STM-1 electrical/optical, STM-4 optical, STM-16 optical, STM-64 optical.

In the OMSN, STM-16 is treated as server signal and client signals are lower order STM signals multiplexed through 4×STM board. Also, client signals FDDI (1.25 Mbps), ETH/FE (10/100 Mbps), ESCON (200 Mbps), digital video (270 Mbps), fibre channel (1.0625 Gbps), FICON (1.0625 Gbps) and GE(1.25 Gbps) can be directly mapped onto STM-16 signal.

A.2.2 Alcatel's 1660 SM equipment

Alcatel's OMSN is a 1660 SM, also known as STM 64 multi-service metro node. Alcatel further identifies this product by release number and version number. 'Release' stands for a set of hardware components and a software package which, as a whole, identify the possible network applications and the equipment performance. Version stands for a further developmental step in a 'release' to improve or add some performance (mainly software) with respect to the previous version. In this section, features of 1660 SM Rel.5.2, Version B are described in brief. (Its predecessor was 1660 SM Rel.4.)

1660 SM Rel.5.2 equipment rack is shown in Fig. A.2. The lower bay of the rack is known as the basic area and the upper bay as the access area. Below the basic area, fan units sub-rack comes as attachment (not shown in Fig. A.2), which helps in preventing overheating of the equipment.

Slot-pairs 25–26, 28–29, 30–31 and 37–38 are known as enhanced slots. In release 4, any STM-16 module occupies one slot-pair and allows the user to access up to 2.5 Gbps throughput. In total, 4 such modules could be plugged-in. In release 5, slim SFP module 'O16ES' of STM-16 rate occupies only one slot in the pair. Such a module can be plugged anywhere from slot 24 to slot 39, but in total, 8 such modules can be used. Remaining slots can be used for other purposes.

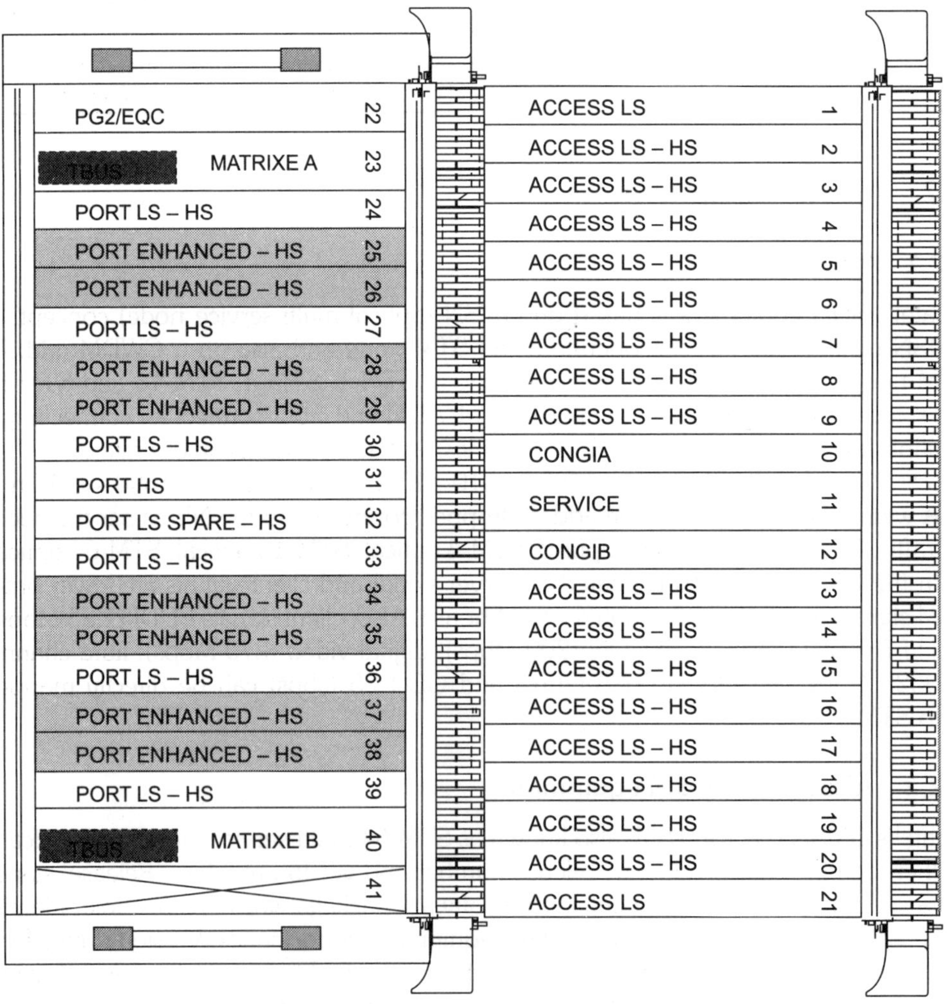

Fig. A.2 *1660 SM Rel.5.2 equipment rack*

Identification of various modules and their functions are described in brief below.

1. Access-LS stands for low speed access modules, viz. E1 tributary modules.
2. Access-HS stands for high speed access modules, viz. E3, E4, STM-1 o and STM-1 e tributary modules.
3. ISA stands for Ethernet access card.
4. Port-HS stands for high speed aggregates, viz. STM-4, STM-16, STM-64 SFP modules.
5. CONGI stands for control and general interface card. It provides power supply derivations, housekeeping and remote alarms and interface with NMS for transporting alarm information.
6. MATRIXE stands for matrix enhanced card. It performs high-order path cross-connection, low-order path cross-connection, synchronisation function and shelf controller function.
7. PQ2/EQC stands for EQUICO card. It performs equipment controller function and provides interface to LCT and NMS.
8. SERVICE stands for service unit which provides auxiliary channels, order-wire and 2 MHz input/output.
9. T-BUS stands for termination bus card. It is a mandatory unit used to provide voltage logical reference to all control and auxiliary bus.

4 × ANY HOST C sub-system

A very special feature of 1660 SM is a 4×ANY HOST C sub-system. This can be plugged into any enhanced slot pair, viz. 25–26 or 28–29 or 34–35 or 37–38. This subsystem performs time division multiplexing/de-multiplexing of client channels. The client streams are mapped in 16 virtually concatenated VC-4s according to a proprietary mapping algorithm. Provisioning parameters and alarm collections are managed through two control interfaces.

Each 4×ANY HOST C sub-system or module that plugs into any enhanced slot has 4 ports for the client signals which can be OL (optical low speed, viz. FE, FDDI, ESCON, digital video) or OH (optical high speed, viz. GE, FICON, fibre channel). Optical wavelength can be 1310 nm (OL-IN, OH-IN) or 850 nm (OL-MM, OH-MM).

Port cards and access cards supported and slots to be used

1660 SM supports following the port cards (placed in basic area of sub-rack) for deriving the bandwidth for the required interface and distributes it through access cards (placed in access area of the sub-rack).

Port card	Purpose	Slot for port card	Access card	Slot for access card	Remarks
P63E1	63 E1 interface	24, 27, 30, 33, 36, 39	Three of A21E1s	1 to 9, 13 to 21	1, 2, 3 correspond to P63E1 in slot 24 and so on; 13, 14, 15 to P63E1 in slot 33 and so on.
P63 E1 N-M4	63 x 2 Mbps ISDN PRI	24, 27, 30, 33, 36, 39	Three of A21E1s	1 to 9, 13 to 21	1, 2, 3 correspond to P63E1N-M4 in slot 24 and so on, 13, 14, 15 to P63E1 N-M4 in slot 33 and so on.
P3E3T3	3 x E3 / T3 interface	24 to 39	A3E3 / A3T3	2 to 9, 13 to 20	Access card in slot 2 is for port card in slot 24 and so on
P4E4N	2 x E4 interface	24 to 39	A2S1	2 to 9, 13 to 20	Access card in slot 2 is for port card in slot 24 and so on
P4S1N	2 x STM-1	24 to 39	A2S1	2 to 9, 13 to 20	Access card in slot 2 is for port card in slot 24 and so on
P4ES1N	4 x STM-1 electrical interface	24 to 39	A4ES1	2 to 9, 13 to 20	Access card in slot 2 is for port card in slot 24 and so on
P16 S1-4E	16 x STM-1 or 4 STM-4	24 to 39	A12OS1E (can derive 12 STM1 or 3 STM4)	Pair slots 2&3, 3&4… 19&20	Access card pair in 2&3 for port card in slot 24 and so on
P16 S1-4E	16 x STM-1 or 4 STM-4	24 to 39	A12OS1SE (can derive 12 STM1 or 3 STM4)	2 to 9, 13 to 20	Access card in slot 2 is for port card in slot 24 and so on
P16 S1-4E	16 x STM-1 or 4 STM-4	24 to 39	A16OS1SE (can derive 16 STM1 or 4 STM4)	2 to 9, 13 to 20	Access card in slot 2 is for port card in slot 24 and so on
ETH-MB	For 11 100Mbps interface	24 to 39	ETH-ATX (can derive 14 FEs)	2 to 9, 13 to 20	Access card in slot 2 is for port card in slot 24 and so on
ETH-MB	For 11 100 Mb interface	24 to 39	GETH-AG (can derive 2 GE)	2 to 9, 13 to 20	Access card in slot 2 is for port card in slot 24 and so on

(Continues…)

Port card	Purpose	Slot for port card	Access card	Slot for access card	Remarks
ISA ES-16	To make available 2.5 Gbps throughput for Ethernet	24 to 39	ETH-ATX (can derive 14 FEs)	2 to 9, 13 to 20	Access card in slot 2 is for port card in slot 24 and so on
ISA ES-16	To make available 2.5 Gbps throughput for Ethernet	24 to 39	GETH-AG (can derive 2 GE)	2 to 9, 13 to 20	Access card in slot 2 is for port card in slot 24 and so on
ISA-PR	Ethernet packet ring can be created using the card on NEs on a ring	Slot pair 25&26 or 26&27... so on 38&39	16FEA–PR or 2GBA–PR	2–3, 3–4, 4–5 or 3–4, 4–5, 5–6… so on 18–19, 19–20, 20–21	16 FE access or 2 GE access can be provisioned on the Ethernet ring; useful for VPN implementation.

Booster and pre-amplifier modules

OMSN BOOSTER +10dBm, +15 dBm and +17 dBm, identified as BST10, BST15 and BST17 respectively, are available to enhance the optical signal strength which can be put in slot 2 to 8 and 13 to 19. Connector options are FC/PC and SC/PC. Pre-amplifier for 2.5 Gbps (STM-16 signal) is available, identified as PR16, which can be plugged in slot 2 to slot 8 and slot 13 to slot 19. A maximum of 8 can be used at a time in the sub-rack.

STM modules supported by 1660 SM

STM-1 optical interface modules supported by 1660 SM are S1.1, L1.1, L1.2, L 4.2 JE. STM-4 optical interface modules supported by 1660SM are S4.1, L4.1, L4.2, L4.2 JE. STM-16 optical interface modules supported by 1660SM are S16.1, L16.1, L16.2, L16.2 JE.

STM–1 SFP optical modules supported are OPTO TRX SFP S–1.1, L1.1, L1.2. STM–4 SFP optical modules supported are OPTO TRX SFP S–4.1, L4.1, L4.2. STM–16 SFP optical modules supported are OPTO TRX SFP S–16.1, L16.1, L16.2. Connector options can be found in[3].

CWDM modules supported by 1660 SM

1660SM equipped with coarse WDM(CWDM) units is called 1660OADM. The cards are

- COADM 1 (2) channel (to add/drop one or two wavelengths)
- MUX/DEMUX (to multiples/de-multiplex eight wavelengths of ITU-T's CWDM grid)

The use of COADM unit for interfacing WDM ring allows the operator to duplicate/ triplicate the starting network capacity based on a single optical channel provided by pure STM–N interfaces.

Wavelengths added/dropped are terminated by STM16 CWDM interfaces operating in compliance with the ITU–T grid. Traffic streams terminated are then cross-connected at SDH level and made available to the relevant client interfaces (SDH/PDH/MPLS/ Ethernet).

A maximum of 8 COADM modules can be used at a time and these can be plugged in slots 1 to 8 or 13 to 20.

Use of MUX/DEMUX unit for interfacing WDM ring allows the operator to terminate the complete pool of optical channels (8 wavelengths or channels in CWDM as per ITU-T grid). MUX/DEMUX unit is required when the traffic matrix is larger than the capacity provided by COADM 1(2) channel/s. MUX/DEMUX module can be plugged into slot 1 to 8 or slot 13 to 20.

CWDM optical SFP modules are identified as SS-162C and SL-162C (for short haul and long haul respectively) and are available to operate at wavelengths 1470 nm, 1490 nm, 1510 nm, 1530 nm, 1550 nm, 1570 nm, 1590 nm and 1610 nm. The details of configurations can be found in [3].

References

[1] Tejas Equipment Manual for TJ100MC1, 2004.

[2] Tejas Equipment Manual for TJ100 MC4/MC-4L/XPE4.

[3] *Technical Handbook of Alcatel*, 1660SM STM64 Multiservice Metro Node, Alcatel's Equipment Manual 3AL 91669 AAAA Ed.03, 2007.

Appendix B

In chapter 9, a brief introduction to SIP, H.323, MGCP and H.248/Megaco was provided in the course of discussions on SDH support for NGN. More details of the same are presented here. This book, which is drawn from the author's experience of teaching SDH at IRISET (Indian Railway Institute of Signal Engineering and Telecommunications), follows the approach employed by the author in his classroom teaching, that is, limiting the discussions in the main course to SDH applications and thereafter devoting a few hours to SIP, H.323, MGCP, H.248/Megaco. The discussion on the topic is covered in this appendix*.

B.1 More About H.323

As mentioned in section 9.6.5, H.323 was the first VoIP standard that helped move the VoIP industry away from proprietary solutions towards interoperable solutions. The initial focus of H.323 was on facilitating multimedia conferencing over LANs that do not provide guaranteed QoS.

H.323 network was conceived as a WAN with number of zones, each zone having a single H.323 gatekeeper (GK), a number of H.323 terminal endpoints (TEs), a number of H.323 gateways (GWs) and a number of multipoint control units (MCUs) interconnected via a LAN. LANs in a zone can be in a single location or at different locations (interconnected by SDH network facilitating transportation of Ethernet traffic over VCs). However, each zone contains exactly one GK, which acts as the administrator of the zone. The functions of various entities in H.323 network, viz. TE, GK, GW, MCU and signalling and control protocols of H.323 have been listed in section 9.6.6.

B.1.1 H.323 terminal description

H.323 terminal is shown in Fig. B.1. The diagram shows the user equipment interfaces, video codec, audio codec, telemetric equipment, H.225.0 layer, system control functions and the interface to the packet-based network. All H.323 terminals have a system control unit, H.225.0 layer, network interface and an audio codec unit. The video codec unit and user data applications are optional.

* Readers will be interested to learn that the author's partiality for the topic earned him the nick-name SIP server at IRISET!

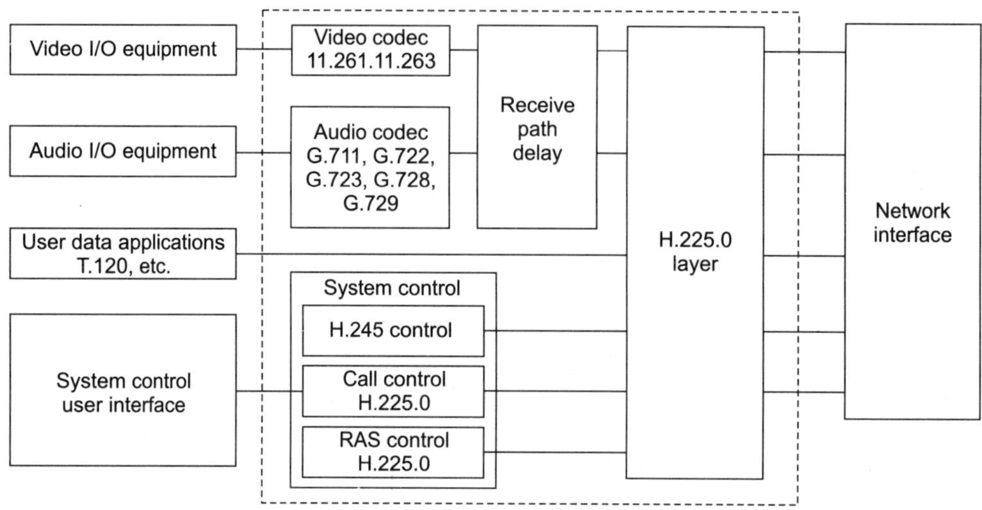

Fig. B.1 *H.323 terminal description*

B.1.2 Call control models in H.323

Two call control models are supported in H.323: GK routed call and direct call.

In the GK-routed call model, all signalling messages are routed through the GK cloud. In this case the use of RAS is necessary. This model allows endpoints to access services provided by the GK cloud, such as address resolution and call routing. It also allows the GKs to enforce admission control and bandwidth allocation over their respective zones. This model is suited to the needs of IP telephony service providers since it allows them to control the network and exercise accounting and billing functions.

GK-routed calls are shown in Fig. B.2. In this figure, the single line arrows represent call signalling channel messages and the double line arrows represent RAS channel messages.

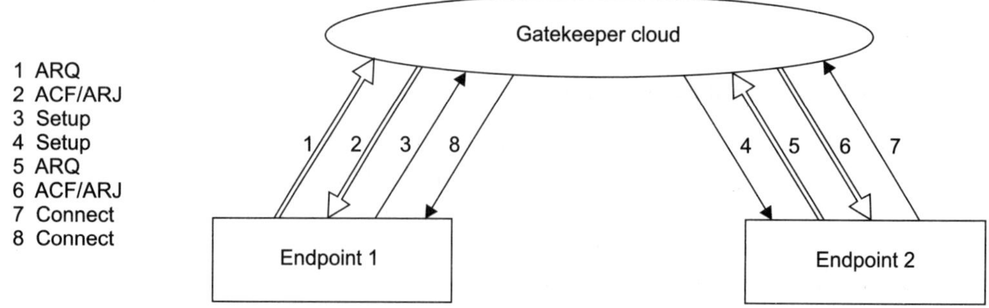

Fig. B.2 *GK routed calls in H.323 network*

In the direct call model, all Q.931 and H.245 signalling messages are exchanged directly between the two endpoints and so are the RTP media streams. As long as the calling endpoint knows the transport address of the called endpoint, it can set up a direct call with the other party. This corresponds to the early PC client model using IP as transport for free Internet phone calls. The GK cloud and RAS channels are optional. When GKs are present, the calling TE (TE1) may request address resolution service from its GK and the called TE (TE2) may ask for permission from its GK to accept the call. This model is unattractive for large-scale carrier deployments because carriers may be unaware of calls being set up, which may prevent them from providing sufficient resources for the call and for charging for it.

Direct calls are shown in Fig. B.3. In this figure, the single line arrows represent call signalling channel messages and the double line arrows represent RAS channel messages.

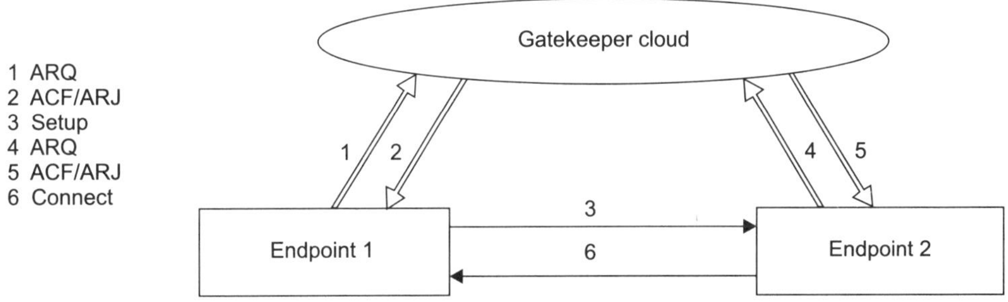

Fig. B.3 *Direct calls*

For simple VoIP calls, H.323 defines *fast connect*, which reduces the phases of a call by combining the Q.931 and H.245 phases.

B.1.3 H.323 services

H.323 v.1 only defined the basic call control and signalling for setting up multipoint multimedia conferences and did not address enhanced services. To enable enhanced services on top of H.323, ITU-T SG16 created the H.450 series which specify supplementary services similar to the features available in the PSTN and PBX. So, we have two sets of services: H.450-based services and non-H.450-based services.

H.450-based services

H.450 is an end-to-end protocol on top of Q.931 and its purpose is to adapt features of PSTN and PBX to IP domain. So, the TEs working on H.450 must understand the service logic in order to make the supplementary service work. The only way this is possible is by making TEs execute most of the service logic. This approach is different from the traditional PSTN enhanced services, where the service logic resides in the switches and not in the endpoints (phones). So, by building more intelligence into TEs via H.450, H.323 is able to offer supplementary services similar to PSTN and PBX on VoIP.

H.323 was not able to offer all supplementary services in one go – it offered them in instalments in various releases:

1. In H.323 v. 1, only the basic call control and signalling for setting up multipoint multimedia conferences are defined. No supplementary services are supported.
2. In H.323 v. 2, three H.450 recommendations were ratified: H.450.1 for generic functional protocol and procedures; H.450.2 for call transfer; and H.450.3 for call diversion, including various flavours of call forwarding and deflection.

H.450.1 defines a generic functional protocol on top of Q.931 for all supplementary services. It also defines the control procedures for the TEs involved in handling protocol messages. The functional protocol defined in H.450.1 is an end-to-end signalling protocol derived from the QSIG protocol for interconnecting private branch exchanges (PBXs).

Each H.450.X with X larger than 1 defines a supplementary service application protocol for a specific service. A supplementary service application protocol data unit (SS-APDU) is encapsulated in the UUIE of a Q.931 message as the H4501 supplementary service parameter. For example, H.450.3 specifies the call diversion services which include call forwarding (unconditional), call forwarding (busy), call forwarding (no reply), and call deflection. These services roughly correspond to various call forwarding features in PSTN. For each service, a set of procedures and the corresponding message flows are defined, such as activation, deactivation, interrogation, registration, and invocation.

3. In H.323 v.3, which was approved in September 1999, five more supplementary services are defined: H.450.4 for call hold; H.450.5 for call park and pickup; H.450.6 for message waiting indication; and H.450.7 for call waiting.

It is fine to have more and more supplementary services on the offer, but it poses significant problems for large-scale deployments for TEs supporting different releases of H.323.

Non-H.450-based services

As already see from above, the disadvantage of H.450-based services is that for all new releases of H.323, TEs needed upgrading before a service is deployed. This slows down the deployment of new services, an undesirable feature in VoIP environment.

The alternative is to implement the services inside GKs. In fact, this alternative comes with a bonus: with GK being used for address resolution, there are many services, such as mobility services, that can be offered to customers. The GK-routed call model allows carriers to introduce more advanced services.

But most services implemented inside the GK are implemented in a proprietary manner. Hence, interoperability and scalability issues remain unsolved.

Realising the limitation of H.323 (the need to upgrade TEs with every release in the case of H.450 based services and service intelligence in GKs, making them proprietary in the case of non-H.450 based services), ITU-T SG16 initiated two new work items in 1999 for version 4. In fact, ITU-T SG16's initiatives are based on the PSTN experience.

(PSTN's similar teething problems got solved later through intelligent network (IN). Concept of IN was introduced in the PSTN to standardise the development of services and locate service logic in a separate platform. In a similar manner, as the VoIP industry matured, the means of integrating IN with GKs also evolved).

The two new initiatives by ITU-T SG16 in H.323 v.4 are:

1. Introduction of an HTTP-based control channel for H.323 devices so that a service provider is able to display web pages to the user with H.323 call-related contents. (This is addressed in Annex K of H.323, which provides a new way to creating new services using a mechanism similar to third party call control).
2. Provision of a new 'stimulus-based' control mechanism for H.323 systems so that a relatively simple H.323 endpoint can rely on the intelligence residing in the network elements such as IN server. (This is addressed in Annex L of H.323, which utilises the 'package' concept for endpoint capability customisation.)

Apart from the above initiatives, H.323 v.4 included five more supplementary services: H.450.8 for name identification; H.450.9 for call completion; H.450.10 for call offer; H.450.11 for call intrusion; and H.450.12 for additional common information network services.

B.1.4 H.323 v.6 enhancements

H.323 v.6 was officially approved in June 2006 with the consent of ITU-T SG16. This study group strived to keep the changes to the base specification to a minimum. There were a few minor additions, but most of the new functionalities being added to H.323 were through extensions that utilise the 'generic' fields and messages in H.225.0 and H.245.

The primary changes to the base H.323 specification is in the introduction of the 'assigned gatekeeper' concept, as described in this section. There were also a number of changes to improve the clarity of certain procedures, including the call clearing procedures, error handling, etc. Enhancements were made to the H.225.0 document to clarify usage of certain fields and to provide a standard mapping of the cause and reason codes for LRJ/ARJ and access reject/ARJ (access reject is defined in H.501), add a new ISUP-number alias type, support the new QoS specification (H.361), and a language field for support of message broadcast (H.460.21). Finally, additions were made to H.323 Annexure G to support the transport of real-time text (ToIP) interleaved with audio to support ITU-T recommendation v.151.

A number of new documents and annexures have been introduced for H.245 that enable H.323 system to utilise various codecs, including GSM, iLBC and H.264. H.245 has been incrementally enhanced with various new features that H.323v.6 systems can utilise.

Security-related documents have also been significantly enhanced and the H.235 document entirely restructured. Now, rather than having a single H.235 document, they are numbered H.235.0, H.235.1, etc. Most importantly, support has been added for SRTP to secure RTP-based media streams.

Most of the new functionality was introduced incrementally over the past two years since the publication of H.323v.5. Some of the newer documents that were approved include:

H.235.0 – H.235.9 - Security for H.323 systems (including SRTP support)
H.239 – Role management
H.241 – Extended video procedures
H.249 – Extended user input indications
H.361 – End-to-end quality of service (QoS) and service priority signalling
H.460.10 – Call party category
H.460.11 – Delayed call establishment
H.460.12 – Glare control indicator
H.460.13 – Called user release control
H.460.14 – Multi-level precedence and pre-emption
H.460.15 – Call signalling transport channel suspension and redirection
H.460.16 – Multiple-message release sequence capability
H.460.17 – Tunnelling RAS through H.225.0
H.460.18 – Traversal of H.323 signalling across network address translators and firewalls
H.460.19 – Traversal of H.323 media across network address translators and firewalls
H.460.20 – Location number for H.323
H.460.21 – Message broadcast for H.323 systems

Assigned gatekeepers

The concept of an 'assigned gatekeeper' is an extension of the alternate gatekeeper mechanism. An 'assigned' gatekeeper is the gatekeeper with which an endpoint should register out of a list of alternate gatekeepers. In the event that the assigned gatekeeper fails, the endpoint would switch to an alternate (as per the alternate gatekeeper procedures). However, either the endpoint or the 'current' gatekeeper will monitor the status of the assigned gatekeeper and, once the assigned gatekeeper is available, the endpoint will transition back to its assigned gatekeeper. This allows the network operator to maintain a preferred provisioned state. While alternate gatekeepers may be geographically dispersed, there is generally a desire to have endpoints register with the alternate gatekeeper that is physically the closest. Whatever the registration preference of the network operator, the assigned gatekeeper concept gives more control to how the resources are utilised.

H.235.1 – H.235.9 - Security for H.323 systems

Most of these documents are enhancements to the former H.235 documents, but a lot of new material has been added, most notably support for SRTP.

H.239 – Role management

H.239 was first introduced with H.323v.5, but got inadvertently omitted from Packetizer's 'What's New in H.323v.5'. H.239 introduces the concept of 'roles' in H.245, which allows, for example, a device to indicate that one video stream is a live video and one a presentation stream.

H.241 – Extended video procedures

This recommendation introduces some extended video procedures for H.323 systems. Most notably, it adds support for H.264 to H.323 systems.

H.249 – Extended user input indication

This recommendation adds the ability to signal new kinds of events like mouse clicks and cursor moves.

H.361 – End-to-end quality of service (QoS) and service priority signalling

This document (formerly called H.323 Annex N) defines how to do end-to-end QoS within H.323 systems.

H.460.10 – Call party category

This document allows transport of the ISUP 'call party category' field within H.323 systems.

H.460.11 – Delayed call establishment

The idea of 'delayed call establishment' is to ensure that media streams or other conditions are met before alerting the user to an incoming call. This allows, for example, a device to ensure that QoS parameters are negotiated or that pinholes can be established through firewalls before ringing the remote party.

H.460.12 – Glare control indicator

This recommendation specifies a mechanism that allows an egress H.323-PSTN gateway to resolve a glare condition (also known as 'dual seizure') that is detected on the gateway when circuit selection is done on the gatekeeper.

H.460.13 – Called user release control

When calling certain numbers (e.g., emergency access numbers), the called party should have some control over how and when the call is released. This recommendation provides that type of functionality.

H.460.14 – Multi-level precedence and pre-emption

This recommendation adds support for MLPP to H.323 systems.

H.460.15 – Call signalling transport channel suspension and redirection

This recommendation allows an intermediary device, for example, to suspend the H.225.0 call signal channel and have it re-routed to another device or, more commonly, point-to-point between the calling and called devices. This allows, for example, a call to be established through a gatekeeper that routes call signalling and, once the call is stable, to have the signalling burden moved off the gatekeeper to the endpoints.

H.460.16 – Multiple-message release sequence capability

Generally, H.323 devices that wish to release the call do so by sending a release complete and closing of the socket. Unfortunately, TCP implementations sometimes throw away messages immediately when a socket is closed, so the remote party does not receive the release complete message. This recommendation defines a handshake to ensure that the release complete message (with the cause code) is properly delivered and received.

H.460.17 – Tunneling RAS through H.225.0

This recommendation is used as part of a NAT/FW traversal solution, along with H.450.19, to allow devices to send and received H.323 calls through a NAT/FW device. The primary idea is to maintain a persistent TCP connection between the endpoint on the private network and the gatekeeper on the public network.

H.460.18 – Traversal of H.323 signalling across network address translators and firewalls

This recommendation defines procedures for traversing a NAT/FW device. It defines new procedures that must be implemented by gatekeepers and endpoints, but also allows for the use of a special 'proxy' device to enable older devices to properly traverse a NAT/FW that do not understand the procedures required to do so autonomously. Like H.460.17, this recommendation relies on H.460.19 for media traversal.

H.460.19 – Traversal of H.323 media across network address translators and firewalls

This recommendation defines the procedures for passing media streams through a NAT/FW device and is used by both H.460.17 and H.460.18.

H.460.20 – Location number for H.323

This recommendation defines how to transport a 'location number' (similar to that found in ISUP) within H.323.

H.460.21 – Message broadcast for H.323 systems

This is a feature wherein H.323 devices may broadcast (using multicast) a message to one or more remote H.323 terminals. Message can be for providing an 'intercom' function for enterprise telephones, a 'paging' service through an enterprise, or a notification system to geographically dispersed terminals. Since standard Internet multicast procedures are made available for H.323 by this feature, it can be used on a wide scale to reach any number of H.323 endpoints in a geographic region or even the entire world.

B.2 SIP (Session Initiation Protocol)

As discussed in Section 9.6, SIP had its origin in pre-commercial stage of VoIP. That has been its bane and though it evolved from the initial from of SIP Internet draft (1996), RFC 2543 (1999) to RFC 3261(2002) with several extensions thereof, it is viewed with suspicion. In this section, the basics of SIP, how it compares with H.323 and how to make best use of SIP are presented.

B.2.1 SIP overview

SIP is a signalling protocol to control the initiation, modification and termination of interactive multimedia sessions. It does so by enabling:

a. Discovery of location of user for establishing or diverting a session
b. Determination of media capabilities of the devices involved in the session
c. Determination of willingness of end user to participate in a session
d. Establishment of session parameters for the parties involved in the session
e. Modification, transfer and termination of an active session
f. Support of user mobility

SIP network has the following elements:

1. *User agent (UA):* It is a logical entity that initiates or responds to SIP transactions. It can be client (UAC) or server (UAS). While UAC initiates SIP requests and accepts SIP responses, UAS accepts SIP requests and sends SIP responses. UA may or may not interact with a human user. UA maintains session or dialog state, i.e., UA is stateful.

2. *Proxy*: It is an intermediate entity in the SIP network that is responsible for forwarding SIP requests to target UAS or proxy, i.e., it does the routing function. It can also enforce policy. Proxies maintain states for the duration of transaction (32 seconds approx.)
3. *Redirect server:* It is a UAS that generates SIP responses to requests received, directing the UAC to contact alternate set of URI (uniform resource identifiers).
4. *Registrar server:* It is a UAS that accepts SIP REGISTER requests and updates the information from the request message into a location database.
5. *Back-to-back user agent (B2BUA):* It is an intermediate entity. It receives incoming requests as UAS and regenerates the request to send on the network as UAC. It maintains dialog state and participates in all transactions in the dialog.

B.2.2 Use of other IETF protocols

SIP uses the following IETF protocols.

1. DNS (domain name service) to resolve host or domain names into routable IP addresses.
2. SDP (session dependent protocol) is used in SIP message body.
3. RTP (real time transport protocol) for transport of packets in real-time.
4. RSVP (resource reservation protocol) to reserve network resources prior to alerting the called party.
5. TLS (transport layer security) as defined in RFC 2246 to provide privacy and integrity of SIP signalling information over the network.
6. STUN (session traversal utilities for NAT, i.e., network address translation) as defined in RFC 5389 to discover the presence and type of NAT between UACs and public internet.

B.2.3 SIP addressing

SIP addresses are typically referred as URI (uniform resource identifier). SIP URI is like e-mail address with a format sip:user@domain:port, sip:user@host:port.

User field can be user name, telephone number within the context of domain or host. Port is an optional field. If not specified, default is 5060.

RFC 3261 specifies secure SIP URI format also, known as SIPS (SIP secure) URI. Its format is sips:user@domain:port, sips:user@host:port. Default port for SIPS URI is 5061.

B.2.4 SIP messages

SIP messages can be broadly grouped into SIP requests and SIP responses. While SIP requests are sent from client to server to invoke a SIP operation, SIP responses are sent from server to a client to indicate the status of a SIP request that the client has sent.

SIP requests

Six SIP requests are defined in RFC 3261 which enable UA and proxy to locate users and initiate, modify and tear down sessions. These are:

INVITE: Through this SIP request, the user or service is invited to participate in a session. The message content has the description of the media session being set up or modified. Called party conveys willingness through 200 OK response.

ACK: This SIP request confirms that the UAC has received the final response to an INVITE request. It is sent end to end for a 200 OK response.

OPTIONS: This SIP request is used by UA to query UAS about its capabilities.

BYE: This SIP request is used by UA to terminate a previously established session.

CANCEL: This SIP request helps UAC and network servers in cancellation of in-progress request.

REGISTER: A client uses this SIP request to register its current location information corresponding to the AOR of the user with SIP servers.

SIP responses

UAC or proxy generates SIP response in response to a SIP request from UAC. SIP responses are numbered from 100 to 699. All 1xx responses are termed provisional responses, implying that they indicate provisional status. Other responses, viz., 2xx to 6xx convey final status, and are hence termed as final responses.

2xx indicates successful processing of SIP request. 3xx indicates the need to re-direct to another UAS for processing. 4xx, 5xx or 6xx indicate failure in processing of SIP request.

Table B.1 summarises SIP responses.

Table B.1 *SIP response table*

Class of response	Status code and meaning
Informational	100-Trying, 180-Ringing, 181-Call being forwarded, 182-Queued,183-Session in progress
Success	200-OK
Redirection	300-Multiple choices, 301-Moved permanently, 302-Moved temporarily, 305-Use proxy, 380-Alternative service

(Continues...)

Table B.1 *Continued*

Class of response	Status code and meaning
Client-error	400-Bad request, 401-Unauthorised, 402-Payment required, 403-Forbidden, 404-Not found, 405-Method not allowed, 406-Not applicable, 407-Proxy authentication required, 408-Request timeout, 410-Gone, 413-Request entity too large, 414-Request URL too large, 415-Unsupported media type, 416-Unsupported URI scheme, 420-Bad extension, 421-Extension required, 423-Interval too brief, 480-Temp. not available, 481-Transaction does not exist, 482-Loop detected, 483-Too many hops, 484-Address incomplete, 485-Ambguious, 486-Busy here, 487-Request terminated, 488-Not accepted here, 491-Request pending, 493-Undecipherable
Server-error	500-Internal server error, 501-Not implemented, 502-Bad gateway, 503-Service unavailable, 504-Server timeout, 505-SIP version not supported, 513-Message too large
Global failure	600-Busy everywhere, 603-Decline, 604-Does not exist anywhere, 606-Not acceptable

B.2.5 SIP message structure

A SIP message consists of start-line, one or more header fields, blank line indicating end of header and the message body (as per requirement).

Start line

The start line for SIP request is a 'request-line' which specifies SIP method, request URI and SIP version. The start line for SIP response is a 'status-line' which consists of SIP-version indicator, SIP-response code and optional reason phrase which is the textual description of the SIP response code.

Header fields

SIP message header fields contain information on signalling and routing for the network entities. The format is similar to HTTP header. Each header field consists of a field name followed by a colon (:) and the field value. Some important SIP headers are described in Table B.2.

Table B.2 *Some important SIP headers and their description*

SIP header	Description
From	Identifies the initiator of SIP request
To	Identifies desired recipient of SIP request
Call-ID	Identifies a series of SIP messages. It is identical for all SIP messages and responses sent by either side UA within a dialog.

(Continues...)

Table B.2 *Continued*

SIP header	Description
Cseq	It has an integer value and method-name. It identifies, orders and sequences SIP requests within a dialog. It also distinguishes message re-transmission and new message.
Via	Identifies the path taken by the request and the location where response needs to be sent
Contact	Identifies a SIP or SIP's URI where UA wants to receive a new SIP request
Allow	Lists the set of SIP methods supported by UA generating the message
Supported	Lists all extensions supported by UA
Require	Identifies extensions required to be supported at remote UA
Content-type	Indicates the type of message body attached to SIP message
Content length	Indicates the size of message body (in decimal) in a SIP message. This is required as SIP messages are ultimately carried over stream-based protocols like TCP.

SIP transactions through messages

A SIP signalling session between two UAs comprises one or more SIP transactions. A SIP transaction occurs between UAS and UAC and involves one or more intermediate servers like proxy, redirect. A SIP transaction comprises all messages that begin with SIP request initiated by UAC, until the final response is received from the UAS. It is identified by call-ID, via-branch, local tag, remote tag, Cseq value.

Ex: SIP register transaction:

UAC sends SIP request 'Register' to 'Registrar'
'Registrar' sends SIP response 100 (Trying) as provisional response
'Registrar' sends SIP response 200 (OK) as final response.

Readers may please note :

1. SIP transaction can result in establishment, modification or termination of session.
2. Establishment of session results in peer-to-peer signalling relationship, known as dialog. Dialog exists throughout duration of a session and between two or more UAs.
3. Multiple SIP transactions can take place within the context of SIP dialog. Each such transaction has increasing integer value in the Cseq header.

Dialog and transaction states are maintained at the SIP UAs or end-points. SIP servers (proxy, redirect) maintain only the transaction state (for at least 32 seconds as per RFC 3261), but not dialog state. This has two advantages:

1. Network servers (proxy/redirect) are free to serve many SIP end points
2. If any server goes out of service, only the transactions in progress are affected but not the established dialogs.

Transport layer protocols for SIP signalling

SIP transactions use either connection oriented transport layer protocol such as TCP (transmission control protocol) or SCTP (stream control transmission protocol) and connectionless protocols such as UDP.

B.2.6 SIP timers

End-to-end reliability in messages/dialog/transactions is guaranteed by use of various timers in SIP. Table B.3 summarises for each SIP timer the default value, the section of RFC 3261 that describes the timer, and the meaning of the timer.

These timers that were hard-coded in the previous releases various vendors' software, can now be configured, thus enabling users to improve the interoperability and performance of their devices and network environment.

Table B.3 *SIP timer's summary*

Timer	Default value	Related sec. of RFC 3261	Meaning
T1	500 ms	17.1.1.1	Round trip time (RTT) estimate
T2	4 s	17.1.2.2	Maximum re-transmission interval for non-INVITE requests & INVITE responses
T4	5 s	17.1.2.2	Maximum duration that a message can remain in the network
Timer A	Initially T1	17.1.1.2	INVITE request retransmission interval, for UDP only
Timer B	64* T1	17.1.1.2	INVITE transaction timeout timer
Timer D	> 32 s for UDP; 0 s for TCP & SCTP	17.1.1.2	Wait time for response retransmissions
Timer E	Initially T1	17.1.2.2	Non-INVITE request retransmission interval, UDP only
Timer F	64*T1	17.1.2.2	Non-INVITE transaction timeout timer
Timer G	Initially T1	17.2.1	INVITE response retransmission interval
Timer H	64*T1	17.2.1	Wait time for ACK receipt
Timer I	T4 for UDP	17.2.1	Wait time for ACK re-transmission

(Continues...)

Table B.3 *Continued*

Timer	Default value	Related sec. of RFC 3261	Meaning
Timer J	64*T1 for UDP; 0 s for TCP & SCTP		Wait time for retransmissions of non-INVITE requests
Timer K	T4 for UDP; 0 s for TCP & SCTP	17.1.2.2	Wait time for response retransmissions

B.2.7 SIP call processing

SIP operation for call processing can be in one of the three basic modes: proxy server, redirect server or B2BUA server.

Proxy server

1. Proxy server accepts INVITE request from UAC.
2. Proxy server contacts the location server to request the address of the called party.
3. Location server identifies the location of called party and provides the address of target server.
4. INVITE request is forwarded to the address so provided. Proxy may add record-route header to INVITE message to ensure all subsequent messages for that dialog are routed via the same proxy. This helps in billing.
5. Called party UA alerts the user. User answers the call.
6. UAS returns 20 OK as response to proxy server.
7. Proxy forwards 200 OK to UAC.
8. UAC confirms receipt of 200 OK by issuing ACK.
9. Proxy forwards the same to called party UA.

Re-direct server

1. Re-direct server accepts INVITE request from the calling party UA.
2. Redirect server contacts location server to get the address of called party.
3. Location server returns the address of called party.
4. Redirect server returns the address directly to the calling party in a 3xx message. Unlike the proxy server, redirect server does not forward the INVITE request.
5. UAC sends ACK to redirect server acknowledging receipt of 3xx message.
6. UAC sends INVITE request direct to the contact address returned by redirect server

7. Called party UA alerts the user and user answers the call.
8. Called part UA provides response 200 OK to the UAC
9. UAC sends ACK to the UAS sending 200 OK response.

B2BUA server

B2BUA is not defined in RFC3261 as a functional entity; it is described as concatenation of UAC and UAS. B2BUA can act as 3PCC (third party call controller) and establish calls between two user agents.

1. B2BUA sends INVITE to user A. This message contains no characteristics of media.
2. User A is alerted; 180 ringing message is sent from user A to B2BUA, but not shown in the message flow. When call is answered, 200 OK is sent to B2BUA. In tihis message, media characteristics are sent.
3. B2BUA sends ACK to user A and INVITE to user B. This INVITE message does not contain media characteristics.
4. User B is alerted and answers the call. 200 OK is sent with offered media characteristics to B2BUA.
5. Using this message, B2BUA creates INVITE message to user A. User A modifies its offered media characteristics if required.
6. User A sends 200 OK with modified offer to B2BUA.
7. B2BUA sends ACK with answer (user A's modified offer) to user B.
8. B2BUA sends an ACK to user A to acknowledge user A's 200 OK response.
9. Call is established between user A and user B.

B2BUA facilitates interworking of SIP devices with other protocols like H.323, MGCP and also transport layer interworking such as TCP-UDP, IPv4-IPv.6 mapping.

Q B.1 *What is the time delay budget of SIP packets?*

According to ITU-T's G.114 recommendation, for good voice quality over VoIP, no more than 150 ms delay should occur end-to-end, one way. G.729 coder look ahead delay of 5 ms, frame processing delay of 10 ms, serialisation delay of 3 ms on the final delivered 64 kbps trunk and lines propagation delay of 32 ms makes a total 60 ms delay. The remaining 90 ms is available for end to end delay and even with 50 ms budget for de-jitter buffer, a comfortable margin of 40 ms is available.

(In fact, on satellite link, a delay of 250 ms earth station to satellite and another 250 ms delay back to earth, i.e., a total of 500 ms are acceptable, as no other alternative exists. ITU-T's G.114 is for the high quality VoIP).

Q B.2 *Are there any products using B2BUA features?*

SIP's B2BUA functions are used to the fullest extent by SIP based IP-PBXs, soft-switches, firewall/NAT traversal applications, call centre applications and conference servers.

B.2.8 H.323 Vs. SIP

Aspect	H.323	SIP
Design approach	H.323 is unified protocol for multimedia communication over IP networks, including audio, video, and data conferencing. It is robust with no interoperability problems	SIP treats call as a session between multimedia streams. As integration of disparate standards is largely left to each vendor, there are interoperability problems. However, SIP has been successful in 'closed environment', where interoperability is facilitated by media gateways
Lineage	H.323 is based on H.324, which in turn was a better H.320. First version of H.323 was issued in the year 1996, Ver.6 in the year 2006.	SIP is frequently allied with the Internet and the World Wide Web by way of HTTP. SIP (RFC 2543) issued in the year 1999 and SIP (RFC 3261) in the year 2006. Though version number remained the same, there was no backward compatibility
Recovery from failure and Reliability	H.323 has 'alternate gatekeepers', 'alternate endpoints' and means of recovering from connection failures. Thus, reliability is high.	SIP has no defined procedures for handling device failure. UA detects the failure of a proxy through timer expiration. It is the responsibility of the user-agent to send a re-INVITE to another proxy, leading to long delays in call establishment.
Message structure	Messages are structured in a standard notation as per ASN.1	Messages are structured as per ABNF (Augmented Backus-Naur Form), which is a syntactical notation, as per IETF's RFC 2234

(Continues...)

Aspect	H.323	SIP
Message encoding	H.323 encodes messages in a compact binary format that is suitable for narrowband and broadband connections (but, not readable, which is not a disadvantage)	SIP messages are encoded in ASCII text format, hence are readable. However, this brings in the disadvantage of long messages and associated bandwidth consumption, delay, and/or processing problems. Sometimes, SIP messages exceed the MTU size when going over WAN links, resulting in delays, packet loss, etc. Hence, in RFC 3485 and 3486, binary encoding of SIP messages has been prescribed.
Media transport	RTP/RTCP, SRTP	RTP/RTCP, SRTP
addressing	H.323 supports URIs, e-mail addresses, and E.164 numbers. H.323 supports the following aliases: E.164 dialled digits, generic H.323 ID, URL, transport address, e-mail address, party number, mobile ISDN/IMSI, ISUP number	SIP interprets only URI-type addresses. It is okay with SIP-SIP devices, but necessitates certain symbol insertions when translating various dialled digits. Example: A '+' sign is inserted in the SIP URI (sip:+18005551212@example.com) in order to indicate that the number is in E.164 format, versus a user ID that might be numeric.
Overlap signalling	H.323 sends overlap signalling digits with no additional overhead, except conveyance of the newly received digits in a single message.	SIP supports overlapped signalling defined in RFC 3578, but additional digit received requires transmission of three messages on the VoIP logical channel (a new INVITE, a 484 response to indicate that the address is incomplete, and an ACK).
Address resolution	H.323 defines an interface between the endpoint and gatekeeper for address resolution using ARQ or LRQ. The H.323 gatekeeper may use any number of protocols to discover the destination address of the callee. The endpoint does not have to be concerned with the mechanics of this process, and the processing requirements for address resolution placed on the gatekeeper by H.323 are for just a single message exchange.	SIP has no address-resolution protocol. However, SIP user agent may route its INVITE message through a proxy or redirect server in order to resolve addresses. The SIP proxy may use various protocols to discover the destination address of the callee. The endpoint does not have to be concerned with the mechanics of this process. Unfortunately, the processing requirements placed on the SIP proxy are higher than with H.323 because at least 3 message exchanges must take place between the SIP device, SIP proxy, and the next hop.

(Continues...)

Aspect	H.323	SIP
Vendor specific extensions	H.323 is extended with non-standard features in such a way as to avoid conflicts between vendors. Globally, unique identifiers help in distinguishing feature elements and data elements	SIP is extended by adding new header lines or message bodies that may be used by different vendors to serve different purposes; thus there is possibility of interoperability problems.
Call signalling	H.323 gatekeeper facilitates address resolution through one RAS message exchange, or may route all call signalling traffic. In large networks, the direct call model may be used so that endpoints connect directly to one another.	When a SIP proxy performs address resolution for the SIP UAC, the proxy is required to handle at least 3 full message exchanges for every call. In large networks, such as IMS networks, the number of messages on the wire may be excessive. (For example, a basic call between two users may require as many as 30 messages on the circuit.)
Statefulness/ Statelessness	H.323 gatekeeper can be stateless using the direct call model.	SIP proxy can be stateless if it does not fork, use TCP, or use multicast.
Call setup and capability negotiation	Simplest call establishment is in 1.5 round trips using UDP, as indicated below: Setup -> <- Connect Ack -> For complex scenarios, entities have to negotiate which channels to open, including audio, video, and data channels. Individual channels may be opened and closed during the call without disrupting the other channels.	A call can be established in as few as 1.5 round trips using UDP: INVITE -> <- 200 OK Ack -> For complex scenarios, SIP entities really do not negotiate, they follow declaration mechanism and follow 'trial and error' approach in case the called party does not support the proposed media.
Call forking	H.323 gatekeeper can control the call signalling and may fork the call to any number of devices simultaneously.	SIP proxies can control the call signalling and may fork the call to any number of devices simultaneously.

(Continues...)

Aspect	H.323	SIP
PSTN interworking	H.323 borrows from traditional PSTN protocols, e.g., Q.931, and integrates with PSTN easily. However, H.323 does not employ the PSTN's circuit-switched technology; H.323 is completely packet-switched.	SIP has no commonality with PSTN and such signalling must be dove-tailed into SIP. Whether PSTN or the 'IMS' variants viz., 3GPP, NGN, 3GPP2, and packet cable, etc., same approach of dove-tailing is needed.
Services	Services may be provided to an endpoint as it places a call, as a call arrives, or during the middle of a call by a gatekeeper or other entity that routes the call signalling. Such services may be provided to the endpoint through a web-browser interface using HTTP or a feature server using Megaco/H.248.	SIP devices can receive service from a SIP proxy as the endpoint places a call, as a call arrives, or during the middle of a call. There is no defined way within SIP of providing services via a web browser or a feature server, as everything is done within the context of a 'session'.
Video and data conferencing	H.323 fully supports video and data conferencing. Procedures to provide control for conference as well as lip synchronisation of audio and video streams are available.	SIP has limited support for video and no support for data conferencing protocols like T.120. SIP has no procedures to control the conference and there is no mechanism within SIP for lip synchronisation.
Packet loss recovery	H.323's 'video fast update' command helps in recovery of packet loss.	There is no standard means of recovering from packet loss in a video stream.
Firewall/NAT support	Firewall/NAT support is provided by H.323 'proxy' or by the endpoint, both in conjunction with a gatekeeper residing in the public network. (defined in H.460.17, H.460.18, 19.)	SIP does not define a firewall/ NAT support, but leaves it to other standards. Example: STUN, TURN, ANAT, and ICE.
Transport protocol	Reliable or unreliable, e.g., TCP or UDP. Most H.323 entities use a reliable transport for signalling.	Reliable or unreliable, e.g., TCP or UDP. Most SIP entities use an unreliable transport for signalling.
Loop detection	Routing gatekeepers can detect loops by looking at the call-identifier and destination-address fields in call-processing messages.	The via header, max-forwards header facilitate loop detection. Details are in IETF's RFC 3261.

(Continues...)

Aspect	H.323	SIP
Multicast signalling	Supported via location requests (LRQ) and auto gatekeeper discovery (GRQ).	Supported through group INVITEs.
Third-party call control	Defined in third-party pause and re-routing. More sophisticated control is defined in the related H.450.x series of standards.	Defined in RFC 3725
Minimum ports for VoIP call	3 (call signalling, RTP, and RTCP.)	3 (SIP, RTP and RTCP.)
Conferencing entity	Yes, an MC is required for this, but it could be co-located in a participating endpoint, or all endpoints could contain an MC.	SIP user agents may perform conferencing themselves.
Media topology	Unicast, multicast, star, and centralised.	Unicast, multicast, star and centralised.
Authentication	Yes, via H.235.	Yes, via HTTP (digest and basic), SSL, PGP, S/MIME or other means.
Encryption	Yes, via H.235 (including use of SRTP, TLS, IPSec, etc.).	Yes, via SSL, PGP, S/MIME, or various other means.
DTMF carriage	H.245 user input indication, RFC 4733, or via the audio stream. The alphanumeric choice of the H.245 user input indication message is the baseline carriage common to all H.323 endpoints; so interoperability is assured.	There is no baseline carriage, which presents issues of interoperability. Transport of DTMF via the INFO method, RFC 4733, KPML, or the audio stream are all options.

B 2.9 Applications of SIP

Above table of comparison seems to be tilting the scale in favour of H.323. But, SIP has its advantages. Consider an NLD (national long distance) network shown in Fig. B.4 with MGs connected on SDH/IP/MPLS cloud. NLD operator carries customer's traffic (STD/ISD) landing on MG ports to UASPs (universal access service providers) i.e., PSTN/PLMN operators. MG ports to UASP connection has to be on E1 links as SS7 signalling is a must for PSTN/PLMN. If customer traffic is also taken on E1s, more MG ports are used up and more hardware costs have to be incurred. Instead, if customer traffic is taken on SIP trunk, E1 ports of MG can be saved and utilised for UASP connection.

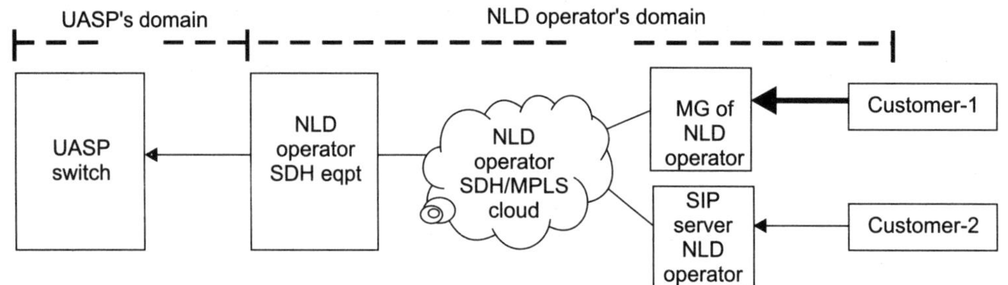

Fig. B.4 *TDM trunks vs. SIP trunk*

Note: Customer-1 is connected on TDM trunk and Customer-2 on SIP trunk to NLD operator. But, traffic from both lands on UASP's switch on TDM only.

Another advantage is smooth operation on networks with BW constraints. There are several codec options with SIP. Their bit rates, sample sizes and mean opinion scores (MOS) are tabulated in Table B.4.

Table B.4 *Voice codec's performance parameters*

Codec/ Compression	Bit rate (kbps)	Sample size (ms)	MOS score
G.711 PCM	64	0.125	4.1
G.726 ADPCM	32	0.125	3.85
G.728 Low delay code excited linear periodic (LD-CELP)	16	0.625	3.61
G.729 CS-ACELP (conjugate structure algebraic code excited linear predictive)	8	10	3.92
G.723.1 ACELP	5.3	30	3.65

With 64 kbps PCM, there is one sample every 125 micro second and with G.711 encoding (8 bit per sample or 8 kbps codec), it is 64 kbps bit rate.

G.729 CS-ACELP codec sends two 10 ms samples per frame with payload of 20 bytes. So, for payload portion, data rate is 20 × 8 bits/20 ms i.e., 8 kbps

Whatever be the codec, addition of IP header (20 byes), UDP header (8 bytes), RTP header (12 bytes) creates an overhead of 40 bytes per frame. With G.711 two 10 ms samples per frame, it means an overhead of 40 bytes in 20 ms or overhead bit rate of 40 × 8bits/20 ms or 16 kbps.

So, with G.729 CS-ACELP, it is 24 kbps per channel. So, on a SIP trunk of FE or 100 Mbps, taking 96 Mbps throughput after allowing for Ethernet frame overheads, we can support 96 Mbps/24 kbps, i.e., 4000 concurrent calls.

There is also the option of RTP header compression in SIP, which reduces IP/UDP/ RTP header from 40 bytes to 4 bytes. Then, per channel BW requirement reduces to 11.2 kbps from 24kbps, which implies more than 8400 concurrent calls on SIP trunk of FE.

Even if the trunk is not FE, with 24 kbps per channel (uncompressed header) or 11.2 kbps per channel (compressed header), channels carried are many more than TDM counterpart of G.711 (64 kbps per channel). This underlines the utility of SIP in BE constraint situations.

B.3 More About ITU-T Rec. H.248 (Megaco)

As discussed in section 9.8.1, a decomposed media gateway separates call control from media conversion. ITU-T Rec. H.248, the media gateway control protocol, also known as Megaco, is for control of elements in a physically decomposed multimedia gateway. It is the product of joint efforts of the IETF and the ITU-T Study Group 16.

Megaco addresses the relationship between the media gateway (MG), which converts circuit-switched voice to packet-based traffic and the media gateway controller (MGC, also called a call agent or soft-switch, which dictates the service logic of that traffic). Megaco instructs an MG to connect streams coming from a packet or cell data network onto a packet or cell stream such as the real-time transport protocol (RTP). Megaco is essentially quite similar to MGCP from an architectural standpoint and the controller-to-gateway relationship, but it supports a broader range of networks, such as ATM.

There are two basic components in Megaco: terminations and contexts. Simply stated, terminations are media connections called physical (time slots) and ephemeral (IP flow) and contexts—associations between terminations which can be added or deleted.

Contexts are created and released by the MG under command of the MGC. A context is created by adding the first termination, and it is released by removing (subtracting) the last termination. In other words, terminations are placed into contexts, which are defined as two or more termination streams being mixed and connected together. The normal, 'active' context might have a physical termination (say, one DS0 in a DS3) and one ephemeral one (the RTP stream connecting the gateway to the network). A termination may have more than one stream, and therefore a context may be a multi-stream context. Audio, video, and data streams may exist in a context among several terminations.

Megaco/H.248 uses a series of commands to manipulate terminations, contexts, events, and signals. The following is a list of the commands:

1. Add – The add command adds a termination to a context. The add command on the first termination in a context is used to create a context.
2. Modify – The modify command modifies the properties, events and signals of a termination.

3. Subtract – The subtract command disconnects a termination from its context and returns statistics on the termination's participation in the context. The subtract command on the last termination in a context deletes the context.
4. Move – The move command atomically moves a termination to another context.
5. Audit-value – The audit-value command returns the current state of properties, events, signals and statistics of terminations.
6. Audit-capabilities – The audit capabilities command returns all the possible values for Termination properties, events and signals allowed by the media gateway.
7. Notify – The notify command allows the media gateway to inform the media gateway controller of the occurrence of events in the media gateway.
8. Service-change – The service change command allows the media gateway to notify the media gateway controller that a termination or group of terminations is about to be taken out of service or has just been returned to service. Service-change is also used by the MG to announce its availability to an MGC (registration), and to notify the MGC of the impending or completed restart of the MG. The MGC may announce a handover to the MG by sending it a service-change command. The MGC may also use service-change to instruct the MG to take a termination or group of terminations in or out of service.

All of these commands are sent from the MGC to the MG, although service-change can also be sent by the MG. The notify command, with which the MG informs the MGC that one of the events the MGC was interested in has occurred, is sent by the MG to the MGC. All Megaco/H.248 messages are in the format of ASN.1 text messages.

Megaco/H.248 connection model

Megaco (H.248) connection model is illustrated in Fig. B.5. As already discussed in the previous section, two principal abstractions of the model are: terminations and contexts.

The three contexts possible are:

1. Call waiting: a context with just one termination, as the caller does not hear anyone else.
2. On-going call between two parties: a context with two terminations is a regular call, as each person hears the other.
3. Conferencing call: a context with more than two terminations is a conference call. All parties hear all others.

The maximum number of terminations in a context is a media gateway (MG) property. MGs that offer only point-to-point connectivity might allow at most two terminations per context. MGs that support multi-point conferences might allow three or more terminations per context.

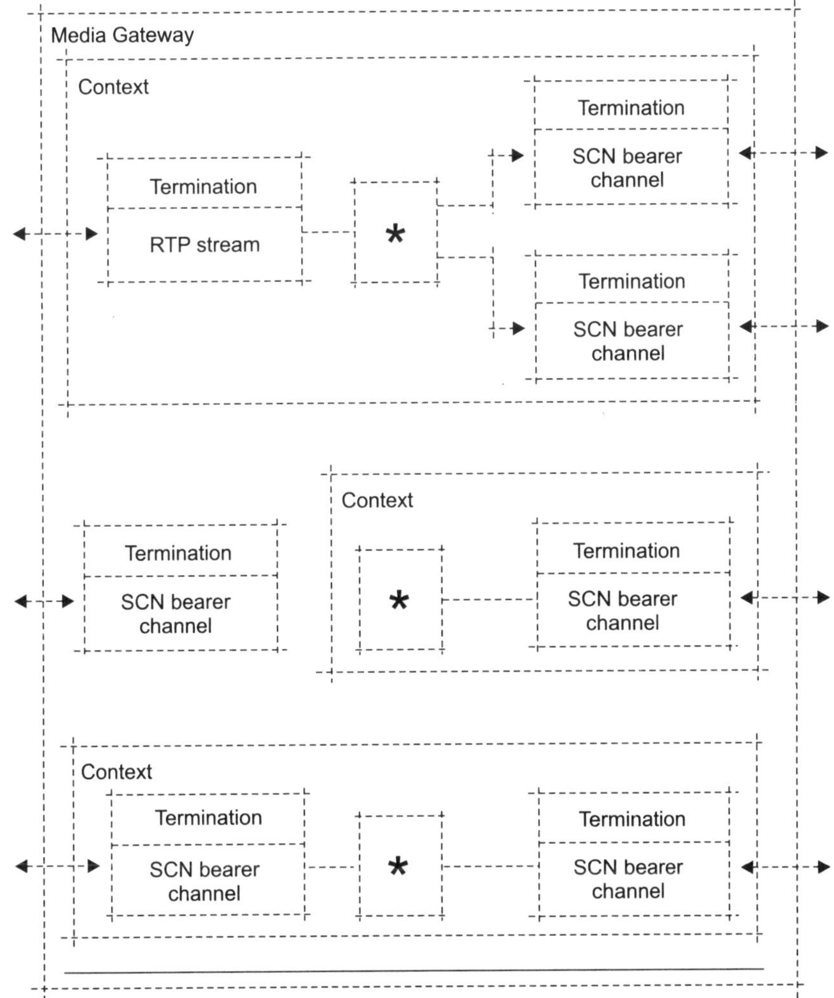

Fig. B.5 *H.248/Megaco connection model for SCN (switched circuit network)*

Note: The asterisk in each box in each of the contexts represents the logical association of terminations implied by the context

The attributes of contexts are:

- Context ID;
- The topology – who hears/sees whom;
- The topology of a context describes the flow of media between the terminations within a context. In contrast, the mode of a termination (send/receive) describes the flow of the media at the ingress/egress of the media gateway.

- The priority is used for a context in order to provide the MG with information about a certain precedence handling for a context. The MGC can also use the priority to control autonomously the traffic precedence in the MG in a smooth way in certain situations (e.g. restart) when a lot of contexts must be handled simultaneously.
- An indicator for an emergency call is also provided to allow preference handling in the MG.

Terminations are referenced by a termination-ID as per a scheme selected by the MG. Termination-IDs of physical terminations are provisioned by the media gateway. The termination-IDs may be chosen to have a structure. For example, a termination-ID may consist of a trunk group and a trunk within the group.

Appendix C

C.1 Migration from IPv4 to IPv6

C.1.1 Need for IPv6

Internet addresses (called IP addresses) are node addresses that identify the location of a node within the Internet. With 32-bit IPv4 address, 4 billion (4,294,967,295, to be precise) addresses are possible. In a study conducted by UN in the year 2004, it was found that with 20% penetration of Internet, assuming 3 persons per IP address, the available IPv4 addresses would be exhausted by the end of year 2008.

IPv6 address has 128 bits. With this the address space would be 3.402823665×1038 or 340,282,366,920,938,463,463,374,607,431,768,211,456, to be exact! Perhaps every square meter of earth's surface can be given 1564 addresses!

There are three types of IPv6 addresses: unicast, anycast and multicast.

a. Unicast means identifying a single interface.

b. Anycast means identifying a set of interfaces and routing to the nearest one.

c. Multicast means identifying a group of interfaces and delivering packets to all in the group.

The prominent five features of IPv6 are already mentioned in section 14.8.4. Additional information about IPv6 addressing architecture is available in RFC 1884, RFC 2073, RFC 2373 and RFC 2374.

Migration from IPv4 to IPv6 without disrupting current operation of existing IPv4 network is a challenging task. The strategy for achieving the same is presented here.

C.1.2 Strategy for migration from IPv4 to IPv6

The infrastructure subsets requiring migration are:

1. Networks and protocols
2. Applications
3. Access networks and customer equipment

The transition strategy for migration of various components in the above sub-sets is to be tested in the test bed. Such test should evaluate operational ease and efficiency in a service provider's environment. Then the implementation is to be taken up on operational networks, with no or minimum disruptions to commercial services. Such a migration strategy involves certain stages comprising a few steps each.

The five stages in the migration from IPv4 to IPv6 are:

1. Fulfilment of prerequisites and setting up of IPv6 test bed
2. IPv4 cloud with IPv4/IPv6 at edges
3. IPv6 core with few IPv4/IPv6 at edges
4. IPv6 cloud interconnecting few leftover IPv4 clouds
5. IPv6 only

Stage 1: Fulfilment of prerequisites and setting up of test bed (Fig. C.1)

1. *Setting up of national internet registry, allocation of contiguous block of IPv6 addresses for the country and IPv6 allocation policy of the country*
   IPv4 addresses are allocated to ISPs and government agencies region-wise. For example, the Asia Pacific Network Information Centre (APNIC) does the allocation for the Asia-Pacific region. IPv6 addresses too can be obtained from APNIC. However, a country should obtain a contiguous block of /32 IPv6 addresses in line with E-164 numbering scheme (for example, '91' belongs to India in E-164 scheme). To take up the issue with APNIC, a National Internet Registry (NIR) is to be set up (the DoT can assume the role). NIR can then allocate /32, /48 and /64 address blocks to ISPs, institutions, government departments, etc.
2. *Upgradation of DNS servers of ISPs (Internet service providers)*
   The DNS (domain name server) translates the name of web site to IP address. So, the DNS of ccTLD domain '.in' should support IPv6 addresses in addition to IPv4 addresses. The URL addresses likewise should be resolvable to both IPv4 or IPv6 and also reachable over IPv4 or IPv6 networks.
3. *IPv6 gateway connectivity to IPv6 Internet*

   A national IPv6 backbone should be created, which will integrate with the existing NIB of IPv4 (NIXI or National Internet Exchange of India in the case of India).
4. *Setting up of test bed for IPv6*
   A carrier grade test bed for IPv6 has to be set up as a national resource to test the following:

   a. Performance of network in IPv6 alone mode/dual mode
   b. Working of DNS

c. NMS and firewalls
d. Billing applications
e. Customer side IPv6 products

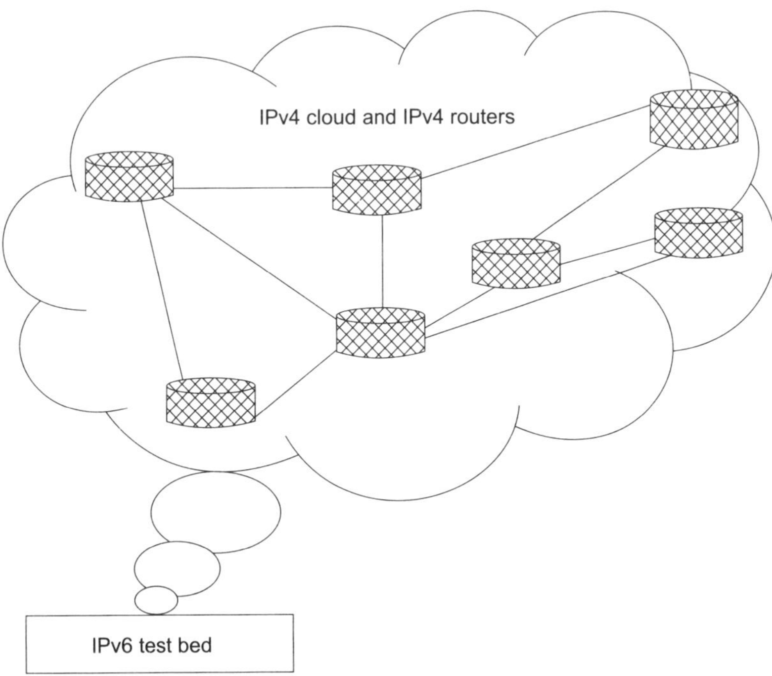

Fig. C.1 *Stage 1 of migration from IPv4 to IPv6*

Stage 2: IPv4 cloud with IPv4/IPv6 at edges

1. *Use of dual-stack routers with tunnelling capabilities at edges:* In this step, only core routers are made dual stack (for supporting IPv4 and IPv6) routers with tunnelling capability of IPv6 over IPv4, i.e., tunnels are formed within IPv4 core to carry IPv6 traffic across the network (refer to Fig. C.2). If the core is MPLS or MPλS-enabled, IPv6 data transfer is greatly facilitated. Else, certain dual-links, i.e., IPv4/IPv6 links are to be developed.

   The stability of IPv4 services is not jeopardized by this; also, the investment at this stage is low since only edge routers are made dual stack. MPLS / MPλS help the cause.

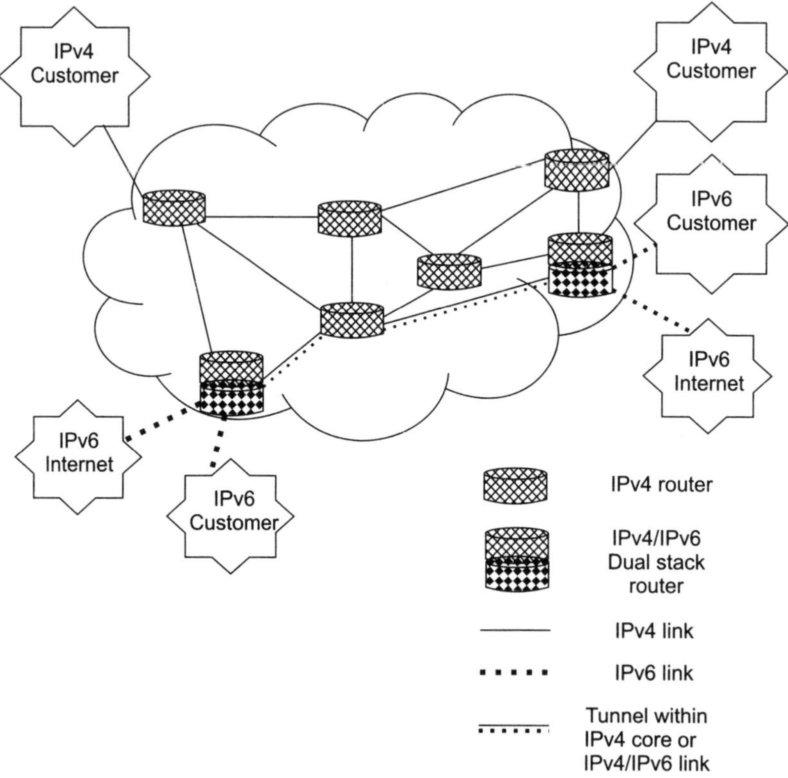

Fig. C.2 *Stage-2 of IPv4 to IPv6 migration (dual stack routers at edges)*

2. *Firewall and NMS for IPv6:* Network management system has to be modified to handle IPv4 and IPv6 links and IPv4/IPv6 links as may be the case while routing packets of IPv4 , IPv6. The firewall protection has to be in place after IPv6 router.
3. *Billing for IPv6:* An interface for IPv6 applications and the mechanism for billing for IPv6 data transfer has to be in place.
4. *IPv6 standardization with respect to other application layer protocols:* The interface to the application layer in IPv6 has to be in place.

Stage 3: Removal of IPv4 from core network (IPv6 only core network)

1. *Core network upgradation to IPv6:* For non-MPLS networks, IPv6 core has to be enabled without IPv4 running. Only few edges have dual-stack routers. This is illustrated in Fig. C.3(a). For MPLS enabled core network, MPLs core below IPv6 core is to be used, as shown in Fig. C.3(b).

2. *New versions of protocols:* Protocols like OSPF (open shortest path first), BGP (border gate-way protocol) are to be run in new versions which support IPv6.
3. *Testing tunnel LSP:* Tunnel label switched paths to be tested for IPv6.
4. *Functionality test of IPv6:* All functionalities of IPv6 are to be tested and implemented and all applications are to be authenticated.
5. *Upgradation of national Internet exchange:* The national internet exchange infrastructure has to be upgragded to support IPv6.

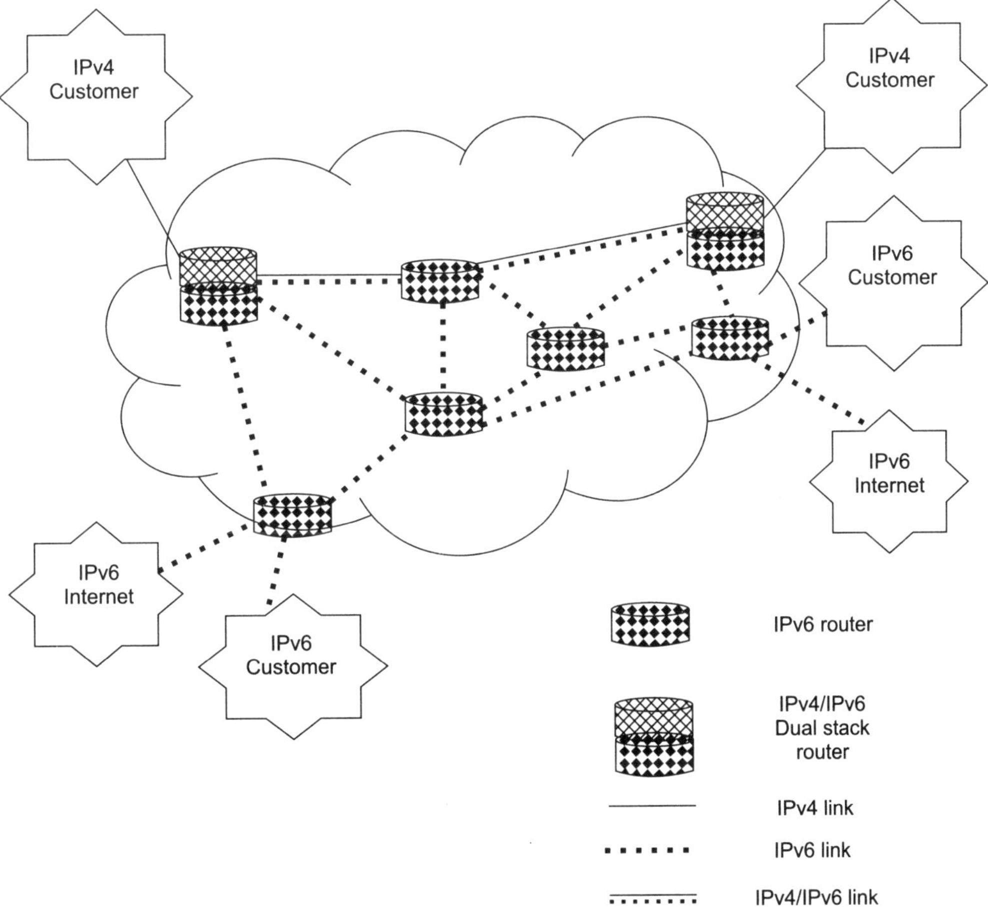

Fig. C.3(a) *Stage-3 of IPv4 to IPv6 migration (non-MPLS enabled core network)*

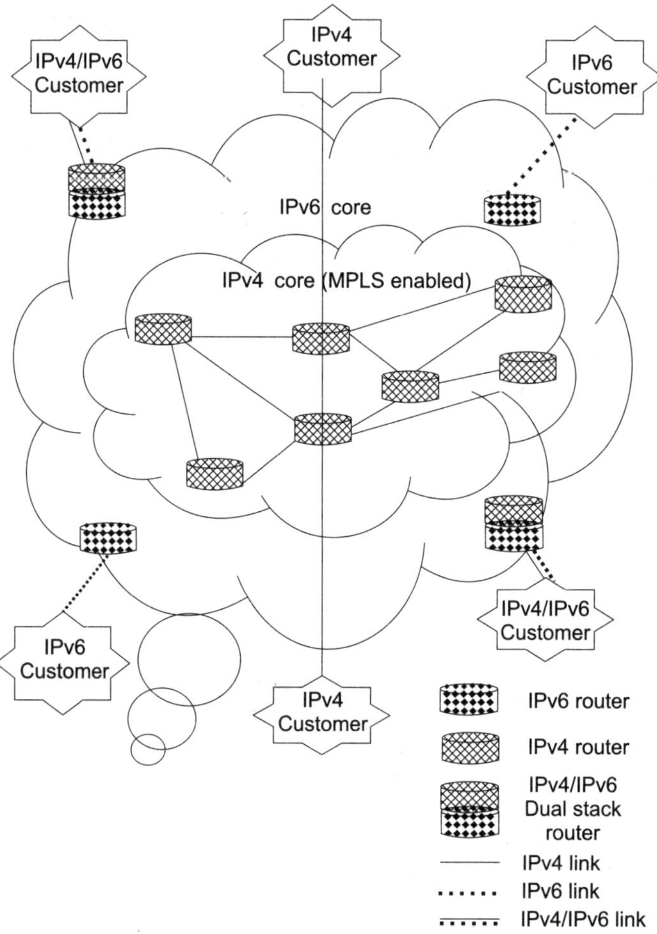

Fig. C.3(b) *Stage-3 of migration from IPv4 to IPv6 (MPLS enabled core network)*

Stage 4: IPv6 only ISP

The interconnectivity of IPv6-only-ISP (Internet service provider) to IPv4 (left out) is shown in Fig. C.4. A translator has to be in place at ASBR (autonomous system boundary router) connecting these two clouds.

Stage 5: IPv6 ISP and IPv6 users (no IPv4)

Stage 5 is illustrated in Fig. C.5. At this stage there are no customer IPv4 nets and all customers are only IPv6.

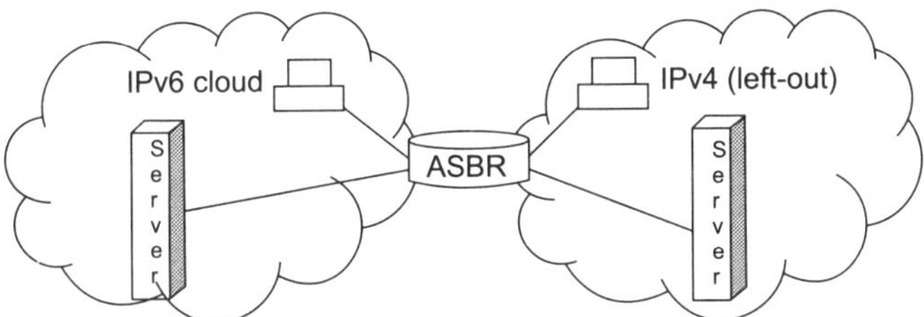

Fig. C.4 *Stage-4 of IPv4 to IPv6 migration*

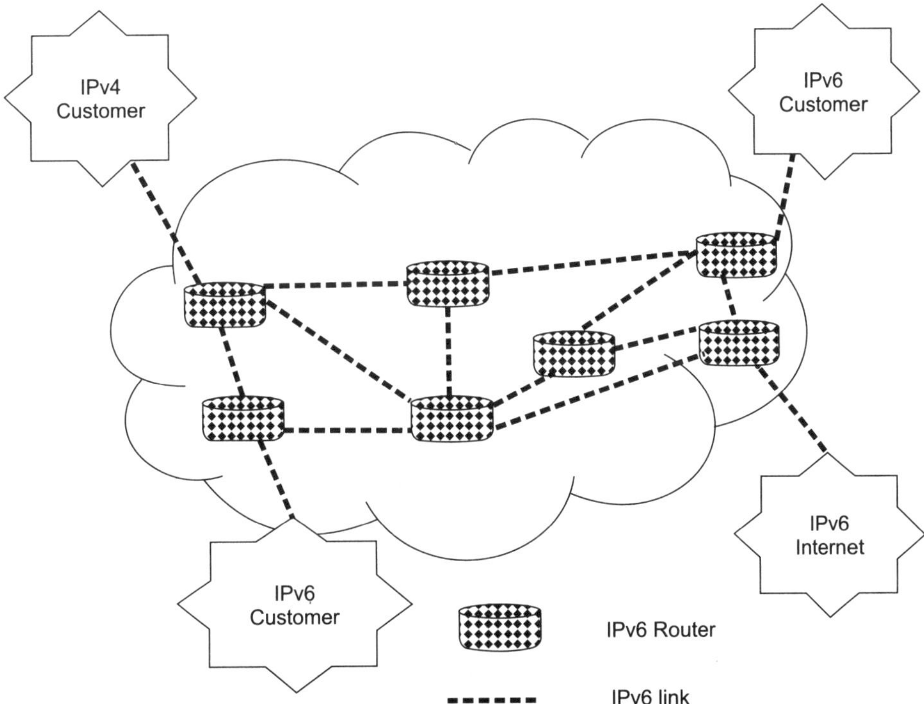

Fig. C.5 *Stage 5 of IPv4 to IPv6 migration (migration is complete)*

Q C.1 *What could be the time-span for migration from IPv4 to IPv6?*

The total time span of IPv4 to IPv6 may last 10 to 15 years, depending on how quickly the migration strategy is implemented. The biggest issue in implementation is the acceptance or readiness of the customer for replacing the access equipment on which they have invested. Another issue would be the availability of security mechanisms like firewalls that support IPv6.

Index

E

F

G

H

P

Q

R

S

T

U

V

W

X

Z